CONVERSION FACTORS

1 atomic mass unit $= 1.66 \times 10^{-27}$ kg

$= 931$ MeV

1 hertz $= 1$ cycle sec^{-1}

1 tesla $= 10^4$ gauss

1 Å $= 10^{-10}$ m

1 eV $= 1.6 \times 10^{-19}$ joule

300 K $= \dfrac{1}{40}$ eV

1 torr $= 1$ mm of Hg

1 curie $= 3.7 \times 10^{10}$ sec^{-1}

1 fermi $= 10^{-15}$ m

1 year $\approx \pi \times 10^7$ sec

MASSES OF PARTICLES

Particle	Symbol	mc^2 (MeV)
Photon	γ	0
Neutrino	ν	0
Electron	e	0.511
Muon	μ	106
Pion	π^+, π^-	140
	π°	135
Kaon	K^+, K^-	494
	$K^\circ, \overline{K}^\circ$	498
Proton	p	938
Neutron	n	940
Lambda hyperon	Λ	1116
Deuteron	d	1876

Introduction
to Atomic and
Nuclear Physics

Aerial view of the National Accelerator Laboratory, Batavia, Illinois. (Photograph courtesy of NAL.)

Introduction to Atomic and Nuclear Physics

HENRY SEMAT   *Professor Emeritus*
The City College of the City University of New York

JOHN R. ALBRIGHT
The Florida State University

HOLT, RINEHART AND WINSTON, INC.
New York   Chicago   San Francisco   Atlanta   Dallas
Montreal   Toronto   London   Sydney

ISBN: 0-03-085402-4

Library of Congress Catalog Card Number: 70-155295

Printed in the United States of America

890 038 098765

To our respective wives

Ray K. Semat

and

Christina B. Albright

Preface

The fifth edition of this book is the product of many years of teaching this subject by the senior author (H.S.) and incorporates the advice, comments, and suggestions of the many teachers, reviewers, and students who have used it. This new edition also incorporates the modern views, teaching experience, and research activities of the junior author (J.R.A.)

The emphasis in this edition is still on the experimental foundations of atomic and nuclear physics with a fair amount of historical material. The major changes consist of (a) the omission of some material, (b) the rearrangement of some of the topics for pedagogical reasons and for a more logical presentation, (c) the updating of all the data, and (d) the addition of many new problems and references. Some details of these changes follow.

The chapter on electricity and magnetism has been omitted and the equations pertaining to these subjects have been written with a constant k so that they can be used with either mks or cgs units as the reader may prefer.

1. In the chapter on special relativity, the material on momentum, energy, and mass has been rewritten so as to emphasize the invariance of mass under Lorentz transformation. This change in approach has been found to enhance the student's problem-solving ability, and it brings his thinking in line with that of physicists who use relativity in their daily work.

2. A chapter on simple Schroedinger quantum mechanics has been added. The calculations are mostly (but not entirely) in one dimension, with no time dependence. The emphasis is on providing a quantitative picture of the origin of (a) the uncertainty principle, (b) quantum numbers for bound states, (c) barrier penetration, and (d) selection rules. The presentation starts from a postulational approach, designed to minimize the amount of effort required to get the student to the point where he can solve problems.

3. A new chapter (11) deals with molecules, statistical mechanics, and solid state physics.

4. The chapter on fundamental particles has been expanded and brought up-to-date by the addition of material on muon-neutrinos, resonances, and the unitary symmetry scheme of Gell-Mann for classifying particles.

5. A new chapter on detecting devices and beam-transport devices has been added.

6. New mathematical appendixes have been added; tables with numerical values of atomic and nuclear properties have been reorganized and brought up-to-date.

Answers to most problems are given at the end of the book. Practically all the other problems contain the answers in the statements of the problems. A separate Solutions Manual containing the solutions to all the problems is available to instructors only.

The book is intended for a one-year course in atomic physics at the sophomore-junior level. With proper selection (Chapters 1–11) it can be used for a one-semester course in atomic physics at this level. It can also be used for a one-semester course in nuclear physics at the junior-senior level (Chapters 3, 11–18).

Appropriate acknowledgments have been made in the earlier editions for the generous help given by many teachers and physicists in preparing those editions. Many of their ideas and suggestions are contained in this new edition. We wish to thank them once again for their help.

We wish to thank Professor Carl H. Poppe of the University of Minnesota, Professor Margaret C. Foster of the State University of New York at Stony Brook, and Professor A. E. Walters of Rutgers, the State University of New Jersey, who were kind enough to read the entire manuscript of this edition. Their suggestions and criticisms have been very valuable and most of them have been incorporated in the book.

Our greatest thanks and appreciation must go to our respective wives, Ray K. Semat and Christina B. Albright, for typing and retyping the manuscript and for their continued encouragement to complete this book.

New York   H. S.
Tallahassee, Florida   J. R. A.
September 1971

Contents

APPENDIXES

Introduction
to Atomic and
Nuclear Physics

PART ONE

Foundations of Atomic and Nuclear Physics

1

Atoms, Ions, and Electrons

1-1 Introduction

The decade from 1895 to 1905 may be termed the beginning of modern physics. During this period J. J. Thomson succeeded in demonstrating the existence of the electron, a fundamental unit of negative electricity with very small mass, Becquerel discovered the phenomenon of natural radioactivity, and Roentgen discovered x-rays. To these discoveries must be added the bold hypothesis put forth by Planck to explain the distribution of energy in the spectrum of the radiation from a blackbody, namely, that electromagnetic radiation, in its interaction with matter, is emitted or absorbed in whole units, called quanta of energy; each quantum has an energy $\mathcal{E} = h\nu$, where ν is the frequency of the radiation emitted or absorbed and h is a constant now called the Planck constant. Einstein (1905), in a treatment of the photoelectric effect, extended Planck's hypothesis by showing that electromagnetic radiation, in its interaction with matter, behaves as though it consists of particles, called photons, each photon having an energy $\mathcal{E} = h\nu$. N. Bohr (1913) used this concept in his very successful and beautiful theory of the hydrogen atom. All these ideas led to one of the most important developments of twentieth-century physics, the quantum theory; this theory includes quantum mechanics and the quantum theory of radiation. During this same period (1905–1913) an equally important change in the foundation of physics resulted from the development of the theory of relativity by Einstein: one part, which deals with systems in motion with uniform velocity relative to one another, is called the special or restricted theory of relativity and will be of major interest here; a second part, in which these restrictions are removed, is called the general theory of relativity and will be of less interest to us.

It is the aim of this book to present the important experimental data on which are based our present ideas of the structure of the atom. An atom is to be regarded not as a static structure composed of particles in fixed positions but

3

rather as a dynamic structure that changes in response to outside agencies, affecting them and, in turn, being affected by them. It is by examining the phenomena that occur during these changes that we get our information concerning the structure of the atom as well as an insight into the nature of the particles and the phenomena that produce or are the result of these changes.

The development of atomic and nuclear physics has not taken place in isolation; it has grown out of the centuries of experience in all of physics. Nor has it been a one-sided affair: the knowledge gained about atomic structure and atomic processes has been fed back into the other branches of physics and has enabled us to get a better understanding of the properties of matter and a deeper insight into such fundamental concepts as mass, energy, linear momentum, angular momentum, and the various conservation laws that govern all of physics.

It will be assumed that the reader is familiar with the fundamental concepts and laws of the traditional subdivisions of physics: mechanics, wave motion, heat, electricity and magnetism, and optics. Many of these concepts and laws will be discussed in terms of their roles in atomic and nuclear physics. The equations used and developed in this book can, in general, be used with cgs and mks units. Each branch of physics tends to employ units that are especially useful; this is also the case in atomic and nuclear physics. These units will be defined and used at appropriate places in the book. Tables of conversion factors will be found in Appendix I.

1-2 The Avogadro Number

The concept of atoms was introduced into chemistry as a hypothesis by Dalton (1802) to explain the formation of compounds from simpler substances called elements. The atoms of any one element were assumed to be identical in all their properties, including mass. Dalton also formulated the law of multiple proportions which states that if two elements combine in more than one proportion to form different compounds the masses of one of the elements which unite with identical amounts of the second element are in the ratio of integral numbers; for example, 16 grams of oxygen combine with 12 grams of carbon to form carbon monoxide, CO, whereas 32 grams of oxygen can combine with 12 grams of carbon to form another compound, carbon dioxide, CO_2. Shortly afterward the law of definite proportions was established. This law states that in any compound the proportion by mass of the constituent elements is a constant; for example, when mercuric oxide is decomposed, for every 16 grams of oxygen liberated, 200.6 grams of mercury are liberated. When oxygen and mercury are combined to form mercuric oxide, their proportions by mass are always in the ratio of 16 to 200.6.

Many of the elements and compounds were studied in the gaseous state:

it was found that the volumes of the elements and compounds involved in a reaction, in which both the initial and final constituents were gases, were connected by very simple relationships such as the laws of Boyle and Charles. To explain the behavior of gases Avogadro (1811) put forth the bold hypothesis that equal volumes of different gases, under the same conditions of temperature and pressure, contain equal numbers of molecules.

In the combination of hydrogen and oxygen to form water vapor, for example, it is found that, when measured at the same temperature and pressure, two liters of hydrogen and one liter of oxygen combine to form two liters of water vapor. The interpretation of this result on the basis of Avogadro's hypothesis is that two molecules of hydrogen and one molecule of oxygen combine to form two molecules of water vapor. The chemical equation expressing this result is

$$2H_2 + O_2 = 2H_2O$$

This equation also states that in the free state hydrogen and oxygen molecules are composed of two atoms each and that water vapor molecules consist of three atoms, two hydrogen atoms and one oxygen atom. From a determination of the combining weights it has been found that 16 grams of oxygen combine with 2.016 grams of hydrogen to form 18.016 grams of water vapor. Since one atom of oxygen combines with two atoms of hydrogen to form water vapor, the relative atomic masses of oxygen and hydrogen are in the ratio of 16 to 1.008.

Since a molecule of oxygen in the free state contains two atoms, the molecular weight of oxygen is 32.00. The molecular weight of hydrogen, H_2, is 2.016. A quantity of any substance whose mass, in grams, is numerically equal to its molecular weight is called a *mole*. The volume occupied by a mole of any gas is called gram molecular volume. At 0°C and 76 cm pressure *the gram molecular volume* of any gas is 22.4 liters. On the basis of Avogadro's hypothesis, every mole of a substance contains the same number of molecules. This number, usually referred to as the *Avogadro number*, will be denoted by the letter N_0.

From the above discussion it is readily seen that, *in the case of an element, a mass in grams equal numerically to its atomic mass must contain N_0 atoms;* for example, 16 grams of oxygen contain $N_0/2$ molecules, hence contain N_0 atoms. The same holds true for any other diatomic molecule. For a monatomic molecule such as helium or neon it is obvious, of course, that there are N_0 atoms in a gram-atomic mass of the element. Therefore, if the Avogadro number N_0 can be determined, the mass in grams of any atom will be known.

The Avogadro number N_0 can be determined by several methods. Two of these methods will be discussed in this chapter. One method is based on a study of electrolysis; the other is based on a study of the Brownian motion of particles suspended in a fluid.

1-3 The Avogadro Number and the Electronic Charge

Solutions of acids, bases, and salts are known to be conductors of electricity. The conductivity of these solutions is due to the presence of *ions* in the solution. An ion is an atom or a group of atoms charged electrically. In the modern theory of the structure of the atom (Section 4-6) an atom consists of a positively charged *nucleus* surrounded by a number of negatively charged particles called *electrons*. In the neutral atom the sum of all the electronic charges is numerically equal to the positive charge of the nucleus. An atom is said to be ionized if the total electronic charge is not equal numerically to the positive charge of the nucleus; similarly, a group of atoms is said to be ionized if the total electronic charge differs numerically from the total positive charge of their nuclei. If there is a deficiency of electrons, the ion is positively charged, and if there is an excess of electrons it is negatively charged. It has been found empirically that the charge on any ion can be expressed in terms of integral multiples of a fundamental quantity of electricity that has been shown to be equivalent to the charge of an electron. If the charge on an ion is equivalent to one electron, it is said to have a valence of unity; if the charge on an ion is equivalent to two electrons, it has a valence of two, and so on.

As a typical example of electrolysis, let us consider a chemical cell containing a solution of silver nitrate, $AgNO_3$, a silver anode connected to the positive terminal of a battery, and a copper cathode connected to the negative terminal of the battery, as shown in Figure 1-1. The silver nitrate in the solution is dissociated into silver ions, Ag^+, and nitrate ions, NO_3^-. Under the action of the electric field between the electrodes the Ag^+ ions migrate toward the cathode and the NO_3^- ions migrate toward the anode. If the chemical cell is examined after the current has been flowing in the circuit for some definite time, it will be found that a mass M of metallic silver Ag has been deposited on the cathode and an equal mass M of metallic silver has been removed from the anode, leaving the concentration of the solution unchanged. A simplified analysis of the action of this cell is as follows: when a silver ion, Ag^+, reaches the cathode, it acquires an electron from it, thus forming a neutral atom which adheres to the cathode. At the anode a silver atom breaks up into a silver ion

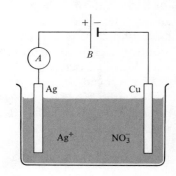

Figure 1-1 Electrolysis

and an electron; the ion, Ag^+, goes into solution to replace the one that was removed and the electron is forced through the external part of the circuit. Thus an atom is removed from the anode for each atom deposited on the cathode. The silver ions are transferred through the solution and the electrons are transferred through the external metallic portions of the circuit.

The behavior of electrolytic cells can be summarized in terms of two laws first formulated by Faraday. The first law states that the quantity of any substance liberated from the solution depends only on the total charge passing through the circuit, or

$$M = kQ \tag{1-1}$$

where M is the mass of the material liberated at one electrode, Q is the quantity of charge transferred, and k is a factor of proportionality called the *electrochemical equivalent* of the substance. From this equation it is evident that k is the mass liberated per unit charge transferred, usually expressed in grams per coulomb.

Faraday's second law states that, for any substance, the mass liberated by the transfer of a quantity of electrical charge Q is proportional to the *chemical equivalent* of the substance, or

$$M = \frac{A}{v} \frac{1}{F} Q \tag{1-2}$$

where A/v, the ratio of the atomic mass to the valence of the element, is the chemical equivalent of the element and F is a constant of proportionality known as the *Faraday constant*.

From Equations (1-1) and (1-2) it will be noted that

$$F = \frac{A}{kv}$$

The value of F can be determined from the results of the experiments on electrolysis. For the case of silver, where $k = 0.0011180$ grams per coulomb, $A = 107.88$ grams per gram-atomic mass, and v is unity, we get

$$F = \frac{107.88}{0.0011180} = 96{,}500 \text{ coulombs/gram-atomic mass}$$

Thus the transfer of 96,500 coul of charge will deposit a gram-atomic mass of a monovalent element.

Since the valence of silver is unity, for every atom of silver deposited on the cathode a charge equivalent to one electron has been transferred through the solution. If e is the charge of one electron, then $N_0 e$ is the total charge transferred when one gram-atomic mass of silver is deposited on the cathode, or

$$F = N_0 e = 96{,}500 \text{ coulombs/gram-atomic mass} \tag{1-3}$$

The value of the Avogadro number can be determined with the aid of

Equation (1-3) by substituting the value of the electronic charge determined by independent methods. The present accepted value of the electronic charge is

$$e = 1.6022 \times 10^{-19} \text{ coul}$$

and the accepted value of the Faraday constant F is

$$F = 96{,}487 \text{ coulombs/gram-atomic mass}$$

so that the Avogadro number N_0 is

$$N_0 = 6.0222 \times 10^{23} \text{ atoms/gram-atomic mass}$$

If the Avogadro number N_0 could be determined independently with the same degree of precision as the electronic charge e, it would be a check on the determination of e as well as additional confirmation of Avogadro's hypothesis. The first direct determination of the Avogadro number was made by Perrin in 1908 in an investigation of the motion and distribution of very small particles suspended in a fluid.

The accepted values of N_0 and e are not the result of any single direct measurement. Rather, the most accurate determination results from combining data from a variety of experiments, each one of which is a measurement of a cluster of fundamental constants (including the experimental error in the measurement). A computer program is used to perform a least-squares calculation of the constants of nature. The values tabulated in Appendix I are the result of this sort of compilation and adjustment.

1-4 Atomic Masses

On the basis of the hypothesis introduced by Dalton (Section 1-2) that all the atoms of any one element were identical in all respects, including that of mass, the numbers representing the masses of elements would give the relative masses of their respective atoms. Others suggested that the atoms of any one element need not be identical in mass, that the numbers representing the atomic masses are only average values of the different weights of the atoms of the particular element. The basis of this suggestion was the hope that the weights of all the atoms could be expressed by integral numbers. Prout (1815) formulated this idea in his hypothesis that the atoms of all the elements are made up of hydrogen atoms. If this hypothesis were correct, the masses of all the atoms would be expressed as integers, with that of hydrogen as unity. This was the system of atomic masses used in the nineteenth century. Accurate measurements showed, however, that the atomic masses of many elements were not integral numbers relative to that of hydrogen. Oxygen, for example, had an atomic mass of 15.99 and that of nitrogen was 13.99. At the beginning of the twentieth century the atomic mass of oxygen was adopted as a standard and assigned the value 16 exactly. This choice was governed partly by the fact that combinations of other elements with oxygen were often used to determine atomic masses.

With the discovery of radioactivity at the end of the nineteenth century and the study of the radioactive elements produced in the process of natural radio-active disintegration, it became evident that many of these elements were identical with known nonradioactive elements except that their atomic weights were different. Thus the group of radioactive elements that were chemically identical with nonradioactive elements would occupy the same place in the periodic table of elements. F. Soddy suggested the name *isotopes* for those elements that occupy the same place in the periodic table. Any one isotope of an element thus consists of atoms that are identical in all respects, including mass. It is interesting to note that the early determination of the atomic masses of the isotopes usually yielded whole numbers. The search for isotopes among nonradioactive elements was begun by J. J. Thomson about 1910. The first element successfully investigated was neon. It was the lightest element whose atomic mass, 20.2, differed appreciably from an integral number. Thomson found two isotopes, one of atomic mass 20, the other of atomic mass 22. The method he used was the determination of the ratio of the charge to the mass of the positive ions formed in an electrical discharge tube containing neon gas. This method is sometimes referred to as positive ray analysis. Since this pio-neering work, instruments of very high precision, known as mass spectrom-eters, have been developed for determining isotopic masses.

The masses of these isotopes were expressed on a physical scale of atomic masses, in which the most abundant isotope of oxygen, $^{16}_{8}O$, was used as a base and its mass assigned the value 16 exactly. Thus, until recently, two systems were in use, one the chemical system of atomic masses based on naturally occurring oxygen, $O = 16$, exactly, the other the physical system of atomic masses based on $^{16}O = 16$ exactly. The mass of an atom of ^{16}O was 16 amu, where 1 amu = 1 atomic mass unit = $\frac{1}{16}$ the mass of an atom of ^{16}O. In 1961 the International Union of Pure and Applied Chemistry recommended the adoption of a unified system of atomic weights based not on oxygen but on the most abundant isotope of carbon, ^{12}C, and assigned the value $^{12}C = 12$ exactly to this isotope. The new table of atomic masses is given in Appendix II.

The new unified unit of atomic mass, represented by the symbol u, is one-twelfth of the mass of an atom of ^{12}C. Since N_0 atoms of ^{12}C have a mass of 12 grams,

$$1 \text{ u} = \frac{1}{12} \times \frac{12 \text{ gm}}{N_0} = \frac{1 \text{ gm}}{6.022169 \times 10^{23} \text{ atoms}}$$

so that

$$1 \text{ u} = 1.66053 \times 10^{-24} \text{ gm}$$

1-5 Brownian Motion

If fine particles suspended in a fluid are examined in the field of a micro-scope, it will be observed that they are in constant haphazard motion. This random motion continues indefinitely and is found to depend on several fac-tors such as the size of the particles, the viscosity of the fluid in which they are immersed, and the temperature of the system. The motion of these particles was first observed by Brown in 1827. Many observers recognized that these particles behave in the same way that molecules of an ideal gas are supposed to behave. The random motion of these particles may be likened to the thermal motion of gas molecules.

The explanation of the Brownian motion, first given by Einstein and Smoluchowski, is based on the assumption that the particles in suspension are continually bombarded by the molecules of the fluid and that this bombardment produces an unbalanced force which accelerates the particle. The motion of the particle through the fluid is opposed by another force due to the viscosity of the fluid. From this theory the distribution of particles in a field of force and their displacement in the course of time can be calculated. Perrin performed these two different types of experiments on Brownian motion, one on the vertical distribution of the particles in the fluid, the other on the displacement of the particles in a given time interval.

1-6   Vertical Distribution of Particles

If the particles suspended in a fluid behave like large-sized gas molecules, then their distribution in the vertical direction should be similar to the distribution of the air in the atmosphere; that is, the density of particles should be greatest at the bottom of the fluid and should decrease exponentially with increasing height. Consider a cylinder of gas of cross-sectional area A and height h. If ρ is the density of the gas, then the mass of gas in a small cylindrical element of height dh is $\rho A\, dh$ and its weight is $\rho Ag\, dh$, where g is the acceleration due to gravity. The weight of this gas is balanced by the upward force due to the difference in pressure on the two circular surfaces of the cylinder. If dp is this difference in pressure, then the upward force is $A\, dp$, from which

$$dp = -\rho g\, dh$$

The minus sign indicates that as the height increases the pressure decreases. Now the density $\rho = M/v$, where M is the mass of a mole of the gas and v is the volume occupied by the mole, so that

$$dp = -\frac{M}{v} g\, dh$$

The general gas law for 1 mole of a gas is

$$pv = RT$$

where T is the absolute temperature of the gas and R is the universal gas constant. Eliminating v from the last two equations, we get

$$dp = -\frac{Mgp\, dh}{RT}$$

or

$$\frac{dp}{p} = -\frac{Mg\,dh}{RT}$$

Integrating this equation (assuming the temperature to be the same throughout the gas), we get

$$\log p = -\frac{Mgh}{RT} + \log C$$

Let $p = p_0$ when $h = 0$; then $\log C = \log p_0$. Hence

$$p = p_0 \exp\left(-\frac{Mgh}{RT}\right)$$

Since the pressure is directly proportional to the density or the number of molecules per unit volume, this equation can be written as

$$n = n_0 \exp\left(-\frac{Mgh}{RT}\right) \qquad (1\text{-}4)$$

where n is the number of molecules per unit volume at a height h above the reference plane and n_0 is the number of molecules per unit volume at the reference plane. This is the well-known law of atmospheres. If n_0 represents the number of molecules per unit volume at the surface of the earth, then n gives the number of molecules per unit volume at any height h above the earth's surface.

The law of atmospheres may be put in a more useful form if we replace the mass M of a mole of gas with $N_0 m$, where m is the mass of a molecule and N_0 is the Avogadro number. It then becomes

$$n = n_0 \exp\left(-\frac{N_0 mgh}{RT}\right) \qquad (1\text{-}5)$$

In applying this law to the determination of the distribution of particles suspended in a fluid, the buoyant force due to the fluid must be taken into consideration. If ρ is the density of the substance and ρ_0, the density of the fluid, then the effective weight of a particle of this substance is $mg(1 - \rho_0/\rho)$, so that the number of particles per unit volume at any height h above the reference plane is given by

$$n = n_0 \exp\left(-\frac{N_0 mg(1 - \rho_0/\rho)h}{RT}\right) \qquad (1\text{-}6)$$

For particles showing observable Brownian motion h is so small that the entire exponent becomes very small in comparison with 1 and the exponential factor may be expanded in a series, of which the first two terms are

$$\frac{n}{n_0} = 1 - \frac{N_0 mg}{RT}\left(1 - \frac{\rho_0}{\rho}\right)h \qquad (1\text{-}7)$$

The higher powers in this series may be neglected.

Figure 1-2 Perrin's method of observing the vertical distribution of particles in a fluid.

Perrin prepared emulsions from gamboge and mastic and with the aid of a centrifuge obtained particles of nearly uniform size. In order to be able to apply Equation (1-7), it is necessary to determine both the density and the volume of the grains in the emulsion. The density was determined in three ways: (a) by the specific gravity bottle method, (b) by measuring the density of the fused mass remaining after the water was boiled off, and (c) by adding potassium bromide to the emulsion until the density of the fluid was equal to that of the particles and then determining the density of the solution.

The sizes of the particles were determined in several ways. One method was to allow a drop of very dilute emulsion to evaporate from a cover glass. It was found that the particles arranged themselves neatly in rows. The sizes of the particles could then be determined by counting the number of particles in a row of known length or the number in a known area. Another method was to time the rate of fall of one of the particles through the fluid and then calculate the size of the particle with the aid of Stokes' law for the velocity of spherical particles falling through a viscous medium.

The experimental determination of the number of particles at any level in the emulsion was made by placing a drop of emulsion in a glass cell about 0.1 mm deep, covered by a microscope cover glass to give a plane surface, as shown in Figure 1-2. To prevent evaporation the edges of the cover glass were treated with paraffin. Under a high-power microscope, the grains in a very thin horizontal section, about 0.001 mm in thickness were sharply defined. Raising or lowering of the microscope made the grains in other sectional layers visible. The method of observation consisted in narrowing the field of view so that not more than five or six particles were visible at any one time. A shutter, placed in the path of light illuminating the emulsion, opened at regular intervals, and the number of grains observed on each such occasion was noted. About 200 observations were made at one level and then the microscope was moved vertically through a known distance and a similar set of observations was made. In this way the ratio n/n_0 was determined. Perrin also used the photographic method, taking photographs of the emulsion at equidistant levels and then counting the number of images of grains appearing on the photographic plate.

Many experiments were performed under a variety of conditions of temperature, with grains of different sizes suspended in several different liquids. As a result of these experiments Perrin obtained $N_0 = 6.88 \times 10^{23}$ as the value of the Avogadro number. When this value was combined with the results of the experiments on electrolysis, the value of the charge of an electron was determined as

$$e = 4.25 \times 10^{-10} \text{ stcoul}$$

1-7   Displacement of Particles in Brownian Motion

The second type of experiment performed by Perrin for the determination of the Avogadro number consisted in observing one of the particles in suspension and measuring the displacement of this particle in a given time interval. The fundamental assumption used in the derivation of the equation for the displacement of such a particle is that the particles suspended in a fluid have a mean kinetic energy equal to the mean kinetic energy of gas molecules at the same temperature. The displacement equation, derived first by Einstein in 1905 (see Appendix V-2), is

$$\overline{x^2} = \frac{2RT}{N_0 K} t \tag{1-8}$$

where x is the displacement of the particle in the x direction in a given time interval t, and $\overline{x^2}$ is the average of the squares of the displacements of the particle in the x direction in equal time intervals t. K is a factor of proportionality determined by the viscosity of the medium.

Perrin tested Einstein's equation in several ways. He used grains of mastic and gamboge in various solutions and at various temperatures. The displacements were determined by plotting the position of the particle after fixed time intervals. A typical trajectory is shown in Figure 1-3. The positions of the particle are represented by dots and the lines joining the dots represent the displacements in the given time interval, say 30 sec. The x component of each

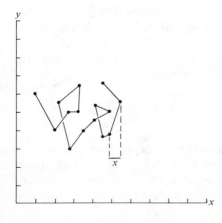

Figure 1-3   Trajectory of a particle in Brownian motion. Each line connecting two dots represents the displacement of the particle in a fixed time interval.

of these displacements can be obtained by projecting them on the x axis; $\overline{x^2}$ is then obtained by squaring and averaging the squares of these displacements.

For the determination of N_0 by the displacement method the microscope was focused at a given level and the first particle that appeared in the center of the field of view was followed for a little while, the positions being noted at given time intervals. The emulsion was then displaced laterally and again the trajectory of the first particle that appeared in the center was traced. From a measurement of 1500 displacements of gamboge particles of 0.367×10^{-4} cm radius Perrin obtained $N_0 = 6.88 \times 10^{23}$ for the value of the Avogadro number. As a result of many determinations of N_0 under different experimental conditions, Perrin adopted

$$N_0 = 6.85 \times 10^{23}$$

as the best value for the Avogadro number, which yielded the value

$$e = 4.2 \times 10^{-10} \text{ stcoul}$$

for the elctronic charge.

Later work by Millikan and Fletcher on Brownian motion in a gaseous medium such as air yielded the value $N_0 = 6.03 \times 10^{23}$, whereas Westgren (1915), working with colloidal gold, silver, and selenium particles, obtained the value $N_0 = 6.05 \times 10^{23}$. Avogadro's hypothesis was thus confirmed experimentally about a century after it was formulated.

In spite of the great improvement in experimental techniques, the precision possible with Brownian motion experiments is not so high as is desirable, in view of the accuracy with which other atomic constants are known, particularly those closely related to the Avogadro number. Modern values of N_0 are obtained by x-ray methods involving determinations of the structure of crystals and the distance d between atomic planes. It is readily shown (see Section 5-6) that the distance d between the atomic planes of a cubic crystal is given by

$$d = \left(\frac{M}{2\rho N_0}\right)^{1/3} \tag{1-9}$$

in which M is the molecular weight of the crystal and ρ is its density.

Crystals of NaCl, KCl, and LiF of very high chemical purity can now be grown artificially. The distance d, also known as the *grating space*, can be determined by diffracting x-rays of known wavelength with the crystal. The value of N_0 can then be calculated by using Equation (1-9).

1-8 Determination of the Charge of an Electron

The earliest experiments in the determination of the electronic charge were performed by Townsend (1897) and J. J. Thomson (1898). Their method consisted in allowing water vapor to condense on ions, thus forming a cloud, and then determining the charge carried by the cloud. The number of individual droplets in the cloud was then computed by weighing the water condensed from the cloud and dividing it by the average weight of a single droplet. The latter was determined by measuring the rate of fall of these droplets through air, assuming Stokes' law to hold. On the assumption that each droplet was condensed on a single ion carrying charge e, they obtained values for e of the order of 3×10^{-10} stcoul.

A very important by-product of this type of investigation was the discovery by C. T. R. Wilson (1897) of a method for producing these clouds. If air that is saturated with water vapor and ionized by some agency is expanded suddenly, the air is cooled and the water vapor condenses on the ions in the air.

H. A. Wilson (1903) made a decided improvement in the method for determining the electronic charge by forming clouds between two capacitor plates which could be connected to the terminals of a battery. This experimental procedure was first to determine the rate of fall of the top surface of the cloud under the influence of gravity alone and then to produce a second cloud and charge the capacitor plates so that the droplets would be urged downward by both gravity and the force due to the electric field. The numerical results for the electronic charge were about the same as those obtained by Townsend and Thomson. All of these experiments suffered from the fact that the weight of a drop of water did not remain constant during the time of observation. Further, the exact number of ions on which each drop was condensed was not known.

Millikan, while repeating the experiments of H. A. Wilson, found that single drops of water could be held stationary between the two plates of the capacitor by adjusting the voltage between the plates so that the weight of the drop could be balanced by the force due to the electric field between the plates. While working with these "balanced drops" he noticed that occasionally the drop would start moving up or down in the electric field. This drop had evidently captured an ion, positive in one case, negative in the other case. This made it possible to determine the charge carried by an ion irrespective of the original charge carried by the drop of water. To avoid the errors due to evaporation Millikan decided to use drops of oil instead of water.

The apparatus consisted essentially of two brass plates A and B about 22 cm in diameter and about 1.5 cm apart, as sketched in Figure 1-4. These plates were placed in a large metal box to avoid air currents. In this experiment small drops of oil are sprayed into the box with an atomizer. After a while one of these drops drifts through the pinhole C in the top of plate A and can then be observed with a telescope. Indirect illumination of the drop is provided by a lamp on the side. When there is no electric field between the plates, the forces acting on the oil drop are its weight $m_1 g$, the buoyant force, $m_0 g$, of the air,

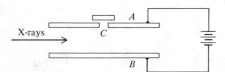

Figure 1-4

and a resisting force due to the viscosity of the medium. Let us set

$$m_1 g - m_0 g = mg$$

where mg is the effective weight of the oil drop in the air. From experiment it is known that the resisting force is proportional to the velocity of the oil drop. Equilibrium will be reached when the velocity of the drop reaches such a value that the resisting force becomes equal to the effective weight of the oil drop. The oil drop will then continue to move downward with uniform velocity. Calling this value of the velocity v, we can write

$$mg = Kv \qquad (1\text{-}10)$$

where K is a factor of proportionality. For very small drops this limiting velocity is reached very quickly. If the capacitor plates are now connected to the battery so that plate A is positive, the velocity of the oil drop will be changed suddenly. This change is due to the fact that the oil drop is charged when it comes from the atomizer. If q represents the charge on the oil drop, the force on it due to the electric field is

$$F = \frac{V}{d} q \qquad (1\text{-}11)$$

where V is the difference of potential between the plates and d is the distance between the plates. If the charge on the oil drop is negative, then the force due to the electric field will be upward and the new velocity v_1 will be given by

$$F - mg = Kv_1 \qquad (1\text{-}12)$$

where v_1 is considered positive in the upward direction. Combining Equations (1-11) and (1-12), we get

$$\frac{V}{d} q - mg = Kv_1 \qquad (1\text{-}13)$$

In Millikan's experiment the air between the capacitor plates was ionized by various methods, such as allowing x-rays or the radiations from radioactive substances to pass through it. The oil drop occasionally acquired an additional ion, either positive or negative, and its velocity in the electric field was observed to change or it may have lost an ion in its passage through the ionized air. If v_n represents its new velocity after the acquisition of an ion of charge q_n, then from Equation (1-13)

$$\frac{V}{d} (q + q_n) - mg = Kv_n \qquad (1\text{-}14)$$

Solving Equations (1-13) and (1-14) for q_n, we get

$$q_n = \frac{d}{V} K(v_n - v_1) \qquad (1\text{-}15)$$

The experiment consists in determining q_n, the charges on the ions captured or lost by the oil drop during the time of observation, which in some cases lasted for several hours. The velocity of the oil drop was measured by timing the passage of its image between two cross hairs in the telescope a known distance apart. The difference of potential V was of the order of several thousand volts. The only quantity left to be evaluated in the determination of q_n is the factor K. This factor K can be expressed in terms of the radius a of the oil drop and the coefficient of the viscosity η of the air as

$$K = 6\pi\eta a \qquad (1\text{-}16)$$

if these drops obey Stokes' law for the velocity of spheres falling through a viscous medium. The radius a may be determined experimentally by measuring the limiting velocity v and substituting this value into Equation (1-10) modified as follows:

$$mg = \tfrac{4}{3}\pi a^3(\rho_1 - \rho_0)g = 6\pi\eta a v \qquad (1\text{-}17)$$

where ρ_1 is the density of the oil and ρ_0 is the density of the air.

A series of experiments was performed to verify the accuracy of Stokes' law. It was found that for very small drops, Stokes' law had to be modified, the correction being given with sufficient accuracy if K is written

$$K = \frac{6\pi\eta a}{1 + b/pa} \qquad (1\text{-}18)$$

where p is the pressure of the air in centimeters of mercury and b is an empirically determined constant. The value of the coefficient of viscosity was determined in another series of experiments, and the value used by Millikan is $\eta = 0.0001825$ dyne sec/cm².

From many determinations of q_n it was found that q_n could always be represented by

$$q_n = ne \qquad (1\text{-}19)$$

where n is an integer and e represents the elementary charge equivalent to that of an electron. The value of the charge of an electron from Millikan's work (1917) is

$$e = 4.770 \times 10^{-10} \text{ stcoul}$$

It must be emphasized that the value of e is the same for both positive and negative charges, since q_n, the ionic charge captured or lost by the oil drop, could be either positive or negative, depending on chance.

Later experimental evidence, particularly from the study of x-ray wavelengths (Section 5-12), indicated that this value of e was too small. Shiba, in 1932, pointed out that the error in Millikan's determination of e was *probably* due to an error in the determination of η. Many experiments have since been performed for the redetermination of this constant. The weighted average of seven different determinations of the coefficient of viscosity of air at 23°C is

$$\eta = 1832.5 \times 10^{-7} \text{ dyne sec/cm}^2$$

When this value of η is used with Millikan's measurements, the value of the electronic charge becomes

$$e = 4.8071 \times 10^{-10} \text{ stcoul}$$

The precision in the determination of the electronic charge by the oil-drop method is limited by the accuracy with which the coefficient of viscosity of air can be determined. This is considerably less than the accuracy with which other associated atomic constants are known. The currently accepted value of the electronic charge is that calculated from the least-squares adjustment of all the fundamental constants, from which

$$e = 4.803 \times 10^{-10} \text{ stcoul}$$

$$= 1.602 \times 10^{-19} \text{ coul}$$

1-9   Electric Discharge through Gases

The phenomenon of the discharge of electricity through gases has been known for many years and has been utilized for the study of many problems connected with atomic structure. In its simplest form a discharge tube consists of a long glass tube with a circular electrode sealed into each end, as shown in Figure 1-5. A smaller side tube is sealed into it so that the pressure of the gas in the tube may be controlled by connecting it to a pumping system.

Let us consider the phenomena when there is air in the tube. Electrode A is connected to the positive side of a source of high potential, such as an induction coil, and electrode C is connected to the negative side. When the pressure of the air inside the tube is reduced to a few millimeters of mercury, the electric discharge fills the entire space between the electrodes with a pink or reddish glow. If the light from this tube is examined with a spectroscope, it

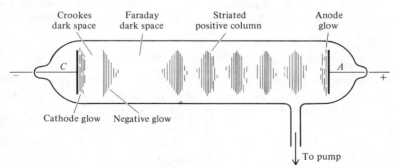

Figure 1-5   Appearance of the electric discharge when the pressure of the air in the tube is about 0.1 mm of mercury.

will be found to consist of a series of lines characteristic of the gases within the tube. This type of discharge is frequently used in studying the spectra of various substances that can be obtained in the form of a gas or vapor.

When the pressure of the air within the tube is reduced to about 0.1 mm of mercury, the appearance of the discharge is approximately as follows: there is a bluish velvety glow around the cathode *C*, called the cathode glow; then a dark space, called the Crookes dark space; then the negative glow, followed by the Faraday dark space; then, filling the rest of the tube up to the anode, the striated positive column; and just around the anode, the anode glow.

The electric field is not uniform along the length of the tube but varies widely as we pass from one electrode to the other. The field is most intense in the Crookes dark space. The width of this dark space depends on the pressure. As the pressure is reduced, this dark space widens out until at a pressure of about 0.001 mm it fills the entire tube. At this pressure the positive column and negative glow have disappeared.

The phenomena observed in the discharge tube may be explained qualitatively by assuming that neutral molecules or atoms are ionized by collisions with ions or free electrons and also that positive ions recombine with free electrons or with negative ions to form neutral atoms or molecules. Under the action of the electric field the positive ions are accelerated toward the cathode, whereas the negative ions and electrons are accelerated toward the anode and thus acquire considerable kinetic energy. A charged particle will lose some of its energy in a collision with a neutral atom or molecule if this collision results in the ionization of the neutral particle, since in the process of ionization work must be done to separate the positive from the negative charges. On the other hand, energy will be released when an electron and an ion recombine to form a neutral atom or molecule. Some of this energy may be emitted in the form of light characteristic of the gas in the tube.

When the pressure of the gas is comparatively low, about 0.001 mm of mercury, the mean free path of the ions and atoms is very large, so that an ion or electron will make very few collisions in traversing the length of the tube. The positive ions, for example, will thus have a great deal of kinetic energy when they strike the cathode. The result of this bombardment of the cathode by the positive ions is that the cathode emits particles called *cathode rays*. These cathode rays travel away from the cathode in nearly straight lines perpendicular to the surface of the cathode since the direction of motion is determined almost exclusively by the very intense electric field in the immediate neighborhood of the cathode.

These cathode rays can be deflected by electric and magnetic fields, and the direction of the deflection shows that they are negatively charged. Certain substances such as glass and zinc sulfide emit fluorescent radiations when bombarded by cathode rays. Cathode rays will also affect a photographic plate. These effects may be used to detect and measure the cathode rays.

1-10 Determination of e/m for Cathode Rays

J. J. Thomson (1897) first successfully determined the nature of the cathode rays and showed that they are *electrons;* he was also among the first to measure the ratio of the charge of the cathode ray to its mass, denoted by the symbol e/m.

A cathode-ray tube typical of the kind used by Thomson is shown in Figure 1-6. It consists of a circular disk C as the cathode and a cylindrical anode A with a small circular hole bored through it along the axis of the cylinder. Two parallel plates of length L separated a distance d are placed behind the anode and a zinc sulfide screen is placed at the end of the tube. A gas discharge is maintained between the anode and the cathode by means of some source of high potential. Most of the cathode rays coming from C strike the anode. Some cathode rays, however, pass through the hole in the anode and proceed with uniform velocity v until they strike the fluorescent screen S at O, producing a bright spot.

If a difference of potential V is applied between the plates PP', the cathode rays will be deflected upward toward the positive plate by a force F, given by

$$F = \frac{V}{d}e = ma \qquad (1\text{-}20)$$

in which e is the charge on a cathode ray and m is its mass. Since the electric field between the plates is uniform, the path of the cathode rays will be parabolic in the region between the plates. After they leave this space they will continue with uniform rectilinear motion until they strike the screen at O'. To determine e/m the acceleration of the particle must be measured. This can be done indirectly by noting that the amount of the deflection y parallel to the electric field between the plates is given by

$$y = \tfrac{1}{2}at^2 \qquad (1\text{-}21)$$

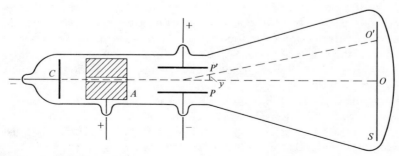

Figure 1-6   A cathode-ray tube.

and that the time t during which the particle is accelerated is given by

$$t = \frac{L}{v} \tag{1-22}$$

where L is the length of the plates and v is the velocity of the rays parallel to the plates. The velocity of the cathode rays can be found very simply with a magnetic field applied at right angles to the electric field and acting over the same length of path L. This magnetic field may be supplied by an electromagnet having its N pole on the side toward the reader and its S pole on the side of the tube away from the reader. If the flux density of the magnetic field is B, the cathode ray will experience an additional force F_1 given by

$$F_1 = Bev \tag{1-23}$$

The direction of the deflection due to the magnetic field will be downward; this field can be adjusted so that the cathode rays are not deviated from their original path, as evidenced by the return of the fluorescent spot to point O. This will occur when the forces due to the electric and magnetic fields are equal, in which case

$$\frac{V}{d}e = Bev \tag{1-24}$$

from which

$$v = \frac{V}{dB} \tag{1-25}$$

The expression for e/m thus becomes

$$\frac{e}{m} = 2\frac{V}{dB^2L^2}y \tag{1-26}$$

All the quantities on the right-hand side of the equation are measurable quantities. The deflection y at the end of the path in the electric field is proportional to the distance OO', which is measured on the fluorescent screen.

The value of e/m will depend on the system of units used. If V is expressed in volts, B in webers per square meter, and the distances in meters, then e/m will be expressed in coulombs per kilogram. Measurements of e/m with electrons from different types of sources such as heated filaments, photoelectric surfaces acted on by light, or radioactive sources all yield the same value of e/m for electrons whose velocities are small relative to the velocity of light. The present accepted value of e/m for electrons is

$$\frac{e}{m} = 1.7588 \times 10^{11} \text{ coul/kg}$$

or

$$\frac{e}{m} = 1.7588 \times 10^8 \text{ coul/gm}$$

or

$$\frac{e}{m} = 1.7588 \times 10^7 \text{ abcoul/gm} \qquad \text{(emu)}$$

1-11 Mass of an Electron

The mass of an electron may now be calculated from the measured values of e/m and its charge e. Using the accepted values of these constants in mksa units, we get

$$m = \frac{e}{e/m} = \frac{1.602 \times 10^{-19} \text{ coul}}{1.7588 \times 10^{11} \text{ coul/kg}}$$

$$m = 9.1096 \times 10^{-31} \text{ kg}$$

or

$$m = 9.1096 \times 10^{-28} \text{ gm}$$

It is instructive to compare the mass of an electron with that of the lightest atom, that is, hydrogen. Since a gram atomic weight of hydrogen contains N_0 atoms and the atomic mass of ^{1}H is 1.00783, the mass of an atom of hydrogen is

$$M_\text{H} = \frac{1.00783}{6.0225 \times 10^{23}} \text{gm}$$

from which

$$M_\text{H} = 1.673 \times 10^{-24} \text{ gm}$$

The ratio of the mass of a hydrogen atom to that of an electron is thus

$$\frac{M_\text{H}}{m} = 1837$$

Since the hydrogen atom ^{1}H consists of a proton and an electron, almost its entire mass is concentrated in the proton. The hydrogen atom is electrically neutral, so that the charge of the proton is $+e$ and that of the electron is $-e$.

Problems

1-1. The following data were recorded during a performance of an oil drop experiment:

plate distance	1.60 cm
distance of fall	1.021 cm

potential difference	5085 volts
viscosity of air	1.824×10^{-4} dyne sec/cm²
density of oil	0.92 gram/cm³
average time of fall	11.88 sec
successive times of rise	22.37 sec
	34.80 sec
	29.25 sec
	19.70 sec
	42.30 sec

Calculate the successive changes in charge on the oil drop and obtain an average value of e from these data. Assume that $b/pa = 0.03$.

1-2.   (a) Prove that Y, the distance between O and O' on the fluorescent screen of Figure 1-6, is given by

$$Y = y \cdot \frac{2D}{L}$$

where D is the distance from the center of the plates to the screen.
The following are the important dimensions of a cathode-ray tube:

distance from anode to screen	33.0 cm
length of plates	7.8 cm
distance between plates	2.4 cm

The plates are placed close to the anode. (b) In the balance method of determining e/m the voltage across the plates was 2800 volts and the magnetic field was 8.20 gauss. When only the magnetic field was on, the deflection on the fluorescent screen was 2.40 mm. Calculate e/m. (c) With the same tube but using the magnetic deflection only, when the accelerating potential between the anode and the cathode was 32,500 volts, a magnetic field of 5.6 gauss produced a displacement of 2.10 cm on the fluorescent screen. Calculate the value of e/m.

1-3.   An electron emitted from a heated filament is accelerated to the anode by a difference of potential of 300 volts between the filament and the anode. Calculate (a) its kinetic energy in ergs, (b) the velocity of the electron when it reaches the anode.

1-4.   An electron moving with a kinetic energy of 5000 eV enters a uniform magnetic field of 200 gauss perpendicular to its direction of motion. Determine the radius of the path of this electron.

1-5.   Determine the velocity of an electron whose magnetic rigidity, when determined by the curvature of its path in a magnetic field, is quoted at 250 gauss cm.

1-6.   A stream of electrons is projected with a velocity v from a point source at an angle θ to the direction of a uniform field of strength B. (a) Describe the paths of these electrons. (b) If electrons are emitted in the form of a hollow cone at a semivertex angle θ, where the magnetic field is along the axis of the cone, show that the electrons will meet (or be focused) at the end of each cycle of their motion.

References

Millikan, R. A., *Electrons + and −*. Chicago: The University of Chicago Press, 1947.
Richtmyer, F. K., E. H. Kennard, and J. N. Cooper, *Introduction to Modern Physics,* 6th ed. New York: McGraw-Hill Book Company, 1969.
Shamos, M. N., *Great Experiments in Physics*. New York: Holt, Rinehart and Winston, 1969.
Taylor, B. N., W. H. Parker, and D. N. Langenberg, *Rev. Mod. Phys.* **41**, 375 (1969).

2 | The Special Theory of Relativity

2-1 Introduction

One of the greatest intellectual achievements of the twentieth century was the formulation of the special theory of relativity in 1905 by Albert Einstein (1879–1955). This theory is an outgrowth of nineteenth century physics—particularly of the synthesis of electromagnetic phenomena into a unified electromagnetic theory in 1864 by James Clerk Maxwell (1831–1879). One of its consequences was the prediction of the existence of electromagnetic waves which travel with the speed of light.

The next step taken by Maxwell was the formulation of the electromagnetic theory of light. Up to this time it had been thought that light waves were similar to the transverse waves that are propagated through a solid. In this older theory the medium through which light was supposed to be propagated was called the *luminiferous ether*. It was imagined that this ether filled all of space and that it was perfectly elastic. It was thus logical to assume that the ether was the medium through which the electromagnetic waves were propagated.

The greatest impetus for the acceptance of the electromagnetic theory of light came in 1887 when Heinrich Hertz (1857–1894) succeeded in producing electromagnetic waves by means of an oscillating current and detecting them with an appropriate circuit containing the proper inductance and capacitance. Modern wireless telegraphy, radio, radar, and television are some of the practical developments based on the work of Maxwell and Hertz.

Attempts were being made at that time to learn more about the luminiferous ether. It might be surmised that the motions of the planets and stars would in some way be affected by the medium through which they were moving, but astronomical data, accumulated over centuries, showed no such effects. Another approach was the design of experiments to determine whether electromagnetic phenomena could be influenced by the motion of the ether. Such an effect could be expected if the ether filled all of space, including the interstices of ponderable matter. It is well known, for example, that the velocity of a wave through an elastic medium is affected by the motion of the medium relative to the observer. The velocity of a sound wave in the atmosphere depends on the motion of the air relative to the observer. To attempt an analogous experiment with electromagnetic waves it is necessary to produce a relative motion of the ether great enough to make accurate measurements of this effect. The velocity of light is 3×10^5 km/sec; the earth moves around the sun with a velocity of about 30 km/sec—about 1/10,000 that of light. An effect produced by such a small relative motion was just within the capabilities of the experimenters of that era. One of the most famous and decisive of these experiments was that performed by A. Michelson and H. Morley in 1887. This experiment had such a profound influence on the development of physics that it will be considered in some detail.

2-2   The Michelson-Morley Experiment

Michelson and Morley (1887) utilized a modification of the Michelson interferometer in their experiment to determine the effect on the velocity of light of the motion of the earth through the ether. The interferometer, shown schematically in Figure 2-1, consists of four pieces of glass, two of them, A and B, made from a single piece of parallel plate glass to be as nearly identical as possible. C and D are silvered on their front surfaces; A is lightly silvered on the back face so that the intensities of the reflected and transmitted beams will be equal. The distances AC and AD are made equal. The entire assembly is mounted on a rigid platform (not shown) capable of rotation about an axis perpendicular to the plane of the paper.

Suppose that the platform is set so that a beam of monochromatic light

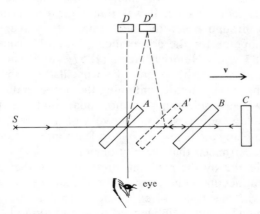

Figure 2-1   Michelson-Morley interferometer adjusted so that arm *ABC* is parallel to the velocity *v* of the earth in its orbit around the sun.

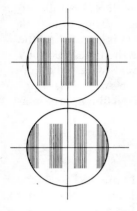

Figure 2-2 Typical interference pattern showing parallel interference fringes. Lower figure shows a shift of the fringes produced by a shift of mirror C by $\frac{1}{4}$ wavelength.

traveling from the source S toward C is moving parallel to the velocity v of the earth in its orbit about the sun. Plates A and B are parallel to each other and at an angle of 45 degrees to the direction of this beam. Let us trace a narrow beam through the interferometer. The beam, on reaching the lightly silvered face of A, is partly reflected to D and partly transmitted through B to C. The two beams are reflected from the front surfaces of mirrors C and D and meet at the rear surface of A, where they produce an interference pattern visible to an observer at O. A typical interference pattern is sketched in Figure 2-2. Those rays of the beam that meet in phase produce bright regions and those rays that meet out of phase by half a wavelength (or half a period) produce dark regions. The presence of the glass plate B serves to make the optical paths A to C and A to D identical. We shall call the length of each path L.

Suppose that the speed of light in a stationary ether is c. The speed of the light along AC relative to the apparatus is $c - v$, and on the return from C to A its speed is $c + v$. The total time t_1 to go from A to C and back to A is

$$t_1 = \frac{L}{c - v} + \frac{L}{c + v}$$

from which

$$t_1 = 2L\frac{c}{c^2 - v^2} \tag{2-1}$$

In going from A to D and back to A the actual path through the ether is along the path $AD'A'$ shown in dotted outline; this is due to the motion of the apparatus through the ether. Its speed along this path is c, so that its speed along the path AD and also along DA is $(c^2 - v^2)^{1/2}$. The time t_2 taken by the light to go from A to D and back to A will differ from the time taken to go from A to C and back to A. The time t_2 is

$$t_2 = \frac{2L}{(c^2 - v^2)^{1/2}} \tag{2-2}$$

Having noted the positions of the interference fringes with AC parallel to the motion of the earth, we then rotate the apparatus through 90 degrees about an axis perpendicular to the plane of the paper so that AD is parallel to the velocity v of the earth and AC is perpendicular to it. Since the times taken by the two beams to travel their respective paths will now equal twice the difference between t_1 and t_2, this will introduce a path difference $\Delta L = 2c(t_1 - t_2)$ and will produce a change in the interference pattern at A. To a first approximation this change in path is

$$\Delta L = 2L \frac{v^2}{c^2} \qquad (2\text{-}3)$$

The effective path length L in this experiment was about 11 meters (produced by multiple reflections) and the path difference should have been about 2.2×10^{-5} cm, sufficient to produce a shift of the interference fringes that could readily have been detected. No change in the interference pattern (other than that which could be ascribed to experimental error) was observed.

The results of this experiment had a profound effect on the foundations of physics in the latter part of the nineteenth century and the first half of the twentieth century. This experiment has since been repeated by D. C. Miller (1925) who claimed to have observed some residual effects, but they were subsequently shown to be due to experimental errors. It will be noted that the quantity to be observed, ΔL, depends on the second power of v/c; C. H. Townes (1958) used the *maser,* an instrument capable of measuring extremely short time intervals with very great accuracy, for a new investigation of the Michelson-Morley type of experiment but one in which the quantity measured depends on the first power of v/c. The results are in complete agreement with those of Michelson and Morley—that is, the speed of light is not affected by the motion of the observer through the ether.

2-3   Newtonian Relativity; Inertial Systems

Newton's formulation of the laws of dynamics is the most widely used set of principles for the formal development of that subject. These laws are three in number. The first law, sometimes also called the *Galilean principle* or the *law of inertia,* states that a body at rest will remain at rest and that a body in motion will continue in motion with constant speed in a straight line unless acted on by an unbalanced force. A coordinate system in which this law holds is called a *Galilean coordinate system* or an *inertial system of reference.*

Newton's second law states that if an unbalanced force acts on a particle the rate of change of momentum of the particle is proportional to the force acting on it. With an appropriate choice of units, such as cgs or mks, this law may be written in the form of an equation:

$$F = \frac{d}{dt}(mv)$$

in which F is the unbalanced or resultant force acting on the particle, m, its mass, v, its velocity, and mv, its momentum. If the mass of the particle is constant, this equation becomes

$$F = m\frac{dv}{dt} = ma$$

where a is the acceleration of the particle.

Newton's third law refers to two particles that interact with one another and states that for every action of one particle on the other there is an equal and opposite reaction produced by the second particle on the first one.

A very important consequence of the above laws is that *the total momentum of a system of particles remains constant if there is no net external force acting on them.* This is the principle of the conservation of momentum. It must be remembered that force, displacement, velocity, acceleration, and momentum are vector quantities.

Experiments have shown that Newton's laws of motion hold to a high degree of approximation when referred to a system of axes fixed on the earth (after making allowance for the earth's rotation). Suppose we perform some dynamical experiment in a train moving with uniform velocity v in the positive x direction and inquire what form Newton's dynamical equations will take when referred to the moving train. Let us call the coordinate system fixed to the earth $OXYZ$ and that fixed to the train $O'X'Y'Z'$. Let us assume that the primed axes are moving in the positive x direction with velocity v and that O and O' coincide at time $t = 0$ (see Fig. 2-3).

The equations of transformation from the primed axes to the unprimed axes are

$$x' = x - vt$$
$$y' = y$$
$$z' = z$$

These equations are sometimes called the Galilean transformation.

If we differentiate these equations with respect to time, using the Newton

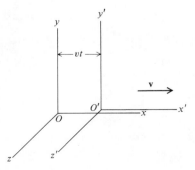

Figure 2-3   The primed coordinate system is moving with uniform velocity v in the x direction. The x and x' axes coincide but are drawn separately for convenience.

notation in which a dot over a symbol represents differentiation with respect to time, we get

$$\dot{x}' = \dot{x} - v$$

$$\dot{y}' = \dot{y}$$

$$\dot{z}' = \dot{z}'$$

These are the equations of transformation for the components of the velocity of a particle as determined by observers in the two coordinate systems.

If we perform a second differentiation with respect to time, we get

$$\ddot{x}' = \ddot{x}$$

$$\ddot{y}' = \ddot{y}$$

$$\ddot{z}' = \ddot{z}$$

as the equations of transformation for the components of the acceleration of the particle as determined by observers in the two coordinate systems. It will be noted that the acceleration of the particle is the same in the two inertial systems. It therefore follows that Newton's second law, $F = ma$, is equally valid when transformed to a set of axes moving with uniform velocity with respect to the original one. Another way of stating this is that Newton's laws of motion remain unchanged in form or are invariant under a Galilean type of transformation.

We are thus led to the conclusion that an observer within the inertial system cannot detect his motion by any dynamical experiment performed wholly within the moving system. This result has long been known and is called Newtonian or classical relativity.

Although Newton conceived of the existence of an absolute space with respect to which the absolute motions of all bodies could be determined, classical dynamics is unable to furnish any criteria for determining this unique reference frame. With the development of the wave theory of light and Maxwell's electrodynamics, attention shifted to the luminiferous ether as a unique frame of reference. Several optical and electrical experiments were designed and performed to determine some effect that could be attributed to the existence and properties of this ether, all without success. The time was ripe for a form of relativity theory that would include electromagnetic as well as dynamical phenomena.

2-4 Fundamental Postulates of Einstein's Special
       Theory of Relativity

Einstein's special or restricted theory of relativity is restricted to inertial systems that are moving with uniform translatory motion relative to each other. Any one observer may regard himself as at rest and express the fundamental laws of physics with respect to a coordinate system at rest with respect to him. The term *stationary system* will be defined as the coordinate system containing this observer and his measuring apparatus.

Einstein's special theory of relativity has at its foundation two fundamental postulates. The first is a generalization of classical relativity to include all fundamental physical laws and may be stated as follows:

Fundamental physical laws should have the same mathematical forms in all inertial systems.

Another way of stating this is that the mathematical formulations of the fundamental laws governing physical phenomena remain unaltered, whether referred to one or the other of two systems of coordinates in uniform translatory motion with respect to one another.

The second postulate is based on the results of the Michelson-Morley and similar experiments. It may be stated as follows:

The speed of light is a constant and is independent of the motion of the source and of the observer. No signal or energy can be transmitted with a speed greater than the speed of light.

The speed of light is defined as the distance traversed by the beam of light divided by the time required to traverse this distance measured by any observer in a given reference frame.

2-5   The Einstein-Lorentz Transformations

Let us consider two observers who are moving with respect to each other with uniform velocity v. Let us choose two rectangular coordinate systems S and S' for these observers and let S' move with velocity v in the x direction relative to S (see Fig. 2-3). An event observed in the S system will occur at coordinates x, y, z at a time t; the same event observed in the S' system will have coordinates x', y', z' and will be observed at a time t'. It should not be assumed that t and t' are necessarily the same, even though the clocks used by the two observers are identical in all respects and have been properly synchronized at the instant they passed each other at the origin of coordinates. This is one of the important differences between the special theory of relativity and the classical theory.

There are various ways of deriving the equations for the transformation of space and time variables appropriate to the special theory of relativity. We shall choose one involving a physical event, the emission of a pulse of light from a source, and consider the description of this event by the two observers. Suppose that the source of light is at O, the origin of the coordinate system S, and suppose that O', the origin of the coordinate system S', coincides with O at the instant that the flash of light is emitted, and let $t = t' = 0$ at this instant. Since the speed of light is constant and independent of the motion of the observer, each one will note a spherical wave spreading out from his origin of coordinates. This is also consistent with the first postulate. If c is the speed of light, the position of the spherical wave will be given by

$$x^2 + y^2 + z^2 = c^2 t^2 \qquad (2\text{-}4)$$

in the S system and by

$$x'^2 + y'^2 + z'^2 = c^2 t'^2 \qquad (2\text{-}5)$$

in the S' system.

The equations of transformation from (x, y, z, t) to $(x'\, y'\, z'\, t')$ that are being sought are those that will transform Equation (2-5) into Equation (2-4). We can use the Galilean transformation as a guide and assume that the equations will be linear and of the form

$$
\begin{aligned}
x' &= K(x - vt) \\
y' &= y \\
z' &= z
\end{aligned} \qquad (2\text{-}6)
$$

and

$$t' = At + Bx$$

where K, A, and B are to be determined. Substitution of the set of Equations (2-6) into Equation (2-5) yields

$$K^2(x - vt)^2 + y^2 + z^2 = c^2(At + Bx)^2$$

Expanding the above equation and collecting terms, we get

$$(K^2 - B^2 c^2)x^2 + y^2 + z^2 + (K^2 v^2 - A^2 c^2)t^2 - 2(ABc^2 + K^2 v)xt = 0$$

Now

$$x^2 + y^2 + z^2 - c^2 t^2 = 0$$

so that we can write

$$(K^2 - B^2 c^2)x^2 + (K^2 v^2 - A^2 c^2)t^2 - 2(ABc^2 + K^2 v)xt = x^2 - c^2 t^2$$

Comparing coefficients of like terms, we get

$$
\begin{aligned}
K^2 - B^2 c^2 &= 1 \\
A^2 c^2 - K^2 v^2 &= c^2 \\
ABc^2 + K^2 &= 0
\end{aligned}
$$

The solution of these three equations yields

$$K = A = \left(1 - \frac{v^2}{c^2}\right)^{-1/2}$$

and

$$B = -\frac{Kv}{c^2}$$

so that the equations of transformation become

$$x' = K(x - vt) \qquad \text{(a)}$$

$$y' = y \qquad \text{(b)}$$

$$z' = z \qquad \text{(c)}$$

$$t' = K\left(t - \frac{vx}{c^2}\right) \qquad \text{(d)} \qquad\qquad (2\text{-}7)$$

$$K = \left(1 - \frac{v^2}{c^2}\right)^{-1/2} \qquad \text{(e)}$$

These equations are known as the *Lorentz-Einstein equations of transformation* of space and time coordinates.

The above equations can also be written as

$$x = K(x' + vt') \qquad \text{(a)}$$

$$y = y' \qquad \text{(b)}$$

$$z = z' \qquad \text{(c)} \qquad\qquad (2\text{-}8)$$

$$t = K\left(t' + \frac{vx'}{c^2}\right) \qquad \text{(d)}$$

This follows immediately from the fact that the S system is moving with a velocity $-v$ relative to the S' system. The value of K, however, remains undisturbed.

2-6 Relativity of Length

Analyses of the Lorentz-Einstein equations lead to extremely interesting and important ideas concerning the fundamental concepts of length and time. Let us consider the measurement of the length of a rod of convenient size and made of some rigid material such as steel. If the rod is at rest with respect to the observer, any one of several well-known methods can be used and its length, L_0, determined. Suppose that this rod is in the S' system and the x' axis is chosen parallel to the length of the rod. The length of the rod can then be expressed as

$$L_0 = x_2' - x_1'$$

where x_1' and x_2' are the coordinates of the two ends of the rod.

The determination of the length of this rod when it is moving with velocity v in the x direction presents a more difficult problem but one that is encountered frequently. There are many ways of doing this but essentially they amount to the determination of the coordinates x_1 and x_2 of the ends of the rod at a given time t by an observer in S. This may be accomplished by using beams of light to record the position of the rod at any instant. The length L of the rod determined by the observer in S will then be

$$L = x_2 - x_1$$

Using Equation (2-7), we find that

$$x_2' - x_1' = K(x_2 - x_1)$$

or

$$L_0 = KL$$

from which

$$L = L_0\left(1 - \frac{v^2}{c^2}\right)^{1/2} \tag{2-9}$$

Thus the measured value of the length of the rod will be less when it is moving parallel to its length than when it is at rest with respect to an observer.

For more than half a century it was generally assumed that the observer would actually *see* that the rod had been shortened by an amount $1/K$. This has been shown by J. Terrell (1959) not to be so. The reason is that the act of *seeing* involves the simultaneous reception of light signals from the different parts of the object. This means that the light received from the end of the rod nearest the observer left the rod at an earlier time than the light received from the end of the rod farthest from the observer. In the derivation of Equation (2-9) it was assumed that the light from the two ends of the rod was emitted simultaneously and recorded by the measuring instruments. When the eye is used as the recording instrument, however, it sees the various parts of the rod at different positions because light reaching the eye from different parts of the rod were emitted at different times. This is just sufficient to cancel the Lorentz contraction.

The above discussion does not affect Equation (2-9). The value of L that appears in this equation must be used wherever appropriate—for example, in determining the velocity of the object and in other calculations.

Before the appearance of Einstein's theory of relativity H. A. Lorentz and G. F. Fitzgerald independently arrived at an explanation of the null result of the Michelson-Morley experiment by making the assumption that the length of a body contracts in the direction of its motion through the ether. The amount of this contraction is just that now given by the relativity theory. An object that has the shape of a sphere of radius R when viewed by a stationary observer will be an ellipsoid of revolution when referred to a set of axes moving with a velocity v with respect to it. The *image* seen by the eye, however, will be circular in all systems and for all directions of motion. The images of objects of other geometrical shapes, although not exhibiting the Lorentz contraction, may appear distorted because the light forming the image was emitted at different times and from different positions of the object in space.

2-7 Relativity of Time

It is apparent from a glance at the equation

$$t' = K\left(t - \frac{vx}{c^2}\right) \qquad\qquad (2\text{-}7\text{d})$$

that the measurement of time and of time intervals by two different observers in uniform relative motion with respect to one another will depend on this motion even though the same or identical clocks are used in this measurement. Let us first consider the case in which a clock is at a given place in the S system—that is, $x_1 = x_2$ during a time interval $\Delta t = t_2 - t_1$. The time interval measured by an observer in S' will be $\Delta t' = t_2' - t_1'$, which from Equation (2-7d) is

$$\Delta t' = K\Delta t$$

so that

$$\Delta t' > \Delta t$$

since

$$K = \left(1 - \frac{v^2}{c^2}\right)^{-1/2}$$

That is, the period of a clock that is moving relative to an observer is longer than when it is measured by an observer who is stationary with respect to it. Stated another way, a clock moving relative to an observer always appears to run slower than it does when it is stationary relative to an observer.

The question of the rate of a clock is of more than academic interest. In addition to the common types of clock with which we are all familiar, there are atomic and nuclear processes that can and are being used for measuring time intervals. Among them are the emission and absorption of radiation and nuclear disintegration. The particles, atoms and nuclei, are usually in motion, and a measurement of a time interval, particularly when $v \approx c$, will be greatly influenced by its velocity. One of these events which has been measured very carefully is the disintegration, or decay, of a mu particle or muon (see Chapter 18). Some of these muons decay when at rest or moving with very small velocities; others decay in flight with velocities very close to the speed of light. The half-life T of muon decay is 1.52×10^{-6} sec when the muon is at rest. It is easy to see that the half-life T', measured when the velocity of the muon is very great, say $v = 0.99c$, will be about seven or eight times as long. This effect, called time dilation, has been checked experimentally, both with cosmic-ray muons and with artificially produced particles.

Another problem of importance is the *ordering* of events as seen by observers who are in uniform relative motion with respect to each other. This also involves the concept of simultaneity. Two events that appear as simultaneous to one observer may not be simultaneous when seen by an observer moving with respect to the first one. Suppose we consider two events that occur with coordinates (x_1, t_1) and (x_2, t_2) in the S system and coordinates (x_1', t_1') and (x_2', t_2') in the S' system. Suppose that they occur simultaneously in the S system; then $t_1 = t_2$. Using Equation (2-7d), we find that

$$t_1' - t_2' = -K \frac{v}{c^2} (x_1 - x_2)$$

The events obviously are *not* simultaneous in the S' system. Similarly, events that are simultaneous in the S' system will not be simultaneous in the S system.

One should next inquire whether there can ever be a crossover in the order of events. Suppose that the two events discussed above occur in the order 1, 2, so that $t_1 > t_2$. Writing the explicit form for $t_1' - t_2'$, we get

$$t_1' - t_2' = K \left[(t_1 - t_2) - \frac{v}{c^2} (x_1 - x_2) \right]$$

If the order of events observed in S' is not to appear reversed in time, then $t_1' > t_2'$ or

$$t_1' - t_2' > 0$$

which implies that

$$(t_1 - t_2) - \frac{v}{c^2} (x_1 - x_2) > 0$$

or

$$\frac{x_1 - x_2}{t_1 - t_2} < \frac{c^2}{v}$$

The above will be true as long as

$$\frac{x_1 - x_2}{t_1 - t_2} < c$$

Thus the order of events will remain unchanged if no signal can be transmitted with a speed greater than the speed of light. The speed of light thus acts as a limiting speed for the transmission of signals and information; this is one part of the fundamental postulate of the special theory of relativity. Hence the order of events will never be interchanged.

2-8 Relative Velocity

Let us consider a particle moving with velocity w in the S' system parallel to the x' axis and assume that the motion is in the $X'Y'$ plane. According to Newtonian relativity, the velocity u of this particle relative to the S system is given by the well-known equation

$$u = v + w$$

where v is the velocity of the S' system relative to S. This is no longer the case when either v or w is comparable to c.

To determine the relativistic form of the equation relating u, v, and w, we must use the Lorentz-Einstein equations of transformation. Now

$$u = \frac{dx}{dt}$$

and

$$w = \frac{dx'}{dt'}$$

From Equation (2-8a) we have

$$x = K(x' + vt')$$

so that

$$dx = K(dx' + v \, dt')$$

and from Equation (2-8d)

$$t = K\left(t' + \frac{v}{c^2} x'\right)$$

so that

$$dt = K\left(dt' + \frac{v}{c^2} dx'\right)$$

hence

$$u = \frac{K(dx' + v \, dt')}{K\left(dt' + \frac{v}{c^2} dx'\right)}$$

or

$$u = \frac{\dfrac{dx'}{dt'} + v}{1 + \dfrac{v}{c^2}\dfrac{dx'}{dt'}}$$

which yields

$$u = \frac{w + v}{1 + vw/c^2} \tag{2-10}$$

Equation (2-10) is the relativistic form of the equation for the composition of parallel velocities. This reduces to the Newtonian form when v and w are small in comparison with c, in which case wv/c^2 is negligible in comparison with unity. In the special case in which $w = c$ we have

$$u = \frac{v + c}{1 + vc/c^2}$$

so that

$$u = c$$

This is in agreement with the fundamental assumption that the speed of light is a constant and independent of the motion of the source or the observer.

2-9 Momentum and Energy in Special Relativity

The classical concepts of momentum and energy are so important that their relativistic forms need to be considered. The classical expression for momentum (mass times velocity) immediately points to a source of difficulty, since velocity does not have simple properties in relativity. It was seen in Equation (2-10) that velocities do not add in the usual way even when they are parallel. The reason for this difficulty is made clearer by consideration of the form

$$v = \frac{dx}{dt} \tag{2-11}$$

A space interval dx transforms according to the Lorentz-Einstein equations, and this fact does not lead to difficulty because dx is in the numerator. The difficulty arises because dt in the denominator also transforms according to the Lorentz-Einstein equations; thus the quotient has complicated transformation properties.

The most appropriate way to resolve this problem is to seek expressions for momentum and energy which will behave just the same as x and t, respectively, under Lorentz-Einstein transformation. As a start toward this goal we introduce the concept of *proper time*. An object with mass m, moving with speed u relative to the laboratory, will be at rest in some frame of reference. In that frame let τ represent time; t will continue to represent time as measured in the laboratory. From Section 2-7 the relation between t and τ is

$$t = \gamma\tau \tag{2-12}$$

where $\gamma = (1 - u^2/c^2)^{-1/2}$. The proper time τ is not affected by the Lorentz-Einstein transformation because it is measured in a frame of reference that is not dependent on any experimental situation. We use γ here to denote the quantity previously called K. In the material to follow we use K when referring to relative motion of axes and γ when considering a moving body.

Let us next recognize that x and t are intimately connected in the Lorentz-Einstein equations. In one dimension Equation (2-7) becomes

$$x' = K\left(x - \frac{v}{c}\, ct\right)$$

$$ct' = K\left(ct - \frac{v}{c}\, x\right) \qquad (2\text{-}13)$$

We have written ct together to make the physical dimensions the same throughout. The quantities x and ct can be written as an ordered pair: (x, ct). This ordered pair, in a more general treatment, is usually called a four-vector because the displacement is, in general, a vector with three components and time provides the fourth component. Next let us differentiate each member of the pair (x, ct) with respect to proper time.

$$\frac{d}{d\tau}(x, ct) = \left(\frac{dx}{d\tau}, c\,\frac{dt}{d\tau}\right)$$

$$= \left(\gamma\frac{dx}{dt}, c\gamma\right)$$

$$= (\gamma u, c\gamma)$$

The process of differentiation does not alter the behavior of the ordered pair under a Lorentz-Einstein transformation. Further, we can multiply each member by m, the mass of the moving object, because m is independent of the frame of reference. Then $(m\gamma u, mc\gamma)$ has exactly the same transformation properties as (x, ct). Last, let us multiply by c; again the transformation properties are unchanged, because all of special relativity is based on the idea that c does not depend on the frame of reference. The ordered pair is now $(m\gamma uc, m\gamma c^2)$.

We define the momentum and the total energy of the object as

$$p = m\gamma u$$

$$\mathcal{E} = m\gamma c^2 \qquad (2\text{-}14)$$

The ordered pair can now be written as $(cp, \mathcal{E})$. It is clear that these quantities transform according to the equations

$$cp' = K\left(cp - \frac{v}{c}\,\mathcal{E}\right)$$

$$\mathcal{E}' = K\left(\mathcal{E} - \frac{v}{c}\,cp\right) \qquad (2\text{-}15)$$

It will be recognized that these definitions have been made in order to preserve simple behavior for the concepts of momentum and energy. We have not yet proved that these definitions are physically useful. Experience has shown that they are indeed useful as extensions of the nonrelativistic concepts of momentum and energy. In the remaining sections of this chapter we demonstrate their utility by showing that these definitions preserve the laws for con-

servation of momentum and energy as well as by showing that our new definitions agree with the classical definitions in the low-velocity limit.

2-10   Conservation of Momentum and Energy

Using the relativistic definitions for momentum and total energy, we next
show that if these quantities are conserved in a frame of reference S they will
also be conserved as measured in a different system S' which moves at a speed
v with respect to S. An object with initial momentum P and energy $\mathcal{E}$ breaks
up into two parts with momenta p_1 and p_2 and energies $\mathcal{E}_1$ and $\mathcal{E}_2$. Conservation
of momentum and energy in the system S is expressed by

$$P = p_1 + p_2$$
$$\mathcal{E} = \mathcal{E}_1 + \mathcal{E}_2 \tag{2-16}$$

In the system S', the energies and momenta can be calculated from Equation (2-15).

$$cP' = K\left(cP - \frac{v}{c}\mathcal{E}\right) \quad \text{(a)}$$

$$\mathcal{E}' = K\left(\mathcal{E} - \frac{v}{c}cP\right) \quad \text{(b)}$$

$$cp_1' = K\left(cp_1 - \frac{v}{c}\mathcal{E}_1\right) \quad \text{(c)}$$

$$\mathcal{E}_1' = K\left(\mathcal{E}_1 - \frac{v}{c}cp_1\right) \quad \text{(d)} \qquad (2-17)$$

$$cp_2' = K\left(cp_2 - \frac{v}{c}\mathcal{E}_2\right) \quad \text{(e)}$$

$$\mathcal{E}_2' = K\left(\mathcal{E}_2 - \frac{v}{c}cp_2\right) \quad \text{(f)}$$

If we add the final momenta, or, in other words, add (c) + (e), we get

$$c(p_1' + p_2') = K\left[c(p_1 + p_2) - \frac{v}{c}(\mathcal{E}_1 + \mathcal{E}_2)\right]$$

$$= K\left(cP - \frac{v}{c}\mathcal{E}\right)$$

$$= cP'$$

so that $P' = p_1' + p_2'$. Adding the energies (d) + (f), we get

$$\mathscr{E}_1' + \mathscr{E}_2' = K\left[(\mathscr{E}_1 + \mathscr{E}_2) - \frac{v}{c}\,c(p_1 + p_2)\right]$$

$$= K\left(\mathscr{E} - \frac{v}{c}\,cP\right)$$

$$= \mathscr{E}'$$

Thus we find that conservation of momentum and energy, as observed in S, implies that an observer in S' will also conclude that momentum and energy are conserved.

The results obtained show that our definitions for momentum and energy have preserved the respective conservation laws. In more advanced work the interplay between these two laws is recognized by referring to them collectively under the name *conservation of four-momentum.*

2-11   Rest Energy, Kinetic Energy, and Total Energy

In the derivation of the Lorentz-Einstein transformation the quantity $c^2t^2 - (x^2 + y^2 + z^2)$ assumes special significance, since it is invariant under a change of reference system. In the case of one-dimensional motion the invariant becomes simply $c^2t^2 - x^2$. It is advisable to inquire whether an invariant exists using $(pc, \mathscr{E})$ which is analogous to the one obtained from (x, ct). Since pc and $\mathscr{E}$ transform in the same way as x and ct, it is clear that the invariant is $\mathscr{E}^2 - p^2c^2$. We may use Equation (2-14), the definition of p and $\mathscr{E}$, to evaluate the invariant quantity.

$$\mathscr{E}^2 - p^2c^2 = m^2\gamma^2c^4 - m^2\gamma^2u^2c^2$$

$$= m^2\gamma^2c^4\left(1 - \frac{u^2}{c^2}\right)$$

Since $\gamma = (1 - u^2/c^2)^{-1/2}$ by definition, we find that

$$\mathscr{E}^2 - p^2c^2 = m^2c^4 \tag{2-18}$$

It is evident that the mass of a particle is a relativistic invariant. The quantity mc^2, which is also invariant, is frequently referred to as the rest energy, since the total energy approaches mc^2 as the momentum goes to zero. The important idea is that an object possesses an intrinsic energy

$$\mathscr{E}_0 = mc^2 \tag{2-19}$$

just because it has mass. The amount of energy involved is quite impressive; for example, if $m = 1$ gram and $c = 3 \times 10^{10}$ cm/sec, we find

$$\mathscr{E}_0 = 1 \text{ gm} \times (3 \times 10^{10} \text{ cm/sec})^2$$

$$= 9 \times 10^{20} \text{ ergs}$$

Similarly, a mass of 1 kg is equivalent to

$$\mathcal{E}_0 = 1 \text{ kg} \times (3 \times 10^8 \text{ m/sec})^2$$
$$= 9 \times 10^{16} \text{ joules}$$

Equation (2-18) is perhaps easier to remember if it is written in the form

$$\mathcal{E}^2 = p^2c^2 + m^2c^4 \qquad (2\text{-}20)$$

a result that looks like the theorem of Pythagoras (see Fig. 2-4). In this form we can also see how the energy of a body depends on its momentum; Figure 2-5 illustrates the fact that the curve is a rectangular hyperbola for bodies of finite mass. For massless objects clearly $\mathcal{E} = pc$, a result which is approached asymptotically for massive particles at very high energies.

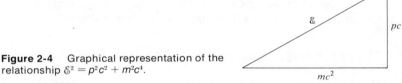

Figure 2-4   Graphical representation of the relationship $\mathcal{E}^2 = p^2c^2 + m^2c^4$.

It may perhaps be surprising that the low-energy limit of Equation (2-20) did not lead to the kinetic energy expression, $\mathcal{E}_K = \frac{1}{2}mu^2$, which is so useful in nonrelativistic mechanics. The fact is that kinetic energy must be defined with care. We can determine the correct expression by calculating the work done in increasing the speed of a particle from zero to its final value u. Suppose that a force F acts parallel to the displacement ds of the particle; then the work done on it is given by

$$dW = F \, ds$$

From Newton's second law

$$F = \frac{dp}{dt}$$

so that

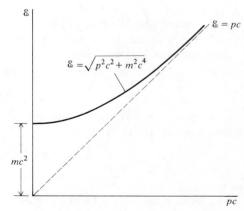

Figure 2-5   Total energy $\mathcal{E}$ as a function of momentum for an object of mass m.

$$dW = \frac{dp}{dt} \, ds = u \, dp$$

Using the fact that $p = m\gamma u$, where $\gamma = (1 - u^2/c^2)^{-1/2}$, we get

$$W = \int_0^{m\gamma u} u \, dp = m \int_0^{\gamma u} u d\left(\frac{u}{\sqrt{1 - u^2/c^2}}\right) \tag{2-21}$$

Integrating this expression and putting in the limits of integration yield

$$W = mc^2\left[\frac{1}{(1 - u^2/c^2)^{1/2}} - 1\right] \tag{2-22}$$

Since the work W was done to increase the velocity of the particle from zero to u, it produced a change in its kinetic energy from zero to $\mathcal{E}_K$, so that

$$\mathcal{E}_K = mc^2(\gamma - 1) \tag{2-23}$$

The relativistic expression for the kinetic energy of a particle will reduce to the more familiar nonrelativistic expression when $u \ll c$. This fact can be seen by expanding the first term in the bracket of Equation (2-23) by the binomial theorem

$$\left(1 - \frac{u^2}{c^2}\right)^{-1/2} = 1 + \frac{1}{2}\frac{u^2}{c^2} + \frac{3}{8}\frac{u^4}{c^4} + \cdots$$

neglecting terms in u/c above the fourth power. Then Equation (2-23) becomes

$$\mathcal{E}_K = mc^2\left(1 + \frac{1}{2}\frac{u}{c^2} + \frac{3}{8}\frac{u^4}{c^4} + \cdots - 1\right)$$

or

$$\mathcal{E}_K = \frac{1}{2}mu^2 + \frac{3}{8}m\frac{u^4}{c^2} + \cdots \tag{2-24}$$

It now becomes apparent that our definition for total energy was reasonable; the kinetic energy $\mathcal{E}_K$ reduces to $\frac{1}{2}mu^2$, when $u \ll c$. From the equations

$$\mathcal{E} = m\gamma c^2 \tag{2-14}$$

$$\mathcal{E}_0 = mc^2 \tag{2-19}$$

and

$$\mathcal{E}_K = mc^2(\gamma - 1) \tag{2-23}$$

it follows that

$$\mathcal{E} = \mathcal{E}_0 + \mathcal{E}_K \tag{2-25}$$

In other words, the total energy of an object is the sum of its rest energy and its kinetic energy.

It should be mentioned that the concept of kinetic energy is less useful in relativity than might be expected. The reason is that the kinetic energy has very complicated properties under a Lorentz-Einstein transformation, being the difference of two quantities that transform in different ways; $\mathcal{E}$ behaves like time and $\mathcal{E}_0$ is invariant. The result is that momentum is the more convenient quantity at relativistic speeds just because its transformation properties are simpler. However, the use of the concept of kinetic energy is sufficiently widespread that it is important to know how to convert from it to momentum and vice versa. The derivation of these conversions is obtained by equating the two expressions for total energy:

$$\mathcal{E} = \mathcal{E}_K + mc^2$$

$$\mathcal{E} = \sqrt{p^2c^2 + m^2c^4}$$

Therefore

$$\mathcal{E}_K = \sqrt{p^2c^2 + m^2c^4} - mc^2 \qquad (2\text{-}26)$$

Solving this equation for p yields

$$p = \frac{1}{c}\sqrt{\mathcal{E}_K{}^2 + 2\mathcal{E}_K\, mc^2} \qquad (2\text{-}27)$$

2-12 Relativistic Mass

One of the important concepts developed in the original formulation of the theory of relativity was that the mass of a particle, as measured by an observer, was a function of its velocity u relative to the observer. The term *relativistic mass*, m_R, was used for the mass of a particle moving with relativistic speed, that is, a speed comparable to the speed of light. The relativistic mass m_R was given in terms of the mass m_0 of the particle at rest by the equation

$$m_R = \frac{m_0}{(1 - u^2/c^2)^{1/2}}$$

The quantity m_0 was usually referred to as the rest-mass of the particle. One of the consequences of this equation is that the relativistic mass approaches infinity as the velocity u approaches the velocity of light c. Stated another way, no material particle can move with a velocity greater than c.

The above equation has been amply verified by many experiments on the measurements of e/m_R for high-speed electrons, beginning with experiments by Kaufmann in 1906 who used high-speed electrons emitted by radioactive substances and repeated with greater accuracy by many others since then.

There are several methods for deriving the equation for the relativistic mass of a particle. One method is to consider an elastic collision between two identical spheres and assume that the principle of conservation of momentum holds in all inertial systems. It is then shown that if such a collision is viewed in two inertial systems moving relative to each other with constant velocity linear momentum will be conserved only if the momentum of each particle is given by

$$p = m_R u$$

where m_R is the relativistic mass.

In modern theoretical physics it is preferred to keep the mass m of a particle constant and to define momentum by

$$p = m\gamma u \qquad\qquad (2\text{-}14)$$

We shall adhere to this description throughout this book and no longer use the terms relativistic mass and rest-mass. This discussion is presented here because these terms will still be found in the literature of science.

2-13   Relativity and Gravitation

The special theory of relativity and the results obtained from it have had a profound effect on the development of atomic and nuclear physics, and conversely, the results of investigations in atomic and nuclear physics have amply verified many of the conclusions drawn from this theory. These conclusions will be developed and discussed at appropriate places in succeeding chapters.

In 1911 Einstein extended the fundamental postulate to noninertial systems —that is, to systems with arbitrary relative motions. This postulate is that the mathematical formulation of a fundamental law of physics should be one that remains invariant when transformed from one reference system to another. This general theory of relativity has, until recently, remained on the periphery of the development of twentieth-century physics. The outstanding success of this theory has been in the development of a theory of gravitation. Until recently it has played no part whatever in the development of atomic and nuclear physics. It may be expected to play a more important role in nuclear physics as the precision of measurement of nuclear processes is increased. Since 1960 a discovery concerning the emission and absorption of gamma rays by identical nuclei, known as the *Moessbauer effect* (see Section 17-16), has been used to examine one of the predictions of the general theory of relativity, the so-called *red-shift*, or shift to the longer wavelengths of radiation from an atom in a gravitational field.

The mathematics needed to discuss the subject of general relativity are beyond the scope of this book. However, it is worth mentioning one important idea concerning gravitation that is fundamental to this theory. This idea is sometimes called the *principle of equivalence;* it expresses the fact that in any given location it is impossible to distinguish between the effects of accelerated motion and those of a gravitational field. Another way of stating this is that the effects of a gravitational field in a given region may be simulated by an appropriate acceleration. The reader may be reminded of the elementary problem of determining the weight of an object that is in the cab of an elevator that is being accelerated in a vertical direction. If the value of the intensity of the gravitational field at a given place is g (also called the acceleration due to gravity), the weight of the object is mg, where m is its mass. If this object is in an elevator being accelerated upward with an acceleration a, the force required to give it this acceleration is

$$m(g + a)$$

If this object is hanging from a spring balance in the elevator cab, its "weight" will be $m(g + a)$. An experimenter in the elevator can conclude that he is in a gravitational field of intensity $g + a$. If the experimenter cannot look out of the elevator cab and is thus not aware of his motion, he will not be able to distinguish between an acceleration of the cab and the presence of a gravitational field by any experiment performed wholly inside the cab. If the elevator cab is accelerated downward with an acceleration a, the "weight" of the object inside will be $m(g - a)$. In the special case in which $a = g$, that of free fall, the object will be "weightless."

Implicit in the above discussion is the equivalence of *inertial* mass and *gravitational* mass—that is, between the mass of a particle as it appears in Newton's laws of motion and the mass of the same particle as it appears in Newton's law of universal gravitation. This equivalence was confirmed to a high degree of precision in an experiment performed about 80 years ago by Eötvös. R. H. Dicke (1964) and his co-workers at Princeton have repeated the Eötvös experiment with greatly increased precision and have obtained results that are consistent with this equivalence.

Problems

2-1. A muon decays with a half-life T when at rest. Muons in motion are observed to decay with a half-life of $6T$; determine their velocity.

2-2. Protons are moving in a circular path with a linear velocity of $0.7c$ relative to the laboratory system. Determine the relative velocities of the protons at points that are diametrically opposite each other.

2-3. Derive the equations for the x, y, and z components of the relative velocity of a particle moving with velocity w in the x', y', z' system; the components of w are w_x, w_y, w_z.

2-4. Derive the equations of transformation for the acceleration of a particle by differentiating the equations derived in Problem 2-3 and using the Lorentz-Einstein equations. Show that for the special case in which the particle is momentarily at rest in the S system, that is,

$$w_x = 0 = w_y = w_z$$

these equations reduce to

$$\ddot{x} = \frac{\ddot{x}'}{K^3} \qquad \ddot{y} = \frac{\ddot{y}'}{K^2} \qquad \ddot{z} = \frac{\ddot{z}'}{K^2}$$

2-5. A relativistic particle whose rest mass is m moves with a velocity of $0.9c$. Determine (a) its total energy and (b) its kinetic energy.

2-6. (a) Show that the velocity of a relativistic particle is given by

$$u = c \, \frac{p}{(p^2 + m^2c^2)^{1/2}}$$

where p is its momentum. (b) Show that the velocity of a relativistic particle of total energy $\mathcal{E}$ is given by $u = c[1 - (mc^2/\mathcal{E})^2]^{1/2}$.

2-7. A relativistic particle of mass m undergoes uniform circular motion with speed u. Show that the centripetal force is given by $m\gamma u^2/R$.

2-8. What velocity is necessary in order to yield $\gamma = 1.1$?

2-9. On a clear night a laser rotated at an angular velocity of 1 radian sec^{-1} produces a spot of light that is visible on Jupiter, 600×10^6 km away. How fast does the spot of light move across Jupiter? Does this violate the theory of relativity? Explain.

2-10. Over what range of velocities can one use the formula $\mathcal{E}_K = \frac{1}{2}mu^2$ with an error no worse than one percent for a particle of mass m?

2-11. Over what range of velocities can one use the formula $\mathcal{E}_K = pc$ with an error no worse than one percent for a particle of mass m?

2-12. Prove that $\mathcal{E}_K \neq \frac{1}{2}m\gamma u^2$.

2-13. Prove that $\mathcal{E}_K \neq p^2/2m\gamma$.

2-14. Prove that $pc/\mathcal{E} = u/c$.

2-15. Continue the binomial expansion to obtain the next term in Equation (2-24), the term of order u^6.

2-16. Find the velocity, momentum, kinetic energy, and total energy for each of the following situations:
(a) An electron in a vacuum tube is accelerated through a potential difference of 100 volts.
(b) An electron in a Van de Graaff accelerator is accelerated through a potential difference of 3 million volts.

2-17. An object S' is moving at a speed of $0.9c$, as measured by an observer in S (at rest). An object S'' also moves (in the same direction as S') at $0.9c$, as viewed by S'. What is the speed of S'' as viewed by S?

2-18. Physicists have often considered building storage devices for high-energy electrons to make colliding beams. If two electrons, each with speed $0.5c$, collide head-on, what will their relative velocity be?

2-19.   Referring to Problem 2-18, what is the total energy of two such colliding electrons, as viewed in the laboratory?

2-20.   Use the binomial theorem to derive Equation (2-3).

Problems for Students Who Know Something about Matrix Algebra

2-21.   If we write the column vector $\begin{pmatrix} x \\ ct \end{pmatrix}$ to represent the ordered pair (x, ct) described in the text, show that the Lorentz-Einstein transformation can be represented by the matrix

$$\begin{pmatrix} K & K\beta \\ K\beta & K \end{pmatrix},$$

where $K = (1 - \beta^2)^{-1/2}$ and $\beta = v/c$.

2-22.   Obtain Equation (2-10), the law for addition of velocities, by multiplying two matrices of the form given in Problem 2-21.

2-23.   Find the inverse of the Lorentz-Einstein transformation matrix and interpret the result physically.

References

Bergmann, P. G., *Introduction to the Theory of Relativity.* New York: Prentice-Hall, 1942, Part I.

Born, M., *Einstein's Theory of Relativity.* New York: Dover Publications, 1962.

Dicke, R. H., *The Theoretical Significance of Experimental Relativity.* New York: Gordon and Breach, 1964.

Einstein, A., *The Meaning of Relativity.* Princeton, N.J.: Princeton University Press, 1953.

————, *Relativity.* New York: Hartsdale House, 1947.

Helliwell, T. M., *Introduction to Special Relativity.* Boston: Allyn and Bacon, 1966.

Kilmister, C. W., *The Environment in Modern Physics.* London: The English Universities Press, 1965.

Leighton, R. B., *Principles of Modern Physics.* New York: McGraw-Hill Book Company, 1959, Chapter 1.

Panofsky, W. K. H., and M. Phillips, *Classical Electricity and Magnetism.* Reading, Mass.: Addison-Wesley Publishing Company, 1955, Chapter 14.

Stephenson, G., and C. W. Kilmister, *Special Relativity for Physicists.* New York: Longmans, Green and Co., 1958.

Taylor, E. F., and J. A. Wheeler, *Spacetime Physics.* San Francisco: W. H. Freeman and Company, 1966.

3 | The Nuclear Atom

3-1   Discovery of Natural Radioactivity

The discovery of radioactivity by Henri Becquerel in 1896, shortly after the discovery of x-rays by Roentgen in 1895, marked the beginning of the modern approach to the study of the structure of the atom. Becquerel was interested in determining whether there was any relationship between the phosphorescence of certain salts after irradiation by ordinary light and the fluorescence of the glass of an x-ray tube which was emitting x-rays. One of the salts used was the double sulfate of uranium and potassium. He wrapped a photographic plate in very thick black paper, placed a crystal of the uranium salt on it, and exposed the whole thing to sunlight. When the plate was developed, a silhouette of the phosphorescent substance appeared in black on the negative, showing that radiations came from the uranium salt. He then varied this experiment by placing a coin, or a metallic screen pierced with an open-work design, between the uranium salt and the photographic plate and, on developing the plate, found the image of each of these objects on the negative.

In attempting to repeat the above experiments, Becquerel ran into some cloudy weather; he put all the materials away in a drawer and waited for a sunny day. A few days later he developed this photographic plate and found that the dark silhouettes again appeared with great intensity, even though the salt had not been exposed to much sunlight. To make certain that this activity goes on without the aid of an external source of light, he built a light-tight box and performed a series of experiments with the photographic plate at the bottom of the box. In one experiment uranium salt crystals were put directly on the photographic plate; he obtained very dark silhouettes of the crystals. In another experiment he put a piece of aluminum between the salt crystals and the photographic plate; he again obtained silhouettes but they were slightly less intense than those obtained without the aluminum plate. He then concluded that the active radiations came from the uranium salt and that the external light had no influence whatever on this activity.

Becquerel then proceeded to experiment with different compounds of uranium, both in crystalline form and in solutions, and found that radiations were emitted by all of them, whether they did or did not phosphoresce. He was thus led to the conclusion that it was the element uranium that was responsible for these radiations. He confirmed this conclusion by repeating these experiments with some commercial powdered uranium. Further experiments showed that the radiations from uranium would also cause the discharge of electrically charged bodies. Shortly thereafter, Rutherford investigated the penetrating power of the radiations from uranium and showed that they were of two types, a very soft radiation easily absorbed in matter, which Rutherford called *alpha rays,* and a more penetrating type of radiation, which he called *beta rays.* It is apparent now that the radiation that affected the photographic plate in Becquerel's experiment consisted of beta rays.

The method used by Rutherford in studying these radiations was an electrical method based on the ionization produced by the radiation in its passage through a gas. The ionization currents so produced can be used for quantitative measurements. Using such a method, Mme Curie showed that the activity of any uranium salt was directly proportional to the quantity of uranium in the salt, thus demonstrating that radioactivity is an atomic phenomenon.

M. and Mme Curie subjected uranium pitchblende to a systematic chemical analysis, and, using an electrical method, measured the activity of the different elements obtained from the pitchblende. In 1898 they succeeded in discovering two new radioactive elements, *polonium* and *radium.* Radium was precipitated in the form of radium chloride. The activity of radium was found to be more than a million times that of a similar quantity of uranium. In 1910 Mme Curie and Debierne obtained pure radium metal by means of electrolysis of the fused salt. The atomic weight of radium as determined by Hoenigschmidt is 225.97. Radium fits in at the end of the second group in the periodic table; it is chemically similar to calcium, strontium, and barium. Many more radioactive substances have been discovered since then, thus filling many of the gaps that existed in the periodic table before the discovery of radioactivity.

3-2   Radiation Emitted by Radioactive Substances

In addition to the alpha and beta rays, naturally radioactive substances emit a third type of radiation called *gamma rays.* The existence of these three distinct types of radiation can be demonstrated very simply. A small quantity of some radioactive salt is placed at the bottom of a long narrow groove in a lead block (Fig. 3-1). A fairly parallel beam will come from the radioactive material R through the slit S. Rays going in all other directions will be absorbed by the lead. This lead block is placed in an airtight chamber and a photographic plate P is placed a short distance above it. To avoid absorption of the rays the air is pumped out of the chamber. A strong magnetic field is applied at right angles to the plane of the paper. After a reasonable exposure, three distinct lines will be

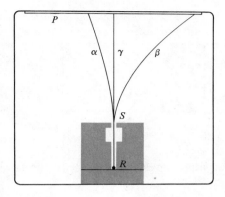

Figure 3-1   Paths of the rays from a radio-active substance R in a magnetic field perpendicular to the plane of the paper. The magnetic field is directed into the paper.

found on the photographic plate. If the magnetic field is directed away from the reader, the positively charged particles or alpha rays will be deflected to the left in the figure, the negatively charged particles or beta rays will be deflected to the right, and the neutral or gamma rays will not be deviated at all.

In very early experiments Henri Becquerel was able to show that the beta particles were similar to cathode rays moving with great velocities. He succeeded in deflecting them first in magnetic fields and then in electric fields, using photographic methods of detection. From the results of these experiments he determined the value of e/m for some of the slower speed beta rays and found this ratio to be about the value then known for electrons. Since then more accurate determinations have been made which show that beta particles emitted by naturally radioactive substances are negatively charged electrons.

3-3   Determination of Q/M for Alpha Particles

One of the best determinations of the ratio of the charge Q to the mass M of the alpha rays was made by Rutherford and Robinson who used the alpha particles emitted by the gas radon ($Z = 86$) and two of its products of disintegration, radium A and radium C', isotopes of polonium of mass numbers 218 and 214, respectively. In the apparatus sketched in Figure 3-2 the alpha particles coming from the radioactive materials contained in the thin-walled glass tube S pass between two silvered glass plates A and B, then through a narrow slit S_1, and strike the photographic plate P placed about 50 cm from the slit. The photographic plate was wrapped in aluminum leaf to protect it from light and from the glow of the source S. With no difference of potential between the plates, a thin stream of undeflected alpha particles passes through the narrow slit S_1 and strikes the photographic plate. When a difference of potential V is maintained between the plates, the alpha particles travel in parabolic paths in the electric field. Only those alpha particles that enter the electric field at some suitable small angle with the plane of the plates will be able to get through the slit S_1. After emerging from this slit they travel in a straight line, tangent to

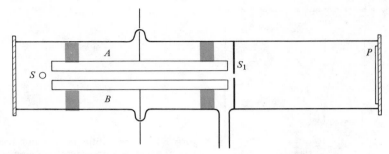

Figure 3-2   Diagram of the apparatus used by Rutherford and Robinson for the determination of Q/M for alpha particles.

this parabola, until they strike the photographic plate. The deflection of a charged particle from its original line of motion produced by the action of an electric field is proportional to Q/Mv^2, where Q is the charge of the particle, M is its mass, and v, its velocity. Hence by measuring the displacement of the line on the photographic plate from the line produced by the undeviated particle the value of Q/Mv^2 can be calculated.

An enlargement of the photograph obtained by the electrostatic deflection method is shown in Figure 3-3. Three distinct lines are visible on each side of the central line. The central line is due to the action of the alpha particles before the field is applied. Lines on both sides of the central line are produced by reversing the electric field. The outermost lines are due to the alpha particles of radon, the next pair of lines, to the alpha particles of radium A, and the innermost pair, to those from radium C'.

To determine the value of Q/M of the alpha particles it is necessary to perform another experiment to determine the velocity v of the alpha particles from the same source. This was done by subjecting the alpha particles from this source to a magnetic field B perpendicular to the direction of their motion, that is, perpendicular to the plane of the figure, and having them strike another photographic plate placed at P. The deflection produced by the magnetic field is proportional to Q/Mv; the latter value can be determined by measuring the

Figure 3-3   Lines on the photographic plate obtained by the electrostatic deflection of the alpha particles. (From Rutherford, Chadwick, and Ellis, *Radiations from Radioactive Substances.* By permission of the Cambridge University Press, publishers.)

Figure 3-4   Lines on the photographic plate obtained by the magnetic deflection of the alpha particles. (From Rutherford, Chadwick, and Ellis, *Radiations from Radioactive Substances.* By permission of The Cambridge University Press, publishers.)

displacement of the line from the line produced by the undeviated particles. Figure 3-4 is an enlargement of the photograph obtained in the magnetic deflection experiment. By reversing the magnetic field, lines were obtained on the other side of the undeviated line. By combining the results of these two experiments the value of Q/M for the alpha particles was computed. The best value was determined from the measurements of the alpha particles of radium C' and yielded

$$\frac{Q}{M} = 4820 \text{ emu/gm}$$

3-4   Nature of the Alpha Particles

The value of Q/M for alpha particles having been accurately determined, a measurement of either Q or M independently would give sufficient evidence of the nature of these particles. One method of determining Q was to count the number of alpha particles, N, emitted by some radioactive substance in a given time interval and then determine the charge NQ carried by these particles. Two methods frequently used for counting alpha particles were (a) the scintillation method and (b) the Geiger counter method.

In the scintillation method alpha particles strike a screen containing a thin layer of powdered zinc sulfide. The energy of the alpha particle is transformed into energy of fluorescent radiation by the small crystals of zinc sulfide. This radiation is in the visible region of the spectrum. Each alpha particle that strikes the screen produces a single scintillation. These scintillations may be viewed with a microscope and the number of alpha particles appearing in the field of view may be counted.

The Geiger counter consists essentially of a cylinder C and a fine wire W mounted parallel to the axis of the cylinder and insulated from it, as shown in Figure 3-5. The cylinder contains a gas, such as air or argon, at a pressure of about 5 to 12 cm of mercury. A difference of potential slightly less than that necessary to produce a discharge through the gas is maintained between the wire and the cylinder wall. Alpha particles can enter the Geiger counter

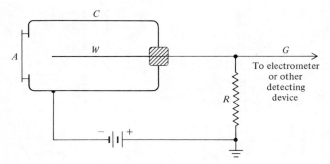

Figure 3-5 Diagram of a Geiger counter.

through the aperture A, which is usually covered with a thin sheet of mica, glass, or aluminum. An alpha particle ionizes the gas along its path; these ions are accelerated by the electric field and produce more ions by collision with neutral atoms and molecules so that the ionization current builds up rapidly. A very high resistance is connected between the wire and ground so that the energy due to the ionization current is rapidly dissipated. The effect is thus the production of a very large current lasting for a very short interval of time. This momentary current registers as a "kick" in an electrometer connected at G. This momentary current may be amplified so that it is capable of operating a loudspeaker or a mechanical counter. By the proper choice of the value of the resistance R, the time constant of the circuit may be made sufficiently small so that each alpha particle that enters the chamber produces a momentary electrical surge which is recorded.

By counting the number of alpha particles entering the aperture A from any source placed a convenient distance away the total number emitted by the source in a given time may be computed. Using this method, Rutherford and Geiger found that 3.57×10^{10} alpha particles/sec are emitted by 1 gram of radium. The present accepted value is 3.70×10^{10} alpha particles/sec.

Knowing the rate at which alpha particles are emitted by a given mass of radioactive material, we may measure the charge on the alpha particles by allowing these particles to charge up a plate connected to an electrometer. Using the alpha particles from radium C, Rutherford and Geiger found the charge Q to be 9.3×10^{-10} stcoul, whereas Regener, using the alpha particles of polonium, obtained the value $Q = 9.58 \times 10^{-10}$ stcoul. Within the limits of experimental error the charge on the alpha particle is equivalent to twice the electronic charge. Using the latter value of Q and the previously determined value of Q/M for computing the mass of the alpha particle, we get

$$M = 6.62 \times 10^{-24} \text{ gm}$$

Comparing this with the mass of the hydrogen atom, we get

$$\frac{M}{M_\text{H}} = \frac{6.62}{1.67} \approx 4$$

The mass of the alpha particle is almost four times that of hydrogen; that is, it has the same atomic weight as helium ($Z=2$). Since the alpha particle carries a charge of $+2e$ and has a mass equal to that of the helium atom, it is probably the nucleus of a helium atom. To make this identification certain Rutherford and Royds (1909) carried out a spectroscopic analysis with the aid of alpha particles emitted by radon.

In this experiment (see Fig. 3-6) some radon was put into the thin-walled glass tube A. This tube was placed in a thick-walled glass tube B that had sealed on to it a capillary tube C into which two electrodes were sealed. Tubes B and C were pumped out and the system was allowed to stand for a few days. The alpha particles emitted by the radon passed through the thin walls of tube A and collected in tube B. The gases collecting in tube B could be compressed and forced into the capillary tube C by letting mercury in through a side tube. After six days enough gas was accumulated and forced into C so that a high voltage across its electrodes produced an electric discharge through the gas. The light coming from this tube was examined with a spectroscope which

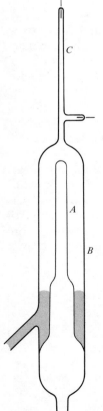

Figure 3-6 Diagram of the tube used to show that alpha particles are helium nuclei.

clearly showed the spectral lines of helium. Control experiments showed that ordinary helium gas could not penetrate through the thin walls of tube A. This spectroscopic evidence proves conclusively that alpha particles are helium nuclei.

3-5 Velocities of the Alpha Particles

The velocities of emission of alpha particles can be measured by allowing them to pass through a magnetic field perpendicular to the direction of motion. In the magnetic spectrograph used by Rosenblum, and sketched in Figure 3-7, the radioactive material R is deposited on a fine wire, and the alpha rays coming through the narrow slit S are bent in a circular path by the magnetic field B perpendicular to the plane of the figure. After traversing a semicircle the alpha rays of any one velocity are focused on the photographic plate P. The air in the chamber is pumped out to avoid loss of velocity by the alpha particles. The radius of the circle is given by the expression

$$BQv = \frac{Mv^2}{r}$$

where the radius of the circle, r, is half the distance from the source to the trace on the photographic plate and v is the velocity of the alpha particle. The magnetic field in this experiment was about 36,000 gauss. Rutherford and his coworkers used a similar apparatus but substituted an ionization chamber for the photographic plate. W. Y. Chang (1948) remeasured the velocities of the alpha particles from polonium and other radioactive sources, using a large, semicircular focusing magnetic spectrograph. The radii of the alpha-particle paths were of the order of 30 cm in a magnetic field of the order of 11,000 gauss, giving greater resolution than in the earlier experiments.

The results of the experiments on the velocities of emission of alpha particles show that these velocities are of the order of 10^9 cm/sec. In many cases the velocity spectrum consists of only a single line—that is, all the alpha particles emitted from this type of element have exactly the same velocity. In another large group of elements the velocity spectrum consists of two or more lines very close together. In a few cases there are several groups of lines

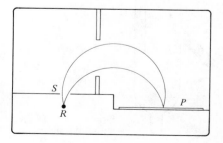

Figure 3-7 Magnetic spectrograph used for measuring the velocities of alpha particles. The magnetic field is perpendicular to the plane of the paper and directed toward the reader.

TABLE 3-1   VELOCITIES OF ALPHA PARTICLES FROM SOME ISOTOPES

Element	Atomic Number Z	Mass Number A	Older Name of the Isotope		Velocity in cm/sec
Po	84	215	Actinium A		1.882×10^9
	84	210	Polonium		1.597
Ra	88	226	Radium	α_1	1.517
				α_2	1.488
Po	84	218	Radium A		1.699
Bi	83	214	Radium C	α_1	1.628
				α_2	1.623
				α_3	1.603
Po	84	214	Radium C'		1.922
Rn	86	222	Radon		1.625
Bi	83	212	Thorium C	α_1	1.711
				α_2	1.705
				α_3	1.665
				α_4	1.645
				α_5	1.642

covering a comparatively large velocity range. A few of the results of these measurements are given in Table 3-1.

3-6   Rutherford's Nuclear Theory of the Atom

Rutherford, in 1911, proposed a nuclear theory for the structure of the atom. He was led to this theory by the results of an experiment by Geiger and Marsden on the scattering of alpha particles by matter. They observed that some of the alpha particles were scattered through angles greater than 90 degrees—that is, they emerged on the side of incidence. To explain such large-angle scattering of fast-moving alpha particles Rutherford assumed that there was an intense electric field within the atom and that the alpha particle was deflected by a single atom. To provide such an intense electric field Rutherford assumed that the entire positive charge of the atom was concentrated in a very small nucleus and that the electrons occupied the space outside the nucleus.

In developing the theory of the scattering of alpha particles by atomic nuclei, Rutherford assumed that both the nucleus and the alpha particle behaved as point charges, that Coulomb's law was valid for such small distances, and that Newtonian mechanics was applicable. To test this theory he instituted a series of experiments on the scattering of alpha particles by thin films of matter. These experiments were performed by Geiger and Marsden in 1913 and repeated with greater accuracy by Chadwick in 1920. They verified Rutherford's nuclear theory of the structure of the atom and showed that the charge on the nucleus of an atom is Ze, where Z is the atomic number of the element and e is the electronic charge.

3-7   Single Scattering of Alpha Particles by Thin Foils

Consider a nucleus of charge Ze stationary at point C and an alpha particle of mass M and charge Q approaching it along the line AB, as shown in Figure 3-8. The original velocity of the alpha particle in the direction of AB is V. There will be a force of repulsion between the two charges given by Coulomb's law,

$$F = k \frac{ZeQ}{r^2} \tag{3-1}$$

where r is the distance between the alpha particle and the nucleus. (It will be recalled that $k = 1$ in the cgs electrostatic system of units and $k = 1/4\pi\epsilon_0$ in the mks system.) Because of this force of repulsion, the alpha particle will be deflected from its original direction and will move in a hyperbolic path with the nucleus at the focus on the convex side of this branch of the hyperbola. When it leaves the region close to the nucleus, the alpha particle will be moving in the direction of OD, making an angle θ with its original direction of motion, AB. It can be shown (see Appendix V-3) that the angle of deflection θ is given by

$$\cot \frac{\theta}{2} = \frac{MV^2}{kZeQ} p \tag{3-2}$$

where p is the distance from the nucleus at C to the original line of motion AB. In the actual experiments on the scattering of alpha particles a large num-

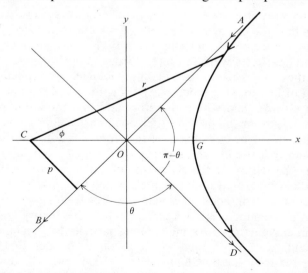

Figure 3-8   The hyperbolic path of an alpha particle in the field of force of a nucleus.

ber of particles were directed against a thin metallic foil. It is therefore necessary to calculate the number of alpha particles scattered through a given angle θ or, what amounts to the same thing, determine the probability that an alpha particle will be deflected through this angle θ. We shall assume that the foil is so thin that the alpha particles suffer no loss in velocity in passing through it; that the probability of multiple deflections of a single alpha particle is negligible.

Suppose that a stream of alpha particles is directed normally on a thin foil of matter of thickness t containing n atoms per unit volume. The number of atoms per unit area of this foil, and also the number of nuclei per unit area of foil, is nt. Any alpha particle whose initial velocity would bring it within a distance p of a nucleus will be deflected through an angle equal to or greater than θ, where θ is given by Equation (3-2). To determine the probability that an alpha particle would come within this distance, imagine a circle of radius p drawn around each nucleus; then the area occupied by all such circles in a unit area of foil is $\pi p^2 nt$. The probability that an alpha particle would come within this distance p of a nucleus is the ratio of the area $\pi p^2 nt$ to unit area. Since every alpha particle that would come within a distance p of a nucleus will be deflected through an angle equal to or greater than θ, the probability that an alpha particle will be deflected through such an angle, or the fraction of the total number of alpha particles that will be deflected through an angle equal to or greater than θ, is given by

$$f = \pi p^2 nt = \pi nt \left(\frac{kZeQ}{MV^2}\right)^2 \cot^2 \frac{\theta}{2} \qquad (3\text{-}3)$$

using the value of p from Equation (3-2).

The probability that the original direction of motion of an alpha particle would fall between the radii p and $p + dp$ or the probability that an alpha particle would be deflected through an angle lying between θ and $\theta + d\theta$ is

$$df = 2\pi pnt \, dp$$

or

$$df = -\pi nt \left(\frac{kZeQ}{MV^2}\right)^2 \cot \frac{\theta}{2} \csc^2 \frac{\theta}{2} \, d\theta \qquad (3\text{-}4)$$

In the experiments on the scattering of alpha particles the scattered particles were detected by the scintillations produced on a fluorescent screen. To compare results for different angles of scattering it is necessary to know the number of alpha particles falling on unit area of the fluorescent screen. To determine this number, consider a stream of alpha particles which are incident normally on the thin foil F as shown in Figure 3-9. The alpha particles that are scattered through an angle θ will travel along the elements of a cone of semivertex angle θ. Similarly, the alpha particles that are scattered through an angle $\theta + d\theta$ will travel along the elements of a cone of semivertex angle $\theta + d\theta$. If

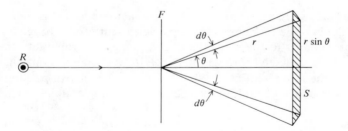

Figure 3-9

the fluorescent screen is to be placed at right angles to the direction of motion of the scattered particles, its shape must be that of a zone of a sphere of radius r and width $r\,d\theta$. Since the radius of this zone is $r\sin\theta$, an element of area dA of the fluorescent screen formed in this way is

$$dA = 2\pi r \sin\theta \cdot r\,d\theta = 2\pi r^2 \sin\theta\,d\theta$$

$$= 4\pi r^2 \sin\frac{\theta}{2}\cos\frac{\theta}{2}\,d\theta \tag{3-5}$$

If N_α is the total number of alpha particles incident on the foil, the number striking unit area of the screen is, from Equations (3-4) and (3-5),

$$N = N_\alpha\,\frac{\pi n t (kZeQ/MV^2)^2 \cot\theta/2\ \csc^2\theta/2\ d\theta}{4\pi r^2 \sin\theta/2\ \cos\theta/2\ d\theta}$$

so that

$$N = \frac{N_\alpha\,k^2 n t (Ze)^2 Q^2}{4r^2 (MV^2)^2 \sin^4\theta/2} \tag{3-6}$$

A study of Equation (3-6) shows that if Rutherford's nuclear theory of the atom is correct the number of alpha particles falling on unit area of a screen at a distance r from the point of scattering must be proportional to

(a) the reciprocal of $\sin^4\theta/2$,
(b) the thickness t of the scattering material,
(c) the reciprocal of the square of initial energy of the particle or $1/(\frac{1}{2}MV^2)^2$,
(d) the square of the nuclear charge, or $(Ze)^2$.

3-8   Experimental Verification of Rutherford's Nuclear Theory of the Atom

Each of the above deductions was tested and verified experimentally in a series of experiments carried out in Rutherford's laboratory. The angular distribution of the alpha particles scattered by a thin foil F, as shown in Figure 3-10, was measured by Geiger and Marsden. The alpha particles from a radon source R passed through a diaphragm D and were scattered by the thin foil F.

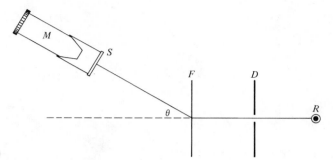

Figure 3-10

The alpha particles that were scattered through an angle θ struck a zinc sulfide screen S and the scintillations were viewed through the microscope M. The microscope and screen were rotated about an axis perpendicular to the plane of the paper and passing through the center of the foil F, and the number of particles striking the screen in a given time was measured at different angles over a range of angles from 5 to 150 degrees. Gold and silver foils were used in the experiment. Some of the results are given in Table 3-2 for gold as the scattering element. If the number of particles N scattered per unit time through an angle θ is proportional to the reciprocal of $\sin^4 \theta/2$, the product $N \times \sin^4 \theta/2$ should be a constant. The last column in Table 3-2 lists the values of these products; these values are approximately constant and lie within the limits of error of the experiment.

To test the dependence of scattering on the thickness of the foil the angle of scattering was kept at about 25 degrees and foils of different thicknesses and also of different materials were used. The alpha-particle source was radium (B + C). The results of the experiments are shown in Figure 3-11, in which the number of particles per minute, N, scattered through an angle of 25 degrees,

TABLE 3-2 SCATTERING OF ALPHA PARTICLES FROM GOLD FOIL

Angle of Deflection θ	$\dfrac{1}{\sin^4 \theta/2}$	Number of Scintillations in Unit Time N	$N \times \sin^4 \theta/2$
150°	1.15	33.1	28.8
135°	1.38	43.0	31.2
120°	1.79	51.9	29.0
105°	2.53	69.5	27.5
75°	7.25	211	29.1
60°	16.0	477	29.8
45°	46.6	1435	30.8
30°	223	7800	35.0
15°	3445	132,000	38.4

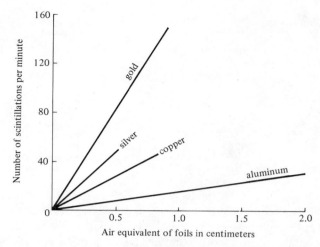

Figure 3-11   Curves showing that the number of alpha particles scattered through a given angle is directly proportional to the thickness of the scattering foil. Each curve is the result of an independent set of measurements.

is plotted as ordinates and the thickness t of the scattering foil of a given material is plotted as abscissae. It is seen that for any one element the number of particles scattered per minute is directly proportional to the thickness of the scattering foil. In the graph the thickness of each foil is expressed in terms of an equivalent length of air path—that is, a thickness of air path that produces the same loss in energy of the alpha particle traversing it as that produced by the material under investigation.

In another series of observations the velocity of the incident alpha particle was changed by placing absorbing screens of mica between the source and the scattering foil. The velocities of the alpha particles were determined empirically by first finding the range R of the alpha particles in air (Section 14-10) and then applying Geiger's rule that

$$R = aV^3 \qquad (3\text{-}7)$$

where a is a constant. The results of the experiment are shown in Table 3-3.

If the number of particles scattered through an angle θ is inversely proportional to the square of the energy of the particles, the product NV^4 should be constant. The values of this product over a wide range of velocities are given in the last column of Table 3-3. These values are considered constant within the limits of error of the experiment.

The experiments of Geiger and Marsden established the essential correctness of Rutherford's nuclear theory of the atom. However, these experiments were not sufficiently accurate to provide a reliable determination of the atomic number Z. It was not until 1920 that Chadwick succeeded in measuring the nuclear charge directly by utilizing alpha-particle scattering techniques. In the meantime Bohr had adopted Rutherford's nuclear hypothesis in his brilliant

TABLE 3-3   VARIATION OF SCATTERING WITH VELOCITY

Range of Alpha Particles	Relative Values of $\frac{1}{V^4}$	Number of Scintillations in Unit Time N	NV^4
5.5	1.0	24.7	25
4.76	1.20	29.0	24
4.05	1.50	33.4	22
3.32	1.91	44	23
2.51	2.84	81	28
1.84	4.32	101	23
1.04	9.22	255	28

work on atomic spectra, and Moseley, from his work on x-ray spectra, was able to determine the nuclear charge and showed that it was equal to Ze.

In Chadwick's experiment alpha particles from the source R, as shown in Figure 3-12, were scattered by a thin foil AA', which was made in the form of an annular ring. The alpha particles that were scattered through an angle θ were counted by the scintillations they produced on the zinc sulfide screen placed at S on the axis of the cone RAA', such that $RA = AS$. The total number of alpha particles falling on the foil AA' could be determined by counting the number reaching S directly from R, since the areas of the screen S and the foil AA' were known. When the scattered rays were investigated, the direct rays from R were cut off by means of the lead plate L. Account was taken of the fact that the annular ring was of finite width. The results of Chadwick's experiments with platinum, silver, and copper foils are listed in Table 3-4.

Within the limits of experimental error, these results are in agreement with Rutherford's nuclear theory of the atom and provide the only direct measurement of the nuclear charge.

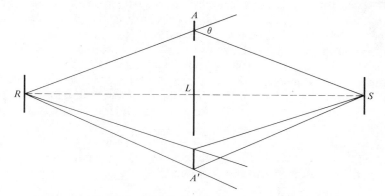

Figure 3-12   Diagram of the arrangement used by Chadwick in his experiments on the scattering of alpha particles.

TABLE 3-4

Element	Nuclear Charge Ze	Atomic Number Z
Cu	$29.3e$	29
Ag	$46.3e$	47
Pt	$77.4e$	78

3-9 Nuclear Sizes: Radii

The results of experiments on the scattering of alpha particles may be used to calculate the distance of approach of the alpha particle to the nucleus (GC in Figure 3-8). The distance of approach will be smallest when the scattering angle θ is greatest—that is, 180 degrees. For this case the direction of the initial velocity V is along the line joining the centers of the nucleus and the alpha particle. At the point of closest approach the alpha particle comes to rest momentarily, so that its energy at this point is simply the Coulomb potential energy $kZeQ/b$, where b is the distance of closest approach. Equating the potential energy to the initial kinetic energy of the alpha particle, we get

$$\frac{kZeQ}{b} = \tfrac{1}{2}MV^2$$

from which

$$b = \frac{2kZeQ}{MV^2} \tag{3-8}$$

In the actual experiments the angle of scattering did not exceed 150 degrees; for this angle the distance of approach was 3.2×10^{-12} cm for gold and about 2×10^{-12} cm for silver. The distance of closest approach b would thus be somewhat smaller than these values. If we consider the process of scattering as a type of collision between two particles, the alpha particle and the nucleus, the distance of closest approach gives an upper limit to the size of the nucleus. Thus the radius of the gold nucleus is less than 3.2×10^{-12} cm. Rutherford also performed experiments on the scattering of alpha particles by lighter nuclei such as aluminum and helium. For helium the closest distance of approach calculated from the results of the experiment yielded a value of about 3×10^{-13} cm.

There are other experimental data from which estimates can be made of the sizes of nuclei. We shall discuss some of these experiments in Part III. These results can be expressed in the following empirical equation for the radius R of a nucleus of an isotope of mass number A:

$$R = r_0 A^{1/3} \tag{3-9}$$

where r_0 is the radius parameter. Numerical values of r_0 vary from about

1.2×10^{-13} to 1.5×10^{-13} cm. The latest data favor the value $r_0 = 1.2 \times 10^{-13}$ cm and will be adopted for all numerical calculations in this book.

3-10   Nuclear Cross Section

The concept of a *nuclear cross section* is a very useful one in nuclear physics. We need not assume that a nucleus is spherical in shape nor that it behaves as a solid elastic sphere of radius R toward projectiles such as alpha particles. We sometimes talk of the "geometrical" cross section given by πR^2, where R is the radius given by Equation (3-9). However, a more useful concept is the *cross section of a nucleus for a given process*, such as the cross section for the scattering of charged particles, the cross section for the absorption of certain particles, or the absorption of radiation such as gamma rays. The term "cross section," used in this sense *is a measure of the probability of the occurrence of the given process*. As an example let us refer once again to the single scattering of alpha particles by nuclei, sometimes called *Coulomb scattering*. It was shown that the probability f that an alpha particle would be scattered by a nucleus through an angle equal to or greater than θ is given by

$$f = \pi p^2 nt \qquad (3\text{-}3)$$

where n is the number of nuclei per unit volume, t is the thickness of the scattering material, and p is the impact parameter. We can now define a nuclear scattering cross section σ_s as

$$\sigma_s = \pi p^2 \qquad (3\text{-}10)$$

hence

$$f = \sigma_s nt \qquad (3\text{-}11)$$

The scattering cross section σ_s is thus proportional to the probability of scattering.

If a beam containing N alpha particles is sent through the substance of thickness t, the number N_s of particles that will be scattered through an angle equal to or greater than θ is

$$N_s = fN = \sigma_s ntN$$

from which

$$\sigma_s = \frac{N_s}{Nnt} \qquad (3\text{-}12)$$

Equation (3-12) can be used to determine scattering cross sections from empirical data.

A large number of nuclear cross sections is of the order of 10^{-24} cm^2. It has been found convenient to use a separate name for this area. The name adopted is *barn;* that is,

$$1 \text{ barn} = 10^{-24} \text{ cm}^2 \qquad (3\text{-}13)$$

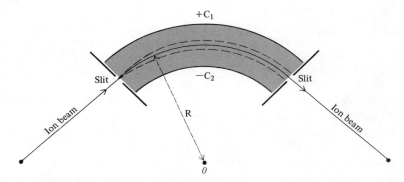

Figure 3-13 Ion beam passing through cylindrical capacitor plates passes through entrance and exit slits after being deviated in electric field.

3-11 Isotopic Masses

The accurate measurement of the masses of the isotopes of elements had an important influence on the development of the nuclear atom and of nuclear physics. These measurements were made originally with mass spectrographs and spectrometers and later through other measurements involving nuclear reactions and nuclear processes. High-precision mass spectrometers were designed and used first by F. W. Aston in England and then by K. T. Bainbridge, A. J. Dempster, and A. O. Nier in the United States and J. H. E. Mattauch in Germany. At first confined to a few laboratories, where they were designed, built, and used for the determination of the relative abundance of the isotopes of an element and the masses of these isotopes, they are now manufactured by many industrial companies and are widely used for many related tasks such as the determination of the products that occur in a complex chemical reaction and the analysis of the constituents of the upper atmosphere.

It will not serve our purpose to describe these mass spectrometers in any detail. It will suffice merely to point out that in most of them a narrow stream of positive ions is sent through a specially shaped electric field between two capacitor plates, as sketched in Figure 3-13, in which they are deflected in such a way that those ions having approximately the same kinetic energy come out in the same direction. These ions then enter a specially designed magnetic field in which the deflection produced is inversely proportional to the momentum of the particle. Ions with the same value of Q/M, where Q is the charge and M is the mass, travel along paths that converge toward a point in the diagram but actually toward a line at right angles to the paper. These lines can be recorded on a photographic plate, as shown in the figure, or the ions may be detected electrically. One method of determining Q/M is to compare the position of this line with that of some known ion or ions whose paths differ from it very slightly. Figure 3-14 is a photograph that shows the separation between the lines of deuterium ^2H and the ions of molecular hydrogen $^1H_2$.

The modern values of a few isotopic masses, given in Table 3-5, are ex-

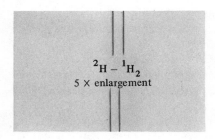

^{2}H – ^{1}H$_2$
5 × enlargement

Figure 3-14   Photograph obtained with the Bainbridge and Jordan mass spectrograph showing the separation of the ions of deuterium, ^{2}H, and molecular hydrogen, ^{1}H$_2$. (From a photograph supplied by K. T. Bainbridge and E. B. Jordan.)

pressed in the unified atomic mass scale with the mass of ^{12}C, the most abundant isotope of carbon taken as the standard, ^{12}C $= 12$ exactly. These values have been obtained from measurements with mass spectrometers and from the results of nuclear reaction experiments. Additional values of isotopic masses can be obtained from the mass differences given in Appendix IV.

3-12   Isotopic Masses and the Constitution of Nuclei

As the result of investigations with the mass spectrograph it has been established that about 280 different isotopes occur in nature. The atomic masses of these isotopes differ very little from whole numbers. The number of naturally

TABLE 3-5

Isotope	Mass in Atomic Mass Units, ^{12}C Scale
n	1.008665
^{1}H	1.007825
^{2}H	2.014102
^{3}H	3.016050
^{3}He	3.016030
^{4}He	4.002603
^{5}Li	5.0125
^{6}Li	6.01512
^{7}Li	7.016004
^{12}C	12.000000
^{13}C	13.003354
^{14}C	14.003242
^{14}N	14.003074
^{16}O	15.994915
^{17}O	16.999133
^{18}O	17.999160
^{235}U	235.043915
^{238}U	238.050770

occurring isotopes per element varies from one for elements fluorine and gold, to 10 for the element tin. Since there are about 100 different elements, there are, on the average, about three isotopes per element.

It is now well established from experiments on the scattering of alpha particles (Section 3-7) and from the study of x-ray spectra (Section 10-1) that an atom consists of a small nucleus with a net positive charge, surrounded by a sufficient number of electrons so that the normal atom is electrically neutral. The results of these experiments lead to the conclusion that the charge of the nucleus of an atom can be represented by Ze, where e is numerically equal to the charge of an electron but positive in sign, and Z is an integer. There are also Z electrons outside the nucleus of a neutral atom. The number Z is called the *atomic number* of the element. The atomic number Z ranges in value from 1 to 105 for presently known elements.

The isotopes of any one element, though they have different atomic masses, all have the same atomic number Z. One conclusion is that the atoms of the various isotopes of any element have exactly the same number of extranuclear electrons. The chemical properties of an element must therefore be ascribed mainly to the arrangement and action of the electrons outside the nucleus. Differences in atomic masses among isotopes of the same element must be due to differences in nuclear structure.

The results of the measurements of the masses of the isotopes show that their masses, based on the unified atomic mass scale, are very nearly integers. This fact suggests the idea that all nuclei are made up of particles, called *nucleons,* each of mass number $A = 1$. At present, two particles of nuclear size and of mass nearly unity are known: these are the proton and the neutron. The proton is the positively charged nucleus of the hydrogen atom of mass number 1. The hydrogen atom consists of a positively charged nucleus, the proton, and an electron outside the nucleus. Since the mass of the hydrogen atom is about 1837 times the mass of the electron, almost the entire mass of the hydrogen atom is due to the proton. The mass of the proton in unified atomic mass units is

$$m_p = 1.00727662 \text{ u}$$

The other particle of unit mass number was discovered experimentally in 1932 by J. Chadwick in experiments on artificial nuclear disintegration (Section 15-4). The neutron has no electric charge and its mass is

$$m_n = 1.0086652 \text{ u}$$

Before the discovery of the neutron, the nucleus was supposed to be made up of protons and electrons. When a tremendous amount of new experimental data on nuclear phenomena became available after the discovery of the neutron, it was necessary to discard the older hypothesis in favor of one in which the nucleus consists of protons and neutrons. There are several arguments against the existence of electrons in the nucleus. Some of these arguments will be presented in appropriate places in the text.

On the hypothesis that a nucleus of an atom consists of protons and neutrons, *the mass number A represents the total number of particles in the nucleus. The atomic number Z is the number of protons in the nucleus,* and *N = A − Z is the number of neutrons in the nucleus; N* is sometimes called the *neutron number* of the isotope. The isotopes of any one element differ only in the number of neutrons in the various nuclei. The convention adopted for representing this information can be illustrated by a typical case, such as oxygen. The atomic number of oxygen is 8; there are three known stable isotopes of oxygen of mass numbers 16, 17, and 18, which are represented by the symbols $^{16}_{8}O$, $^{17}_{8}O$, and $^{18}_{8}O$, respectively, the atomic number being written as a subscript on the lower left-hand side of the chemical symbol of the element, and the mass number as a superscript on the upper left-hand side of the chemical symbol. The atomic number is sometimes omitted, since the chemical symbol uniquely determines it. Thus the isotopes of oxygen will sometimes be designated as ^{16}O, ^{17}O, ^{18}O.

3-13   Mass of a Nucleus

It might at first be supposed that the mass of a nucleus should be the sum of the masses of its constituent particles. A survey of the data, however, shows that the mass of a nucleus is, in general, less than the sum of the masses of its constituent particles in the free state. To account for this difference in mass use is made of the *principle of equivalence of mass and energy,* a principle first developed by Einstein in his theory of relativity. Einstein's principle states that a mass m is equivalent to an amount of energy $\mathcal{E}$, and the equation relating these quantities is

$$\mathcal{E} = mc^2 \tag{3-14}$$

where c is the speed of light.

If Δm is the decrease in mass when a number of particles combine to form the nucleus of an atom, the principle of equivalence of mass and energy states that an amount of energy

$$\Delta\mathcal{E} = \Delta mc^2 \tag{3-15}$$

is released in this process. This amount of energy represents the *binding energy* of the particles in the nucleus.

To illustrate the relationship between mass and energy and the meaning of the *binding energy of a nucleus* let us consider the isotope of lithium of mass number $A = 7$. Its atomic number $Z = 3$; hence it is composed of three protons and four neutrons, plus three electrons outside the nucleus. Let us compare the sum of the masses of the constituent particles in the free state with the mass of the lithium nucleus. The numerical values are taken from Table 3-5. Except for the neutron, these values are all atomic masses; that is, there will be one electron with each proton and three electrons with the lithium nucleus. Thus

$$4 \, {}_0^1 n = 4 \times 1.008665 = 4.034660$$

$$3 \, {}_1^1 \text{H} = 3 \times 1.007825 = 3.023475$$

$$4 \, {}_0^1 n + 3 \, {}_1^1 \text{H} = \overline{7.058135}$$

$${}_3^7 \text{Li} = 7.016004$$

$$\Delta m = \overline{0.042131 \text{ u}}$$

In the subtraction the masses of the electrons cancel out. The difference in mass between these constituent nucleons in the free state and in the nucleus of lithium is

$$\Delta m = 0.042131 \text{ u}$$

This difference in mass is equivalent to the binding energy of the ^{7}Li nucleus. This nucleus thus has a stable structure. In any attempt to disrupt this nucleus energy will have to be supplied from some external source.

The use of atomic masses rather than nuclear masses in calculations of nuclear energy changes is correct in all cases except the one in which a nuclear transformation involving the emission of a positron takes place. (Section 15-8).

3-14   Energy and Mass Units

Energy units other than the erg and joule have been found to be more convenient and more useful in atomic and nuclear physics. Because of the principle of equivalence of mass and energy, the unified atomic mass unit u is sometimes used as an energy unit. This unit has already been defined as one-twelfth of the mass of an atom of ^{12}C; its value in grams is

$$1 \text{ u} = 1.660531 \times 10^{-24} \text{ gm}$$

Using the precise value of $c = 2.997925 \times 10^{10}$ cm/sec in the equation

$$\mathscr{E} = mc^2$$

we get

$$1 \text{ u} = 1.492 \times 10^{-3} \text{ erg}$$

or

$$1 \text{ u} = 1.492 \times 10^{-10} \text{ joule}$$

Another convenient and widely used unit of energy is the electron volt (eV). It is defined as the kinetic energy acquired by a particle with a charge of one electron when it is accelerated in an electric field by a difference of potential of one volt. Since the work done on an electronic charge e by a difference of potential V is eV,

$$1 \text{ electron volt} = 1 \text{ eV} = 1.602 \times 10^{-19} \text{ coul} \times 1 \text{ volt}$$
$$= 1.602 \times 10^{-19} \text{ joule}$$
$$= 1.602 \times 10^{-12} \text{ erg}$$

Multiples of this unit of energy and their symbols are

$$1 \text{ kiloelectron volt} = 1 \text{ keV} = 1000 \text{ eV}$$
$$1 \text{ megaelectron volt} = 1 \text{ MeV} = 10^6 \text{ eV}$$
$$1 \text{ gigaelectron volt} = 1 \text{ GeV} = 10^9 \text{ eV}$$

[*Note.* In the United States 10^9 is called a billion and 10^9 eV, designated as BeV, is called a billion electron volts. In Great Britain, however, 10^9 is a thousand million, whereas 10^{12} is a billion. To avoid confusion the prefix giga (from gigantic) is now used for 10^9.] The conversion factor from unified atomic mass units to MeV is

$$1 \text{ u} = 931.48 \text{ MeV}$$

Thus in the preceding example the binding energy of ^{7}Li, which was found to be 0.04213 u, becomes much more meaningful and impressive when expressed as 39.25 MeV.

The mass of an electron m_e, which is

$$m_e = 9.1091 \times 10^{-28} \text{ gm}$$
$$= 9.1091 \times 10^{-31} \text{ kg}$$

is frequently expressed in terms of its equivalent in electron volts; thus

$$m_e = 0.511 \text{ MeV}$$

or in unified atomic mass units

$$m_e = 5.486 \times 10^{-4} \text{ u}$$

Similarly the proton mass in MeV is $m_p = 938.259$ MeV and the neutron mass is $m_n = 939.553$ MeV.

Problems

3-1. (a) Show that the path of an alpha particle in the electrostatic deflection experiment performed by Rutherford for the determination of Q/M (Section 3-3) is a parabola given by

$$y = x \tan \theta - \tfrac{1}{2}\left(\frac{V}{d}\frac{Q}{M}\right)\frac{x^2}{v_0^2 \cos^2 \theta}$$

where θ is the angle made by the initial velocity v_0 with the line joining the source S and the slit S_1 and V is the difference of potential between the plates. (b) Show that only those alpha particles that are directed initially at an angle θ given by

$$\sin 2\theta = \frac{V}{d} \frac{Q}{M} \frac{L}{v_0^2}$$

will get through the slit S_1; L is the distance from S to S_1. (c) Calculate the value of the angle θ in this experiment for the alpha particles from radon when $V = 2000$ volts; the distance between the plates is 4 mm and the length of each plate is 35 cm.

3-2.   Alpha particles from polonium, $A = 210$, are directed normally against a thin sheet of gold of thickness 10^{-5} cm. The density of gold is 19.32 gm/cm³. Determine the fraction of the incident alpha particles scattered through angles greater than 90 degrees.

3-3.   (a) Protons of 1 MeV energy are scattered by nuclei of gold. Assuming the Rutherford type of scattering, calculate the distance of closest approach. (b) Compare this distance with the radius of the proton. (c) Determine the distance of closest approach of deuterons of the same energy.

3-4.   Using the value of the radius of a nucleus given by Equation (3-9), determine the average density of nuclear matter.

3-5.   Alpha particles from ^{214}Po (ThC′) are deflected in a strong magnetic field. Their magnetic rigidity, expressed in terms of the field strength B and radius of curvature r of the path, is found to be $Br = 427.07$ kilogauss cm. Determine (a) the kinetic energy of these alpha particles and (b) their velocity.

3-6.   Alpha particles from a polonium ($A = 210$) source are deflected in a magnetic field $B = 25,000$ gauss and travel in a circular path of radius r, such that $Br = 331.76$ kilogauss cm. Determine (a) the radius of the path and (b) the kinetic energy of the alpha particles.

3-7.   Bismuth $A = 212$ (Thorium C) emits alpha particles of two different velocities. When traversing a magnetic field of strength $B = 20,000$ gauss they travel in circular paths of radii r_1 and r_2 such that $Br_1 = 354.34$ kilogauss cm and $Br_2 = 355.51$ kilogauss cm. Assuming that these alpha particles start from a point source and travel in semicircular paths, (a) determine the separation of the two alpha-particle lines in this spectrum and (b) calculate their kinetic energies.

3-8.   Singly charged lithium ions of mass numbers 6 and 7, liberated from a heated anode, are accelerated by means of a difference of potential of 400 volts between the anode and the cathode and then pass through a hole in the cathode into a uniform magnetic field perpendicular to their direction of motion. If the magnetic flux density is 800 gauss, determine the radii of the paths of these ions.

3-9.   Uranium isotopes of mass numbers 235 and 238 are to be separated from a piece of uranium by using a mass spectrometer that will deflect them through 180 degrees into two collectors 4.0 cm apart. If the singly charged ions have energies of 2000 eV when entering the magnetic field, calculate (a) the magnetic flux density necessary to achieve this separation and (b) the radii of the paths of the ions.

3-10.   Using the data on the masses of isotopes from Table 3-5, calculate the binding energy of a neutron in a ^7_3Li nucleus. Express the result in both u and MeV.

3-11.   A *velocity-selector* for positive ions used in some mass spectrographs is shown schematically in Figure 3-15. It consists of a parallel-plate capacitor and a magnet which provides a field of uniform flux density **B** throughout the region between the plates of the capacitor. The direction of **B** is at right angles to the direction of the electric field intensity **E** between the plates. A stream of positive ions of charge q and mass M is sent through this region in a direction at right angles to both **E** and **B**. Derive an equation for those ions that go through the space between the capacitor plates without deviation.

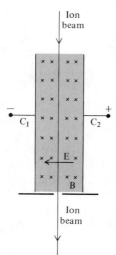

Figure 3-15   Velocity selector for positive ions. Magnetic flux density B directed into plane of figure, at right angles to electric field E and to velocity of ions.

3-12.   A narrow stream of positive ions, having passed through a velocity selector, enters a uniform magnetic field of flux density **B** in a direction at right angles to **B**. (a) Determine the radii of their paths. (b) Show that the ratio M/q of these ions is directly proportional to the radii of their paths.

3-13.   A beam of protons, with kinetic energy equal to 10 MeV, impinges on a copper target of thickness 1 mg/cm². The beam current is 10^{-6} amp, and the total reaction cross section is 100 millibarns. How many reactions occur per second?

References

Conn, G. K. T., and H. D. Turner, *The Evolution of the Nuclear Atom*. London: Iliffe Books, 1965.

Cork, J. M., *Radioactivity and Nuclear Physics*. New York: D. Van Nostrand Company, 1950, Chapters 1–4.

Jayaram, R., *Mass Spectrometry*. New York: Plenum Press, 1966.

Kaplan, I., *Nuclear Physics*. Reading, Mass.: Addison-Wesley Publishing Company, 1955, Chapter 10.

Reed, R. I., *Mass Spectrometry*. New York: Academic Press, 1965, pp. 1–92.

Rutherford, E., J. Chadwick, and C. D. Ellis, *Radiations from Radioactive Substances*. London: Cambridge University Press, 1930, Chapters II, VIII.

Tralli, N., *Classical Electromagnetic Theory*. New York: McGraw-Hill Book Company, 1963.

4

Some Properties of Electromagnetic Radiation

4-1 The Nature of Light

Ever since the days of Newton and Huygens—that is, since the latter part of the seventeenth century—there have been two fundamentally different theories concerning the nature of light. Newton proposed a corpuscular theory of light without specifying definitely the nature of these corpuscles; Huygens proposed a wave theory to explain exactly the same phenomena that were explained by Newton in his corpuscular theory. In one important case, that of the velocity of light in material media, deductions from these two theories led to divergent results. On the basis of Newton's corpuscular theory, light should travel faster in the denser medium, whereas on the basis of the wave theory light should travel slower in the denser medium. It was not until the middle of the nineteenth century that the velocity of light in a dense medium (water) was determined by Foucault and Fizeau and found to be less than that in a vacuum. By this time the wave theory had become fairly well established, mainly because of the work of Young, Fresnel, and Arago on the phenomena of interference and diffraction of light. Further, the polarization of light by reflection and by double refraction through crystals could be explained on the assumption that light waves were transverse waves. The wave theory of light was the only one that could explain satisfactorily all the optical phenomena then known.

Although the wave theory of light was definitely established, the nature of the waves remained a puzzle. It was thought at first that these waves were similar to the transverse waves which are propagated through an elastic solid. A luminiferous ether, filling all of space and having some of the properties of an elastic solid, was postulated as the medium through which light waves were propagated. Maxwell, in his work on electricity and magnetism (1864), showed that a disturbance consisting of transverse electric and magnetic fields should be propagated through the ether with the speed of light. Hertz (1887) succeeded in producing electromagnetic waves by means of an oscillatory current and showed the correctness of Maxwell's theory. Modern radio, radar, television and microwaves are some of the practical developments based on the work of Maxwell and Hertz.

Serious difficulties arose in connection with the properties of the luminiferous ether through which these waves were assumed to be propagated. Einstein's theory of relativity (1905) resolved these difficulties by showing that such an ether was not necessary for the propagation of electromagnetic waves. As a consequence of this theory, light waves are now regarded as electromagnetic oscillations, consisting of variations in the intensities of transverse electric and magnetic fields, each of which may exist in free space—that is, space completely devoid of matter.

In spite of the successes of the electromagnetic theory, there were many phenomena that could not be explained by this theory. Among them are atomic spectra, both in emission and absorption, thermal radiation or blackbody radiation, and the photoelectric effect. The explanation of these phenomena, all of which involve the interaction of radiation and matter, led to the development of the quantum theory of radiation, one of the most important contributions of twentieth-century physics to our understanding of nature. The theory had its origin in the explanation of the distribution of energy in blackbody radiation by Max Planck (1900) who postulated that in its interaction with matter electromagnetic radiation behaved as though it consisted of particles of energy, each particle of energy, or *quantum,* having an energy proportional to the frequency ν of the radiation, or in the form of an equation

$$\mathscr{E} = h\nu \tag{4-1}$$

where h is a constant of proportionality now called the Planck constant.

The greatest impetus to the development and acceptance of this theory came from the explanation of the photoelectric effect by A. Einstein (1905). This was followed by the theory of the hydrogen atom by N. Bohr (1913) who succeeded in explaining the origin of the spectral lines of hydrogen and started the development of the Bohr-Rutherford theory of the nuclear atom.

The above, and other important phenomena to be discussed in this book, showed that electromagnetic radiation had a dual character—that of transverse waves and of particles, called *photons,* each photon having an energy proportional to its frequency. Beginning in 1925, the work of L. De Broglie, E. Schroedinger, and P. A. M. Dirac started the formulation of a new quantum theory, known as *quantum mechanics* or *wave mechanics,* which included the older quantum theory, and vastly extended it to new phenomena, particularly to the waves associated with material particles and to nuclear physics. Some of these developments will be discussed in later chapters.

4-2 Radiation from an Accelerated Charge

On the basis of classical electrodynamics it is shown that an accelerated charge radiates energy in the form of electromagnetic waves. Referring to Figure 4-1, if a charge q moving with a velocity $v \ll c$ is given an acceleration

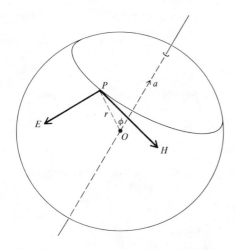

Figure 4-1   The electric intensity E and the magnetic intensity H are tangent to the surface of the sphere.

a when it is at point O, a transverse electromagnetic wave will travel outward in all directions from this point as a center. At any point P a distance $r = ct$ from the center, with ϕ the angle between OP and the acceleration a, the electromagnetic wave will consist of an electric field of intensity E and a magnetic field of intensity H at right angles to each other in a plane at right angles to OP, the direction of motion. The magnitude of the electric and magnetic field intensities at the time $t = r/c$ are given by the following equations:

In cgs units	In mksa units

$$E = \frac{qa}{rc^2} \sin \phi \qquad \text{(4-2a)}$$

$$E = \frac{qa}{4\pi\epsilon_0 rc^2} \sin \phi \qquad \text{(4-2c)}$$

$$H = \frac{qa}{rc^2} \sin \phi \qquad \text{(4-2b)}$$

$$H = \frac{B}{\mu_0} = \frac{qa}{4\pi rc} \sin \phi \qquad \text{(4-2d)}$$

In the cgs system used in Equations (4-2a) and (4-2b) the charge q is to be expressed in statcoulombs, r in centimeters, a in centimeters per second^2, and c in centimeters per second, E will then be in dynes per statcoulomb and H in oersteds. With this choice of units the numerical values of E and H will be identical. In the mksa (meter, kilogram, second, ampere) system q is expressed in coulombs, ϵ_0 is the permittivity of the vacuum, and μ_0 its magnetic permeability; E will be in newtons per coulomb, H in amperes per meters, and B in webers per meter^2 (or in teslas).

Because of the simplicity and symmetry of the equations in cgs units, the rest of the discussion in this chapter will be based on Equations (4-2a) and (4-2b). A glance at these equations shows that the maximum values of E (and also H) occur at points for which $r = OP$ is at right angles to the direction of the acceleration a and are zero when $\phi = 0$. At any particular point P, with r and ϕ constant, E is directly proportional to the acceleration a. Thus, if a varies harmonically with time, E will also vary harmonically.

In electrodynamics it is shown that the energy per unit volume in an electric field in free space is given by

$$w_E = \frac{E^2}{8\pi} \tag{4-3}$$

Similarly the energy per unit volume in a magnetic field in free space is given by

$$w_M = \frac{H^2}{8\pi} \tag{4-4}$$

Hence the total energy per unit volume w in the electromagnetic field is given by

$$w = \frac{E^2}{8\pi} + \frac{H^2}{8\pi} = \frac{E^2}{4\pi} \tag{4-5}$$

Since this energy is propagated with velocity c, the amount of energy I which flows in unit time through unit area perpendicular to the direction of propagation is

$$I = \frac{cE^2}{4\pi} \tag{4-6}$$

Hence I represents the instantaneous value of the intensity of the electromagnetic wave at any point in space. It will be noted, by referring to Equations (4-2a) and (4-2b), that the intensity of the electromagnetic wave produced by a charge which is accelerated along the axis is a maximum in a direction at right angles to the acceleration, and is zero in a direction parallel to the acceleration.

The rate at which energy is radiated from an accelerated charge can be calculated by integrating the intensity I over the surface of a sphere of radius r

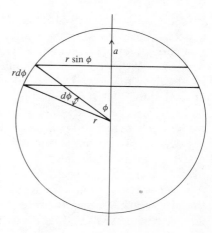

Figure 4-2

containing the accelerated charge at its center. Referring to Figure 4-2, consider a small element of surface area included between two small circles of radii $r \sin \phi$ and $r \sin (\phi + d\phi)$ and of area

$$dA = 2\pi r \sin \phi \cdot r \, d\phi$$

The amount of energy dS, which passes through this element of surface in unit time, is

$$dS = I \, dA = \frac{cE^2}{4\pi} \cdot 2\pi r^2 \sin \phi \, d\phi$$

Substituting the value of E from Equation (4-2a) and integrating over the whole surface, we get

$$S = \int_0^A I \, dA = \frac{1}{2} \frac{a^2 q^2}{c^3} \int_0^\pi \sin^3 \phi \, d\phi$$

which yields

$$S = \frac{2}{3} \frac{a^2 q^2}{c^3} \tag{4-7}$$

In this equation S represents the rate at which energy is radiated from the accelerated charge; it depends on the square of the acceleration and also on the square of the charge. It is thus independent of the direction of the acceleration and of the sign of the charge.

The simplest type of electromagnetic wave is a plane wave in which the electric field intensity is in one direction only, say the y direction, which vibrates with a single frequency and which has a single wavelength λ. Such a wave is called a *linearly polarized monochromatic* electromagnetic wave; the direction of polarization is taken as the direction of vibration of the vector representing the electric field intensity. In the example above the direction of polarization is the y direction. Since an electromagnetic wave is a transverse wave, its direction of propagation is at right angles to the direction of vibration of the electric vector that represents the intensity of the electric field. A linearly polarized monochromatic electromagnetic wave consists not only of an electric field but a magnetic field as well. The magnetic field is always at right angles to the electric field, varying periodically and always in time phase with the electric field. Such a linearly polarized electromagnetic wave traveling, say, in the x direction, can be represented by the following two equations:

$$E = E_0 \sin \frac{2\pi}{\lambda} (x - ct) \tag{4-8a}$$

$$H = H_0 \sin \frac{2\pi}{\lambda} (x - ct) \tag{4-8b}$$

where E is the electric intensity at any point in the path of the wave at any instant of time, H is the intensity of the magnetic field at the same point at the

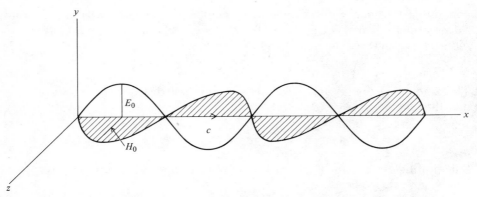

Figure 4-3 Graph showing the values of the intensities of the electric and magnetic fields in a plane electromagnetic wave at a given instant of time.

same time, but always at right angles to the electric vector E. Both E and H are in a plane at right angles to the direction of propagation. In Figure 4-3, E is chosen parallel to the y axis and H, parallel to the z axis; E_0 and H_0 are the amplitudes of the electric and magnetic intensities, respectively. This plane wave is propagated in the x direction with the velocity c.

The average intensity of the wave is the average value of the energy flowing through unit area in unit time. Since the rate at which electromagnetic energy flows through unit area is given by $I = c(E^2/4\pi)$, the average intensity can be obtained by averaging I over a convenient interval of time, say half a period, to yield

$$I_{av} = \frac{1}{T/2} \int_0^{T/2} I \ dt = \frac{2}{T} \int_0^{T/2} \frac{c}{4\pi} E_0^2 \sin^2 \frac{2\pi}{\lambda} (x - ct) \ dt$$

or

$$I_{av} = \frac{cE_0^2}{8\pi} \tag{4-9}$$

The average intensity of the electromagnetic wave is thus proportional to the square of the amplitude of the electric intensity, which, of course, is equal to the square of the amplitude of the magnetic intensity.

4-3 Polarization of Light

The analysis of electromagnetic radiation has a twofold interest: one concerns the properties and nature of the radiation and the other the information that can be obtained from this analysis about the nature, behavior, and origin of the radiation and about the systems that interact with it. The information may be obtained from measurements of wavelengths or frequencies composing the radiation, the distribution of intensities among these wavelengths, and their states of polarization. As far as the optical region is concerned, methods of determining the wavelengths of the radiation—that is, its spectrum—are well known from elementary physics and are merely mentioned here. The instrument used for this analysis is usually a spectroscope in which the dispersing device is either a prism or a diffraction grating. Measurement of the intensity of the radiation is usually a difficult procedure and involves the conversion of the radiation into other forms of energy. Among the devices used are photographic plates in which the density of the line is related to the intensity of the radiation, thermocouples for the conversion of the radiation into heat, other calorimetric devices, photovoltaic and photoelectric devices, and others to be described later.

There are several methods for determining the state of polarization of radiation and also for changing it. In the optical region the devices and methods commonly used are the reflection of the light from a glass plate set at the polarizing angle, transmission of the light through a sheet of Polaroid or through a Nicol prism, or simply the scattering of the light by some convenient substance. It will be recalled that when a beam of unpolarized light is reflected from a glass plate at the polarizing angle, such that

$$\tan i_p = \mu \qquad (4\text{-}10)$$

where i_p is the polarizing angle and μ is the index of refraction of the glass, the reflected beam is linearly polarized with its direction of polarization parallel to the surface of the glass and perpendicular to the direction of propagation. The direction of polarization is taken as the direction of vibration of the vector representing the intensity of the electric field of the electromagnetic radiation; for example, if the graph of Figure 4-3 represents the linearly polarized beam of light, its direction of polarization is parallel to the y axis.

Although the methods listed above are all suitable for the polarization and analysis of optical wavelengths, only scattering has been found useful in the region of very short wavelengths. Polarization by scattering can be demonstrated by sending a beam of unpolarized light vertically down a tube of water containing some fine particles in suspension as shown in Figure 4-4; soapy water will do. Since light is propagated as a transverse wave, all of the vibra-

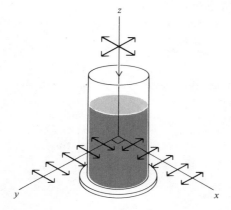

Figure 4-4   Polarization by scattering.

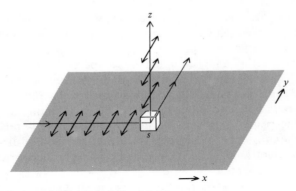

Figure 4-5   Scattering of linearly polarized light.

tions are in the horizontal or x-y plane. If we now analyze the light that is scattered through 90 degrees with the aid of a sheet of Polaroid or a Nicol prism, it will be found that the light is linearly polarized. If the light is viewed along the x axis, its direction of vibration will be found to be parallel to the y axis; conversely, the light traveling in the y direction is polarized with the direction of vibration parallel to the x axis. A simple explanation can be obtained from the fact that the vibrations of the electric vector of the original beam were parallel to the x-y plane. In its passage through the solution the beam of light excited the charges in the medium and produced accelerations in the horizontal plane. We can resolve these accelerations into x and y components. As shown by Equation (4-2), the radiation from charges accelerated in the y direction has maximum intensity in the x direction, with zero intensity in the y direction, and conversely for the radiation from the x component of the acceleration; the maximum intensity of its radiation will be in the y direction.

A substance that polarizes light by scattering can also be used as an analyzer. Suppose, for example, that linearly polarized light traveling in the x direction with its vibrations in the y direction is scattered by S, as shown in Figure 4-5. When the intensity of the light is measured in the y-z plane, it will be found that the intensity is a maximum in the z direction and nearly zero in the y direction.

4-4   Circular and Elliptic Polarization

An interesting and important type of polarization is circular polarization, a special case of elliptic polarization. Since linear vibrations of a charged particle give rise to linearly polarized radiation, it is to be expected that circular motion may be thought of as the sum of two mutually perpendicular linear vibrations of equal amplitude but out of phase by 90 degrees; thus

$$x = R \cos \omega t \qquad y = R \cos \left(\omega t + \frac{\pi}{2} \right) \qquad (4\text{-}11)$$

are the equations for the x and y coordinates of a particle moving in a circular path of radius R with uniform angular velocity ω with the origin at the center of the circle. If the amplitudes of the x and y displacements are not equal, the particle is moving in an elliptic path.

Circularly polarized light and elliptically polarized light can be generated from linearly polarized light with the aid of a thin section of a doubly refracting crystal known as a *quarter-wave plate*. Such a plate is made by cutting a thin section of a uniaxial crystal such as calcite, quartz, or mica so that the *optic axis* is parallel to the crystal faces. When a plane monochromatic wave is incident normally on the face of such a plate, the ordinary ray and the extraordinary ray are refracted normally so that there is no angular separation between the two rays. These two rays, it will be recalled, are linearly polarized at right angles to each other and travel through the crystal with different velocities. If the thickness t of the crystal is such that the difference in optical path lengths between the two rays produces a difference in phase of $\pi/2$ on their emergence, the crystal is called a quarter-wave plate. If the phase difference between the two rays on emergence is π, the crystal is called a *half-wave plate*.

To illustrate the method by which a quarter-wave plate produces circularly polarized light, let us start with a beam of unpolarized monochromatic light and send it through a polarizer with its axis at 45 degrees to the vertical as shown in Figure 4-6. Only linearly polarized light with its direction of vibration parallel to the axis of the polarizer will be transmitted through it. This linearly polarized light is now allowed to pass through a quarter-wave plate set with its optic axis in the vertical direction. We can imagine the linearly polarized light resolved into two equal components: one, marked E, parallel to the optic axis, and the other, marked O, perpendicular to the optic axis. These two components, the extraordinary ray E and the ordinary ray O, travel through the quarter-wave plate with different velocities and emerge out of phase by $\pi/2$—that is, the E ray has maximum amplitude when that of the O

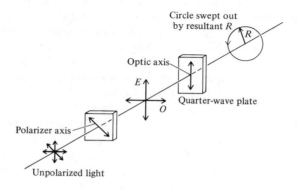

Figure 4-6   Quarter-wave plate used to produce circularly polarized light.

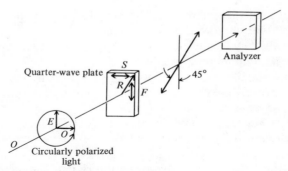

Figure 4-7   Quarter-wave plate changes circularly polarized light to linearly polarized light and shows the direction of rotation of the incident beam.

ray is zero, and the O ray has maximum amplitude when the E ray has zero amplitude. The sum of these two vectors is a vector that rotates so that its terminus sweeps out a circle; hence the name circularly polarized light.

If this circularly polarized light is now allowed to pass through a sheet of Polaroid or a Nicol prism used as an analyzer, and the analyzer is rotated about the beam as an axis, it will be found that the intensity of the transmitted beam remains constant. If a second quarter-wave plate is placed behind the first one so that their optic axes are parallel, the difference in phase between the E and O rays will become π and the transmitted beam will become linearly polarized.

In analyzing circularly polarized light, it is important to know whether the resultant electric intensity vector is rotating clockwise or counterclockwise as viewed by the observer. This is particularly necessary when the source of the circularly polarized radiation is an atomic or nuclear system instead of the optical system described above. Since the action of a quarter-wave plate depends on the difference in velocities of the E and O vibrations passing through it, it is first necessary to determine by optical means which of them has the greater velocity. This information can be marked on the plate. Suppose that in the particular plate at hand the vibrations parallel to the optic axis move faster than those perpendicular to it; these directions are marked F and S, respectively, as shown in Figure 4-7. Suppose further that a source emits light which is circularly polarized in a counterclockwise direction, as viewed from the analyzer. This means that the ray marked O reaches its maximum a quarter of a period ahead of ray E. If, now, this light goes through the plate, the O ray will be retarded by a quarter of a period and be brought into phase with the E vibrations, thus producing linearly polarized light with its vibrations at an angle of $+45$ degrees with S. Conversely, the incidence of light that is polarized in a clockwise direction will produce a resultant linear polarization with its direction of vibration at an angle of -45 degrees with S. It remains merely to determine the direction of vibration of the linearly polarized light coming through the quarter-wave plate with an analyzer to decide whether the original light was polarized in a clockwise or counterclockwise direction.

4-5 Blackbody Radiation

All bodies radiate electromagnetic energy. The amount and the character-istics of this radiation from any body depend on two factors: (a) its temperature T and (b) the nature of its surface. The same body will also absorb radiation falling on it. A body with a polished surface is a poor radiator and a poor ab-sorber of radiation. A body with a rough or blackened surface is a good radia-tor and a good absorber of radiation. A surface coated with lampblack is an excellent absorber and emitter of radiation. The terms *absorptivity* and *emissivity* are used to characterize the nature of the surface; a surface that can absorb all of the radiation falling on it has an absorptivity of 1; its emissivity is also 1. A body with such a surface is called a *blackbody*. The absorptivities and emissivities of other surfaces are all less than unity.

For experimental purposes a blackbody consists of a hollow box with a very small aperture O in one face, as shown in Figure 4-8. Any radiation that reaches the aperture can enter it; that is, it is completely absorbed. Once in-side the box the radiation will strike the walls and will interact with some of the electrons, atoms, and molecules of the material. Some of the radiation will be scattered without change in wavelength; and some of the energy of the electromagnetic wave will be absorbed by the constituents, thereby producing changes in their energy. At this point it is necessary to make some assumptions concerning this process. Classically, one assumes that these constituents can absorb arbitrary amounts of energy without any restrictions. The electrons, for example, are accelerated by the electromagnetic field of the incident radia-tion and in turn radiate electromagnetic waves according to Equations (4-2a and b). The atoms and molecules may be set into vibration and emit radiation in various directions. The scattered and emitted radiation travels to other parts of the enclosure where it produces further energy changes and a very small fraction passes through the aperture O. The temperature of the walls may change during this time, but ultimately equilibrium will be established between the radiation in the enclosure and the surrounding walls at some temperature T. The radiation coming through the aperture O is a sample of the radiation

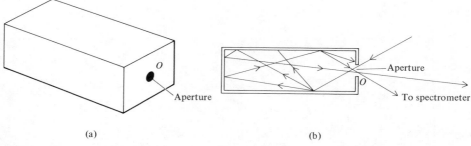

(a)   (b)

Figure 4-8   (a) Insulated hollow enclosure with aperture O in one face. (b) Longitudinal section of hollow enclosure. O is the blackbody.

in the hollow enclosure. Since the aperture O is a blackbody, this radiation is usually called blackbody radiation at temperature T.

Several important results have been obtained empirically from examinations of blackbody radiation. One of these, known as the *Stefan-Boltzmann law,* is that the rate R at which energy is radiated per unit area by a blackbody at temperature T is

$$R = \sigma T^4 \qquad (4\text{-}12)$$

where σ, known as the *Stefan-Boltzmann constant,* has the value

$$\sigma = 5.6697 \times 10^{-5} \text{ erg/cm}^2 \text{ sec K} \qquad (4\text{-}13)$$

with the temperature T expressed in kelvins (K). [*Note.* The symbol for the absolute temperature in Kelvin degrees is now simply K instead of °K.]

The composition of the radiation from blackbodies has been analyzed carefully and accurately by means of spectrometers which utilize either rock-salt prisms or diffraction gratings. In the curves of Figure 4-9 the intensity I of the radiation at a particular wavelength λ is plotted as ordinate and the wavelength λ as abscissa. Each curve represents the complete spectrum taken at a particular temperature. At lower temperatures most of the wavelengths are in the infrared region. At higher temperatures the intensities at all wavelengths increase and include wavelengths in the visible region (4000–7800 Å); and at sufficiently high temperatures the radiation extends into the ultraviolet region (<4000 Å).

It will be noted that each curve passes through a maximum; that is, the intensity I is a maximum at a particular wavelength λ_m for a given temperature of the radiation. Further, as the temperature of the radiation is increased, the particular wavelength λ_m at which the intensity is a maximum is shifted toward lower values. It is found empirically that λ_m varies inversely as the temperature T of the blackbody radiation, or, in the form of an equation,

$$\lambda_m T = b \qquad (4\text{-}14)$$

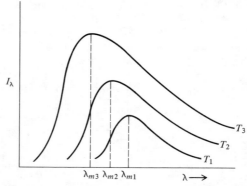

Figure 4-9   Intensity I_λ of radiation as a function of wavelength λ for blackbody radiation at different temperatures; $T_3 > T_2 > T_1$.

where b is a universal constant whose value is

$$b = 2.8978 \times 10^{-1} \text{ cm K} \qquad (4\text{-}15)$$

Equation (4-14) is known as *Wien's displacement law*. This equation can be used to determine the temperature of a blackbody spectroscopically by measuring the wavelength λ_m at which the intensity of the radiation is a maximum. This method has been used extensively in determining stellar temperatures.

In any actual experiment for measuring the intensity I_λ of a particular wavelength λ the measurement is always made over a small wavelength interval $d\lambda$ between λ and $\lambda + d\lambda$. By performing the integration

$$\int_0^\infty I_\lambda \, d\lambda$$

either graphically or from a knowledge of I_λ as a function of λ we can arrive at the Stefan-Boltzmann law.

4-6   The Planck Radiation Law

The radiation coming from the aperture O of the blackbody of Figure 4-8 is characteristic of the radiation inside the hollow enclosure. In an attempt to explain the distribution of energy among the wavelengths it is more convenient to focus our attention on the radiation inside the enclosure. Let ρ represent the energy density of the radiation—that is, the amount of energy per unit volume of the enclosure. As stated earlier this radiation interacts with the electrons, atoms, and molecules of the material of the walls to set the electrons in vibration. Attempts to derive the distribution of energy among the various frequencies (or wavelengths) of the radiation by using classical electrodynamics, thermodynamics, and statistical mechanics were not very successful. The best-known result, called the *Rayleigh-Jeans law,* is given by the equation

$$\rho_\nu d\nu = \frac{8\pi\nu^2}{c^3} kT \, d\nu \qquad (4\text{-}16)$$

where k is the Boltzmann constant. For comparison with experiment this equation can be put in terms of the wavelength λ by using

$$\nu\lambda = c \qquad (4\text{-}17)$$

and

$$d\nu = -\frac{c}{\lambda^2} \, d\lambda \qquad (4\text{-}18)$$

Now the energy $\rho_\nu d\nu$ contained in a frequency interval $d\nu$ between ν and $\nu + d\nu$ is equal to that contained in a corresponding wavelength interval $d\lambda$ between $\lambda + d\lambda$; that is, using Appendix V-9,

$$\rho_\nu \, d\nu = \rho_\lambda \, d\lambda$$

hence

$$\rho_\lambda \, d\lambda = \rho_\nu \frac{c}{\lambda^2} \, d\lambda$$

which yields

$$\rho_\lambda \, d\lambda = \frac{8\pi}{\lambda^4} kT \, d\lambda \tag{4-19}$$

as another form of the Rayleigh-Jeans equation.

This equation agrees with experimental results only at very long wavelengths at any given temperature T, as illustrated in Figure 4-10. At shorter wavelengths, and thus at higher frequencies, the Rayleigh-Jeans equation predicts values that are much too large; ρ_λ approaches infinity as $\lambda \to 0$, contrary to all experimental evidence.

M. Planck (1900) resolved this difficulty by making a radical assumption that the energy of the oscillators (in the walls of a blackbody) cannot have any arbitrary values but are limited to sets of discrete values. According to Planck's hypothesis, if a particle is oscillating with a frequency ν, its energy can have only the values that are given by the equation

$$\mathcal{E} = nh\nu \tag{4-20}$$

where n is an integer and h is a constant of proportionality now called the *Planck constant*. The quantity $h\nu$ is called a *quantum of energy* and n is called a *quantum number* and can have the values

$$n = 0, 1, 2, 3, \ldots \tag{4-21}$$

In the classical theory the oscillators in a substance at temperature T can have a continuous range of energies, but in Planck's theory these oscillators are limited to a set of discrete values of the energy which differ by a single quantum of energy, as shown in Figure 4-11. For high frequencies and small wave-

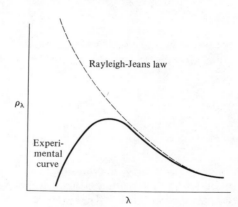

Figure 4-10   Rayleigh-Jeans law agrees with experiment at long wavelengths; $\rho_\lambda \to \infty$ as $\lambda \to 0$.

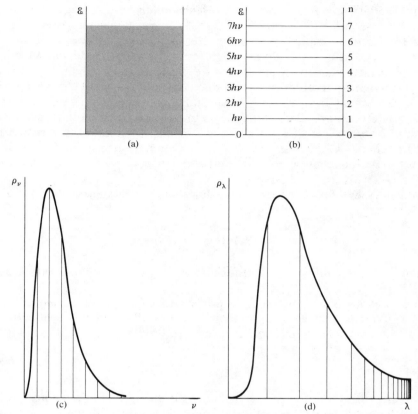

Figure 4-11   (a) Classical—continuous distribution of energy of oscillators at temperature *T*. (b) Planck's hypothesis—discontinuous set of energy levels of oscillators at temperature *T*. (c) Frequency distribution of energy levels. (d) Wavelength distribution of energy levels.

lengths these energy states are widely separated; for low frequencies and long wavelengths these energy states are closer together, approaching the continuous or classical distribution of energy states as $\nu \to 0$ or $\lambda \to \infty$.

One of the most important differences between the classical and the quantum descriptions of the energy states of the oscillators is in the values of the average energies of these systems. One of the hypotheses of statistical mechanics is the *equipartition of energy theorem* which states that the *average kinetic energy* per degree of freedom of a system of particles in thermal equilibrium at temperature *T* is $\frac{1}{2}kT$, where *k* is the Boltzmann constant. Thus in the case of a monatomic ideal gas whose atoms have three degrees of freedom of translational motion the average kinetic energy per particle is $\frac{3}{2}kT$. For the case of a linear harmonic oscillator which has one degree of freedom its average kinetic energy is $\frac{1}{2}kT$. Since the total energy of such an oscillator is twice its average kinetic energy, its average total energy is kT. (See Problem 4-12.)

The Rayleigh-Jeans equation for the distribution of the energy in blackbody radiation in equilibrium with the material of the walls at temperature T contains the factor kT which we can interpret as the average energy of the oscillators in equilibrium with the radiation. In the derivation of this equation it was assumed that the walls of the enclosure were made of conducting materials and that when equilibrium was established the radiation would consist of standing waves of all frequencies. These standing waves would have nodes at appropriate positions, depending on the modes of vibration in these standing (or stationary) waves. To each of these modes of vibration was assigned an average energy kT. It was then a simple matter to determine the number of modes in a frequency interval $d\nu$ between ν and $\nu + d\nu$ or a wavelength interval $d\lambda$ between λ and $\lambda + d\lambda$.

The number of modes per unit volume in the frequency interval $d\nu$ was found to be

$$N(\nu) \, d\nu = \frac{8\pi\nu^2}{c^3} \, d\nu \tag{4-22}$$

which, when multiplied by kT, leads directly to the Rayleigh-Jeans law, Equation (4-16).

It is obvious that if the energies of the oscillators are restricted to integral values of $h\nu$, the average value of their energies will in general be different from kT. Planck calculated this average value (see Problem 4-14) and found it to be

$$\mathcal{E} = \frac{h\nu}{\exp\,(h\nu/kT) - 1} \tag{4-23}$$

Assuming this to be the average value of the energies of the various modes of oscillation in blackbody radiation, Planck obtained the equation

$$\rho_\nu \, d\nu = \frac{8\pi\nu^2}{c^3} \, \frac{h\nu}{\exp\,(h\nu/kT) - 1} \, d\nu \tag{4-24}$$

for the distribution of the energy in blackbody radiation. In terms of the wavelengths of the radiation this equation is

$$\rho_\lambda \, d\lambda = \frac{8\pi hc}{\lambda^5} \, \frac{1}{\exp\,(hc/kT\lambda) - 1} \, d\lambda \tag{4-25}$$

The agreement with experiment is excellent.

4-7   Einstein's Derivation of Planck's Radiation Law

A very instructive and simplified derivation of the Planck radiation law was given by Einstein in 1917 in which he introduced an important concept concerning the probability of the emission and/or absorption of radiation. Suppose we focus our attention on just two of the possible energy states of the oscil-

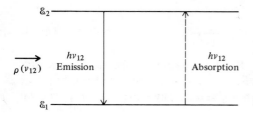

Figure 4-12

lators, $\mathcal{E}_1$ and $\mathcal{E}_2$, in equilibrium with the radiation field at temperature T. As shown in Figure 4-12, let $\mathcal{E}_2 > \mathcal{E}_1$, with

$$\mathcal{E}_2 - \mathcal{E}_1 = h\nu_{12} \qquad (4\text{-}26)$$

where ν_{12} is the frequency of the radiation emitted when an oscillator goes from energy state $\mathcal{E}_2$ to energy state $\mathcal{E}_1$ and is also the frequency of the radiation absorbed by an oscillator when it is raised from the energy state $\mathcal{E}_1$ to state $\mathcal{E}_2$. Let $\rho(\nu_{12})$ represent the energy density of the radiation of frequency ν_{12} in the blackbody radiation.

Suppose that there are n_2 oscillators in energy state $\mathcal{E}_2$; there is a certain probability that some of these oscillators will go to state $\mathcal{E}_1$ spontaneously and emit radiation of frequency ν_{12}. Let us call this probability of spontaneous emission A_{21}. Since there are n_2 oscillators in state $\mathcal{E}_2$, the number emitting radiation of frequency ν_{12} in any given time interval is n_2A_{21}. The action of the radiation field of density $\rho(\nu_{12})$ may stimulate some of the other oscillators in state $\mathcal{E}_2$ to emit radiation. Let B_{21} represent the probability of *stimulated emission* under the influence of the radiation field of density $\rho(\nu_{12})$. The number of oscillators going from state $\mathcal{E}_2$ to state $\mathcal{E}_1$ by stimulated emission in the same time interval will be proportional to the energy density $\rho(\nu_{12})$. Hence for the number of oscillators that will emit stimulated radiation of frequency ν_{12}, we can write the value $n_2B_{21}\rho(\nu_{12})$.

Oscillators in energy state $\mathcal{E}_1$ can be raised to energy state $\mathcal{E}_2$ only by the absorption of radiation of frequency ν_{12}. Let us call the B_{12} the probability of absorption of radiation under the influence of the radiation field of density $\rho(\nu_{12})$. Since there are n_1 oscillators in this energy state, the energy absorbed in this same time interval will be $n_1B_{12}\rho(\nu_{12})$.

Since the radiation field is in equilibrium with the oscillators, the number of atoms per unit time, N_{21}, going from energy state $\mathcal{E}_2$ to $\mathcal{E}_1$ will equal the number N_{12} going from state $\mathcal{E}_1$ to $\mathcal{E}_2$. Now

$$N_{21} = n_2A_{21} + n_2B_{21}\rho(\nu_{12}) \qquad (4\text{-}27a)$$

and

$$N_{12} = n_1B_{12}\rho(\nu_{12}) \qquad (4\text{-}27b)$$

so that

$$n_1B_{12}\rho(\nu_{12}) = n_2[A_{21} + B_{21}\rho(\nu_{12})] \qquad (4\text{-}28)$$

from which

$$\frac{n_2}{n_1} = \frac{B_{12}\,\rho(\nu_{12})}{A_{21} + B_{21}\,\rho(\nu_{12})} \tag{4-29}$$

In order to evaluate the ratio n_2/n_1, Einstein assumed, just as Planck did, that the energy states of the oscillators followed a Boltzmann distribution law; that is, the number of oscillators $n(\mathcal{E})$ in a given energy state $\mathcal{E}$ is proportional to the factor

$$\exp\left(-\frac{\mathcal{E}}{kT}\right)$$

The Boltzmann distribution has already been derived for a special case in Section 1-6. (A more general discussion will be found in Chapter 11.)
Hence we can write

$$\frac{n_2}{n_1} = \exp\left[-\frac{(\mathcal{E}_2 - \mathcal{E}_1)}{kT}\right]$$

$$= \exp\left(-\frac{h\nu_{12}}{kT}\right) \tag{4-30}$$

Substituting this value into Equation (4-29) and solving for $\rho(\nu_{12})$, we get

$$\rho(\nu_{12}) = \frac{A_{21}\,\exp\,(-h\nu_{12}/kT)}{B_{12} - B_{21}\,\exp\,(-h\nu_{12}/kT)} \tag{4-31}$$

or

$$\rho(\nu_{12}) = \frac{A_{21}}{B_{12}\,\exp\,(h\nu_{12}/kT) - B_{21}} \tag{4-32}$$

In order to evaluate the Einstein coefficients A_{21}, B_{12}, and B_{21}, we can compare Equation (4-32) with experimental results. If we assume that

$$B_{12} = B_{21}$$

then Equation (4-32) becomes

$$\rho(\nu_{12}) = \frac{A_{21}}{B_{21}}\,\frac{1}{\exp\,(h\nu_{12}/kT) - 1} \tag{4-33}$$

which is the correct form of the radiation law. A comparison of the above equation with the Planck radiation law shows that

$$\frac{A_{21}}{B_{21}} = \frac{8\pi h\nu_{12}{}^3}{c^3} \tag{4-34}$$

and the radiation law becomes

$$\rho(\nu_{12}) = \frac{8\pi h\nu_{12}{}^3}{c^3}\,\frac{1}{\exp\,(h\nu_{12}/kT) - 1} \tag{4-35}$$

The Einstein coefficients play very important roles in the discussion of intensities of radiation and can be derived with the aid of quantum mechanics.

4-8 Photoelectric Effect

The photoelectric effect was first observed by Hertz in his work on the production of electromagnetic waves. Hertz noted that the air in the spark gap became a better conductor when it was illuminated by ultraviolet light from an arc lamp. Hallwachs (1888) found that when ultraviolet light was incident on a negatively charge zinc surface the surface lost its charge rapidly. If the surface was positively charged, there was no loss of charge under the action of the light. A neutral surface became positively charged when illuminated by ultraviolet light. It is evident that only negative charges are emitted by the surface because of the action of the ultraviolet light. Measurements of e/m for these negative charges by the usual electric and magnetic deflection methods show that these charges are electrons.

A typical arrangement for the study of the photoelectric effect is shown in Figure 4-13. A glass tube has a quartz window W sealed onto it to permit ultraviolet light to enter the tube, P is the photoelectric surface to be investigated, and C is a hollow cylinder that collects the electrons emitted by P. A small hole in the base of the cylinder permits light to reach the plate P. In this type of work it is very important that the surface of the plate P be as clean as possible. Not only must the air be pumped out of the tube, but the entire tube must be baked during the pumping operation to get rid of as much gas as possible if the results of the experiment are to have any quantitative significance. Further, to ensure that no electrons come from the cylinder C because of the action of scattered light, C is usually coated with copper oxide or some other substance that is comparatively insensitive photoelectrically. However, when

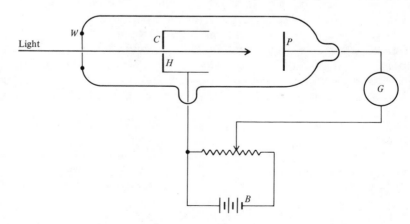

Figure 4-13   Experimental arrangement for measuring the photoelectric current.

C and *P* are made of different materials, it is found that a *contact* difference of potential exists between them. This contact difference of potential may be of the order of one or two volts. If the contact difference of potential between *C* and *P* is enough to make *C* negative with respect to *P*, it will oppose the motion of the electrons. In all photoelectric experiments correction must be made for this contact difference of potential. In the discussion that follows the values of the potential difference between *C* and *P* have been corrected for this effect.

When light from some source such as a quartz mercury arc lamp is incident on the plate *P*, the electrons emitted by the plate are collected by the cylinder *C*. A difference of potential is maintained between *P* and *C* by means of a potentiometer arrangement, and the photoelectric current is measured by a sensitive galvanometer *G*. The photoelectric current is found to depend on two factors: (a) the intensity of the incident light, and (b) the wavelengths of the incident beam. To determine the effect of each of these factors monochromatic light of known wavelength must be used. There are two points of interest in the photoelectric effect: one is the velocity with which the electrons are emitted by the surface; the other is the number of electrons emitted under known conditions.

4-9 Velocity of the Photoelectrons

If monochromatic light of wavelength λ and intensity I is incident on the surface *P*, the electrons emitted from the surface will be acted on by the electric field between the plate *P* and the collecting cylinder *C*. With a potentiometer arrangement the electric field can be varied by varying the difference of potential between *P* and *C*. If *C* is made positive with respect to the plate, the electrons will be accelerated toward *C*; if *C* is made negative, the electrons will be retarded. The current, as registered by the galvanometer, is proportional to the number of electrons per second reaching the cylinder. If the photoelectric current i is plotted against the difference of potential V between *C* and *P*, as shown in Figure 4-14, it is found that for all positive values of *V* the current is constant, but as *C* is made negative with respect to *P* the current decreases rapidly and becomes zero at some value V_0. If the intensity of the monochromatic beam of light is increased from I_1 to I_2 and the experiment is repeated, the photoelectric current is increased in the same ratio for all values of *V*. As

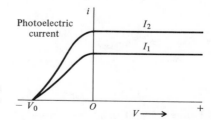

Figure 4-14 Curves of the photoelectric current produced by monochromatic beams of light of two different intensities.

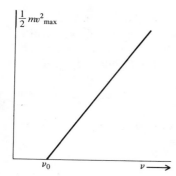

Figure 4-15 Dependence of the stopping potential on the frequency of the incident radiation.

V is made negative, the photoelectric current decreases sharply and reaches zero at the same value of the voltage V_0, which is called the *stopping potential* for this particular wavelength λ.

There are two results of great importance that have been obtained in this experiment. The direct proportionality between the maximum current and the intensity of the light indicates that the number of electrons per second emitted by the surface P is directly proportional to the intensity of the incident beam of light. The fact that the stopping potential V_0 is independent of the intensity of the beam can be interpreted only by assuming that the kinetic energy of the electrons emitted by the surface does not exceed a certain maximum value given by

$$V_0 e = \tfrac{1}{2} m v_{\max}^2 \tag{4-36}$$

Those electrons that leave the surface with kinetic energies less than the maximum are stopped by smaller values of the potential difference. This explains the decrease in the current when the potential difference between C and P is made negative.

The dependence of the stopping potential on the wavelength of the light was investigated by Millikan (1916) in a series of very careful experiments with sodium and potassium as the photoelectric surfaces. The surfaces were illuminated by light of different wavelengths. The stopping potential V_0 was determined for each particular wavelength. The results of Millikan's experiments can best be represented in a graph (Fig. 4-15), in which $V_0 e$ is plotted against the frequency ν of the light incident on the given surface. The curve is a straight line given by the equation

$$V_0 e = \tfrac{1}{2} m v_{\max}^2 = h(\nu - \nu_0) = h\nu - h\nu_0 \tag{4-37}$$

where h is the slope of the line and ν_0 is the smallest frequency that can cause the emission of an electron from the surface. The frequency ν_0 is known as the *threshold frequency* and depends on the nature of the surface. The slope h, however, is a constant that is independent of the nature of the surface.

This quantity h is Planck's constant, which we have already introduced. In atomic phenomena h plays a role whose importance is comparable to that of e (electron charge), m (mass of the electron) or c (speed of light). The photoelectric effect provides a way of measuring the ratio h/e. From his value of e, Millikan determined the value of h to be 6.55×10^{-27} erg sec. The most recent work, by Parker, Langenberg, Denenstein, and Taylor (1969), leads to the result

$$h = 6.6262 \times 10^{-27} \text{ erg sec}$$

$$= 6.6262 \times 10^{-34} \text{ joule sec} \tag{4-38}$$

4-10   Einstein's Photoelectric Equation

The direct dependence of the energy of the photoelectron on the frequency of the incident light cannot be explained by the electromagnetic wave theory of light. In this classical theory there should be a relationship between the intensity of the incident light and the energy of the photoelectron. The intensity of an electromagnetic wave depends on the square of the amplitude of the electric vector and is independent of the frequency of the light. To explain the photoelectric effect Einstein (1905) made use of the concept of a *quantum of energy,* a concept introduced into physics by Planck (1900), in order to explain the distribution of energy among the various wavelengths in the radiation from a blackbody at temperature T. According to Planck's theory, whenever radiation is emitted or absorbed by such a body, the energy is emitted or absorbed in whole *quanta,* where a *quantum of energy* is given by

$$\mathcal{E} = h\nu \tag{4-39}$$

in which ν is the frequency of the radiation and h is the Planck constant. Such a quantum of energy has since received the name *photon*. The energy of a quantum or a photon is directly proportional to the frequency of the radiation. In Einstein's explanation of the photoelectric effect the entire energy of a photon is transferred to a single electron in the metal, and when the electron comes out of the surface of the metal it will have an amount of kinetic energy given by

$$\tfrac{1}{2}mv^2 = h\nu - W \tag{4-40}$$

where W is the work required to remove the electron from the metal. Equation (4-40) is known as Einstein's photoelectric equation. For those electrons that come out with maximum kinetic energy Equation (4-40) becomes identical with Equation (4-37).

The photoelectric effect is not confined to the action of light on metallic surfaces. It can occur in gases and liquids as well as in solids. The nature of the radiation that produces the photoelectric effect includes the whole range of electromagnetic waves, from the very short gamma and x-rays to the regions of the ultraviolet, the visible, and the infrared wavelengths. By measuring the kinetic energies of the emitted photoelectrons, when the substance is bombarded by photons of energy $h\nu$, we can obtain valuable information concerning the origin of these electrons and the properties of the systems from which they came. We shall have occasion to discuss many of these phenomena throughout this book. In this chapter we shall present a brief discussion of the properties of the conduction electrons and the information obtained about conductors from the photoelectric effect.

4-11   Photoelectrons and Conduction Electrons

The distribution of the energies of the conduction electrons in a solid is shown in Figure 4-16, in which the number of electrons $N(\mathcal{E})\,d\mathcal{E}$ with energies in the range of $\mathcal{E}$ to $\mathcal{E} + d\mathcal{E}$ is plotted against the energy $\mathcal{E}$. The equation of this curve can be derived on the basis of reasonable assumptions by using quantum statistical mechanics (see Section 11-6) and will be shown to be $N(\mathcal{E})\,d\mathcal{E} = C\mathcal{E}^{1/2}\,d\mathcal{E}$, where C is a constant. At the absolute zero of temperature this curve ends abruptly at a value of $\mathcal{E} = \mathcal{E}_F$, called the *Fermi energy* and given by

$$\mathcal{E}_F = \frac{h^2}{2m}\left(\frac{3\rho}{8\pi}\right)^{2/3} \tag{4-41}$$

where h is the Planck constant, m is the mass of an electron, and ρ is the number of conduction electrons per unit volume.

This distribution differs completely from that to be expected from classical theory; for example, at absolute zero of temperature, assuming equipartition of energy, all electrons should have the same energy, since $kT = 0$. The modern theory involves two new concepts: (a) the Pauli exclusion principle and (b) a new type of statistics developed by Fermi and Dirac for particles subject to the Pauli exclusion principle.

The Pauli exclusion principle applies to all particles that have an intrinsic angular momentum or *spin* equal to $\frac{1}{2}h/2\pi$. It will be shown that electrons, protons, and neutrons, among others, have spin angular momenta of $\frac{1}{2}h/2\pi$. These particles are called *fermions*. The Pauli exclusion principle states that in a closed system no two identical fermions can have the same energy or,

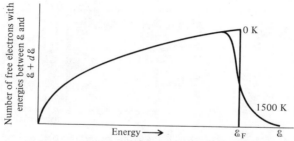

Figure 4-16   Distribution of energies among the conduction electrons of a metal at 0 K and at 1500 K.

stated another way, in a closed system no two electrons can be in the same energy state.

The conduction electrons are treated as though they were particles of a gas, an electron gas, moving freely in the conductor and exerting very little force on each other. The electrostatic forces due to the nuclei essentially cancel each other, except for those at the surface of the metal. The electrons can thus be assumed to be moving in a potential energy well. Particles moving in such a well are limited to discrete energy values. In the quantum mechanical solution given in Section 7-7 the electron is assumed to be moving in a three-dimensional well of infinite height. Fermi-Dirac statistics is then applied to determine the distribution of the electrons at these energy levels, yielding Equation (4-41). Figure 4-17 is a schematic representation of this distribution. Although the original solution was for a potential well of infinite height, the potential energy at the surface of the metal is taken at a height $\mathscr{E}_s$ above the lowest energy of the conduction electrons, taken as the zero level.

The electrons will normally fill the lowest levels first, and because of the exclusion principle these levels will be filled quickly, and at zero degrees kelvin the highest filled energy level will be $\mathscr{E}_F$, the Fermi energy. At high temperatures there is only a slight change in the energies and their distribution among the electrons in the higher energy levels, as shown in the distribution curve for the conductor at 1500 K.

In the quantum mechanical solution quoted above no account was taken of the spin of the electrons. It will be shown later that an electron can have only

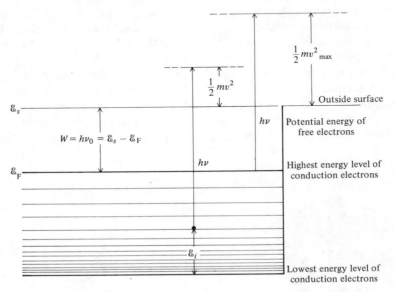

Figure 4-17   Energies of conduction electrons inside a metal lie between 0 and $\mathscr{E}_F$. A photon of energy $h\nu$ ejects an electron with kinetic energy $\frac{1}{2}mv^2$; another photon of equal energy ejects another electron with maximum kinetic energy.

two orientations of its spin axis at a given energy level. There thus can be two electrons in each of these levels with their axes of spin oriented in opposite directions.

For the purposes of this discussion we may consider the distribution of energy among the free electrons in a conductor at room temperature to be almost the same as one at absolute zero. Referring to Figure 4-17, we note that $\mathcal{E}_S$ is the energy required to remove a conduction electron from the lowest energy level.

$\mathcal{E}_i$ is the energy of any one of the conduction electrons; it can have any value from zero to $\mathcal{E}_F$. Suppose that a photon of energy $h\nu$ is incident on the metallic surface and ejects a photoelectron. This electron will leave the surface of the metal with an amount of kinetic energy given by

$$\tfrac{1}{2}mv^2 = h\nu - (\mathcal{E}_S - \mathcal{E}_i) \tag{4-42}$$

Thus there can be a wide range of values for the kinetic energies of the electrons ejected by the action of photons of frequency ν. The electrons ejected with maximum kinetic energy are those that had energies $\mathcal{E}_i = \mathcal{E}_F$ inside the metal; thus

$$\tfrac{1}{2}mv^2_{\max} = h\nu - (\mathcal{E}_S - \mathcal{E}_F) \tag{4-43}$$

A comparison of Equations (4-40) and (4-43) shows that

$$W = \mathcal{E}_S - \mathcal{E}_F = h\nu_0 \tag{4-44}$$

The quantity W is usually called the work function of the metal; its numerical value is of the order of a few electron volts for most conductors and is the same as the work function for thermionic emission from these metals. At higher temperatures the distribution of energies among the free electrons lacks a sharp maximum; hence the curve for the stopping potential should not cut the V axis at V_0 but should approach this axis asymptotically. This has been observed in careful measurements on the photoelectric effect.

4-12 Phototubes

The simple photoelectric tube, which consists of a photosensitive surface used as a cathode and a metal in the form of a wire or other geometrical shape acting as the anode to collect the electrons emitted by the cathode, has come into common use for a great variety of purposes. The photoelectric current may be amplified with the aid of appropriate circuits so that very weak light signals can produce measurable effects. Special phototubes, known as *photomultiplier tubes,* have been developed in which the amplification takes place within the phototube itself. The design of a photomultiplier tube is shown schematically in Figure 4-18, in which P is the photosensitive cathode. Light striking P causes the ejection of photoelectrons from it; these electrons are then attracted to a metal surface called a *dynode* and labeled 1 in the figure. It is known that

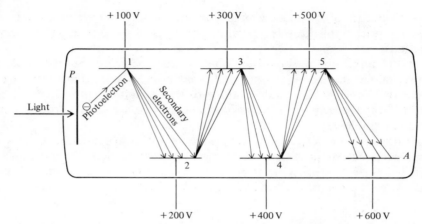

Figure 4-18 Schematic diagram illustrating the operation of a photomultiplier tube. Four secondary electrons are assumed to be emitted for each electron striking a dynode.

an electron striking a metal surface with sufficient velocity can cause the ejection of one or more electrons from the surface. This process is called *secondary emission.*

Suppose that a photoelectron striking dynode 1 produces R electrons by secondary emission; these secondary electrons are then directed toward a

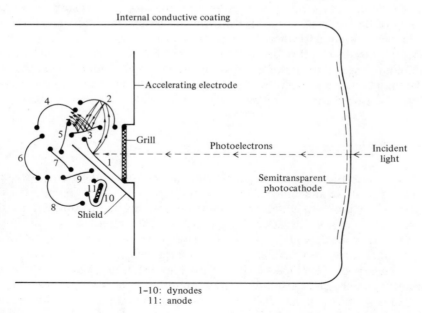

1–10: dynodes
11: anode

Figure 4-19 Schematic diagram showing the arrangement of the electrode structure of a photomultiplier tube. (Courtesy of RCA, Victor Division.)

second dynode (2), by making its potential higher than that of dynode 1. Suppose that R electrons are again ejected by secondary emission for each incident electron; then for each electron emitted by the photosensitive surface there are now R^2 electrons. If the tube consists of n dynodes, each successive dynode at a suitably higher potential than the preceding one, the multiplication factor of the tube will be R^n. Modern photomultiplier tubes have 10 to 16 dynodes. The electrons coming from the nth dynode go to the anode A, from which they proceed to some detecting circuit. An idea of the magnification produced by a photomultiplier tube can be obtained by setting $R = 4$ and the number of dynodes, $n = 10$. Then the multiplication factor is 4^{10} or 10^6. The photomultiplier tube is thus capable of detecting single electrons emitted by the photocathode, since with a multiplication factor of 10^6 a measurable charge is produced. Some high-gain tubes have been developed with multiplication factors up to 10^9; details of such a tube are given in Figure 4-19.

Phototubes are widely used as detectors of electromagnetic radiation. They can detect flashes of light caused by the passage of charged particles through matter (see Sections 13-7 and 13-8). They are used with computing machinery to detect the holes in punched paper tape or in punched cards. They find an everyday use in the "electric eye" that causes doors to open.

4-13   Pressure and Momentum of Radiation

The existence of radiation pressure was first shown by Maxwell (1871) to be a consequence of the electromagnetic theory of light. Experimental evidence of the existence of radiation pressure was first demonstrated by Lebedew (1900) in Russia and independently by Nichols and Hull (1901) in the United States. Essentially they showed that if light falls normally on a surface it exerts a force on that surface. It can be shown that if light falls on a black or completely absorbing surface the pressure P on the surface is given by

$$P = w \qquad (4\text{-}45)$$

where w is the energy density per unit volume of the radiation. Using the expression for the energy density in terms of the electric field intensity E given by Equation (4-5), we get

$$P = \frac{E^2}{4\pi} \qquad (4\text{-}46)$$

If the radiation falls normally on a surface that reflects a fraction r of the incident radiation, the pressure on the surface will be increased by this amount so that

$$P = (1 + r)w \qquad (4\text{-}47)$$

It can also be shown, on the basis of electromagnetic theory, that radiation possesses momentum. If the beam consists of plane electromagnetic waves, the amount of momentum per unit volume p_v is given by

$$p_v = \frac{w}{c} \tag{4-48}$$

where c is the velocity of light.

The equation for the pressure of radiation can also be derived on the basis of quantum theory and special relativity. It will be recalled that the relativistic equation for the energy $\mathcal{E}$ and the momentum p of a particle of mass m is

$$\mathcal{E}^2 = p^2 c^2 + (mc^2)^2 \tag{2-20}$$

Since a photon has zero rest-mass, its momentum is given simply by

$$p = \frac{\mathcal{E}}{c} \tag{4-49}$$

The energy of a photon whose frequency is ν is given by

$$\mathcal{E} = h\nu \tag{4-39}$$

where h is the Planck constant; hence the momentum of the photon can be written as

$$p = h\frac{\nu}{c} \tag{4-50}$$

or, using the relationship

$$\nu\lambda = c \tag{4-17}$$

we get

$$p = \frac{h}{\lambda} \tag{4-51}$$

for its momentum in terms of the wavelength λ of the radiation.

Each photon absorbed by a surface transfers momentum h/λ to it. If we consider a parallel beam of light as consisting of n photons per unit volume traveling in the x direction with velocity c, the number of photons per second striking a surface perpendicular to the x axis is ncA, where A is the area of the surface. If the surface absorbs this radiation completely, the rate of change of momentum of this surface will be

$$ncA\,\frac{h}{\lambda} = nh\nu A$$

By Newton's laws this is simply the force acting on it. The pressure produced by this radiation will then be

$$P = nh\nu \tag{4-52}$$

which is the same as

$$P = w \qquad (4\text{-}45)$$

Any system that emits or absorbs radiation will undergo a change in momentum equal to the momentum of the radiation emitted or absorbed by it. This applies to nuclear, atomic, and molecular systems as well as to macroscopic systems.

4-14   Angular Momentum of Radiation

In addition to linear momentum, electromagnetic radiation may transport angular momentum. In the wave theory of radiation angular momentum should be associated with circular polarization, whereas in the quantum theory angular momentum should be associated with the spin of a photon. In the latter theory the angular momentum of a photon due to its spin is assigned the value $h/2\pi$ (sometimes written as $\hbar$), where h is the Planck constant. Exchanges of angular momentum between radiation and matter on the atomic and subatomic scale will play a very important role in the phenomena to be discussed in the rest of this book; these phenomena provide ample justification for the assignment of spin angular momentum $\hbar$ to the photon.

Exchanges of angular momentum between radiation and matter on the macroscopic scale were first demonstrated by R. Beth in 1936. It was shown by J. H. Poynting (1909) that the angular momentum G transmitted per unit area per second by a circularly polarized beam of light is $\lambda/2\pi$ times the linear momentum, or

$$G = \frac{\lambda}{2\pi} p_v = \frac{\lambda w}{2\pi c} \qquad (4\text{-}53)$$

This transfer of angular momentum will produce a torque in its passage through a doubly refracting crystal, which, though extremely small, may be measurable. A simplified explanation of the method by which a torque is produced by the passage of light through a doubly refracting crystal is that the electric polarization produced in the crystal is not in the same direction as the electric field intensity producing this polarization. The torque L per unit volume can be shown to be given by

$$L = P_E E \sin (P_E, E) \qquad (4\text{-}54)$$

where E is the electric field intensity, and P_E is the electric polarization; P_E is a vector that is the sum of the electric dipole moments per unit volume.

In the experiment performed by Beth, a beam of light was first linearly polarized by a Nicol prism and then circularly polarized light by a quarter-wave plate. This beam was sent through a half-wave plate H. (See Figure 4-20.) The half-wave plate changes the direction of polarization of the beam so that

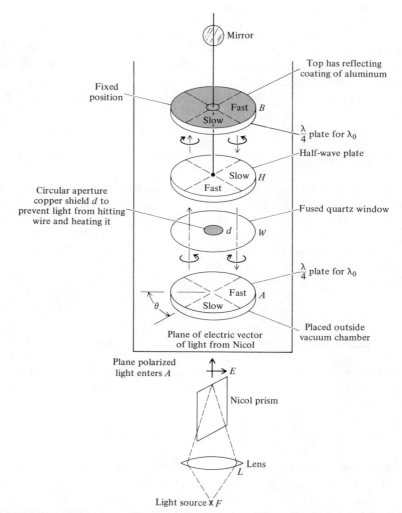

Figure 4-20   Schematic diagram of apparatus for the measurement of the angular momentum of light. [After R. A. Beth, *Phys. Rev.*, **50**, 115 (1936).] The half-wave plate, *H*, is subjected to a small torque, causing the mirror to rotate.

the emergent beam is polarized opposite to the entering beam—that is, a positive or counterclockwise polarization is changed to a negative or clockwise polarization. This is twice the effect that would have been produced if the beam had been completely absorbed by the crystal. Another factor of 2 was obtained by sending the emergent beam into a quarter-wave plate *B* which had a reflecting coating of aluminum on its top side. The effect on the beam emerging from *B* after reflection was equivalent to that of a half-wave plate, so that the beam going downward and passing through *H* produced a torque equal to that of the beam traveling upward.

The half-wave plate H was suspended in a vacuum by means of a fiber pass-ing through a hole in plate B; the other end of the fiber was attached to a rigid support. The torque produced in the half-wave plate produced a twist of the fiber; the angle of twist was measured by the deflection of a beam of light from the small mirror M.

Various tests were carried out to check theoretical predictions, such as the effect on the torque of changing the spectral distribution of the incident light; also the effect on the torque of changing the angle θ between the axis of crystal A and the direction of the electric field intensity E. The experimental results compared very favorably with theoretical predictions. For the intensities used in these experiments the torques were of the order of 10^{-9} dyne cm.

The equation for the angular momentum G of the radiation can be written in terms of the angular momentum of the photons by noting that the energy per unit volume w is given by

$$w = nh\nu \tag{4-55}$$

where n is the number of photons per unit volume. Equation (4-53) becomes

$$G = \frac{nh\lambda\nu}{2\pi c}$$

so that

$$G = n\hbar \tag{4-56}$$

Thus each photon of energy $h\nu$ has an angular momentum due to its spin equal to $\hbar$. The quantity $\hbar$ is frequently used as a unit of angular momentum; the photon is then said to have a spin of 1 in units of $\hbar$.

Problems

4-1. An electron with kinetic energy of 2000 eV strikes a metallic surface and is stopped in a distance of 3×10^{-5} cm. (a) Determine its accelera-tion. (b) Calculate the electric field intensity at a point 40 cm from the point of incidence on the metal and at right angles to the direction of the acceleration. Express the value in cgs and mksa units. (c) Calculate the energy per unit volume at this position.

4-2. Using the data in Problem 4-1, calculate the rate at which energy is radiated from this electron.

4-3. When a copper surface is illuminated by the radiation of wavelength $\lambda = 2537\text{Å}$ from a mercury arc, the value of the stopping potential is found to be 0.24 volts. (a) Calculate the wavelength of the threshold value for copper. (b) On the assumption that there are two free electrons per atom of copper, calculate the value of $\mathscr{E}_F$. (c) Using the above values, calculate the work done in taking an electron of lowest energy through the surface of the copper.

4-4. The photoelectric threshold of tungsten is 2300 Å. Determine the energy of the electrons ejected from the surface by ultraviolet light of wavelength 1800 Å.

4-5. Assuming the sun to be a blackbody with a surface temperature of 6000 K, (a) calculate the rate at which energy is radiated from it. (b) Determine the loss in mass per day due to this radiation. (c) At what rate does radiation reach the upper atmosphere of the earth? Compare this rate with the value of the solar constant (2 cal cm^{-2} min^{-1}).

4-6. Using Wien's displacement law, calculate the wavelength at which the intensity of the solar spectrum is a maximum. Compare it with the measured value.

4-7. A student observes that a certain phototube requires 0.25 volt to serve as the stopping potential for light with a wavelength of 5000 Å. If the light has a wavelength of 3750 Å, the stopping potential is one volt. Calculate h/e from the above data.

4-8. The work function for hydrogen is about 13.6 eV. (a) For photons of wavelength equal to 10 Å, incident on hydrogen, what is the maximum kinetic energy of the photoelectrons which are emitted? (b) What is the ratio of the maximum speed of these photoelectrons to the speed of light?

4-9. Three circular plates are arranged vertically as shown in Figure 4-21. The plates are held in place by a perfect thermal insulator and the volume between the plates is evacuated. The two outer plates are maintained at constant temperatures of 0° and 100°C, respectively. What is the equilibrium temperature of the center plate?

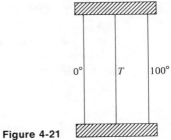

Figure 4-21

4-10. In the experiments of Nichols and Hull the measured value of the pressure of the radiation on a reflecting surface was found to be 7 × 10^{-5} dyne/cm^2. (a) Calculate the average value of the energy density of the radiation. (b) Determine the average value of the electric-field intensity of this radiation. (c) Suppose that the radiation consists entirely of light of wavelength 6000 Å; determine the number of photons per second striking unit area of the surface.

4-11. (a) Show that the Planck constant h has the dimensions of angular momentum. (b) How many photons per second will be needed to produce a torque of 4×10^{-9} dyne cm on passing through a half-wave plate if, initially, they are all polarized in the positive direction? Assume that $\lambda = 6000$ Å.

4-12. The acceleration a of a linear harmonic oscillator is given by

$$a = -\omega^2 x$$

(a) Show that its velocity v can be written as

$$v = v_0 \cos \omega t$$

where v_0 is the velocity at $x = 0$ when $t = 0$. (b) Show that the average value of its kinetic energy $\overline{\mathcal{E}}_K$ is one-half its maximum kinetic energy. (c) Show that its total energy $\mathcal{E}_T$ is equal to its maximum kinetic energy. (d) Show that its total energy at any instant is $2\overline{\mathcal{E}}_K$.

4-13. Assume that a system consists of a very large number of linear harmonic oscillators that interact slightly and are in equilibrium at temperature T. Assume further that there is a continuous distribution of energy among these oscillators given by the Boltzmann distribution [see Equation (11-32)]

$$f(\mathcal{E}) = C \exp\left(-\frac{\mathcal{E}}{kT}\right)$$

where k is the Boltzmann constant. Hence the probability $P(\mathcal{E})\, d\mathcal{E}$ that an oscillator will have an energy between $\mathcal{E}$ and $\mathcal{E} + d\mathcal{E}$ is given by

$$P(\mathcal{E})\, d\mathcal{E} = C \exp\left(-\beta \mathcal{E}\right)$$

where $\beta = 1/kT$, and C is a constant. The average value $\overline{\mathcal{E}}$ of these oscillators will then be given by

$$\overline{\mathcal{E}} = \frac{\int_0^\infty \mathcal{E}\, P(\mathcal{E})\, d\mathcal{E}}{\int_0^\infty P(\mathcal{E})\, d\mathcal{E}}$$

Show that the average value of the energy of an oscillator is

$$\overline{\mathcal{E}} = \frac{1}{\beta} = kT$$

Note. When the equation of $\overline{\mathcal{E}}$ is written explicitly, it will be recognized that the integrand of the numerator is the negative derivative with respect to β of the integrand of the denominator. Hence

$$\overline{\mathscr{E}} = -\frac{d}{d\beta} \ln \int_0^\infty \exp{(-\beta\mathscr{E})}\, d\mathscr{E}$$

4-14.   Referring to Problem 4-13, assume that the linear harmonic oscillators have no continuous distribution of energy but are limited to discrete values of the quantum of energy $h\nu$ such that an energy state is given by

$$\mathscr{E}_n = nh\nu$$

where $n = 0, 1, 2, 3, \ldots .$ Show that for such a system the average value $\overline{\mathscr{E}}$ of the energy of the oscillators is given by

$$\overline{\mathscr{E}} = \frac{h\nu}{\exp{(h\nu/kT)} - 1}$$

Note. Follow the steps indicated in Problem 4-13 but replace the integrand with the infinite sum; that is,

$$\overline{\mathscr{E}} = \frac{\displaystyle\sum_{n=0}^\infty nh\nu \exp{(-\beta nh\nu)}}{\displaystyle\sum_{n=0}^\infty \exp{(-\beta nh\nu)}}$$

Write the denominator in its explicit form as an infinite series and compare it with the series obtained by the binomial expansion of the function $(1 - x)^{-1}$.

References

Most of the references listed at the end of the first three chapters contain material pertinent to this chapter; in addition, we list:

Born, M., *Atomic Physics*. New York: Hafner Publishing Company, 1959.
Lorrain, P., and D. R. Corson, *Electromagnetic Fields and Waves*. San Francisco: W. H. Freeman and Company, 1970, Chapter 11.

5 | X-Rays

5-1   Discovery of X-Rays

X-rays were discovered by Roentgen in 1895. He found that the operation of a cathode-ray tube produced fluorescence in a platinum-barium-cyanide screen placed at some distance from the tube. The source of the rays causing this fluorescence was traced to the walls of the cathode-ray tube. In further experiments he found that the interposition of various thicknesses of different substances between the screen and the tube reduced the intensity of the fluorescence but did not obliterate it completely. This showed that these x-rays, as Roentgen called them, had very great penetrating power. It was also found that these rays could blacken a photographic plate and could ionize a gas.

The x-rays, or Roentgen rays, traveled in straight lines from the source and were not deflected in passing through electric or magnetic fields. They were thus not charged particles. Roentgen tried to reflect and refract them but without success. Almost a quarter of a century later it was shown that x-rays can be refracted and reflected. One reason why Roentgen was unable to find these properties is that the index of refraction of x-rays is very small and slightly less than one (see Section 5-11).

Haga and Wind, in 1899, sent a beam of x-rays through a narrow aperture; they actually succeeded in getting a diffraction pattern, but the effect was so small that their results were not generally accepted as conclusive. It was not until 1912 that the wave nature of x-rays was definitely established by Laue's experiments on the diffraction of x-rays by crystals. Barkla's experiments (1906) on the polarization of x-rays established the fact that these rays were transverse waves similar to light waves.

5-2   Production of X-Rays

The most common method of producing x-rays is to direct a beam of fast-moving electrons toward a substance, usually called a target. In Roentgen's experiment the cathode rays struck the walls of the tube so that the glass wall

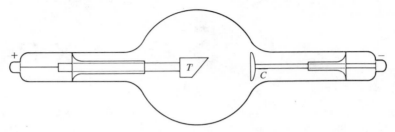

Figure 5-1 Gas-filled type of x-ray tube; *C* is the cathode and *T* is the target.

became the source of the x-rays. The gas-filled type of x-ray tube is a modification of the cathode-ray tube. Instead of allowing the cathode rays to strike the walls of the tube, the cup-shaped cathode *C* focuses them onto a metal target *T*, as shown in Figure 5-1. The gas pressure in the tube is of the order of 0.001 torr (1 torr = 1 mm of Hg), and the difference of potential between the cathode and target is usually of the order of 30,000 to 50,000 volts. The electrons from the cathode are stopped by the target, which then becomes the source of x-rays. These x-rays proceed in all directions from the target.

The Coolidge type of x-ray tube is a thermionic tube in which the cathode is a tungsten filament; one modern design of a Coolidge tube is shown in Figure 5-2. When the filament is heated to incandescence by means of a current supplied either by a storage battery or by a step-down transformer, electrons are emitted by the filament. These electrons are accelerated to the target by a difference of potential maintained between them. The filament is placed inside a metallic cup in order to focus the electrons onto the target. The tube must be highly evacuated so that no electric discharge can take place in the residual gas under normal operating conditions. One great advantage of the Coolidge type of tube is that the emission, hence the current in the tube, can be controlled by varying the temperature of the filament. In general the Coolidge tube is more stable in operation than the gas type of tube.

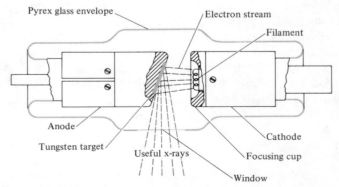

Figure 5-2 A modern Coolidge type of x-ray tube. (Courtesy of the General Electric X-ray Corp.)

Coolidge x-ray tubes have been designed to operate at voltages that range from a few hundred to about one million volts. Various types of high-voltage sources are used in operating x-ray tubes. The most common type of high-voltage source is the step-up transformer with its secondary coil well insulated from the primary coil. If the a-c voltage from the secondary is applied directly across the cathode and the target of the x-ray tube, the tube acts as its own rectifier—that is, current flows in the tube only during that half of the cycle in which the target is positive with respect to the cathode. In those experiments in which it is necessary to have a constant direct current through the tube the transformer terminals are connected to a rectifier circuit consisting of two or more high-voltage rectifiers together with a large capacitor and inductance coils. The rectified, constant d-c voltage is then applied to the x-ray tube.

When an x-ray tube is to be operated continuously with comparatively large amounts of power, special arrangements must be made for cooling the target. One common method of doing this is to mount the target material on a hollow copper tube and to circulate water through this tube. Almost any substance can be used as the target of an x-ray tube, depending on the problem under investigation. The targets of general-purpose x-ray tubes are usually made of tungsten or molybdenum because these metals have high melting points.

An entirely different type of x-ray source was developed by D. W. Kerst in 1941; this tube is called a *betatron*. In the older type of x-ray tubes the electrons which strike the target acquire their energy by the application of a high voltage between the filament and the target. In the betatron the electrons acquire their energy by the action of the force exerted on them by the electric field which accompanies a changing magnetic field. Some betatrons now in operation accelerate electrons so that they have energies up to 100 MeV when they strike the target and produce x-rays. These x-rays have been used in nuclear experiments as described in Chapter 15. The mode of operation of the betatron will be described in greater detail in Chapter 12 (Particle Accelerators).

5-3   Measurement of the Intensity of X-Rays

The intensity of a beam of x-rays may be measured by any one of its effects, such as the blackening of a photographic plate, the rise in the temperature of a piece of lead which absorbs these rays, or the ionization produced in a gas or vapor. Several of these effects will be described in this and other chapters of the book.

The *ionization chamber* is commonly used for measuring the intensity of x-rays; it makes use of the ionization produced in a gas or vapor by the passage of x-rays through it. A typical ionization chamber is shown in Figure 5-3. It consists of a metal cylinder *C* containing some convenient gas or vapor such as air or methyl bromide at about atmospheric pressure. A metal rod *R*, insulated from the cylinder, runs parallel to the axis of the cylinder. The x-ray beam

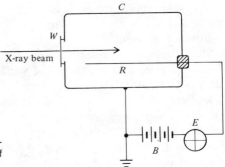

Figure 5-3 Ionization chamber and electrometer for measuring the intensity of x-rays.

enters the chamber through a thin window W, usually made of mica or thin aluminum, and ionizes the gas in the chamber. A battery B maintains a difference of potential between R and C so that the ions are set in motion toward C and R just as soon as they are formed. This ionization current is measured by the electrometer E. Experiment shows that the ionization current is directly proportional to the intensity of the x-ray beam.

The intensity of the x-rays coming from any tube has been found to depend on the element used as the target, the power supplied to the tube, and the difference of potential between the target and cathode.

The Geiger counter tube, originally developed for the counting of alpha particles (Section 3-4), has also been adapted for measuring the intensity of x-rays. Another device that has come into great use, particularly for the measurement of x-rays from high-voltage sources, is the *scintillation counter* (Section 13-7). In this device the x-rays are allowed to penetrate a transparent crystal or liquid in which the energy of the x-radiation is transformed into visible fluorescent radiation. This radiation is then incident on the sensitive surface of a photomultiplier tube. The photoelectric current thus produced can be amplified and measured with appropriate circuits.

5-4 Diffraction of X-Rays

The explanation of the origin of x-rays on the basis of classical electrodynamics is that x-rays are emitted in the form of electromagnetic pulses or groups of waves when the electrons are stopped by the target of an x-ray tube. In the process of being stopped, the incoming electrons interact with the electrons and nuclei of the atoms of the target. Some of the incoming electrons are accelerated in the Coulomb fields of the nuclei and electrons; as a result, they radiate energy in the form of x-rays. The effects of these interactions on the target atoms will be considered in later sections.

The existence of a wave motion can be definitely established only by diffraction and interference phenomena. The conditions under which these phenomena occur are well known for waves in the visible region of the spectrum;

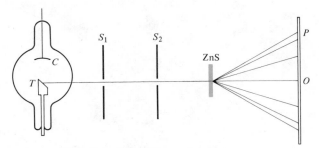

Figure 5-4   Arrangement of apparatus for producing a Laue diffraction pattern.

for example, if yellow light is allowed to pass through a diffraction grating having about 6000 lines to the centimeter, a diffraction pattern is obtained consisting of a central image or undeviated line, then a first-order image which is deviated by about 20 degrees from the original direction, and farther on a second-order image, and so on. An analysis of the action of the grating shows that the spacing between the lines ruled on the grating should be of the order of magnitude of the wavelength of the light used. The results of the early experiments on the diffraction of x-rays indicated that their wavelengths were of the order of 10^{-8} or 10^{-9} cm. It occurred to M. von Laue (1912) that the ordered arrangement of atoms or molecules in crystals fulfilled all the conditions essential for the diffraction of such short wavelengths. The spacing between these atoms or molecules was known to be of the order of 10^{-8} cm. A crystal differs from an ordinary diffraction grating in that the diffracting centers in the crystal are not all in one plane; the crystal acts as a space grating rather than as a plane grating.

Figure 5-5   Photograph of Laue diffraction pattern of rocksalt. (From photograph by J. G. Dash.)

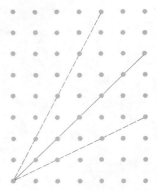

Figure 5-6 Orientations in a crystal of some planes that are rich in atoms.

Following Laue's suggestion, Friedrich and Knipping (1912) carried out the following experiment. A narrow pencil of x-rays was allowed to pass through a thin crystal of zinc blende (ZnS). The emergent beam fell on a photographic plate P, as shown in Figure 5-4. The diffraction pattern obtained consisted of the central spot at O and a series of spots arranged in a definite pattern about O.

Figure 5-5 is a photograph of the Laue diffraction pattern obtained by passing a narrow pencil of x-rays through a thin crystal of rock salt perpendicular to its *cleavage* planes; these cleavage planes are parallel to a surface along which a crystal can be readily split or broken. From a knowledge of the structure of the crystal some of the wavelengths in the incident radiation could be computed. A simple interpretation of the diffraction pattern was given by W. L. Bragg. He assumed that the diffraction spots were produced by x-rays scattered from certain sets of parallel planes within the crystal which contained large numbers of atoms. That some planes are richer in atoms than others can be seen by considering a two-dimensional array of points and drawing lines through these points, as in Figure 5-6. These lines correspond to planes in a three-dimensional crystal.

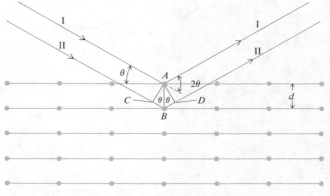

Figure 5-7 "Reflection" of x-rays from atomic planes.

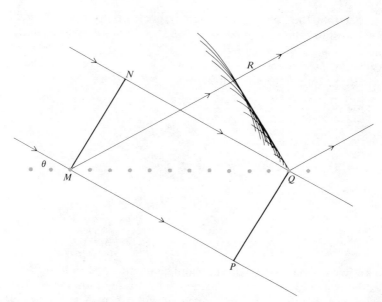

Figure 5-8 Plane wave front *MN* advancing toward *PQ* is reflected by atoms between *M* and *Q* to form the reflected wave front *RQ*-Huygens construction. The angle of incidence and the angle of reflection are equal.

It is easy to derive the condition under which the x-rays scattered from a series of atomic planes will produce an intense spot. Consider a series of parallel atomic planes spaced a distance d apart. Suppose a narrow beam of x-rays from some source is incident on these planes at an angle θ, as shown in Figure 5-7. The beam will be scattered in all directions by the atoms in the various atomic planes. Let us first consider the x-rays scattered by any one plane. Following the usual Huygens construction, it can be seen that the scattered beam will have a maximum intensity at an angle θ to this plane, which is equal to the angle of incidence. Figure 5-8 shows a portion *MN* of the incident wave front advancing to *PQ*, thus causing the atoms between *M* and *Q* to send out wavelets in all directions. These wavelets produce a wave front *QR*, the reflected wave, which advances along the direction *MR*. This is identical with ordinary optical reflection and is independent of the wavelength of the x-rays or of the spacings of the atoms in the plane. However, if we now consider the effect of the reflections from a set of parallel planes spaced a distance d apart, the wavelets from the atoms in the different planes will reinforce each other and produce maximum intensity only if they meet in phase. The condition under which reinforcement occurs can be derived by considering a small portion of the beam that is incident at an angle θ to the planes of the crystal and then reflected at this same angle. Consider two rays such as I and II which are scattered by two particles *A* and *B* in adjacent planes, as shown in Figure 5-7. Ray II travels a longer path than ray I; this difference in path is evidently $(CB + BD)$. The spacings are greatly enlarged; actually these two rays are so close together that they produce a single impression on a photographic plate. Whenever the difference in path $(CB + BD)$ is a whole wavelength, λ, or a whole multiple of the wavelength, $n\lambda$, the waves will reinforce one another and produce an intense spot.

The condition for reinforcement is thus

$$(CB + BD) = n\lambda$$

and from the figure

$$CB = d \sin \theta$$

and

$$BD = d \sin \theta$$

hence

$$n\lambda = 2d \sin \theta \qquad\qquad (5\text{-}1)$$

which is known as the *Bragg equation* and gives the condition for the *reflection* of x-rays from a series of atomic planes. The Bragg equation is essentially a restrictive condition on the appearance of intense maxima in the scattered beam of x-rays. A more extensive analysis would show that no other intense maxima exist than those predicted by the Bragg equation. If the distance between atomic planes is known, the wavelength of the x-rays which produce intense maxima on reflection by the crystal can be calculated by using the Bragg equation; or, conversely, using x-rays of known wavelengths, distances between atomic planes can be computed.

In the Bragg equation n is always an integer. When $n = 1$, the difference in path between waves reflected from any two adjacent planes is one wavelength. For this case

$$\lambda = 2d \sin \theta_1$$

gives the condition for the first-order reflection of wavelength λ from the crystal. For $n = 2$ the equation becomes

$$2\lambda = 2d \sin \theta_2$$

that is, the second-order reflection of the same wavelength will occur at θ_2, a larger angle of incidence and reflection.

The analysis used in deriving the Bragg equation can now be used to explain the Laue diffraction pattern. The x-rays that penetrate the crystal are scattered from the different atomic diffraction centers. We can construct sets of parallel atomic planes inside the crystal as shown in Figure 5-9 and apply the Bragg equation to each set of parallel planes. The value of the incident angle will be different for each set of planes and each set of parallel planes will have its own particular value for the distance d, but these distances will be related in a simple way to the distances between planes that are parallel to the cleavage face because of the geometry of the crystal. The x-rays scattered from any set

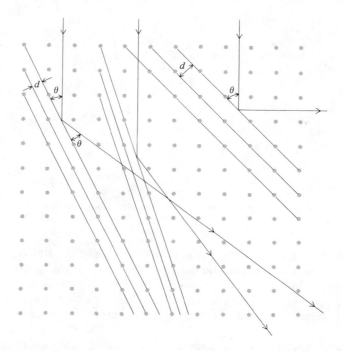

Figure 5-9   Reflection of x-rays from sets of atomic planes within a crystal to produce the
Laue diffraction pattern.

of parallel planes will reinforce each other only if the particular wavelength
that will satisfy the Bragg equation is present in the incident beam. Furthermore, an examination of Figure 5-9 will show that there are comparatively
few sets of parallel planes that are sufficiently rich in atoms to produce intense
diffraction spots. Hence, even if the incident radiation contains a wide range
of wavelengths, which is the most common method of producing a Laue pattern, the number of intense diffraction spots produced will be small in spite of
the large number of atomic diffraction centers in the crystal.

5-5   Single Crystal X-Ray Spectrometer

The Laue diffraction patterns are complex and difficult to interpret. Instead
of using a crystal as a transmission grating, Bragg set up the crystal as a reflection grating. A typical experimental arrangement is shown in Figure 5-10.
Two lead slits, S_1 and S_2, define a narrow beam of x-rays coming from the
target T. This beam of x-rays strikes the crystal C at some angle θ and is reflected by it to the photographic plate P. The crystal C is mounted on a spectrometer table and can be rotated so that the glancing angle θ may be varied.

At each particular setting of the crystal only the particular wavelength λ
that satisfies the Bragg equation $n\lambda = 2d \sin \theta$

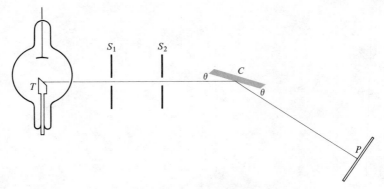

Figure 5-10   Single crystal x-ray spectrometer, with photographic plate.

will be reflected to the photographic plate. As the crystal is rotated, other wavelengths will be reflected by the crystal. In this way a spectrum of the incident beam is obtained. The spectrum may consist of several orders. If, in the Bragg equation ($n\lambda = 2d \sin \theta$) n has the value of unity, the spectrum is said to be a first-order spectrum; if $n = 2$, a second-order spectrum is obtained, and so on.

In many experiments the photographic plate is replaced by an ionization chamber, a Geiger counter, or a scintillation counter which can be rotated about the same axis as the crystal.

The choice of crystal to be used in an x-ray spectrometer is determined by several factors, such as the range of wavelengths to be examined, the ease with which a good surface can be obtained, and the reflecting power of the crystal. Crystals most commonly used are calcite, quartz, and rocksalt. It is obvious that for use in measuring x-ray wavelengths the distance d between atomic planes in the crystals must be known from other data. Sufficient data can be obtained from crystallographic studies for some of the simpler crystals, such as rocksalt and calcite, for an independent determination of the spacing between atomic planes.

5-6   The Grating Space of Rocksalt Crystals

The distance between atomic planes, or the *grating space* of a crystal, can be calculated when the structure of the crystal is known. Rocksalt, one of the best known and simplest of crystals, has the geometrical structure of a simple cube, with ions of sodium and chlorine arranged alternately at the corners of a cube, as shown in Figure 5-11. The distance between atomic planes can be found by determining the volume of one of these elementary cubes. If M is the molecular weight of NaCl and ρ is its density, the volume of one mole of NaCl is

$$v = \frac{M}{\rho}$$

Since there are $2N_0$ ions or diffracting centers in a mole, the volume associated with each ion will be

$$V = \frac{M}{2\rho N_0}$$

where N_0 is the Avogadro number. The distance d between ions will therefore be

$$d = \sqrt[3]{\frac{M}{2\rho N_0}} \tag{5-2}$$

The molecular weight of sodium chloride is 58.45 and its density is 2.164 grams/cm³. Substitution of the value of N_0 yields $d = 2.814 \times 10^{-8}$ cm. This was the value known in 1913. The measurements of x-ray wavelengths soon achieved high precision. Siegbahn, from whose laboratory came a great many of the precise determinations of x-ray wavelengths, adopted the value

$$d = 2.81400 \times 10^{-8} \text{ cm at } 18°C$$

for the grating space of rocksalt. Large, good, single crystals of rocksalt of the type needed for x-ray measurements are not readily available and, when available, have the defect of being hygroscopic; calcite is a much better crystal for such measurements and is more readily obtainable as large single crystals. Careful comparisons between calcite and rocksalt crystals made with monochromatic x-rays led to the adoption of the value

$$d = 3.02945 \times 10^{-8} \text{ cm at } 18°C$$

for the grating space of calcite.

Until recently (1967) all measurements of x-ray wavelengths were based

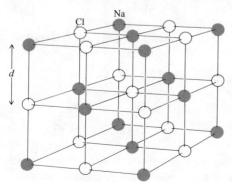

Figure 5-11   Arrangement of sodium ions and chlorine ions in a sodium chloride crystal.

on this value of the grating space of calcite. Advances in design of x-ray spectrometers, some of which will be discussed in subsequent sections, led to much greater accuracy in the determinations of x-ray wavelengths. Discrepancies began to appear in these measurements which indicated that different calcite crystals had slightly different values of the grating space constant d. J. A. Bearden undertook an investigation of this problem in 1960 and found that more consistent and precise results could be obtained by measuring x-ray wavelengths with reference to a standard of wavelength. He recommended (1967) that a very intense characteristic line emitted by a tungsten target, known as the $W_{K_{\alpha 1}}$ line, be used as the primary standard; its measured value is

$$\lambda(W_{K_{\alpha 1}}) = (0.2090100 \pm 5 \text{ ppm}) \text{ Å}$$

(ppm means parts per million.)

As a standard, it is given an exact value, and this value is set at

$$\lambda(W_{K_{\alpha 1}}) = 0.2090100 \text{ Å*}$$

The difference between 1 Å and 1 Å* is exceedingly small, except for extremely precise measurement, and for most of the discussions in this book the two units may be considered identical. However, for completeness and for the convenience of those who may have need to convert from Å* to Å, the recommended conversion factor is

$$\frac{\lambda(\text{Å})}{\lambda(\text{Å*})} = 1.000197(56)$$

with an accuracy of 5.6 ppm.

The $K_{\alpha 1}$ lines of Cr, Cu, Mo, and Ag are recommended as secondary standards. The wavelengths of these and other lines of the K series of selected elements are listed in Table 10-1.

5-7 Typical X-Ray Spectra

With a knowledge of the grating space of a crystal and the use of the Bragg equation, it is possible to measure the wavelengths of the x-rays emitted by the target of an x-ray tube. When resolved by means of a crystal spectrometer, the heterogeneous beam of x-rays from a target is found to consist of two distinct types of spectra:

1. A continuous spectrum.
2. A sharp line spectrum superposed on the continuous spectrum.

Typical x-ray spectra, obtained by Ulrey, are shown in the curves in Figure 5-12, in which the intensity of a given wavelength is plotted against the wavelength in angstrom units. The curve for tungsten shows the continuous spectrum of the x-rays coming from the tungsten target of an x-ray tube operated at a voltage of 35 kV. The curve for molybdenum was obtained under similar

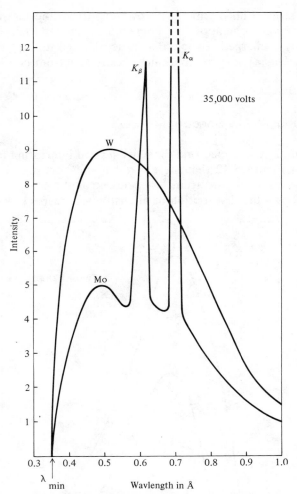

Figure 5-12   Typical x-ray spectra from tungsten and molybdenum targets, respectively.

conditions and shows two sharp lines characteristic of the element molybdenum superposed on the continuous spectrum. These lines are known as the
K_α and K_β lines of molybdenum. To get the K_α and K_β lines of tungsten, the
tube would have to be operated at about 70 kV (Section 10-3).

The two distinctively different types of spectra from the same target must
have different origins. It will be shown in Chapter 10 that the sharp line spectrum is produced by the energy changes which take place as a result of the
rearrangement of the electrons in the various electronic levels of the atom
following the transfer of energy to the atom by the impinging electron. The
continuous spectrum, on the other hand, results from the radiation emitted by
the electrons which are accelerated in the Coulomb field of force of the nuclei

of the target atoms. The term *bremsstrahlung* (from the German: *bremse*—brake and *strahlung*—radiation) has come to be accepted to describe the radiation emitted by a charged particle that is accelerated in the Coulomb field of force of a nucleus. Thus the continuous x-ray spectrum is a type of bremsstrahlung.

5-8   Continuous X-Ray Spectrum

The continuous x-ray spectrum presents several interesting features. It will be noted from Figure 5-12 that there is a definite short wavelength limit λ_{min} to the continuous x-ray spectrum independent of the material of the target. Figure 5-13 shows the distribution of intensity with respect to wavelength for

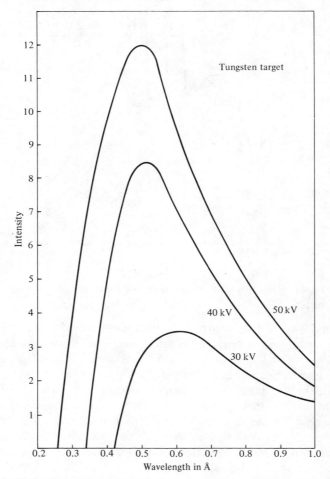

Figure 5-13   The continuous x-ray spectrum from a tungsten target showing the dependence of the short wavelength limit on the voltage across the tube.

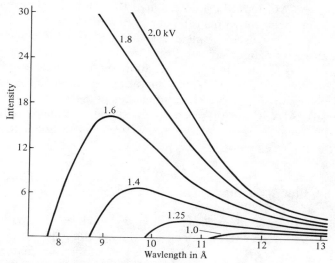

Figure 5-14   Intensity of the continuous x-ray spectrum as a function of wavelength at comparatively low voltages. [Stephenson and Mason, *Phys. Rev.*, **75**, 1713 (1949).]

different accelerating potentials. When the voltage across the tube is increased, the short wavelength limit λ_{min} is shifted toward smaller values. Duane and Hunt showed that the short wavelength limit of the continuous spectrum varies inversely as the voltage across the tube. Put in terms of frequencies,

$$\nu_{max} = \frac{c}{\lambda_{min}} \tag{5-3}$$

and

$$Ve = h\nu_{max} = \frac{hc}{\lambda_{min}} \tag{5-4}$$

where V is the voltage across the tube, e is the electronic charge, and h is the Planck constant. Ve represents the energy with which an electron strikes the target. One may look at this phenomenon as the inverse of the photoelectric effect, since the maximum kinetic energy of an electron striking the target is

$$\tfrac{1}{2}mv_{max}^2 = Ve = h\nu_{max} \tag{5-5}$$

Duane and Hunt carried out a series of careful experiments on the determination of the short wavelength limit of the continuous x-ray spectrum for various voltages across the x-ray tube and determined the value of the Planck constant h to be

$$h = 6.556 \times 10^{-27} \text{ erg sec}$$

This was in good agreement with the value of h determined by means of the photoelectric effect.

Stephenson and Mason (1949) determined the intensity distribution of the continuous x-ray spectrum with a tube operated at comparatively low voltages of 1 to 2.0 kV. They used a single crystal spectrometer placed under an evacuated bell jar and detected the x-rays with a Geiger counter. The distribution in intensities at a fixed voltage, as a function of the wavelength, is essentially similar to that obtained at higher operating voltages, as shown in Figure 5-14. They also compared the relative intensities of the x-rays emitted by three different targets, Al, Cu, and W, when operated under similar conditions. They found that the total intensity of the x-rays was proportional to the atomic number of the target element.

5-9 Wavelengths of Gamma Rays

In examining the radiations from radioactive materials, it was found that gamma rays were not affected by electric and magnetic fields. It is now known that gamma rays are short electromagnetic waves of the same nature as x-rays, except that they are emitted from the nucleus of the atom. The wavelengths of some of the gamma rays have been measured by x-ray crystal diffraction methods.

Rutherford and Andrade (1914) made use of the Bragg spectrometer for determining the wavelengths of gamma rays from radium B. A narrow beam of gamma rays from the radioactive preparation R, after passing through a fine slit in a lead block, was reflected by the rocksalt crystal C onto the photographic plate P, as shown in Figure 5-15. By rotating the crystal from 0 to about 15 degrees a series of sharp lines was obtained corresponding to the gamma-ray wavelengths emitted by radium B. This method has been used for measuring the wavelengths of the gamma rays from many other radioactive

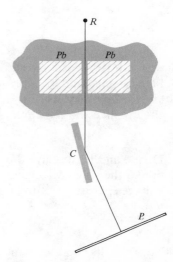

Figure 5-15 Bragg-type crystal reflection method of determining gamma-ray wavelengths.

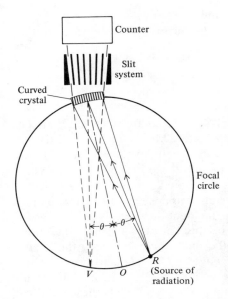

Figure 5-16 Schematic diagram of the Du Mond curved crystal spectrometer used in measuring short wavelength gamma rays and x-rays. R is the source of the radiation, V is its virtual image, and θ is the Bragg angle.

substances. The smallest gamma-ray wavelength measured with the Bragg type of crystal spectrometer is $\lambda = 0.016$ Å, one of the lines emitted by radium C. In order to produce x-rays of the same wavelength, the x-ray tube would have to be operated at a difference of potential of 770,000 volts.

The very small angles involved in the measurement of very short wavelengths limit the accuracy of the Bragg type of single crystal spectrometer. A new design, suggested by Du Mond (1927) and first constructed by Mlle Cauchois (1932) with a curved crystal, has since been developed into an instrument of high precision for the measurement of gamma rays and x-rays of very short wavelength. Figure 5-16 is a schematic diagram of one design used by Du Mond in some of his experiments. The crystal, originally a flat lamina of quartz 2 mm thick and 50 mm wide, is curved by placing it between two cylindrical steel surfaces which have windows in them to permit the radiation to pass through the crystal. The radius of curvature of the crystal is 2 meters. The atomic planes, which are perpendicular to the curved surface, converge when produced toward the point O. A circle through O and through the curved crystal with a diameter equal to the radius of curvature of the crystal is called the *focal circle*. A small source of gamma rays placed at point R on this focal circle sends a narrow beam of radiation at an angle θ to the crystal planes; it is reflected from these planes and transmitted through the crystal in such a manner as to form a virtual image on the focal circle at V. The precision with which the angle θ is measured depends on the sharpness of the virtual image V. A sufficiently large Geiger counter or scintillation counter is placed behind the crystal to receive this beam. A lead collimator is placed between the counter and the crystal to stop the direct transmitted beam but allows the reflected beam to pass through it. The Bragg angle θ is determined from the position of the source R on the focal circle. Du Mond and his co-workers have measured wavelengths smaller than 0.01 Å with an accuracy of 10^{-5} Å with the curved crystal spectrometer.

5-10   X-Ray Powder Crystal Diffraction

Comparatively few substances are available in the form of large single crystals for use in the Bragg type of spectrometer. Most substances exist as aggregates of very small crystals. Methods have been developed for studying the crystal structure of a substance, even though these crystals may be very minute. If the substance is made into a very fine powder, it may be considered as made up of a multitude of very small crystals. Suppose a very narrow pencil of monochromatic x-rays of wavelength λ is incident on the powder at S, as in Figure 5-17. Since these small crystals are oriented at random, there undoubtedly will be some crystal oriented at just the right angle to satisfy the Bragg equation ($n\lambda = 2d \sin \theta$) for this particular wavelength λ. The scattered beam will have maximum intensity at an angle 2θ with respect to the incident beam of x-rays, and because of the random orientation of the crystals these maxima will lie on a cone of central angle 4θ. If the x-rays scattered by the powder are incident on a photographic plate placed perpendicular to the incident beam, concentric circles will be registered on it corresponding to the different orders of reflection. Another method is to bend a thin strip of photographic film in the form of a circle with the powder at the center of this circle. Small holes are cut in this film to permit the direct beam of x-rays to enter and leave the camera without blackening the photographic film.

Frequently, instead of a monochromatic beam, x-rays from a target that emits a few intense characteristic lines are used for these powder photographs. Since the relative intensities of these characteristic lines are known, it is not difficult to determine which circles correspond to the different wavelengths. Typical powder crystal diffraction patterns are shown in Figure 5-18. This

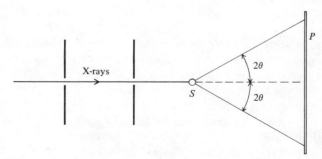

Figure 5-17   Method of obtaining x-ray diffraction patterns from a substance in powder form.

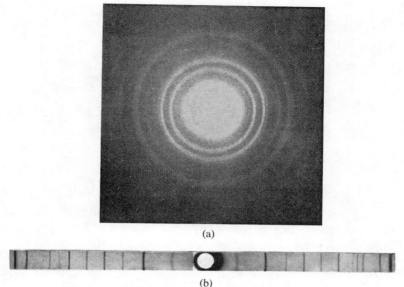

(a)

(b)

Figure 5-18   (a) X-ray powder diffraction pattern of aluminum. (Reproduced with the permission of A. W. Hull.) (b) X-ray powder diffraction pattern of tungsten obtained with a photographic film bent in the form of a circular cylinder. Radiation from a copper target was used in making this pattern. (From a photograph by L. L. Wyman and supplied by A. W. Hull.)

method of crystal analysis was first developed by Debye and Scherrer and by Hull and is one of the most valuable aids in studying crystal structure.

5-11   Refraction of X-Rays

Several of the early experimenters, including Roentgen and Barkla, tried to find out whether x-rays are refracted when they enter a material medium. The phenomenon of refraction, however, was not discovered until 1919, when Stenström observed that Bragg's law ($n\lambda = 2d \sin \theta$) did not yield the same value for λ in the different orders of reflection of a monochromatic beam from a crystal. The explanation of this apparent failure of Bragg's law is that x-rays are refracted when they penetrate the crystal. Because of refraction, the wavelength in the crystal is different from that outside. If μ is the index of refraction, λ, the wavelength in a vacuum, and λ', the wavelength in the crystal, then

$$\mu = \frac{\lambda}{\lambda'} \qquad (5\text{-}6)$$

Further, the angle of incidence on the inner crystal planes is not the angle between the incident beam and the surface of the crystal but the angle between

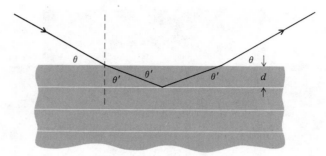

Figure 5-19   Refraction of x-rays on penetrating a crystal.

the refracted ray and the atomic planes in the crystal. The deviations from Bragg's law show that the angle of refraction θ' is less than the grazing angle angle of incidence θ—that is, the x-rays are bent away from the normal on penetrating the crystal. This means that the index of refraction for x-rays is less than unity. In Figure 5-19, if θ is the angle between the incident ray and the surface of the crystal and θ', the angle of refraction, then Snell's law of refraction takes the form

$$\mu = \frac{\cos \theta}{\cos \theta'} \qquad (5\text{-}7)$$

The angle θ' is not only the angle of refraction but also the angle that the ray makes with the atomic planes, so that inside the crystal Bragg's law is

$$n\lambda' = 2d \sin \theta' \qquad (5\text{-}8)$$

However, the angle that is actually measured is the angle θ and the wavelength usually desired is the wavelength λ in a vacuum. A modified form of Bragg's law containing λ, θ, and μ can be obtained by combining Equations (5-6) and (5-7) with (5-8). Remembering that

$$\sin \theta' = (1 - \cos^2 \theta')^{1/2}$$

and substituting the values of cos θ' and λ' from Equations (5-6) and (5-7), we find that Equation (5-8) becomes

$$\frac{n\lambda}{\mu} = 2d \left(1 - \frac{\cos^2 \theta}{\mu^2} \right)^{1/2}$$

$$n\lambda = 2d(\mu^2 - \cos^2 \theta)^{1/2}$$

or

$$n\lambda = 2d(\mu^2 - 1 + \sin^2 \theta)^{1/2}$$

and by factoring out sin θ we have

$$n\lambda = 2d \sin \theta \left[1 + \frac{\mu^2 - 1}{\sin^2 \theta}\right]^{1/2} \tag{5-9}$$

Equation (5-9) is a modified form of Bragg's law. For purposes of calculation this equation can be simplified by expanding the quantity in brackets by means of the binomial theorem; thus

$$\left[1 + \frac{\mu^2 - 1}{\sin^2 \theta}\right]^{1/2} = 1 + \frac{1}{2}\frac{\mu^2 - 1}{\sin^2 \theta} + \cdots$$

$$= 1 + \frac{1}{2}\frac{(\mu + 1)(\mu - 1)}{\sin^2 \theta}$$

Experiment shows that μ does not differ appreciably from unity and to a very close approximation $\mu + 1 = 2$, so that the brackets can be written as

$$\left[1 + \frac{\mu^2 - 1}{\sin^2 \theta}\right]^{1/2} = 1 - \frac{1 - \mu}{\sin^2 \theta}$$

and the modified form of Bragg's law is then

$$n\lambda = 2d\left[1 - \frac{1 - \mu}{\sin^2 \theta}\right]\sin \theta \tag{5-10}$$

The more common method of expressing the results is to give the quantity $\delta = 1 - \mu$, which shows how the index of refraction differs from unity; for example, the index of refraction of calcite for the wavelength $\lambda = 0.708$ Å differs from unity by the amount

$$\delta = 1 - \mu = 1.85 \times 10^{-6}$$

Two interesting conclusions can be drawn from the fact that the index of refraction of x-rays in material media is less than unity. One is that when x-rays pass through a prism the beam will be deviated in a direction opposite to that for ordinary light. The path of an x-ray beam through a prism is shown in Figure 5-20. The other conclusion is that at a certain critical angle θ_c the x-ray beam should not be refracted into the medium at all but should be totally reflected. At this critical angle of incidence $\cos \theta' = 1$ and

$$\mu = \cos \theta_c \tag{5-11}$$

For all angles smaller than θ_c there should be no refraction at all, only total reflection of the x-ray beam.

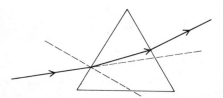

Figure 5-20   Path of a beam of x-rays refracted through a prism. The beam is bent away from the normal on entering the prism.

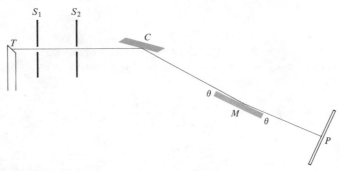

Figure 5-21   Method of reflecting monochromatic x-rays from a mirror *M.*

An experimental method for the determination of this critical angle is shown in Figure 5-21. A narrow beam is obtained by passing the x-rays from the target T through two slits S_1 and S_2. The crystal C reflects a particular wavelength λ onto the plane surface M of the material. If the grazing angle of incidence θ is less than the critical angle θ_c, the beam will be totally reflected to the photographic plate P and the surface M will act as a mirror. As the angle of incidence is increased, the beam continues to be reflected until $\theta = \theta_c$, after which the beam is refracted into the material medium and the intensity of the reflected beam becomes almost zero. Figure 5-22 is a photograph of a beam of x-rays reflected from a palladium surface.

According to the classical theory of dispersion, when an electromagnetic wave of frequency v passes through a substance containing n free electrons per unit volume, the index of refraction of the substance is given by

$$\mu = 1 - \frac{ne^2}{2\pi mv^2} \tag{5-12}$$

provided that the natural frequency of the electrons is small in comparison with v and that absorption is negligible. Equation (5-12) can be rewritten as

$$1 - \mu = \delta = \frac{ne^2}{2\pi mv^2}$$

Setting

$$v = \frac{c}{\lambda}$$

we get

$$\delta = \frac{ne^2\lambda^2}{2\pi mc^2} \tag{5-13}$$

Now

$$\mu = \cos\theta_c$$

Since θ_c is small, $\cos\theta_c$ can be expanded in terms of θ_c to yield

$$\cos\theta_c = 1 - \frac{\theta_c^2}{2} + \cdots$$

so that

$$\delta = 1 - \mu = \frac{\theta_c^2}{2}$$

or

$$\theta_c = \sqrt{2\delta} = \sqrt{\frac{ne^2}{\pi mc^2}}\,\lambda \tag{5-14}$$

Measurements of the critical angle show that it is directly proportional to the wavelength of the x-rays and to the square root of the density of the material, except in the neighborhood of an absorption limit (Section 10-3). This is in good agreement with Equation (5-14). However, the classical theory is inadequate to explain refraction in the neighborhood of an x-ray absorption limit. A quantum theory of dispersion has been developed that predicts results in fair agreement with available experimental data. A discussion of the quantum theory of dispersion, however, is beyond the scope of this book.

The index of refraction of a medium for an electromagnetic wave is defined as

$$\mu = \frac{c}{w} \tag{5-15}$$

where c ($=3\times10^{10}$ cm/sec) is its velocity in a vacuum and w is the velocity of the wave in the medium. Since the index of refraction for x-rays is less than unity, its velocity w in a material medium is greater than c. At first sight this may appear to be in contradiction to the fundamental postulate of the special theory of relativity which states that electromagnetic waves are always propagated with the same velocity c. Actually there is no contradiction. On the

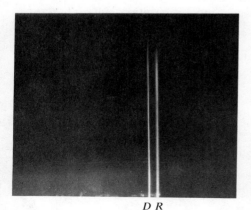

$D\ R$

Figure 5-22   Reflection of monochromatic x-rays (AgKγ line) from a palladium mirror near the critical angle. D is the direct beam and R is the reflected beam. (From photograph by H. Semat.)

microscopic point of view, a material body is not a continuous medium but consists of nuclei and electrons separated by distances that are large relative to their sizes. These particles may be imagined as being situated in empty space. The incident wave, in traversing this space, sets the charges into vibration, causing them to emit electromagnetic waves. Two sets of waves therefore travel through this space: the incident wave and the waves produced by the forced vibrations of the electrons. These elementary waves travel through the space with velocity c; but, by superposition of the elementary waves, a new wave form is produced, which, on the macroscopic (large-scale) point of view, consists of a train of waves traveling through the material body with a wave velocity w that may be greater or smaller than c (see Section 6-7). The energy carried by the incident wave and the energy radiated by the forced vibrations of the electrons always travel with velocity c.

The refracted wave thus consists of a train of waves traveling with a velocity w whose magnitude depends upon the binding forces acting on the electrons. The dielectric constant k of the material medium is determined by the magnitude of these binding forces, and, it can be shown, on the basis of Maxwell's electromagnetic theory of light, that the index of refraction, μ, is given by the equation

$$\mu = \sqrt{k} = \frac{c}{w} \tag{5-16}$$

It is this equation that is used to derive the expression for the index of refraction given by Equation (5-12).

5-12 Measurement of X-Ray Wavelengths by Ruled Gratings

An extremely important application of the total reflection of x-rays is the measurement of x-ray wavelengths by using a ruled grating with a known number of lines per millimeter. As long as the grazing angle of incidence θ is less than the critical angle $\theta_c = \sqrt{2\delta}$ the x-ray beam will be totally reflected from a polished surface. This surface can be made into a diffraction grating by ruling

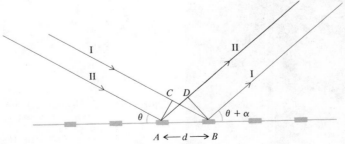

Figure 5-23 Reflection of x-rays from a ruled grating.

a series of uniformly spaced lines a distance d apart. The spaces between the rulings will act as diffracting centers to produce various orders of diffraction on either side of the regularly reflected line. In Figure 5-23, if the rays are incident at an angle θ to the surface of the ruled grating, an intense maximum will be produced at some other angle $(\theta + \alpha)$, provided the difference in path between two adjacent rays is a whole number of wavelengths. The difference in paths between rays I and II is

$$CB - AD = d \cos \theta - d \cos (\theta + \alpha)$$

so that the condition for an intense maximum at this angle of reflection is

$$d[\cos \theta - \cos (\theta + \alpha)] = n\lambda \qquad (5\text{-}17)$$

where n is an integer and may be either positive or negative. The regular reflection at the angle θ is called the zero order; the other orders of reflection are called positive if α is positive and negative if α is negative.

Diffraction gratings for x-rays have been made of glass, speculum metal, silver, and gold and the number of lines per millimeter has ranged from 50 to 600. Since the grating space d is accurately known and the angles θ and $(\theta + \alpha)$ can be measured with a high degree of precision, this method can be used as a check on the wavelengths measured with crystal gratings. A comparison of some wavelengths measured by ruled gratings and by crystals is made in Table 5-1.

TABLE 5-1   COMPARISON OF WAVELENGTHS MEASURED BY RULED GRATINGS AND BY CRYSTALS

Line	λ_g by Grating	λ_c by Crystal	$\lambda_g - \lambda_c$ in Percent	Observer
Mo L_α	5.4116Å	5.3950Å	+0.31	Cork
Mo L_β	5.1832	5.1665	+0.33	"
Cu K_α	1.54172	1.5387	+0.20	Bearden
Cu $K_{\beta 1}$	1.39225	1.3894	+0.20	"
Cr K_α	2.29097	2.2859	+0.22	"
Cr $K_{\beta 1}$	2.08478	2.0806	+0.22	"

In every case the wavelength measured with a ruled grating is greater than that measured with a crystal grating. The possible sources of error in the ruled grating measurements were investigated by many physicists but in no case could they account for this large discrepancy—about $\frac{1}{4}$ of 1 percent. This led to a reexamination of the basis on which crystal grating wavelength measurements depend, namely, the grating space d. It will be recalled that this distance, for rocksalt, is given by the equation

$$d = \sqrt[3]{\frac{M}{2\rho N_0}} \qquad (5\text{-}2)$$

and for crystals that are not simple cubes, such as calcite and quartz, this equation becomes

$$d = \sqrt[3]{\frac{M}{2\rho N_0 \phi(\beta)}} \tag{5-2a}$$

where $\phi(\beta)$ is a function of the angles of the crystal lattice. The least accurately known constant entering in the computation of d is the Avogadro number N_0. The Avogadro number is computed from the relationship

$$F = N_0 e$$

where F is the Faraday constant and e is the electronic charge; F is very accurately known from many experiments extending over a long period of time. To get agreement between the ruled grating and the crystal grating measurements of x-ray wavelengths it is necessary to increase the value of e from Millikan's value 4.77×10^{-10} stcoul to about 4.803×10^{-10} stcoul. As stated in Section 1-8, redeterminations of the value of the electronic charge have led to the adoption of the value

$$e = 4.803 \times 10^{-10} \text{ stcoul}$$

Within recent years improved methods have increased the accuracy of the determinations of the densities of crystals. Chemically pure crystals of NaCl, KCl, LiF, and Si have been prepared artificially, and the grating spaces of these crystals have been determined by measuring the Bragg angle for monochromatic x-rays whose wavelengths are known from ruled grating measurements. Using the known molecular weights of these crystals, we can determine values of the Avogadro number with the aid of Equation (5-2). The Avogadro number determined in this manner, together with the known value of the Faraday constant, leads to a determination of the electronic charge.

5-13   Absorption of X-Rays

When a parallel beam of x-rays passes through any material, the intensity of the emergent beam is less than that of the incident beam. The decrease in intensity of the beam when it traverses a small thickness dx of the material (Fig. 5-24) depends on the thickness and the intensity I of the beam, or

$$-dI = \mu I \, dx$$

where $-dI$ represents the decrease in intensity of the beam and μ is a factor of proportionality called the *absorption coefficient*. It represents the fraction of the energy removed from the beam per centimeter of path and is sometimes referred to as the total absorption coefficient of the material.

The above equation can be written as

$$\frac{dI}{I} = -\mu \, dx$$

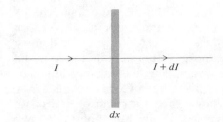

Figure 5-24

and on integration becomes

$$I = I_0 \exp(-\mu x) \qquad (5\text{-}18)$$

where I_0 is the initial intensity of the beam and I is its intensity after it has traversed a thickness x of the material.

There are several ways in which the atoms of a substance can remove energy from an incident beam of x-rays. One process is that of *scattering*, in which the electrons of the atoms are set into forced vibrations by the incident x-ray waves and then radiate electromagnetic waves in all directions. We will devote the remaining sections of this chapter to discussions of various aspects of the scattering process.

Another process that can occur when x-rays are incident on a substance is a type of photoelectric effect in which x-ray photons of sufficient energy eject electrons from different energy levels of the atom. The subsequent electronic transitions within the atoms are accompanied by the emission of x-rays that are characteristic of these atoms. Such radiation is usually referred to as *fluorescent radiation* and is emitted in all directions. We will discuss this process in greater detail in Chapter 10.

With x-rays of higher energy, greater than 1 MeV, several other processes occur; these usually involve the interaction of x-ray photons with the nuclei of atoms. These photonuclear effects will be discussed in Part III of this book. Another effect that is of very great interest is the conversion of x-ray photons of energy greater than 1.02 MeV into electron-positron pairs. The phenomenon of pair production will also be left to Part III.

The absorption coefficient μ may thus be the sum of a variety of different coefficients, each one representing a different process by which energy is removed from the x-ray beam; this may be represented by the equation

$$\mu = \sigma + \tau + \epsilon \qquad (5\text{-}19)$$

where σ represents the scattering coefficient, τ, the fluorescent transformation coefficient, and ϵ, the coefficients of any other processes that may occur. It is sometimes possible to select one of these processes for investigation by suitable choices of substance and incident x-ray energy.

A glance at the equation defining the absorption coefficient μ shows that it can be interpreted as the fraction of the energy absorbed per unit length of substance traversed. Although this is a very useful quantity for many purposes, it does depend on the thermodynamic properties of the substance, such as its particular phase—solid, liquid, gas—and its pressure and temperature. A more useful concept is the *mass absorption coefficient* μ/ρ, where ρ is the density of the substance. The dimensions of μ/ρ are those of area per unit mass; in the cgs system these are square centimeters per gram. The mass absorption coefficient is characteristic of the substance for a given type of x-ray beam; it may be interpreted as the fraction of energy removed from a beam of unit cross section per unit mass of the substance traversed. Using Equation (5-19), we can write

$$\frac{\mu}{\rho} = \frac{\sigma}{\rho} + \frac{\tau}{\rho} + \frac{\epsilon}{\rho} \qquad (5\text{-}20)$$

where σ/ρ is the mass scattering coefficient, τ/ρ, the mass transformation coefficient, and ϵ/ρ, the mass coefficient for the particular process involved.

5-14 Atomic Absorption Coefficient

From the microscopic point of view, an *atomic absorption coefficient*, μ_a, is of greater interest than a mass absorption coefficient. The former can be defined in terms of the latter by the equation

$$\mu_a = \frac{\mu}{\rho} \frac{M}{N_0} \qquad (5\text{-}21)$$

where N_0 is the Avogadro number and M is the mass of one gram-mole of the isotope of the element under discussion. The dimensions of μ_a are those of an area/atom; for example, μ/ρ for copper, for x-rays of 0.70 Å, is about 50 cm²/gram. Taking $M = 63$ gram per gram-atomic mass and $N_0 = 6 \times 10^{23}$ atoms per/gram-atomic mass, Equation (5-21) yields

$$\mu_a = 50 \times 10^{-22} \text{ cm}^2/\text{atom}$$

If we imagine the atom to be a sphere of radius of about 2×10^{-8} cm, we can think of this atom as having a geometrical cross section of approximately 12×10^{-16} cm². The absorption cross section is thus very much smaller than the geometrical cross section.

The atomic absorption coefficient, hence the cross section for absorption, varies with the wavelength of the incident x-rays for any one substance. We can interpret the atomic absorption coefficient as a measure of the probability of the absorption of the given radiation by the atoms of the substance irradiated. As has already been shown, absorption of radiation by the atoms may result in its scattering, which will be represented by an *atomic scattering coefficient* σ_a, or it may result in the transformation of its energy into fluorescent radiation which will be represented by an *atomic fluorescent transformation coefficient* τ_a. The value of each of these coefficients is a measure of the probability for the occurrence of the given type of process and is expressed in units of square centimeters per atom.

5-15   Scattering of X-Rays

When a beam of x-rays passes through a substance, the electrons in this substance are set into vibration and radiate x-rays in all directions. The radiation emitted by these electrons is called *scattered* or *secondary radiation*. If E is the electric intensity of the incident wave, the acceleration of the electron will be

$$a = \frac{Ee}{m}$$

where e is the charge of the electron and m is its mass. Both E and e are in cgs electrostatic units. If the speed of the electron is small in comparison with the speed of light c, then the electric intensity E_ϕ of the scattered wave at a distance r from this electron is

$$E_\phi = \frac{ea \sin \phi}{rc^2} \qquad (4\text{-}2a)$$

where ϕ is the angle between E and r, as shown in Figure 5-25. Substituting the value for the acceleration in the above equation, we get

$$E_\phi = \frac{Ee^2}{rmc^2} \sin \phi \qquad (5\text{-}22)$$

Since the intensity of the wave is proportional to the square of the electric vector, the ratio of the intensity I_ϕ of the scattered radiation to the intensity I of the incident radiation is

$$\frac{I_\phi}{I} = \frac{E_\phi^2}{E^2} = \frac{e^4}{r^2 m^2 c^4} \sin^2 \phi \qquad (5\text{-}23)$$

Choose a set of axes so that the incident or primary beam is propagated parallel to the x axis. The electric intensity E will then be in a plane parallel to the y-z plane. Since the position of the y and z axes can be chosen arbitrarily in this

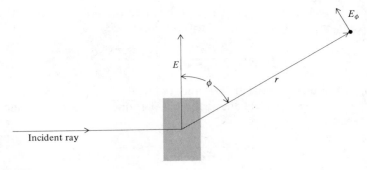

Figure 5-25

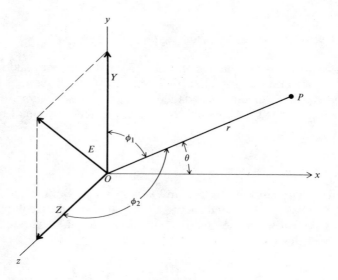

Figure 5-26

plane, let the y axis be taken in the plane determined by the direction of the scattered ray OP and the primary ray (Figure 5-26), and let the electron be at the origin of coordinates. The electric vector E may be resolved into two perpendicular components, Y and Z, such that

$$E^2 = Y^2 + Z^2$$

Since the intensity of the wave is proportional to the square of the electric field intensity, we can write

$$I = I_y + I_z$$

where I_y is the intensity of the y component of the incident wave and I_z is the intensity of its z component. Since the primary ray is unpolarized, the average values of Y^2 and Z^2 are equal; hence the intensities of the y and z components are equal, yielding

$$I_y = I_z = \frac{I}{2}$$

Now the intensity, I_1, of the scattered beam at the point P due to the y component of the incident wave is, from Equation (5-23),

$$I_1 = I_y \frac{e^4 \sin^2 \phi_1}{r^2 m^2 c^4} = \tfrac{1}{2} I \frac{e^4 \cos^2 \theta}{r^2 m^2 c^4} \tag{5-24}$$

where ϕ_1 is the angle between OP and Y and $\theta = \pi/2 - \phi_1$. Similarly, the intensity I_2 of the scattered beam at the point P due to the z component of the incident wave is

$$I_2 = I_z \frac{e^4 \sin^2 \phi_2}{r^2 m^2 c^4} = \tfrac{1}{2} I \frac{e^4}{r^2 m^2 c^4} \tag{5-25}$$

where ϕ_2 is the angle between OP and Z, and is always $\pi/2$. The total intensity I_e of the scattered wave at P due to the energy radiated by a single electron is thus

$$I_e = I_1 + I_2 = I\,\frac{e^4}{2r^2m^2c^4}\,(1 + \cos^2\theta) \qquad (5\text{-}26)$$

If there are n electrons per unit volume of material, and if we make the assumption that each electron is effective in scattering the x-rays independently of all the other electrons, then the intensity of the scattered wave reaching point P, per unit volume of the scatterer, will be given by

$$I_s = nI_e = I\,\frac{ne^4}{2r^2m^2c^4}\,(1 + \cos^2\theta) \qquad (5\text{-}27)$$

This analysis of the intensity of the scattered x-ray beam leads to three conclusions, each of which can be investigated experimentally. One is that a measurement of the total energy scattered should yield the number of electrons per unit volume effective in scattering, hence the number of electrons per atom effective in scattering. Another is that when the scattering angle θ is 90 degrees, the scattered beam should be linearly polarized with the direction of vibration of the electric vector parallel to the z axis. The third is that Equation (5-27) predicts a definite distribution of intensity of the scattered beam as a function of the angle of scattering with a minimum intensity at 90 degrees.

5-16   Determination of the Number of Electrons per Atom

The rate at which energy is scattered by the electrons of a substance can be calculated by integrating I_s over a large sphere of radius r with its center at the scatterer. A convenient element of surface area is one over which I_s has a constant value and, from Figure 5-27, is

$$dA = 2\pi r \sin\theta \cdot r\,d\theta$$

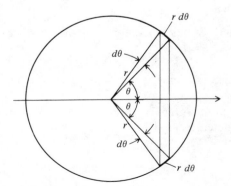

Figure 5-27

The amount of energy W_s that is scattered per unit volume per second is therefore

$$W_s = \int_0^A I_s \, dA$$

$$= \frac{\pi I n e^4}{m^2 c^4} \int_0^\pi (1 + \cos^2 \theta) \sin \theta \, d\theta$$

$$= \frac{8\pi}{3} \frac{n e^4}{m^2 c^4} I \tag{5-28}$$

Let

$$\sigma = \frac{W_s}{I} = \frac{8\pi}{3} \frac{n e^4}{m^2 c^4} \tag{5-29}$$

The quantity σ is called the scattering coefficient of the substance and represents the fraction of the energy removed from the incident beam by the process of scattering per centimeter of path traversed in the substance. A measurement of σ would make possible an evaluation of the number of electrons effective in scattering. One of the earliest determinations of the scattering coefficient was made by Barkla (1911), and later repeated with greater accuracy by Hewlett, using carbon as the scattering element. Hewlett measured the intensity of the scattered beam over a range of angles θ from about 0 to π, thus performing the integration experimentally. The wavelength of the incident beam was 0.71 Å. The result obtained for the mass scattering coefficient is $\sigma/\rho = 0.2$, where ρ is the density of carbon. Solving Equation (5-29) for n, we get

$$n = \sigma \cdot \frac{3 m^2 c^4}{8 \pi e^4}$$

Since n is the number of electrons per cubic centimeter, n/ρ is the number of electrons per gram of material and is given by

$$\frac{n}{\rho} = \frac{\sigma}{\rho} \cdot \frac{3 m^2 c^4}{8 \pi e^4} \tag{5-30}$$

Putting in the experimentally determined values on the right-hand side of Equation (5-30) yields

$$\frac{n}{\rho} = 3 \times 10^{23} \text{ electrons/gm}$$

There are $(6.02 \times 10^{23})/12 = 5.01 \times 10^{22}$ atoms/gram of carbon. Therefore the number of electrons per atom effective in scattering is

$$\frac{3 \times 10^{23}}{5 \times 10^{22}} = 6$$

which is the atomic number of carbon. Hence the number of electrons effective in scattering is equal to the atomic number of the scattering element. This is in excellent agreement with the results of the experiments on the scattering of alpha particles and is a direct determination of the number of electrons outside the nucleus of the atom.

It must be remarked that such good agreement between the results of the experiments on the scattering of x-rays and Thomson's theory has been obtained only with scattering substances of low atomic number and then only with x-rays of wavelengths greater than 0.1 Å. The theory of the scattering of x-rays will be discussed further in Section 5-19.

5-17 Polarization of X-Rays

This classical theory of scattering predicts that the beam scattered at an angle of 90 degrees should be linearly polarized. This can be seen by a glance at Equations (5-24) and (5-25). For $\theta = 90$ degrees the intensity due to the y component of the electric vector will be zero, whereas the intensity at any point in the X-Y plane due to the z component of the electric vector is independent of the angle. Thus in the y direction the scattered beam will be linearly polarized with the direction of vibration parallel to the z axis. To detect this polarization, another scatterer is used as an analyzer. Barkla was the first to

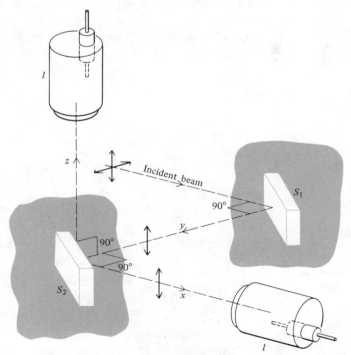

Figure 5-28 Arrangement of apparatus for determining the polarization of x-rays scattered through 90 degrees.

show the polarization of the beam scattered at an angle of 90 degrees. Compton and Hagenow (1924) repeated this experiment with improved apparatus. Figure 5-28 shows a diagram of their apparatus. A narrow beam of x-rays from a tungsten target was scattered by S_1 to a second scatterer placed at S_2 so that the scattered beam made an angle of 90 degrees with the original beam. The scattering materials were blocks of paper, carbon, aluminum, and sulfur. The beam scattered by S_2 was examined in two directions at right angles to each other and perpendicular to the direction S_1S_2. The intensity in the direction S_2-x was found to be a maximum, whereas that in the direction S_2-z was nearly zero within the limits of error of the experiment. This is in complete agreement with the electromagnetic theory, since, if only the z component of the electric vector is present in the beam scattered from S_1 to S_2, then the intensity in the direction S_2-z should be zero for a transverse wave; similarly the beam should have maximum intensity in the direction S_2-x.

5-18 Intensity of the Scattered X-Rays

Many measurements have been made on the distribution of intensity in the scattered x-ray beam as a function of the angle of scattering. For x-ray wavelengths greater than 0.2 Å the distribution follows the $(1 + \cos^2 \theta)$ law fairly well except for small angles of scattering, in which cases the intensity of the scattered beam is much greater than that predicted by Equation (5-27) (see Fig. 5-29). This large intensity of the scattered beam at small angles can be explained on the assumption that the phase differences of the rays scattered at

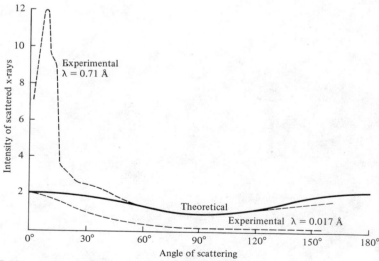

Figure 5-29 A comparison of the theoretical and experimentally determined values of the intensity of the scattered x-rays as a function of the angle of scattering.

these angles are small so that the waves scattered by neighboring electrons reinforce each other.

For wavelengths much smaller than 0.2 Å, however, the intensity falls off rapidly at large angles; the scattering coefficient is much smaller than that predicted by the classical theory. The lowest curve in Figure 5-29 shows the intensity of the rays scattered by iron; the incident rays were hard gamma rays of wavelength 0.017 Å.

Evidently the classical theory of scattering is not completely satisfactory. To account for the small intensity of the beam at large angles of scattering, A. H. Compton proposed a quantum theory of scattering in which the x-rays, considered as photons, cause the ejection of electrons from the atom as a result of the scattering process. As will be shown in the next section, the greater the angle of scattering, the greater the amount of energy removed from the beam by these ejected or *recoil* electrons.

5-19 The Compton Effect

In the experiments on the scattering of x-rays by matter it was noted that for wavelengths of the order of 1 Å the experimental results were in good agreement with the classical theory of scattering, but for shorter wavelengths there was great divergence between the classical theory and experimental results. In working with scattered radiation, A. H. Compton (1923) observed that the wavelength of the radiation scattered by a block of paraffin in a direction at right angles to the incident beam was greater than the wavelength of the incident beam. The theory of this effect was given by Compton and also by Debye at about the same time.

Consider the incident radiation as consisting of photons, or quanta, of energy $h\nu$ traveling in the direction of the primary ray with velocity c. From the relationship between mass, energy, and momentum derived from the theory of relativity a photon of energy $h\nu$ has a momentum of $h\nu/c$. Suppose that this photon strikes a comparatively free electron at rest. If we assume that the principles of conservation of energy and conservation of momentum hold during this process, then, as a result of this collision, the electron will acquire a velocity v in a direction making an angle θ with the direction of motion of the incident photon, and a photon of energy $h\nu'$ will be scattered at an angle ϕ with the original direction as shown in Figure 5-30. From the principle of conservation of energy we get

$$h\nu = h\nu' + mc^2(\gamma - 1) \tag{5-31}$$

where $mc^2(\gamma - 1)$ is the kinetic energy of the electron as derived on the basis of the special theory of relativity, and γ is $(1 - v^2/c^2)^{-1/2}$.

Resolving the momentum vectors into two rectangular components, along and at right angles to the direction of the incident photon, and using the principle of conservation of momentum, we get

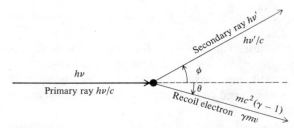

Figure 5-30 The Compton effect.

$$\frac{h\nu}{c} = \frac{h\nu'}{c} \cos \phi + \gamma m\upsilon \cos \theta \qquad (5\text{-}32)$$

$$0 = \frac{h\nu'}{c} \sin \phi - \gamma m\upsilon \sin \theta \qquad (5\text{-}33)$$

The solution of these equations (Appendix V-4) yields the following results:

$$\lambda' - \lambda = \frac{h}{mc} (1 - \cos \phi) \qquad (5\text{-}34)$$

$$\cot \frac{\phi}{2} = (1 + \alpha) \tan \theta \qquad (5\text{-}35)$$

$$\mathcal{E} = mc^2(\gamma - 1) = h\nu \frac{2\alpha \cos^2 \theta}{(1 + \alpha)^2 - \alpha^2 \cos^2 \theta} \qquad (5\text{-}36)$$

where $\alpha = h\nu/mc^2$, $\mathcal{E}$ is the kinetic energy of the recoil electron, and the wavelengths λ' and λ corresponding to the frequencies ν' and ν have been introduced.

Equation (5-34) states that the wavelength of the ray scattered at any angle ϕ should always be greater than the wavelength of the incident radiation. Furthermore, this difference in wavelength should not depend on the nature of the scattering material but should depend only on the angle of scattering.

The factor h/mc with m, the rest mass of the electron, is called the *Compton wavelength of the electron,* and is usually designated by the symbol λ_c. Its numerical value can be determined by substituting the known values of the constants h, m, and c, obtaining

$$\lambda_c = 0.02426 \ \text{Å} \qquad (5\text{-}37)$$

A glance at Equation (5-34) shows that the change in wavelength of the electromagnetic radiation interacting with the electron will never exceed $2\lambda_c$.

Equation (5-35) gives the relationship between the direction of motion of the recoil electron and the scattered photon, whereas Equation (5-36) gives the kinetic energy of the recoil electron in terms of the energy of the incident photon and the angle θ.

A typical experimental arrangement for studying the Compton effect is shown in Figure 5-31. X-rays from some target T, giving out strong characteristic radiation, are scattered in all directions by the body A. The radiation scattered in some direction ϕ is then allowed to fall on the crystal C of an x-ray spectrometer, which analyzes the beam into its component wavelengths. This radiation may be measured with the aid of an ionization chamber B or with a photographic plate in place of B.

The results of a typical set of measurements are shown in Figure 5-32. In this case the K_α line of molybdenum (see Chapter 10) was scattered by graphite, and measurements were made at scattering angles of 45, 90, and 135 degrees. Each scattered beam is seen to consist of two distinct lines. One line, P, has a wavelength corresponding to the wavelength of the incident radiation. The second line, M, has a longer wavelength, λ', which depends on the angle of scattering. The wavelength λ' of this *modified* line is in good agreement with that calculated from Equation (5-34). Similar results were obtained when other substances were used as scatterers. In all cases the wavelength of the modified line was found to depend only on the angle of scattering and not on the nature of the scattering substance.

The presence of a line with the same wavelength as the incident radiation is not predicted by Equation (5-34). In deriving this equation it was assumed that the electron which took part in the scattering was a "free" electron and that it was ejected from the atom. The unmodified line is due to the interaction between the incident quanta and bound electrons. In this case the bound electrons do not receive energy or momentum from the incident quantum and there is no change in the wavelength of the scattered photon. In light atoms, such as Be, C, and Al, the electrons are probably bound more loosely than in the heavier elements. The modified line should be relatively more intense than the unmodified line for light elements, whereas the reverse should be true for the heavier elements. This has actually been confirmed by the scattering experiments.

The results of the experiments on the Compton effect leave no doubt that in

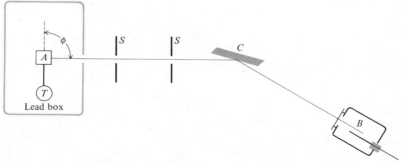

Figure 5-31   Method of determining the wavelength of the scattered x-rays.

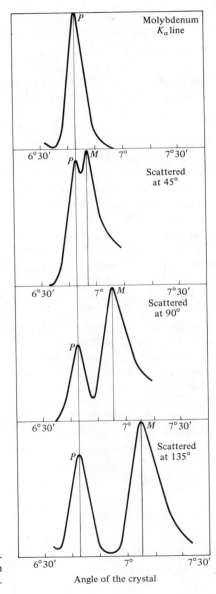

Figure 5-32 Curves showing the displacement of the "modified" line in the Compton effect for three different angles of scattering.

its interaction with matter, radiant energy behaves as though it were composed of particles. Similar behavior was observed in the photoelectric effect. It will be shown later that in the processes of emission and absorption, light behaves as though it consisted of corpuscles, but the phenomena of the interference and diffraction can be explained only on the hypothesis that radiant energy is propagated as a wave motion. We are thus led to the conclusion that radiant energy exhibits a dual character, that of a wave and that of a corpuscle. The relationship between these two concepts, wave and corpuscle, will be examined more fully in the next chapter.

5-20   Compton Recoil Electrons

In addition to a change in wavelength, the theory of the Compton effect predicts that every collision should be accompanied by the ejection of a recoil electron. According to Equation (5-35), these electrons will be ejected at angles θ less than 90 degrees, with those ejected in the forward direction having maximum energy. There are two different modes of approach to the problem: one is the investigation of the energy of the recoil electron as a function of the angle θ, and the other is the investigation of the simultaneity of the appearance of the recoil electron with that of the scattered photon. Both problems were investigated shortly after the discovery of the Compton effect. Compton and Simon (1925) sent a beam of x-rays through a Wilson cloud chamber and measured the maximum range of the recoil electrons. The kinetic energies calculated from the range measurements agreed within about 20 percent with those predicted by theory. The problem of simultaneity was investigated by Bothe and Geiger; they sent a beam of x-rays through hydrogen and detected the recoil electrons and scattered photons in two adjacent Geiger counters. The Geiger counter pulses were recorded on moving photographic film. They observed a reasonable number of coincidences between the scattering of a photon and the ejection of a recoil electron. No great accuracy can be claimed for these experiments, since photons are not directly detected by Geiger counters

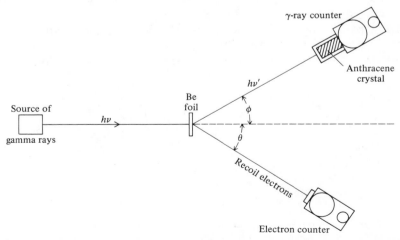

Figure 5-33   Schematic diagram of the apparatus of Cross and Ramsey for investigating the Compton effect.

but are detected only by the recoil electrons produced in the gas in the Geiger counter chamber.

The development of fast-counting crystal scintillation counters and the great improvement in electronic circuits led to a reinvestigation of both problems. In one experiment by Cross and Ramsey (1950) a collimated beam of gamma-ray photons from a RaTh source was allowed to strike a foil of beryllium (see Fig. 5-33). The scattered photons were detected by one scintillation counter and the recoil electrons detected in a second scintillation counter. The pulses from these two counters were fed into a coincidence circuit—that is, only coincidences of the two counters were registered. In the experiment, the gamma-ray counter was kept at a fixed angle $\phi = 30$ degrees. The energy of the gamma-ray photons was 2.62 MeV. The electron recoil scattering angle for this case can be calculated from Equation (5-35) and shown to be $\theta = 31.3$ degrees. In one part of the experiment, both counters remained fixed at the appropriate angles while the beryllium scatterer was moved. The coincidence rate showed a maximum at the proper position of the scatterer, that is, the position for which $\phi = 30$ degrees and $\theta = 31.3$ degrees. In another part of the experiment, the beryllium foil and the gamma-ray detector were kept fixed in position but the electron detector was moved about 2 degrees on either side of the correct position. The curve of coincidence counts against angle (see Fig. 5-34) showed a maximum at 31.3 degrees. The resolving time of the circuits used was 0.3 microsecond; these results thus confirm the fundamental assumptions of the Compton effect and also show that the scattered photon and recoil electron are emitted simultaneously within the above time interval.

The question of simultaneity in the Compton effect was investigated in great detail by Hofstadter and McIntyre (1950), using the somewhat different arrangement shown in Figure 5-35. Gamma-ray photons from a small rod of radioactive ^{60}Co were collimated by means of a thick lead shield and allowed to strike a crystal of stilbene. This stilbene crystal has two functions: (a) it scatters a photon $h\nu'$ which then travels to another larger stilbene crystal mounted on a photomultiplier tube; (b) the first stilbene crystal also acts as the detector of the recoil electron released in it by converting its energy into

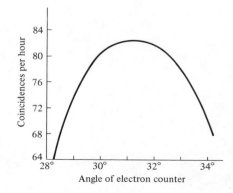

Figure 5-34   Graph of coincidences against angle of the electron counter with gamma-ray counter held at 30 degrees. [After *Phys. Rev.* **80**, 933 (1950).]

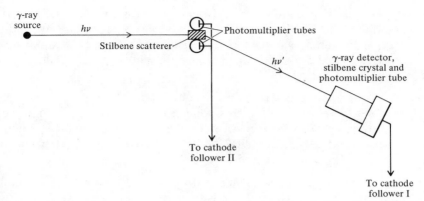

Figure 5-35   Schematic diagram of the apparatus used by Hofstadter and McIntyre to investigate simultaneity in the Compton effect.

light, which is detected by two adjacent photomultiplier tubes connected in parallel. The pulses from the gamma-ray detector and electron detector are fed to opposite plates of a cathode-ray tube and the pulses produced are photographed. The timing of the cathode-ray pulses is checked by impressing two known simultaneous pulses on the two detecting circuits.

In the above experiment the gamma-ray detector was set at scattering angles of 30, 50, 70, and 90 degrees, and simultaneous pulses were observed from the recoil electron and scattered photon detectors. The accuracy of the experiment was such that the pulses were considered to be simultaneous within 1.5×10^{-8} sec.

Problems

5-1.   The mass absorption coefficient of silver is 38 cm²/gm for x-rays of wavelength 0.4 Å and 11 cm²/gm for a wavelength of 0.5 Å. (a) Determine the atomic absorption coefficients of silver for these wavelengths. (b) Compare these values with the geometrical cross section of silver atoms.

5-2.   The mass absorption coefficient of lead for x-rays of wavelength 0.5 Å is 60 cm²/gm; for 2.25 Å, 500 cm²/gm. (a) Determine the atomic absorption coefficients of lead for these wavelengths. (b) Compare these values with the geometrical cross section of lead atoms.

5-3.   The absorption coefficient of aluminum is 0.73 cm⁻¹ for x-rays of wavelength 0.20 Å. Determine (a) the mass absorption coefficient and (b) the atomic absorption coefficient. Compare the latter with the geometrical cross section of aluminum atoms.

5-4.   The mass absorption coefficient of x-rays of wavelength $\lambda = 0.7$ Å is

5 gm^{-1} cm^2 for aluminum and 50 gm^{-1} cm^2 for copper. The density of aluminum is 2.7 gm/cm^3 and that of copper is 8.93 gm/cm^3. What thickness in centimeters of each of these substances is needed to reduce the intensity of the x-ray beam passing through it to one-half its initial value?

5-5.   An x-ray spectrometer has a crystal of rocksalt set at an angle of 14 degrees to the beam of x-rays coming from a tube operated at a constantly increasing voltage. An intense x-ray line first appears when the voltage across the tube is 9045 volts. (a) Calculate h/e and (b) calculate h, using the accepted value of e.

5-6.   Calculate the grating space d of calcite from the following data: molecular weight $M = 100.091$, density $\rho = 2.71029$ gm/cm^3, and $\phi(\beta) = 1.09594$.

5-7.   The radiation from an x-ray tube operated at 40 kV is analyzed with a Bragg x-ray spectrometer, using a calcite crystal cut along its cleavage plane. (a) Calculate the short wavelength limit of the x-ray spectrum coming from this tube. (b) What is the smallest angle between the crystal planes and the x-ray beam at which this wavelength can be detected?

5-8.   The wavelength of the K_α line of silver is 0.563 Å. The radiation from a silver target is analyzed with a Bragg spectrometer, using a calcite crystal. Determine (a) the angle of reflection for the first order and (b) for the second order. (c) What is the highest order for which this line may be observed?

5-9.   Monochromatic x-rays are reflected in the first order from a calcite crystal set with its cleavage planes at an angle of 13 degrees with respect to the x-ray beam. These x-rays are allowed to fall on a silver mirror at a very small angle to the plane of the mirror. The mirror is then rotated until the critical angle is reached. (a) Calculate the wavelength of the x-rays incident on the mirror. (b) Calculate the index of refraction of silver for this x-ray beam. (c) Determine the critical angle. The density of silver is 10.5 gm/cm^3.

5-10.   A ruled glass surface covered with a thin layer of gold forms a diffraction grating with 200 lines/mm. A very narrow beam of the copper K_α radiation, $\lambda = 1.541$ Å, is incident on the grating at an angle of 20 min to its surface.
(a) Show that for small angles Equation (5-17) can be put in the form

$$n\lambda = d\left(\alpha\theta + \frac{\alpha^2}{2}\right)$$

by expanding the cosine functions.
(b) Calculate the angle between the first-order and zero-order beams.

5-11. (a) Write the equations for the Compton effect for the case in which the velocity of the recoil electron is small in comparison with the velocity of light so that relativistic effects may be neglected. (b) Solve these equations for the change in wavelength as a function of the angle of scattering. Compare this result with Equation (5-34).

5-12. The K_α radiation from a molybdenum target, $\lambda = 0.709$ Å, is scattered from a block of carbon, and the radiation scattered through an angle of 90 degrees is analyzed with a calcite crystal spectrometer. (a) Calculate the change in wavelength produced in the scattering process. (b) Determine the angular separation in the first order between the modified and unmodified lines produced by rotating the crystal through the required angle.

5-13. (a) Calculate the angle between the direction of motion of the recoil electron and the incident photon in Problem 5-12. (b) Determine the energy of the recoil electron.

5-14. Monochromatic x-rays of wavelength $\lambda = 0.124$ Å are scattered from a carbon block. (a) Determine the wavelength of the x-rays scattered through 180 degrees. (b) Determine the maximum kinetic energy of the recoil electrons produced in this scattering process.

5-15. The value σ/n obtained from Equation (5-29) is the scattering cross section per electron. Show that the "radius" of the electron calculated from the above value is $(\frac{8}{3})^{1/2} a_c$, where a_c is the classical electron radius.

5-16. A photon of frequency ν is scattered through an angle of 90 degrees by an electron which was at rest. (a) Write the relativistic equations for conservation of momentum and energy. (b) Working from these equations, show that the scattered frequency is given by

$$\nu' = \frac{\nu m c^2}{h\nu + mc^2}$$

5-17. A photon of frequency ν is backscattered at 180° by an electron which was at rest. (a) Write the equations for conservation of momentum and energy for this process. (b) From these equations calculate the momentum of the recoil electron in terms of ν and m.

5-18. The absorption coefficient for 10 keV x-rays in a certain material is 2 cm^{-1}. What fraction of incident photons is present at a depth of 1 cm in this material?

References

Bacon, G. E., *X-Ray and Neutron Diffraction*. New York: Pergamon Press, 1966. (Contains excerpts from many original papers.)

Bragg, W. H., and W. L. Bragg, *X-Rays and Crystal Structure*. London: George Bell & Sons, 1925, Chapters II, III, IV, and X.

Clark, G. L., *Applied X-Rays*. New York: McGraw-Hill Book Company, 1955, Chapters I–X.

Compton, A. H., and S. K. Allison, *X-Rays in Theory and Experiment*. Princeton, N.J.: D. Van Nostrand Company, 1935, Chapters I–III, IX.

Guinier, A., and D. L. Dexter, *X-Ray Studies of Materials*. New York: Wiley-Interscience, 1963.

Nuffield, E. W., *X-Ray Diffraction Methods*. New York: John Wiley & Sons, 1966, Chapters 1–3, 5, 7.

Richtmyer, F. K., E. H. Kennard, and J. N. Cooper, *Introduction to Modern Physics*. New York: McGraw-Hill Book Company, 1969.

Siegbahn, M., *The Spectroscopy of X-Rays*. New York: Oxford University Press, 1925, Chapters I–III, VII.

Worsnop, B. L., *X-Rays*. New York: John Wiley & Sons, 1950.

6 | Waves and Particles

6-1 De Broglie's Hypothesis

In order to explain the results of some of the experiments involving the interaction between radiant energy and matter, such as blackbody radiation, the photoelectric effect, and the Compton effect, it was necessary to assign to radiant energy some properties characteristic of particles rather than waves. The amount of energy assigned to such a particle of radiant energy, or photon, is given by

$$\mathcal{E} = h\nu \tag{6-1}$$

where h is the Planck constant and ν is the frequency of the radiation. The frequency ν is usually computed from measurements of the wavelength λ of the radiation, using the relationship

$$\nu = \frac{c}{\lambda} \tag{6-2}$$

where c is the velocity of light. The wavelength λ can be measured only by some experiment that involves interference or diffraction, phenomena characteristic of wave motion. In spite of the fact that radiation possesses this dual character, it never exhibits both characteristics in any one experiment. In a given experiment it behaves either as a wave or a corpuscle.

According to a hypothesis introduced by L. De Broglie (1924), this dual character, wave and particle, should not be confined to radiation alone but should also be exhibited by all the fundamental entities of physics. On this hypothesis, electrons, protons, atoms, and molecules should have some type of wave motion associated with them. De Broglie was led to this hypothesis by considerations based on the special theory of relativity and on the quantum theory.

Let us return for a moment to a consideration of electromagnetic radiation. We have already shown, on the basis of the relativistic equation relating energy and momentum, that the momentum p of a photon is given simply by

$$p = \frac{\mathcal{E}}{c} \tag{6-3}$$

so that

$$p = \frac{h\nu}{c} \tag{6-4}$$

or

$$p = \frac{h}{\lambda} \tag{6-5}$$

The quantity $1/\lambda$ is commonly used by spectroscopists and is known as the *wave number*, or the number of waves per centimeter in the monochromatic beam of light. If we let

$$k = \frac{1}{\lambda} \tag{6-6}$$

then

$$p = hk \tag{6-7}$$

Since momentum is a vector quantity whose magnitude is given by p, and h is a scalar quantity, k may be interpreted as the magnitude of a vector quantity whose direction is that of the direction of propagation of the light waves of length λ.

De Broglie carried these considerations over into the dynamics of a particle. On De Broglie's hypothesis a wavelength λ is associated with each particle and is related to the momentum p of the particle by the equation

$$p = \frac{h}{\lambda} \tag{6-5}$$

If m is the mass of the particle and v is its velocity, then

$$p = \gamma m v \tag{6-8}$$

so that

$$\lambda = \frac{h}{\gamma m v} \tag{6-9}$$

with

$$\gamma = \left(1 - \frac{v^2}{c^2}\right)^{1/2} \tag{6-10}$$

Equation (6-9) gives the relationship between the wavelength λ associated with a particle and the mass m and velocity v of the particle. The existence of these waves was demonstrated experimentally by Davisson and Germer (1927) and G. P. Thomson (1928).

6-2   Electron Diffraction Experiments of Davisson and Germer

De Broglie's hypothesis that material particles should exhibit a dual character, that of a wave and that of a corpuscle, has led to many interesting and far-reaching consequences. The wavelength associated with any particle of mass m moving with velocity v, small in comparison with c, so that $\gamma = 1$, is

$$\lambda = \frac{h}{mv} \tag{6-9a}$$

where h is the Planck constant. If the particle is an electron which has acquired its velocity v under the action of a difference of potential V, its kinetic energy, if v is small in comparison with c, is

$$\tfrac{1}{2}mv^2 = Ve \tag{6-11}$$

and the wavelength associated with it can be expressed as

$$\lambda = \frac{h}{\sqrt{2mVe}} \tag{6-12}$$

For a difference of potential of 100 volts, for example,

$$\lambda = 1.22 \times 10^{-8} \text{ cm} = 1.22 \text{ Å}$$

This wavelength is of the order of magnitude of the distances between atomic planes in crystals. This fact immediately suggests the possibility of showing the existence of these waves by using crystals as diffraction gratings for electrons in a manner analogous to their use with x-rays. Such a series of experiments was first carried out by Davisson and Germer (1927).

The experimental arrangement used by Davisson and Germer is shown in

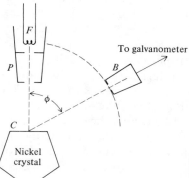

Figure 6-1   Outline of the experimental arrangement in the electron diffraction experiments of Davisson and Germer.

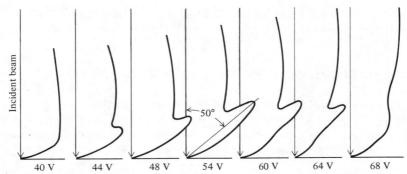

Figure 6-2 Curves, plotted in polar coordinates, showing the intensity of the scattered beam at different angles of scattering. The incident beam in each case is perpendicular to the face of the nickel crystal.

Figure 6-1. Electrons from a hot tungsten filament are accelerated by a difference of potential V between the filament F and the plate P. Some of these electrons emerge from a small opening in the plate P and strike the surface of a nickel crystal at normal incidence. The electrons are scattered in all directions by the atoms in the crystal. The intensity of the electron beam scattered in any given direction is determined by allowing the electrons to enter a chamber or bucket, B, set in the appropriate position, and then measuring the deflection produced by a galvanometer connected to the bucket. By rotating the bucket B about a line in the face of the crystal through the point of incidence as an axis, the intensity of the scattered beam can be measured as a function of the angle of scattering. The results of one such set of measurements are shown in the set of curves of Figure 6-2. These curves are plotted in polar coordinates; the length of the radius vector is proportional to the intensity of the scattered beam and the angle between the radius vector and the y axis is the angle of scattering. The crystal is held in a fixed position throughout this set of measurements. When the difference of potential between F and P is 40 volts, the curve is fairly smooth. At 44 volts a distinct spur appears on the curve at an angle of about 60 degrees. The measurement of the distribution of intensity in the scattered beam is repeated at higher voltages. The length of the spur increases until it reaches a maximum at 54 volts at an angle of 50 degrees, then decreases and disappears completely at 68 volts at an angle of about 40 degrees.

The selective reflection of the "54-volt" electrons at an angle of 50 degrees can be explained as due to the constructive interference—that is, reinforcement—of the electron waves from the regularly spaced atoms of the nickel crystal. Consider a regular array of atoms such as that shown in Figure 6-3. Several sets of parallel planes rich in atoms can be drawn through this array in a manner identical with that used in the explanation of the Laue x-ray diffraction patterns. The parallel lines drawn in the figure represent the traces of one such set of planes perpendicular to the plane of the figure. A beam of electrons incident on the crystal will make some angle α with the normal to these planes. If the wavelength of the De Broglie waves associated with these electrons satisfies Bragg's law (see Section 5-4)

$$n\lambda = 2d \sin \theta = 2d \cos \alpha \qquad (6\text{-}13)$$

then the waves scattered from these planes will have the correct phase relationships to reinforce one another and will produce an intense beam reflected at an equal angle α to the normal. The Bragg equation can be put into a more useful form by noting that the distance d between atomic planes is related to the distance D between atoms in the surface layer by the simple equation

$$d = D \sin \alpha$$

Substitution of this value of d in Equation (6-13) yields

$$n\lambda = 2D \sin \alpha \cos \alpha = D \sin 2\alpha$$

from which

$$n\lambda = D \sin \phi \qquad (6\text{-}14)$$

where

$$\phi = 2\alpha$$

The above equation can be applied to the measurements made by Davisson and Germer. For the case of the "54-volt" electron beam $\phi = 50$ degrees and $n = 1$; from x-ray data D is known to be 2.15 Å; hence

$$\lambda = 2.15 \text{ Å} \times \sin 50° = 1.65 \text{ Å}$$

This wavelength can be compared with the value obtained by substituting

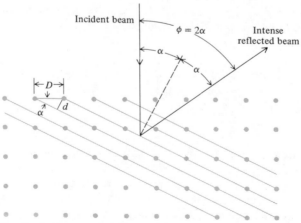

Figure 6-3 Diffraction of electron waves by a crystal.

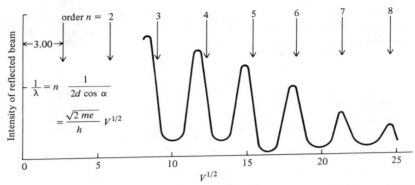

Figure 6-4 Curve obtained when the angle of incidence and the angle of reflection of the electron beam were each kept at 10 degrees, while the energy of the electrons was increased.

$V = 54$ volts into Equation (6-12). The latter yields $\lambda = 1.66$ Å. Similar measurements were made with higher energy electrons and comparable results were obtained in each case. The results of these experiments are in good agreement with De Broglie's hypothesis.

In one variation of this experiment an oblique angle of incidence was used. An examination of the intensity of the scattered beam showed that there was an intense maximum at an angle of "reflection" equal to the angle of incidence, in accord with Bragg's law of reflection from a crystal grating. In this case the atomic planes producing the diffraction pattern are parallel to the surface.

With the angle of incidence kept fixed, measurements were made of the intensity of the beam reflected at this angle when the energy of the incident electrons was varied. Figure 6-4, in which the intensity of the reflected beam is plotted against the square root of the voltage between filament and plate, shows a series of maxima almost equally spaced. Each maximum represents a different order of diffraction. The existence of these maxima follows directly from an application of Bragg's law

$$n\lambda = 2d \cos \alpha \qquad (6\text{-}13)$$

where α is now the angle between the incident beam and the normal to the atomic planes. Combining this with Equation (6-12), we have

$$\frac{1}{\lambda} = \frac{\sqrt{2me}}{h} V^{1/2} = \frac{n}{2d \cos \alpha} \qquad (6\text{-}15)$$

Whenever the wavelength is such as to satisfy Bragg's law, there will be an intense maximum at an angle of reflection equal to the angle of incidence. Since α is kept constant, $2d \cos \alpha$ is also constant; there will therefore be an intense maximum for each new value of n, the order of diffraction, as V is changed. From Equation (6-15), the spacings between maxima should all be the same and proportional to $1/2d \cos \alpha$. Actually the positions of the maxima differ

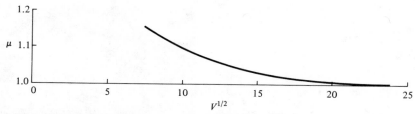

Figure 6-5 The index of refraction of nickel plotted against $V^{1/2}$.

slightly from the calculated positions as shown by the positions of the arrows in Figure 6-4. This discrepancy can be explained as due to the refraction of the electron beam in the crystal, since for the higher voltages the electrons do penetrate the crystal. The index of refraction μ can be calculated with the aid of the modified form of Bragg's law, (see Section 5-11) which is

$$n\lambda = 2d(\mu^2 - \sin^2 \alpha)^{1/2} \tag{6-16}$$

Values of the index of refraction are plotted against $V^{1/2}$ in Figure 6-5. For low energy electrons μ is large and greater than unity; at higher energies μ decreases and approaches unity.

6-3 Electron Diffraction Experiments of G. P. Thomson

G. P. Thomson (1928) was able to secure electron diffraction patterns by passing a narrow beam of cathode rays through very thin films of matter. The cathode rays were produced in a gas-discharge tube operated at potentials varying from 10,000 to 60,000 volts. The cathode rays, after passing through the thin film F, were received on a photographic plate at P, as shown in Figure 6-6. The pattern on the photographic plate consisted of a series of well-defined concentric rings about a central spot, as shown in Figure 6-7. This pattern is very similar in appearance to x-ray powder diffraction patterns.

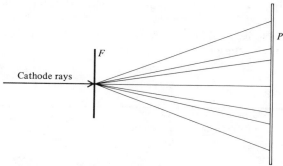

Figure 6-6 Schematic diagram of the electron diffraction experiments of G. P. Thomson.

Figure 6-7 Diffraction pattern obtained by passing a beam of electrons through gold foil. (Reproduced from Thomson, *Wave Mechanics of Free Electrons*, Cornell University Press.)

Ordinary metals, such as gold, silver, and aluminum, are microcrystalline in structure—that is, they consist of a large number of very small crystals oriented at random. If any wave of length λ is incident on a thin film of such a microcrystalline substance, a definite circular diffraction pattern will be obtained; the wavelength λ can be determined if the lattice constant d of the microcrystals is known. In Thomson's experiments a beam of electrons moving with speed v was used instead of the x-ray beam and a circular diffraction pattern was obtained. The wavelength λ associated with the electron can be calculated with the aid of Bragg's formula, using the value of d from x-ray data. Actually Thomson determined the wavelength of the electrons from the De Broglie formula

$$\lambda = \frac{h}{p}$$

and then calculated the spacing d between atomic planes and compared it with x-ray determinations. Table 6-1 gives a few of the results of Thomson's experiments.

The results of these experiments confirm De Broglie's hypothesis that there is a wave motion associated with every moving electron.

The electron diffraction camera has since been developed into an instrument of great precision and utility, particularly for the study of surface phenomena such as corrosion and other chemical changes and for the determination of crystal grating spaces of microcrystals. Figure 6-8 is a photograph taken by Oliver Row and N. R. Mukherjee showing an electron diffraction pattern of gold; the thickness of the gold foil was about 250 Å.

TABLE 6-1   VALUES OF d IN Å

Metal	X-Rays	Cathode Rays	
Aluminum	4.05	4.06	4.00
Gold	4.06	4.18	3.99
Platinum	3.91	3.88	
Lead	4.92	4.99	

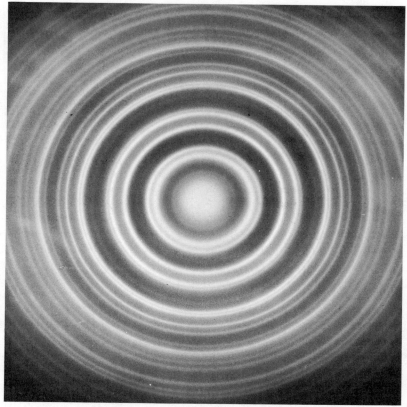

Figure 6-8   Electron diffraction pattern of gold; thickness of the gold film was about 250Å. (From photograph by Oliver Row and N. R. Mukherjee.)

6-4   Waves Associated with Atoms and Molecules

Since De Broglie's formula applies to any mass particle moving with speed v, it should be possible to secure diffraction and interference effects with atoms and molecules as well as with electrons. The diffraction of atoms and molecules by crystals was first clearly demonstrated by Stern and his co-workers. In these experiments they investigated the diffraction of hydrogen and helium molecules, using crystals of lithium fluoride and sodium chloride. A stream of molecules from an oven at a known temperature T was directed against the surface of the crystal at some angle θ. The surface of the lithium fluoride crystal in this case behaves in very much the same manner as a "crossed" diffraction grating does in optical experiments—that is, the diffracting centers are at the corners of squares. The incident beam is scattered in all directions, and intensity maxima for any given wavelength occur only at those points where the waves from the different diffracting centers meet in phase.

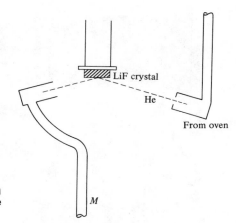

Figure 6-9 Schematic diagram showing the arrangement of apparatus used in the diffraction of atoms and molecules.

In the case of the molecular beam scattered by the crystal, the points of maximum intensity correspond to regions of maximum pressure of the gas. Hence Stern used a very sensitive manometer, M (see Fig. 6-9), to locate the diffraction pattern produced by the crystal. This manometer could be rotated about an axis at right angles to the face of the crystal and could thus measure the intensity of the scattered molecular beam at various points. Figure 6-10 is a typical curve showing the diffraction pattern obtained by reflecting helium from the LiF crystal. The intensity maximum at 0 degrees corresponds to the regularly reflected beam in which the angle of reflection is equal to the angle of incidence. Then, as the manometer is rotated on either side of the reflected beam, the pressure is found to drop to a minimum value at about 5 degrees and to rise again to a maximum on either side at about 12 degrees from the directly reflected beam. These two maxima correspond to the first-order diffraction patterns obtained with a grating.

The wavelength associated with the helium atoms can be calculated, using the known spacing of the atoms in the lithium fluoride crystal and the positions of the first-order diffraction patterns. The wavelengths calculated in this manner are in excellent agreement with those predicted by the De Broglie formula

$$\lambda = \frac{h}{p}$$

where p is the average momentum of the helium atoms calculated from a knowledge of the temperature of the helium in the oven. Many experiments were performed using temperatures varying from 100 to 650 K. Diffraction patterns were obtained in all cases. Similar experiments were performed with hydrogen molecules, H_2, and the wavelengths obtained were found to be in agreement with De Broglie's hypothesis.

A similar experiment was performed by T. H. Johnson in which a beam of hydrogen atoms was reflected from a lithium fluoride crystal and allowed to strike a plate coated with molybdenum oxide. The oxide was reduced to metal-

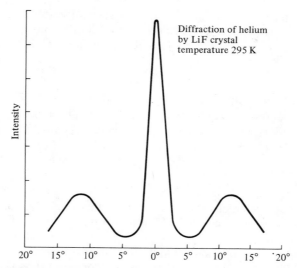

Intensity

Diffraction of helium
by LiF crystal
temperature 295 K

20° 15° 10° 5° 0° 5° 10° 15° ˙20°

Figure 6-10 The diffraction pattern obtained with helium reflected from a lithium fluoride
crystal.

lic molybdenum wherever the hydrogen struck the plate, and the pattern thus
formed was then photographed. The diffraction patterns obtained agreed with
the predictions based on De Broglie's hypothesis and with the results obtained
by Stern.

6-5 Diffraction of Neutrons

The existence of De Broglie waves having been definitely established, we
have one more powerful tool for studying the interaction between different
types of particles. The neutron is unique among these particles in that it has
no charge; hence it will not be influenced by the electrostatic fields of electrons
and nuclei and will penetrate most substances very readily. In the majority
of cases the interaction of neutrons with matter takes the form of elastic collisions with nuclei. The neutron loses a fraction of its energy in each such
collision, and if the substance occupies a sufficient volume, the neutrons will
finally reach an equilibrium velocity depending on the temperature of the
substance.

Another type of interaction between neutrons and nuclei is one in which
the neutron is captured by the nucleus. We shall consider the subject of neutron capture in greater detail in Chapter 15. Here we shall merely make use
of the fact that the capture of a neutron by a nucleus results in an unstable isotope; the latter then usually disintegrates with the emission of some particle
such as a beta particle, or a proton, or an alpha particle. The detection of the
emitted particle is one of the simplest methods of showing that the neutron
has been captured.

When neutrons are in equilibrium with a mass of material at absolute temperature T, they will have a distribution of velocities corresponding to this temperature. Their most probable velocity is that for which the kinetic energy $\mathcal{E}_k$ is given by

$$\mathcal{E}_k = kT \tag{6-17}$$

where k is the Boltzmann constant whose value is $k = 1.38 \times 10^{-16}$ erg/deg. Thus if neutrons come to equilibrium in a block of graphite or paraffin at room temperature, which we can take as approximately 300 K, then the kinetic energy of these neutrons will be

$$\mathcal{E}_{300} = 1.38 \times 10^{-16} \times 300$$

$$= 4.14 \times 10^{-14} \text{ erg}$$

The most probable velocity of a neutron of this energy is about 2.2×10^5 cm/sec. Since 1.6×10^{-12} erg $= 1$ eV, a neutron in equilibrium with matter at 300 K has an energy of about 1/40 eV. Neutrons of such low energy are usually referred to as *slow* neutrons, or *thermal* neutrons.

The De Broglie wavelength of thermal neutrons can thus readily be evaluated, yielding

$$\lambda = 1.82 \text{ Å}$$

which is about the same order of magnitude as x-rays. Hence the experimental techniques developed for x-rays can be applied, with suitable modifications, to the study of the interactions of slow neutrons with matter. Powerful sources of slow neutrons are now available from *nuclear reactors,* sometimes called chain-reacting piles. W. H. Zinn (1947) studied the distribution of wavelengths, hence energies, of a neutron beam from a chain-reacting pile

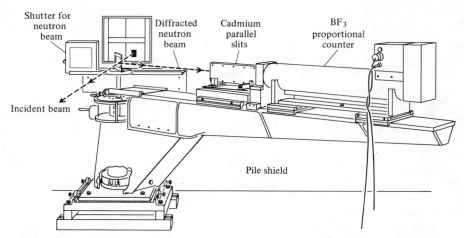

Figure 6-11   Diagram of the neutron spectrometer used by W. H. Zinn in the study of neutron diffraction. (Courtesy of W. H. Zinn, Argonne National Laboratory.)

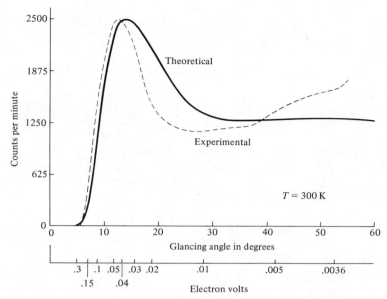

Figure 6-12   Energy distribution among neutrons in thermal column of a graphite reactor as determined with a calcite crystal neutron spectrometer. [Zinn, *Phys, Rev.,* **71**, 755 (1947).]

that had been slowed down to room temperature by passing through a large block of graphite. The neutrons were reflected from a calcite crystal, and the intensity of the reflected beam at a given crystal angle was measured by allowing the neutrons to enter a proportional counter filled to a pressure of 40 cm of Hg with boron trifluoride gas enriched with the ^{10}B isotope, as shown in Figure 6-11. Slow neutrons are easily captured by ^{10}B nuclei, which then emit alpha particles. These alpha particles ionize the gas in the chamber, producing a pulse whose amplitude is proportional to the energy of the alpha particle. The graph of Figure 6-12 shows the distribution of energy among the neutrons in the thermal column.

E. O. Wollan and C. G. Shull have studied the structure of crystalline substances, using the Laue diffraction technique and the powder crystal method. In the latter case they first monochromatized a narrow beam from the Oak Ridge graphite reactor by reflecting it from a NaCl crystal at an angle of about 6.5 degrees. The monochromatic beam of neutrons was scattered by the powdered crystals under investigation; the intensity of the beam scattered at a given angle enters a proportional counter filled with BF_3 gas enriched with the ^{10}B isotope. The counter rotates about an axis through the center of the spectrometer table on which the powder is mounted. The counter readings are recorded automatically. A typical powder diffraction pattern obtained with powdered diamond is shown in Figure 6-13. The numbers above the peaks

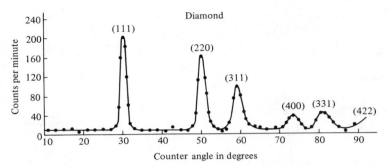

Figure 6-13 Neutron diffraction pattern produced by powdered diamond. The numbers above the peaks indicate the crystal planes which produced them. [Wollan and Shull, *Phys, Rev.,***73**, 834 (1948).]

designate the planes in the diamond effective in producing these intense maxima.

In producing Laue diffraction patterns, such as that shown in Figure 6-14, Wollan, Shull, and Marney sent a narrow beam of neutrons possessing a continuous wavelength distribution of 0.5 to 3.0 Å through a sodium chloride crystal 0.35 cm thick, with the incident beam parallel to one of the axes of the cube. The photographic film was placed 6.4 cm from the crystal. However, since neutrons do not affect a photographic film to any extent, the film was covered with a sheet of indium 0.5 mm thick. Neutrons are readily captured by indium, and the unstable nuclei thus formed disintegrate with the emission of beta particles. The latter do affect the photographic plate. A hole was cut in the center of the indium plate to reduce the blackening due to the undeflected neutron beam. Laue patterns of a variety of crystals have been obtained in this way, utilizing the De Broglie waves associated with neutrons.

Figure 6-14 Neutron Laue diffraction pattern produced by a crystal of rocksalt. (Photograph supplied by E. O. Wollan and C. G. Shull.)

6-6   Velocity of De Broglie Waves

The velocity of the De Broglie waves associated with a particle is not necessarily the same as that of the particle. The relationship between the two velocities can be found very readily. If λ is the length of the De Broglie wave and ν is its frequency, then the velocity w of this wave is given by the usual equation

$$w = \nu\lambda \qquad (6\text{-}18)$$

The momentum p of the particle is related to the wavelength by the fundamental equation

$$p = \frac{h}{\lambda} \qquad (6\text{-}5)$$

Let us now assume that the relationship between the total energy $\mathcal{E}$ of the particle and the frequency ν of the associated wave is given by the usual equation from quantum theory

$$\mathcal{E} = h\nu \qquad (6\text{-}1)$$

that the total energy $\mathcal{E}$ and mass m are given by the Einstein relativity equation

$$\mathcal{E} = m\gamma c^2 \qquad (6\text{-}19)$$

and that the momentum p is given by

$$p = m\gamma v \qquad (6\text{-}20)$$

where v is the velocity of the particle and $\gamma = (1 - v^2/c^2)^{1/2}$. It then follows that

$$w = \mathcal{E}\,\frac{\lambda}{h}$$

or

$$w = \frac{\mathcal{E}}{p} = \frac{\gamma m c^2}{\gamma m v}$$

from which

$$w = \frac{c^2}{v} \qquad (6\text{-}21)$$

Since the velocity v of a material particle is always less than the velocity of light c, the velocity of the De Broglie wave associated with it is always greater than c. There is no contradiction between this fact and the postulate of the theory of relativity that no signal—that is, energy—can be transmitted with a speed greater than c. We have already encountered velocities of waves greater than c—for example, the velocity of x-rays in material media. R. W. Wood has shown that the index of refraction of visible light passing through media in which anomalous dispersion is produced is less than unity—for example, yellow light passing through a prism of sodium. This means that its wave velocity in sodium is greater than c.

The fact that the wave velocity is greater than the particle velocity does not mean that the De Broglie waves get away from the particle. Instead we can think of the particle inside a *wave packet* or group of waves, with the entire group or packet traveling with the particle velocity v, while the individual waves composing the group travel with the wave velocity w. In our discussion of wave and group velocity in the next section we shall prove that the group velocity of the De Broglie waves is equal to the particle velocity.

The wave velocity of the De Broglie waves differs from the velocity of light in one important aspect: the wave velocity is a function of the wavelength even in free space. To show this we can start with the relativistic equation for the total energy $\mathscr{E}$ of a particle of mass m and momentum p:

$$\mathscr{E}^2 = p^2c^2 + m^2c^4 \tag{6-22}$$

and solve it for m^2, obtaining

$$m^2 = \frac{\mathscr{E}^2}{c^4} - \frac{p^2}{c^2} \tag{6-23}$$

Now, since

$$p = \frac{h}{\lambda}$$

and

$$\mathscr{E} = h\nu$$

we can write

$$m = \frac{h}{c} \sqrt{\frac{\nu^2}{c^2} - \frac{1}{\lambda^2}}$$

For De Broglie waves

$$w = \nu\lambda$$

hence

$$m = \frac{h}{c} \sqrt{\frac{w^2}{c^2\lambda^2} - \frac{1}{\lambda^2}} \tag{6-24}$$

from which

$$w = c \sqrt{1 + \frac{m^2c^2}{h^2}\lambda^2} \tag{6-25}$$

This equation shows that for a particle of mass $m > 0$ the wave velocity w is always greater than $c;$ furthermore, the wave velocity of the De Broglie waves is a function of the wavelength even in free space.

As a special case of De Broglie waves, consider those waves that are propagated with a wave velocity $w = c$. This corresponds to the propagation of electromagnetic waves. The velocity of the associated particle, the photon, is therefore also equal to c.

6-7   Wave and Group Velocities

In any dispersive medium, that is, one in which the velocity w of a wave depends on its wavelength, waves of different wavelengths are propagated through the medium as a group with a velocity u which is different from w. To derive the relationship between these two velocities let us consider two waves of slightly different wavelengths traveling together through a dispersive medium. In Figure 6-15 the upper wave ABC has a wavelength λ and a wave velocity w, whereas the wave $A'B'C'$ drawn just below it has a wavelength $\lambda + d\lambda$ and is moving with a velocity $w + dw$ in the same direction as the upper wave.

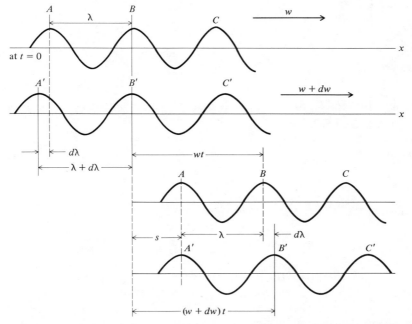

Figure 6-15   Wave and group velocities: t is the time required for A' to get into phase with A; in this time B has traveled a distance wt while the position of maximum amplitude had advanced a distance s with the group velocity u.

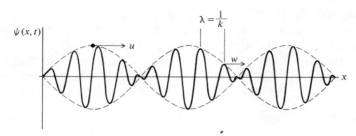

Figure 6-16   Group velocity u and wave velocity w.

The two waves, although drawn separately, are traveling together through the same space. Let us choose an instant when two crests, say B and B', coincide at time $t = 0$. The amplitude of the combined wave will be a maximum at this position. After a certain time t, B' will have moved ahead of B by an amount $d\lambda$ so that now A' coincides with A. The position of maximum amplitude has thus shifted in this time to the place at which A and A' coincide. Let us call this distance s and define the group velocity u by the equation

$$u = \frac{s}{t} \tag{6-26}$$

From Figure 6-15 it can be seen that

$$s = wt - \lambda$$

so that

$$u = w - \frac{\lambda}{t} \tag{6-27}$$

It can also be seen from the figure that the distance $d\lambda$ that B' has moved from B in the same time t is

$$d\lambda = (w + dw)t - wt = t \, dw$$

from which

$$t = \frac{d\lambda}{dw}$$

Substituting this value of t in Equation (6-27) yields

$$u = w - \lambda \frac{dw}{d\lambda} \tag{6-28}$$

Equation (6-28) gives the relationship between the velocity u of a group of waves and the wave velocity w of the individual waves of the group. If there is no dispersion—for example, light traveling in a vacuum—then $dw/d\lambda = 0$ and $u = w$; that is, the group velocity and wave velocity are the same.

The equation for the group velocity u can be put in a more concise form when written in terms of the frequency ν and the wave number $k = 1/\lambda$ as follows:

$$u = \frac{dv}{dk} \tag{6-29}$$

The addition of two sine waves of slightly different wavelengths traveling in the same direction, say the x direction, with slightly different velocities can readily be performed analytically (see Problem 6-10). The result is shown graphically in Figure 6-16; a wave of wavelength λ travels to the right with wave velocity w but with a slowly varying amplitude; the envelope of this wave also moves to the right with a slower velocity u, the velocity of the group of waves.

6-8 Group Velocity and Particle Velocity

It is a comparatively simple matter to show that the group velocity of the De Broglie waves is the same as the velocity of the particle. It has been shown that the wave velocity of the De Broglie waves is given by

$$w = c \sqrt{1 + \frac{m^2 c^2}{h^2} \lambda^2} \tag{6-25}$$

Now

$$\frac{dw}{d\lambda} = \frac{c}{\sqrt{1 + (m^2 c^2/h^2)\lambda^2}} \times \frac{m^2 c^2}{h^2} \lambda$$

Substituting this value of $dw/d\lambda$ into Equation (6-28) and using Equation (6-25) for w, we obtain

$$u = \frac{c^2}{w}$$

But we have shown that

$$v = \frac{c^2}{w} \tag{6-21}$$

hence

$$u = v$$

That is, the group velocity of the De Broglie waves is the same as that of the particle. Stated in another way, the De Broglie waves move with the particle.

6-9 Heisenberg's "Uncertainty Principle"

An interesting interpretation of the duality, wave and particle, of both matter and radiant energy has been given by Heisenberg. The concepts of a particle and of a wave have been built up on the basis of experiments performed on a comparatively large scale; these concepts are mental pictures formed on the basis of such experiments. When applied to experiments involving quantities of the order of magnitude of atomic dimensions, these concepts can have the validity of analogies only. The concept of the electron as a particle, for example, was derived from the results of experiments on the motion of the electron through electric and magnetic fields. The problem of particle dynamics is to predict the position and velocity of the particle at any time t when its initial position and velocity are known. But the experiments on electron diffraction show that this is not always possible. Electrons starting with the same initial conditions are not all scattered through the same angle by the crystals; the result is a diffraction pattern showing a definite distribution of these electrons with respect to both position and momentum. But a diffraction pattern is the best evidence that we are dealing with a wave phenomenon. To apply the wave concept to a single electron, the electron may be pictured as a small bundle or packet of waves extending over some small region of space Δs. The association of a wave packet with an electron means that the position of the electron at any instant of time t cannot be specified with any desired degree of accuracy; all that can be said of the electron is that it is somewhere within this group of waves that extends over a small region of space Δs.

Heisenberg's *uncertainty principle* refers to the *simultaneous* determination of the position and the momentum of the particle and states that the uncertainty Δx involved in the measurement of the coordinate of the particle and the uncertainty Δp_x involved in the simultaneous measurement of its momentum are governed by the relationship

$$\Delta x \cdot \Delta p_x \gtrsim \frac{h}{4\pi} \tag{6-30}$$

where h is the Planck constant. The uncertainty principle is one method of describing the wave-particle duality of all physical entities; it does not in any way depend on the inaccuracies involved in the design and use of measuring instruments.

An examination of a few idealized experiments will serve to show how the wave concept acts as a limitation on the particle concept, giving rise to the uncertainty principle. One such idealized experiment was given by Bohr. Suppose we wish to determine the position of an electron, using some instrument such as a microscope of very high resolving power. It can be shown that the resolving power of a microscope is given by

$$\Delta x = \frac{\lambda}{\sin \alpha}$$

where Δx represents the distance between two points that can just be resolved by the microscope, λ is the wavelength of the light used, and α is the semivertical angle of the cone of light coming from the illuminated object. The uncertainty in the determination of the x coordinate of the position of the electron is represented by Δx. To make Δx as small as possible, light of very short wave-

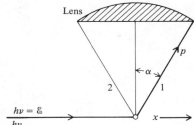

Figure 6-17   Schematic diagram of the gamma-ray microscope experiment.

length must be used, either x-rays or gamma rays. The minimum amount of light that can be used is a single quantum $h\nu$. When the electron scatters this quantum into the microscope, as shown in Figure 6-17, the electron will receive some momentum from the quantum (Compton effect). Since the scattered quantum can enter the microscope anywhere within the semivertical angle α, its contribution to the x component of the momentum of the electron is unknown by an amount

$$\Delta p_x = p \sin \alpha = \frac{h}{\lambda} \sin \alpha$$

where h/λ is the momentum of the quantum. The product of the uncertainties in the determination of the simultaneous values of the position and momentum of the electron is therefore

$$\Delta x \cdot \Delta p_x = \frac{\lambda}{\sin \alpha} \cdot \frac{h}{\lambda} \sin \alpha = h$$

Another illustration of the uncertainty principle is supplied by an experiment in which a beam of electrons passes through a narrow slit and is then recorded on a photographic plate placed some distance away, as sketched in Figure 6-18. Every electron that is registered on the photographic plate must have passed through the slit, and if its width is Δy, then the y coordinate of the electron is indeterminate by an amount Δy. Making this width smaller increases the accuracy in the knowledge of the y coordinate of the electron at the instant it passes through the slit. With a very narrow slit a very definite diffraction pattern is observed on the photographic plate. This diffraction pattern is interpreted to mean that the electron receives additional momentum parallel to the slit at the instant that it passes through the slit. If p is the momentum of the electron, the component in the y direction is $p \sin \theta$, where θ is the angle of deviation. The electron may be anywhere within the diffraction pattern, so that if α is the angular width of the pattern, the uncertainty in the knowledge of the y component of the momentum of the electron is

$$\Delta p_y = p \sin \alpha$$

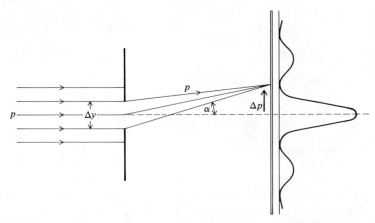

Figure 6-18   Diffraction of electrons by a narrow slit. A graph of the intensity pattern obtained is shown on the right.

The angular width of the diffraction pattern is determined by the slit width Δy and is given by the equation

$$\Delta y \sin \alpha = \lambda$$

Hence the product of the uncertainties in the simultaneous determination of the y coordinate and y momentum of the electron in its passage through the slit is

$$\Delta y \cdot \Delta p_y = p \sin \alpha \cdot \frac{\lambda}{\sin \alpha} = p\lambda$$

From De Broglie's hypothesis

$$p = \frac{h}{\lambda}$$

therefore

$$\Delta y \cdot \Delta p_y = h$$

Another set of variables can be used to express Heisenberg's uncertainty principle. If $\mathcal{E}$ is the energy of the system at time t, it may be shown that

$$\Delta \mathcal{E} \cdot \Delta t \geq \frac{h}{4\pi} \qquad (6\text{-}31)$$

where $\Delta \mathcal{E}$ is the uncertainty in our knowledge of the value of the energy $\mathcal{E}$ and Δt is the uncertainty in the knowledge of the time.

6-10 Probability Concept

The wave-particle parallelism must be extended to include an interpretation of the intensity of light and also the intensity of the electron beam. According to the wave theory of light, the intensity is determined by the square of the amplitude of the electric vector at the point under consideration. On the corpuscular theory the intensity of the beam of light must be determined by the number of photons per second which pass through a unit area perpendicular to their direction of motion, or

$$N \sim E_0{}^2 \qquad\qquad (6\text{-}32)$$

where E_0 is the amplitude of the electric vector and N is the number of photons per unit volume of the beam; also, Nc is the number of photons passing through unit area in unit time. The intensity relationship $N \sim E_0{}^2$, while adequate for intense beams, is no longer satisfactory when dealing with very weak beams for which N is a very small number ($N \ll 1$). In this case it becomes difficult to determine the exact location of each particle in the continuous wave; that is, the expression $N \sim E_0{}^2$ involves an indeterminacy with respect to the position of these photons in space.

An alternative approach to the problem is through a statistical interpretation. The square of the amplitude, $E_0{}^2$, can be thought of as proportional to the *probability* of a photon's crossing unit area perpendicular to the direction of motion in unit time at the point under consideration; for example, if a beam of light passes through a narrow slit and is incident on a photographic plate, a definite diffraction pattern will be obtained showing regions of great intensity alternating with regions of very small intensity. On the statistical interpretation the probability of a photon's striking the photographic plate is very great where the intensity is great and very small where the intensity is small. In the case of a very weak beam of light, say one in which a single photon passes through the slit every minute, it is impossible to predict just where any individual photon will strike the photographic plate. All that can be said is that the probability of the photon's striking a certain portion of the plate is large just where the wave theory predicts large intensity and the probability is small just where the wave theory predicts small intensity. If only a few photons strike the photographic plate, their arrangement will undoubtedly be haphazard; but if a sufficient time is allowed to elapse so that a large number of photons reach the photographic plate, the result will be the diffraction pattern predicted by the wave theory.

The above mode of description can be applied to the diffraction of an electron beam by a narrow slit. The wave associated with the electron is the De Broglie wave. There is a certain probability that an electron, after passing through the slit, will strike a given point on the photographic plate; this probability is proportional to the square of the amplitude of the associated De Broglie wave. It is impossible to predict just where any one electron will strike the plate; yet after an interval of time sufficient to allow a large number of electrons to strike the plate a definite diffraction pattern will be observed, and the intensities at the different points will correspond to the amplitudes of the diffracted waves at those points.

The phenomena of transmission and reflection at a plane surface can be explained in a similar manner. If a system of waves is incident on a plane surface, part of it will be reflected and part transmitted, and their intensities will be proportional to the squares of the amplitudes of the reflected and transmitted waves, respectively. According to the corpuscular theory, the particle associated with the incident wave has a certain probability of being reflected and a certain probability of being transmitted, these probabilities being proportional to the squares of the amplitudes of the corresponding waves.

From the discussion of the last two sections it can be asserted that although there is an indeterminacy in the description of phenomena from the corpuscular point of view, there is no lack of determinacy from the point of view of the wave theory. The wave functions necessary to describe these phenomena are continuous functions of the coordinates and the time. These wave functions are obtained from the solutions of the appropriate wave equations: in general, second-order partial differential equations involving the coordinates and the time. For the case of light, these wave functions are obtained from the solutions of the differential equations that form the basis of the electromagnetic theory of light. For the case of material particles, these wave functions are obtained from the solutions of a wave equation first formulated by Schroedinger and forming the basis of a new division of physics called *wave mechanics*.

6-11 Schroedinger's Equation for a Single Particle

Schroedinger developed an equation describing the motion of a particle and its associated De Broglie wave which has become the foundation of wave mechanics or quantum mechanics. He arrived at this equation by considerations that involved analogies between geometrical optics and physical optics, comparing the path of a particle with that of a ray of light, and the associated wave with the electromagnetic waves.

One can arrive at the Schroedinger equation by starting with a typical wave equation and incorporating De Broglie's hypothesis about the waves associated with material particles in it. If some aspect of a wave, represented by Ψ, is propagated through space with a velocity w, the differential equation representing its motion can be written as

$$\frac{\partial^2 \Psi}{\partial x^2} + \frac{\partial^2 \Psi}{\partial y^2} + \frac{\partial^2 \Psi}{\partial z^2} = \frac{1}{w^2} \frac{\partial^2 \Psi}{\partial t^2} \tag{6-33}$$

In general, Ψ, will be a function of x, y, z, and t. For the case of electromagnetic waves, $\Psi(x, y, z, t)$ may represent one of the components of the electric field or the magnetic field. The solution of this equation, subject to certain restrictions and boundary conditions, will yield the value of Ψ as a function of the space coordinates and the time.

To arrive at the Schroedinger equation for the motion of a single particle of mass m moving with velocity $v \ll c$ in a field of force in which its potential energy is $V(x, y, z)$ and its total energy is $\mathcal{E}$ it is necessary to evaluate the wave velocity w with which the associated wave Ψ is traveling. For the nonrelativistic case ($v \ll c$) the kinetic energy can be written as

$$\tfrac{1}{2}mv^2 = \mathcal{E} - V$$

or

$$mv^2 = 2(\mathcal{E} - V)$$

and the momentum p will be given by

$$p = mv = \sqrt{2m(\mathcal{E} - V)} \tag{6-8}$$

It has been shown that in terms of the wave associated with this particle

$$\mathcal{E} = h\nu \tag{6-1}$$

$$p = \frac{h}{\lambda} \tag{6-5}$$

and

$$w = \nu\lambda \tag{6-18}$$

so that

$$w = \frac{h\nu}{p}$$

from which

$$w = \frac{h\nu}{\sqrt{2m(\mathcal{E} - V)}} \tag{6-34}$$

In general, only those wave functions Ψ that are harmonic functions of the time are of physical significance. Such functions can be expressed as $\sin 2\pi\nu t$ or $\cos 2\pi\nu t$ or combinations of sine and cosine functions represented by the exponential function $\exp(2\pi\nu it)$, where $i = \sqrt{-1}$. If we are dealing with a monochromatic wave of frequency ν, the function Ψ may be represented as the product of two factors as follows:

$$\Psi(x, y, z, t) = \psi(x, y, z)\exp(2\pi\nu it)$$

where $\psi(x, y, z)$ is a function of the space coordinates only. Differentiating the above equation with respect to time only yields.

$$\frac{\partial^2 \Psi}{\partial t^2} = -4\pi^2\nu^2\,\psi \exp(2\pi\nu it) \tag{6-36}$$

Differentiating the above equation for Ψ with respect to each of the coordinates yields

$$\frac{\partial^2\Psi}{\partial x^2} + \frac{\partial^2\Psi}{\partial y^2} + \frac{\partial^2\Psi}{\partial z^2} = \left(\frac{\partial^2\psi}{\partial x^2} + \frac{\partial^2\psi}{\partial y^2} + \frac{\partial^2\psi}{\partial z^2}\right)\exp\,(2\pi\nu it) \qquad (6\text{-}37)$$

We can thus write for the space factor $\psi(x, y, z,)$

$$\frac{\partial^2\psi}{\partial x^2} + \frac{\partial^2\psi}{\partial y^2} + \frac{\partial^2\psi}{\partial z^2} = -\frac{4\pi^2\nu^2}{w^2}\,\psi \qquad (6\text{-}38)$$

Substituting the value of w from Equation (6-34), we get

$$\frac{\partial^2\psi}{\partial x^2} + \frac{\partial^2\psi}{\partial y^2} + \frac{\partial^2\psi}{\partial z^2} = -\frac{8\pi^2 m}{h^2}\,(\mathcal{E} - V)\,\psi \qquad (6\text{-}39)$$

This is the time-independent form of the Schroedinger wave equation for a single particle.

In the special case of a particle whose motion is restricted to one dimension only, say the x direction, and whose potential energy $V(x)$ is independent of the time, the Schroedinger equation reduces to an ordinary differential equation

$$\frac{d^2\psi}{dx^2} = -\frac{8\pi^2 m}{h^2}\left(\mathcal{E} - V(x)\right)\psi \qquad (6\text{-}40)$$

We shall present solutions of this equation for several interesting cases in Chapter 7.

A very important symbolic method of writing and treating the Schroedinger equation can be developed by rewriting the above equation as follows:

$$-\frac{\hbar^2}{2m}\frac{d^2\psi}{dx^2} + V(x)\psi = \mathcal{E}\psi$$

and then

$$\left[-\frac{\hbar^2}{2m}\frac{d^2}{dx^2} + V(x)\right]\psi = \mathcal{E}\psi \qquad (6\text{-}41)$$

The two terms in the brackets can be thought of as an operator H, called the *Hamiltonian,* so that the time-independent Schroedinger equation can be written as

$$H\psi = \mathcal{E}\psi \qquad (6\text{-}42)$$

that is, the Hamiltonian operating on function $\psi(x)$ yields a constant $\mathcal{E}$ multiplying ψ.

The Hamiltonian is actually the total energy of the particle written as a function of the coordinate and its momentum. Thus, if a particle is moving with a velocity v in a field in which its potential energy is $V(x)$, its total energy is

$$\tfrac{1}{2}mv^2 + V(x) = \frac{p^2}{2m} + V(x)$$

where

$$p = m \frac{dx}{dt} = mv$$

Thus

$$H(p, x) = \frac{p^2}{2m} + V(x) \tag{6-43}$$

If we can write the equation for the total energy of a particle in the Hamiltonian form, then the Schroedinger equation can be obtained by replacing

$$p \rightarrow \frac{\hbar}{i} \frac{d}{dx}$$

and

$$x \rightarrow x$$

where

$$i = \sqrt{-1}$$

This operational method will form the starting point of our discussion of quantum mechanics in Chapter 7.

6-12 Electron Optics

Nearly all the phenomena that are usually associated with light and x-rays and which form the subject of optics can also be observed with electrons. Electrons can be reflected and refracted; interference and diffraction phenomena can be produced at will; electrons from a point source may be focused by passing them through properly shaped electric or magnetic fields; such fields play the role of lenses. This has led to the development of a completely new branch of science, known as *electron optics*. Investigations in this subject have led to a better understanding of physical phenomena. Several important instruments have been produced that have wide applications. One of these is the *electron microscope*.

A microscope is used to provide an enlarged image of a small object as well as to show greater detail in its structure. The latter property is determined by the resolving power of the microscope. We have already seen that the limit of the resolving power of a microscope is determined by the wavelength of the incident radiation. In the case of optical microscopes the limit of the resolving power of the optical microscope is of the order of magnitude of the wavelength of visible light, which we may take as 5000 Å. But, since the wavelengths associated with electrons are determined by the relation

$$\lambda = \frac{h}{mv}$$

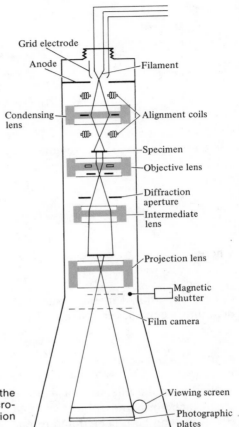

Figure 6-19 A schematic drawing of the column of the RCA EMU-3 electron microscope. (Courtesy of the Radio Corporation of America.)

it is possible to get much smaller wavelengths by using appropriate accelerating voltages on the electrons and thus to produce microscopes with much greater resolving power in the range of about 5 to 10 Å.

Figure 6-19 is a schematic drawing of the essential parts of a compact electron microscope of high resolving power operating in the range of 50 to 100 kV. This microscope uses magnetic fields both for the objective lens and for the projection lens. Either an electromagnet or a permanent magnet may be used for these lenses. The specimen to be investigated has to be very thin so that electrons of about 50 keV energy can be transmitted through it without loss of energy. The instrument is designed so that the image can be focused on a fluorescent screen for visual examination and is also arranged so that a photographic plate can be put in front of the screen to photograph the image. A diffusion pump is connected to the microscope to provide a vacuum of about 5×10^{-5} torr. This instrument can also be provided with an adapter so that it can be used as an electron diffraction camera. Other electron microscopes

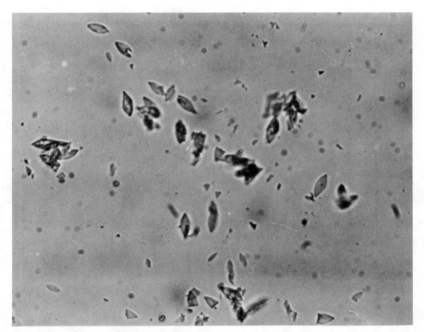

Figure 6-20   Light micrograph of monohydrated aluminum oxide. A representative micrograph that is perhaps a factor of three from the theoretical limit of an optical microscope using visible light. Magnification 1000×. (Courtesy of J. Hillier, RCA Laboratories.)

have been designed for the study of the surfaces of specimens by the reflection of electrons from them.

Electron microscopes are now being operated with accelerating voltages up to about 3 MV. Since the penetrating power of electrons depends on their energy, it is now possible to investigate the properties and structure of comparatively thick samples of material. The high-voltage microscopes probably will have their greatest use in the fields of solid-state physics and metallurgy.

An idea of the meaning of the term *resolving power* as well as the difference in resolving powers between light and electron microscopes can be obtained from the two photographs shown in Figures 6-20 and 6-21, taken by James Hillier. Figure 6-20 is a photograph taken with a light microscope of monohydrated aluminum oxide. This micrograph is estimated to be about a factor of 3 from the theoretical limit of a microscope using visible light. Figure 6-21 is an electron micrograph of the tip of one of the crystals of the same substance.

The limitation on the resolving power of the electron microscope is not the wavelength of the electrons but the aberrations associated with lenses, such as spherical aberration and chromatic aberration. The latter can be easily understood by recalling that electrons, in their passage through matter, may lose

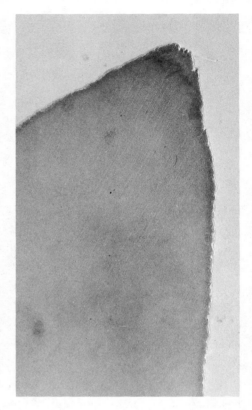

Figure 6-21 Electron micrograph of the tip of one of the plates of the same monohydrated aluminum oxide. The resolving power is about a factor of three from the theoretical limit of the electron microscopes of about 1950. Magnification 68,000×. Compare with Figure 6-20. (Courtesy of J. Hillier, RCA Laboratories.)

energy by inelastic collisions. The wavelengths of such electrons will be increased and the focus will be blurred.

There are variations in the use of electron microscopes. Recently (1970) Crewe, Wall, and Langmore used a scanning-beam technique with an electron microscope operated at 30 kV to detect the presence of heavy atoms such as uranium and thorium in specially prepared solutions and organic compounds. They were able to ascertain the positions of such atoms in the samples. The separations are just within the resolving power of their instrument.

Problems

6-1.   Calculate the wavelength associated with an alpha particle emitted by the nucleus of an atom of radon. Compare this wavelength with the diameter of the nucleus.

6-2.   A stream of electrons moving with a velocity v_1 in a vacuum passes through a surface into a region in which its potential energy is U and its velocity is v_2. Assuming that the index of refraction μ of the waves associated with matter is given by

$$\mu = \frac{w_1}{w_2}$$

(a) Show that the index of refraction of the electrons is given by

$$\mu = \frac{v_2}{v_1} = \sqrt{1 + \frac{U}{\frac{1}{2}mv_1{}^2}}$$

(b) Show that if the electrons enter the surface at an angle of incidence $i > 0$, they will be refracted toward the normal.

6-3.   A stream of electrons of 240 eV energy is incident on the surface of platinum at an angle of 30 degrees to the normal. The electron energy U in platinum is known to be about 12 eV. (a) Calculate the index of refraction of platinum for these electrons. (b) Determine the angle of refraction. (c) Determine the wave velocity of the electron waves in platinum. (d) Determine the velocity of the electrons in platinum.

6-4.   Electrons from a heated filament are accelerated by a difference of potential between the filament and the anode of $30\,\mathrm{kV}$; a narrow stream of electrons coming through a hole in the anode is transmitted through a thin sheet of aluminum. Assuming Bragg's law to hold, calculate the angle of deviation of the first-order diffraction pattern. (b) Determine the velocity of these waves in the aluminum foil.

6-5.   It can be shown that the rate at which a gas at absolute temperature T moves out of an orifice in an oven is the same as that of a gas moving out of the aperture with a uniform velocity equal to $\frac{1}{4}\bar{c}$, where $\bar{c}$ is the average velocity of the molecules. Further, the average kinetic energy of the molecules of a gas is given by

$$\tfrac{1}{2}MC^2 = \tfrac{3}{2}kT$$

where

$$C^2 = \frac{3\pi}{8}\bar{c}^2$$

k is Boltzmann's constant and is equal to 1.37×10^{-16} erg K^{-1}. (a) Derive an expression for the length of the waves associated with a stream of molecules of a gas at temperature T. (b) Calculate the wavelength associated with a stream of helium molecules issuing from an oven at 300 K. (c) Devise an experiment for showing the existence of these waves, giving approximate dimensions of the essential parts of the apparatus.

6-6.   Deep-water waves travel with a wave velocity $w = (g\lambda/2\pi)^{1/2}$, where g

is the gravitational acceleration. Show that the group velocity of such waves is $u = w/2$.

6-7. Ripples on the surface of a liquid travel with a wave velocity $w = (2\pi S/\lambda\rho)^{1/2}$ in which S is the surface tension and ρ is the density of the liquid. Show that the group velocity of the ripples is $u = \frac{3}{2}w$.

6-8. (a) Light of wavelength λ travels through a medium of index of refraction μ with a wave velocity w given by $\mu = c/w$. Show that the group velocity is given by

$$u = w\left(1 + \frac{\lambda}{\mu}\frac{d\mu}{d\lambda}\right)$$

(b) Determine the ratio of the group velocity to the wave velocity of light of wavelength 6000 Å, in a glass, from the following data: when $\lambda = 5890$, $\mu = 1.5682$; when $\lambda = 6235$, $\mu = 1.5663$.

6-9. Using the relationship $w = \nu\lambda$ derive Equation (6-25) from Equation (6-26).

6-10. The equation for a wave traveling in the positive x direction with wave velocity w is

$$y = \sin\frac{2\pi}{\lambda}(x - wt)$$

(a) Write this equation in terms of the frequency ν and wave number k. (b) Write the equation for a similar wave of frequency $\nu + d\nu$ and wave number $k + dk$ traveling with a velocity $w + dw$. (c) Add these two waves using the usual expression for the sine of the sum of two angles and the sine of the difference of the same two angles. (d) Neglecting second-order differentials in comparison with first-order differentials, show that the resultant wave is given by

$$y = 2\cos 2\pi\left(\frac{dk}{2}x - \frac{d\nu}{2}t\right)\sin 2\pi(kx - \nu t)$$

(e) Determine the velocity of the group of waves u by inspection of the above equation.

6-11. An electron's position is known to an accuracy of about 10^{-8} cm. How accurately can its velocity be known?

6-12. A recently discovered meson has a lifetime of about 10^{-23} sec. How accurately can its rest energy be known?

6-13. A photon of frequency ν is scattered by an electron of mass m. The scattered photon of frequency ν' leaves with a direction at 90° to the incident photon. Show that the De Broglie wavelength of the recoiling electron is $\lambda_{\text{recoil}} = c(\nu^2 + \nu'^2)^{-1/2}$.

6-14. Which one of the following particles has the longest wavelength?

(a) An alpha particle with a kinetic energy of 3 eV; (b) a neutron with a kinetic energy of 13 eV; (c) an electron with a kinetic energy of 26 keV; (d) an 80-keV photon.

References

Bacon, G. E., and K. Lonsdale, "Neutron Diffraction," in *Reports on Progress in Physics*. London: The Physical Society of London, 1953, XVI, 1–61.

Bohm, D., *Quantum Theory*. Englewood Cliffs, N.J.: Prentice-Hall, 1951, Parts I, II, and III.

Born, M., *Atomic Physics*. New York: Hafner Publishing Company, 1959.

Feynman, R. P., R. B. Leighton, and M. Sands, *The Feynman Lectures on Physics*, Vol. III. Reading, Mass.: Addison-Wesley Publishing Company, 1965, Chapters 1–3.

Heisenberg, W., *The Physical Principles of the Quantum Theory*. Chicago: The University of Chicago Press, 1930, Chapters I–IV.

Jacob, L., *An Introduction to Electron Optics*. New York: John Wiley & Sons, 1951.

Jammer, M., *The Conceptual Development of Quantum Mechanics*. New York: McGraw-Hill Book Company, 1966.

Pinsker, Z. G., *Electron Diffraction*. London: Butterworths, 1953.

Sherwin, C. W., *Introduction to Quantum Mechanics*. New York: Holt, Rinehart and Winston, 1959.

7 Elements of Quantum Mechanics

7-1 Postulates of Quantum Mechanics

As a result of the material in Chapter 6, it is clear that we need to develop a wave theory in order to provide a quantitative description of matter on the atomic scale. There are many ways of getting started at this task. The method employed here is to make a set of assumptions which is known to lead to the correct results. This technique has the disadvantage of neglecting the historical development of the theory; also, our assumptions will not appear in their most general form. The advantage is that we shall be able to calculate results more quickly without calling on a detailed knowledge of various classical theories.

To further our aim of getting to the subject without full generality we shall impose a few restrictions which will simplify our task considerably. We shall ignore relativistic effects. We shall start with a one-dimensional theory, in spite of the fact that a full treatment would require at least three dimensions. Also, we shall assume that our system is independent of time. Last, we shall restrict ourselves to a one-particle theory.

Subject to the above restrictions, we now state six postulates that will suffice to describe some simple systems. We consider a single particle of mass m whose position is given by the coordinate x. It will perhaps be moving with velocity v and momentum p.

I. *There is a wave function $\psi(x)$; knowledge of this function will enable us to obtain a complete description of the behavior of our particle.* From Chapter 6 it is clear that the uncertainty principle precludes any description that is as definite as the classical Newtonian approach to mechanics. By a "complete description" we mean a description that contains as much information as possible consistent with the uncertainty relation. It will be seen as we proceed that $\psi(x)$ will often be a complex quantity involving $i = \sqrt{-1}$. The meaning and use of ψ will be elucidated by the remaining postulates.

II. *Certain observable (that is, measurable) quantities can be represented as mathematical operators.* In particular, we shall make considerable use of operators for the familiar physical quantities, position and momentum. We shall assume that the position operator x_{op} is simply the coordinate x as a number to be multiplied with some function.

$$x_{op} = x \qquad (7\text{-}1)$$

The momentum operator will be assumed to be

$$p_{op} = \frac{\hbar}{i} \frac{\partial}{\partial x} \qquad (7\text{-}2)$$

where $\hbar = h/2\pi$. This form is manifestly in need of completion; we need to have a function to the right of an operator in order to give meaning to our symbols. If $\phi(x)$ is any function of x, the meaning of our two equations is simply

$$x_{op}[\phi] = x\phi; \qquad p_{op}[\phi] = \frac{\hbar}{i} \frac{\partial \phi}{\partial x} \qquad (7\text{-}3)$$

III. $\psi(x)$ *is continuous.* As we have seen, the function ψ will doubtless be subjected to differentiation and so we expect it to be continuous.

IV. $\psi^*\psi\,dx$ *is the probability of finding the particle in the interval* $(x, x+dx)$. The function $\psi^*(x)$ is the complex conjugate of ψ, the function that is obtained by changing the sign of $\sqrt{-1}$ wherever it appears in $\psi(x)$. The physical interpretation of this postulate has been discussed in Section 6-10. A direct consequence of this postulate is that, given the wave function, we can calculate the probability of finding a particle in a finite interval.

$$\text{Prob [particle is in } (x_1, x_2)] = \int_{x_1}^{x_2} \psi^*\psi\,dx \qquad (7\text{-}4)$$

Since the particle has to be somewhere, we expect that the probability should be unity that we shall find it if we hunt everywhere. Therefore

$$\int_{-\infty}^{\infty} \psi^*\psi\,dx = 1 \qquad (7\text{-}5)$$

V. *If* α_{op} *is the operator associated with some observable quantity* α, *then the average value* $<\alpha>$ *is given by the expression*

$$<\alpha> = \int_{-\infty}^{\infty} \psi^*\alpha_{op}\psi\,dx \qquad (7\text{-}6)$$

This postulate enables us to relate operators and wave functions to reality, but it reminds us that we cannot, in general, predict the outcome of the measurement of a physical quantity. Rather, we must be satisfied to predict the result of averaging answers from many measurements of this quantity.

VI. *The function ψ is a solution of the equation*

$$H_{\mathrm{op}}\psi = \mathcal{E}\psi \qquad (7\text{-}7)$$

where H_{op} is the operator for the total energy of the particle; $\mathcal{E}$ is a constant.
Equation (7-7) is called the *time-independent Schroedinger equation*. It will
typically be a differential equation. The usual procedure in a quantum mechanical problem is (a) to define the energy operator; (b) solve the Schroedinger equation for ψ; (c) in the process learn what values of $\mathcal{E}$ are possible—
they turn out to be just the values of the total energy that the particle can have;
(d) proceed to calculate whatever is needed, using postulates IV and V. Later
we shall illustrate this procedure for several examples.

In all the situations we shall encounter, the energy operator H_{op} can be written as the sum of the kinetic energy operator T_{op} and the potential energy
operator V_{op}.

$$H_{\mathrm{op}} = T_{\mathrm{op}} + V_{\mathrm{op}} \qquad (7\text{-}8)$$

In one dimension the kinetic energy operator becomes

$$T_{\mathrm{op}} = \tfrac{1}{2}m(v_{\mathrm{op}})^2 = \frac{1}{2m}(p_{\mathrm{op}})^2 \qquad (7\text{-}9)$$

Recalling Equation (7-2), we obtain

$$T_{\mathrm{op}} = -\frac{\hbar^2}{2m}\frac{\partial^2}{\partial x^2} \qquad (7\text{-}10)$$

In our treatment it will turn out that the potential energy will always depend on
the position x of the object rather than on such quantities as velocity or momentum, thus appearing as a multiplicative function. The time-independent
Schroedinger equation (7-7) can now be written for one-dimensional problems
in the form

$$-\frac{\hbar^2}{2m}\frac{d^2\psi}{dx^2} + V(x)\psi = \mathcal{E}\psi \qquad (7\text{-}11)$$

We may use ordinary derivatives because ψ depends only on x.

The method of solving this differential equation is very much dependent
on the specific form of the potential energy $V(x)$. Each form of V which one
assumes leads to a completely different problem.

7-2 The Infinite Square Well

Perhaps the simplest useful form of $V(x)$ is that of the infinite square well,
as shown in Figure 7-1, in which

$$V(x) = 0 \qquad 0 < x < a$$

$$V(x) = \infty \qquad \text{elsewhere} \qquad (7\text{-}12)$$

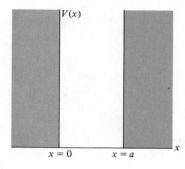

Figure 7-1 Potential energy associated with an infinitely deep square well.

This choice of potential corresponds to restricting our particle to be in the interval (0, a). It follows that *outside* (0, a) we must require that $\psi = 0$, since the particle may never be found there. The task, then, is to work with the Schroedinger equation *inside* (0, a). The correct form is simply

$$-\frac{\hbar^2}{2m}\frac{d^2\psi}{dx^2} = \mathcal{E}\psi \tag{7-13}$$

Rearranging, we get

$$\frac{d^2\psi}{dx^2} + \frac{2m\mathcal{E}}{\hbar^2}\,\psi = 0 \tag{7-14}$$

The general solution of this differential equation is

$$\psi = A \sin\left(\frac{\sqrt{2m\mathcal{E}}}{\hbar}\,x\right) + B \cos\left(\frac{\sqrt{2m\mathcal{E}}}{\hbar}\,x\right) \tag{7-15}$$

where A and B are constants of integration. If we evaluate ψ at $x = 0$, we find that $\psi(0) = B$, but $\psi(0) = 0$ if approached from the left. As a result of Postulate III, ψ must be continuous. Therefore $B = 0$ and

$$\psi = A \sin\left(\frac{\sqrt{2m\mathcal{E}}}{\hbar}\,x\right) \tag{7-16}$$

We can get some very valuable insight into the problem if we use the same continuity condition at the point where $x = a$

$$\psi(a) = A \sin\left(\frac{\sqrt{2m\mathcal{E}}}{\hbar}\,a\right) = 0 \tag{7-17}$$

We now have two choices; either $A = 0$ or $\sin(\sqrt{2m\mathcal{E}}\,a/\hbar) = 0$. We cannot permit A to vanish; if $A = 0$, then $\psi = 0$ everywhere, and we will have $\int_{-\infty}^{\infty} \psi^*\psi \, dx = 0$ in contradiction to Postulate IV. Therefore

$$\sin\left(\frac{\sqrt{2m\mathcal{E}}}{\hbar}\,a\right) = 0 \tag{7-18}$$

The properties of the sine function tell us that

$$\frac{\sqrt{2m\mathcal{E}}}{\hbar}\, a = n\pi \qquad (7\text{-}19)$$

where n is some integer. If we solve this expression for $\mathcal{E}$, we get

$$\mathcal{E} = \frac{n^2\pi^2\hbar^2}{2ma^2}, \qquad n = \text{integer} \qquad (7\text{-}20)$$

Notice that since n appears only as n^2, we can ignore negative integers. Notice also that we cannot have $n = 0$. The reason for this fact is not so obvious until we rewrite Equation (7-16) in the form

$$\psi(x) = A\, \sin\frac{n\pi x}{a} \qquad (7\text{-}21)$$

Then it is clear that if $n = 0$ we obtain $\psi = 0$ everywhere, again contradicting Postulate IV. Therefore $n = 1, 2, 3, \ldots$.

One more quantity, A, must be evaluated before we know ψ completely. Again we call on Postulate IV.

$$1 = \int_{-\infty}^{\infty} \psi^*\psi\, dx = |A|^2 \int_0^a \sin^2\frac{n\pi x}{a}\, dx \qquad (7\text{-}22)$$

$$= |A|^2 \frac{a}{2}$$

Therefore $|A|^2 = 2/a$. It is customary to let A be a real number, and so $A = \sqrt{2/a}$. The final solution, then, is

$$\psi = \sqrt{\frac{2}{a}}\, \sin\frac{n\pi x}{a} \qquad \mathcal{E} = \frac{n^2\pi^2\hbar^2}{2ma^2}, \qquad n = 1, 2, 3, \ldots \qquad (7\text{-}23)$$

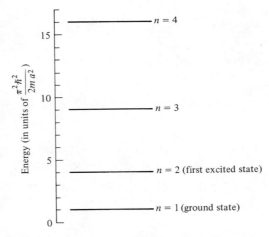

Figure 7-2   Energy-level diagram for the infinite square well.

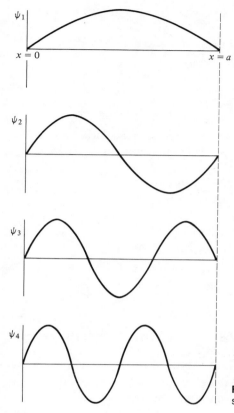

Figure 7-3   Wave functions for the infinite square well.

Figures 7-2 and 7-3 illustrate the energies and wave functions for the first few values of n.

We encounter something here which is typical of quantum mechanical problems involving trapped particles. When an object has insufficient energy to escape from its potential well, the energy will be *quantized*. This particle in a square well cannot possess just any energy. Only those summarized by (7-23) are allowed. The lowest possible energy is *not* zero. This fact is intimately related to the uncertainty principle. By trapping our particle in a region of length a we have gained information about its position that precludes any possibility that the particle might be at rest. To know that a particle is at rest is to know its momentum with infinite precision. The result is that the lowest energy (called the *ground-state* energy) is different from zero.

Another result that is typical of bound-state problems is the appearance of the integer n. This integer is called a *quantum number*. There is only one such integer in this problem because of the restriction to one dimension. A three-dimensional problem would have given rise to three quantum numbers.

It is interesting to note that there is a connection between the De Broglie hypothesis of Chapter 6 and our solution to the infinite square well. A glance at the wave functions for several states (Fig. 7-3) shows that $n/2$ De Broglie waves are capable of fitting into the well if the particle is in the nth quantum state.

7-3   Expectation Values for the Infinite Square Well

We are now ready to put Postulate V to use. We can calculate the average position of a particle in a square well by performing the integral

$$<x> = \int_{-\infty}^{\infty} \psi^* x \psi \, dx \qquad (7\text{-}24)$$

Then

$$<x> = \frac{2}{a} \int_0^a x \sin^2 \frac{n\pi x}{a} \, dx \qquad (7\text{-}25)$$

$$= \frac{2}{a} \frac{a^2}{4} = \frac{a}{2}$$

This result is quite reasonable; the function $\psi^*\psi$ is symmetric about the point $x = a/2$, indicating that the particle will be in the left-hand half of the box just as much as the right-hand half (see Fig. 7-4).

A slightly more sophisticated calculation (with an easier integral) will enable us to find the average momentum of the particle. This time we must recall that the momentum operator is the differential operator $(\hbar/i)(\partial/\partial x)$.

$$<p> = \int_{-\infty}^{\infty} \psi^* \, p_{\text{op}} \psi \, dx = \frac{2}{a} \int_0^a \sin \frac{n\pi x}{a} \frac{\hbar}{i} \frac{\partial}{\partial x} \sin \frac{n\pi x}{a} \, dx$$

$$= \frac{2\hbar}{ia} \int_0^a \sin \frac{n\pi x}{a} \left(\frac{n\pi}{a}\right) \cos \frac{n\pi x}{a} \, dx \qquad (7\text{-}26)$$

$$= 0$$

Again, the result is not startling. It reminds us that this particle will be moving toward the left just as much as it moves toward the right. It does *not* imply a lack of motion. We saw this effect in the preceding section and we can see it again by averaging p^2.

$$<p^2> = \int_{-\infty}^{\infty} \psi^* \, (-\hbar^2) \frac{\partial^2}{\partial x^2} \psi \, dx$$

$$= -\frac{2\hbar^2}{a} \int_0^a \sin \frac{n\pi x}{a} \frac{\partial^2}{\partial x^2} \sin \frac{n\pi x}{a} \, dx$$

$$= \frac{2\hbar^2}{a} \left(\frac{n\pi}{a}\right)^2 \int_0^a \sin^2 \frac{n\pi x}{a} \, dx = \frac{n^2\pi^2\hbar^2}{a^2} \qquad (7\text{-}27)$$

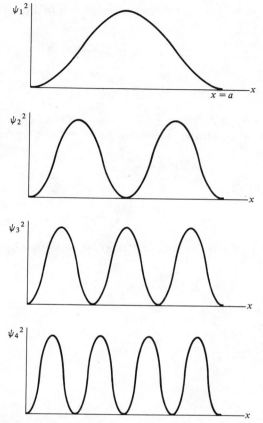

Figure 7-4 $\psi^*\psi$ for the infinite square well.

The uncertainty in x can be defined by the relationship

$$(\Delta x)^2 = <(x - <x>)^2>$$ (7-28)

We can expand the squared deviation from the average value of x to get

$$(\Delta x)^2 = <x^2 - 2x<x> + <x>^2>$$

It follows that

$$(\Delta x)^2 = <x^2> - <x>^2$$ (7-29)

We now use this form to calculate the uncertainty in our knowledge of x. First we need $<x^2>$.

$$<x^2> = \frac{a^2}{3} - \frac{a^2}{2n^2\pi^2}$$

as the student can verify (see Problem 7-6). Then

$$(\Delta x)^2 = \frac{a^2}{12} - \frac{a^2}{2n^2\pi^2}$$ (7-30)

For the uncertainty in the momentum the analog of Equation (7-29) is

$$(\Delta p)^2 = <p^2> - <p>^2 \tag{7-31}$$

So

$$(\Delta p)^2 = \frac{n^2\pi^2\hbar^2}{a^2} \tag{7-32}$$

where we have used $<p^2>$ from (7-27) and $<p>$ from (7-26). We can check to see whether these uncertainties agree with Heisenberg's principle.

$$(\Delta p \; \Delta x)^2 = \frac{n^2\pi^2\hbar^2}{a^2} \left(\frac{a^2}{12} - \frac{a^2}{2n^2\pi^2} \right)$$

$$\Delta p \; \Delta x = \hbar \sqrt{\frac{n^2\pi^2}{12} - \frac{1}{2}} \tag{7-33}$$

The smallest value of this product corresponds to the ground state, $n = 1$, for which

$$\Delta p \; \Delta x \approx 0.567 \; \hbar$$

a result that agrees with

$$\Delta p \; \Delta x \geq \frac{\hbar}{2}.$$

7-4   The Free Particle

Let us consider a one-dimensional particle that is not trapped in any kind of potential well. In other words, the particle is free to move without being acted on by any force. Then $V = 0$, and the Schroedinger equation becomes

$$-\frac{\hbar^2}{2m} \frac{\partial^2\psi}{\partial x^2} = \mathcal{E}\psi \tag{7-34}$$

If we define a quantity k, called the wave number, by

$$k = \sqrt{\frac{2m\mathcal{E}}{\hbar^2}} \tag{7-35}$$

and rearrange, we get

$$\frac{d^2\psi}{dx^2} + k^2\psi = 0 \tag{7-36}$$

the same equation that described the particle within a square well. This time, instead of using sines and cosines, let us write the solution with exponential functions:

$$\psi = C \exp (ikx) + D \exp (-ikx) \tag{7-37}$$

For reasons that will become apparent later, let us set $D = 0$. Then

$$\psi = C \exp{(ikx)} \tag{7-38}$$

If we try to evaluate C, we run into problems. We ought to have

$$\int_{-\infty}^{\infty} \psi^* \psi \, dx = 1$$

however, in this case

$$\psi^* \psi = |C|^2 \tag{7-39}$$

and for any finite value of C we will end up with infinity if we integrate over the entire x axis.

If we consider the physics of the problem, we shall see that indeed $\psi^* \psi$ makes sense. It is independent of x, telling us that the particle can be anywhere at all. We have an infinite amount of uncertainty in its position. The expression $\Delta p \, \Delta x \geq \hbar/2$ tells us that we may then hope to determine the momentum with no uncertainty. Let us try.

$$<p> = \int_{-\infty}^{\infty} \psi^* \frac{\hbar}{i} \frac{\partial}{\partial x} \psi \, dx$$

$$= \frac{\hbar}{i} (ik) \int_{-\infty}^{\infty} \psi^* \psi \, dx = \hbar k \tag{7-40}$$

$$<p^2> = \int_{-\infty}^{\infty} \psi^* (-\hbar^2) \frac{\partial^2}{\partial x^2} \psi \, dx$$

$$= -\hbar^2 (-k^2) \int_{-\infty}^{\infty} \psi^* \psi \, dx = \hbar^2 k^2 \tag{7-41}$$

Thus

$$(\Delta p)^2 = <p^2> - <p>^2 = 0 \tag{7-42}$$

fulfilling our prediction. Notice that $<p>$ is positive, meaning that our particle has a momentum of exactly $\hbar k$ and is moving toward the right. If these calculations had been performed by using $\exp{(-ikx)}$ instead of $\exp{(+ikx)}$, the results would have been the same, except that $<p> = -\hbar k$, meaning that $\exp{(-ikx)}$ describes a particle moving toward the negative end of the x axis.

It is important to note that there is a fundamental difference between the two quantum mechanical problems that we have solved so far. For the particle in a square well the energy was quantized—it could take on only certain discrete values. For the free particle there is no such restriction on the energy. No integers appear to enumerate the energy states available. This distinction between a free particle and a trapped particle is more general than might be guessed from the special nature of these two simple examples. Other problems in which particles are bound (such as electrons trapped in an atom) will give rise to discrete energy levels. Unbound systems (such as two atoms that collide) will give rise to an energy continuum.

7-5 The Step Potential

As a simple example of a quantum-mechanical collision problem let us consider a particle of momentum $\hbar k$ which collides with a stationary wall of finite height. The arrangement is shown in Figure 7-5. The potential energy of the particle is given by

$$V(x) = 0 \qquad x < 0$$
$$V(x) = V_0 \qquad x > 0 \tag{7-43}$$

The Schroedinger equation is easily written down for the two regions separately.

$$\frac{d^2\psi}{dx^2} + \frac{2m\mathcal{E}}{\hbar^2}\psi = 0 \qquad x < 0$$
$$\frac{d^2\psi}{dx^2} + \frac{2m(\mathcal{E} - V_0)}{\hbar^2}\psi = 0 \qquad x > 0 \tag{7-44}$$

Here $\mathcal{E}$ represents the energy that the particle possesses in region 1. Following the procedure of the preceding section, we use the wave number

$$k_1 = \frac{\sqrt{2m\mathcal{E}}}{\hbar} \tag{7-45}$$

to describe the particle in region 1; in region 2 the wave number is by analogy

$$k_2 = \frac{\sqrt{2m(\mathcal{E} - V_0)}}{\hbar} \tag{7-46}$$

The wave function also must be written down in two pieces. In region 1 there will be an incident wave of momentum $+\hbar k_1$; there will also be a reflected wave

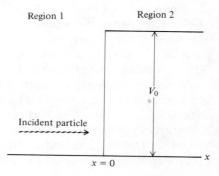

Figure 7-5 The step potential.

of momentum $-\hbar k_1$. In region 2 there will be a transmitted wave of momentum $+\hbar k_2$. The relative amounts of these three waves have not been specified so far; we denote the amplitudes by A, B, and C, respectively.

$$\psi(x) = \underset{\text{incident}}{A \exp(ik_1x)} + \underset{\text{reflected}}{B \exp(-ik_1x)} \qquad x < 0$$

$$\underset{\text{transmitted}}{\psi(x) = C \exp(ik_2x)} \qquad x > 0$$

(7-47)

In order to proceed it will be necessary to eliminate some of the arbitrary constants. We have at our disposal two principles to help us with this task. First of all, ψ must be continuous. Applying this condition at $x = 0$, we get

$$A + B = C \qquad (7\text{-}48)$$

A second condition is that $d\psi/dx$ must be continuous. This condition arises in order that $d^2\psi/dx^2$ may remain finite for use in the Schroedinger equation. In (7-44) $d^2\psi/dx^2$ must not become infinite because all the other terms remain finite. So the condition that $d\psi/dx$ is continuous yields

$$k_1A - k_1B = k_2C \qquad (7\text{-}49)$$

With two equations and three unknowns (A, B, C) it would appear that we were in bad shape; but the thing to do is to solve for B and C in terms of A, since the amplitude of the incident wave is, after all, adjustable by the experimenter. On solving these two equations, we get

$$B = \frac{k_1 - k_2}{k_1 + k_2} A$$

$$C = \frac{2k_1}{k_1 + k_2} A$$

(7-50)

The next item is the physical interpretation of the mathematical formalism. The exact form of the interpretation happens to depend crucially on the relative sizes of $\mathcal{E}$ and V_0.

If $\mathcal{E} > V_0$, we are talking about a particle which, in classical theory, would have possessed enough energy to climb the wall and continue along the positive x axis. In this case k_2 is a real number; the incident, reflected, and transmitted waves are all simply propagating with no attenuation. The only thing unusual about all this is that quantum mechanics does not tell us whether a given projectile will be reflected or not. Classically, a particle with $\mathcal{E} > V_0$ would proceed—slowed down but not reflected. In the quantum description there will always be a reflected wave, since $B \neq 0$ (except for the trivial case in which $k_1 = k_2$ that arises only when $V_0 = 0$). We cannot predict for certain that the particle will be reflected, but we *can* calculate the *probability* that it will be reflected.

The probability of reflection or transmission can be obtained by defining the flux associated with a wave in a given region as the average of the velocity operator:

$$\text{flux} = \int \psi^* \, v_{op} \psi \, dx \tag{7-51}$$

The velocity operator is, of course,

$$v_{op} \equiv \frac{1}{m} \, p_{op} = \frac{\hbar}{im} \frac{\partial}{\partial x} \tag{7-52}$$

Then T, the probability of transmission, is the ratio of the transmitted flux to the incident flux.

$$T = \frac{\int_0^\infty C^* \exp(-ik_2 x) \, (\hbar/im) \, (\partial/\partial x) \, C \exp(ik_2 x) \, dx}{\int_{-\infty}^0 A^* \exp(-ik_1 x) \, (\hbar/im) \, (\partial/\partial x) \, A \exp(ik_1 x) \, dx}$$

$$= \frac{k_2}{k_1} \left| \frac{C}{A} \right|^2 \tag{7-53}$$

Similarly the probability of reflection R is the ratio of flux reflected to incident.

$$R = \frac{\int_0^{-\infty} B^* \exp(ik_1 x) \, (\hbar/im) \, (\partial/\partial x) \, B \exp(-ik_1 x) \, dx}{\int_{-\infty}^0 A^* \exp(-ik_1 x) \, (\hbar/im) \, (\partial/\partial x) \, A \exp(ik_1 x) \, dx} = \left| \frac{B}{A} \right|^2 \tag{7-54}$$

If the expressions for B and C from Equation (7-50) are inserted, we obtain

$$T = \frac{4k_1 k_2}{|k_1 + k_2|^2}$$

$$R = \left| \frac{k_1 - k_2}{k_1 + k_2} \right|^2 \tag{7-55}$$

For the case in which k_2 is purely real it is easy to show that $T + R = 1$, indicating that an incident particle must undergo either reflection or transmission.

If $\mathcal{E} < V_0$, the particle would be reflected, according to classical physics, but in the quantum treatment there will be some transmission because $C \neq 0$ for all finite values of k_2 in Equation (7-50). The physical meaning becomes more lucid if we realize that $\mathcal{E} < V_0$ implies that k_2 is imaginary. So we will write it in the form

$$k_2 = iK \tag{7-56}$$

where K is a real number. Then the transmitted wave becomes $C \exp(-Kx)$, it is no longer a propagating, oscillating wave—it is now purely attenuating. But the fact remains that even for $\mathcal{E} < V_0$ the wave function does not vanish in region 2 of Figure 7-5. Therefore, if an experimentalist buried a detector in the forbidden territory and fired low-energy projectiles at the wall, there would be some nonzero probability that they could penetrate and be detected. This phenomenon of barrier penetration has been observed experimentally for various physical systems. It is a good illustration of a fundamental difference between quantum and classical mechanics. In subsequent chapters of this book this concept will be used repeatedly for varied applications.

7-6 The Harmonic Oscillator

The harmonic oscillator is a system for which the force is linear and acts to oppose displacement:

$$F = -k_s x \tag{7-57}$$

k_s is the spring constant; we shall restrict ourselves to a one-dimensional oscillator with coordinate x. If the particle that is oscillating has mass m, Newton's second law becomes

$$m \frac{d^2 x}{dt^2} = -k_s x \tag{7-58}$$

It is customary to rewrite this equation in the form

$$\frac{d^2 x}{dt^2} + \omega^2 x = 0 \tag{7-59}$$

where

$$\omega = \sqrt{\frac{k_s}{m}} \tag{7-60}$$

ω is the angular frequency of the oscillator. The solution of this classical problem is

$$x = A \sin \omega t + B \cos \omega t \tag{7-61}$$

where A and B are constants.

To consider the problem in a quantum-mechanical fashion, we need to recall that the potential energy is related to the force by the expression $F = -dV/dx$. Then the potential energy for the harmonic oscillator becomes

$$V = \tfrac{1}{2} k_s x^2 \tag{7-62}$$

and the Schroedinger equation, $H\psi = \mathcal{E}\psi$, becomes

$$\frac{\hbar^2}{2m} \frac{d^2 \psi}{dx^2} + \tfrac{1}{2} k_s x^2 \psi = \mathcal{E}\psi \tag{7-63}$$

Again it is customary to rewrite this equation in such a way as to reduce the number of constants. This end can be achieved by substituting

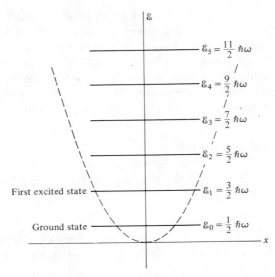

Figure 7-6 Energy levels for the harmonic oscillator. The parabola represents the potential energy $V = \frac{1}{2} k_s x^2$. The energy levels are equally spaced, $\mathcal{E}_n = (n + \frac{1}{2})\hbar\omega$, $n = 0$, 1, 2, ...

$$y = \alpha x \qquad \alpha = \sqrt[4]{\frac{mk_s}{\hbar^2}} \qquad (7\text{-}64)$$

Notice that y is dimensionless. If, in addition, we define another dimensionless quantity,

$$\eta = \frac{2\mathcal{E}}{\hbar\omega} \qquad (7\text{-}65)$$

then Equation (7-63) becomes

$$\frac{d^2\psi}{dy^2} + (\eta - y^2)\psi = 0 \qquad (7\text{-}66)$$

This equation is called Hermite's differential equation, and it is necessary to find solutions to it that also obey the normalization condition, Equation (7-5). Details of this calculation appear in Appendix V-6. We shall simply indicate the results here. Just as in the case of the infinite square well, only certain discrete energies are allowed. They are labeled by a quantum number n. Each energy has associated with it a wave function ψ_n. The situation is summarized in Figures 7-6 and 7-7. We can write down explicit forms for the first few wave functions and a general relationship for the nth quantum state (see Table 7-1 on page 202). The expression for the energy,

$$\mathcal{E}_n = (n + \frac{1}{2})\hbar\omega \qquad n = 0, 1, 2, \ldots \qquad (7\text{-}67)$$

is particularly important and simple: the levels are equally spaced. Since the ground state occurs when $n = 0$, we see that here, as in the case of the infinite square well, there is a nonzero energy in the ground state. The quantity $\mathcal{E}_0 = \frac{1}{2}\hbar\omega$ is often referred to as the zero-point energy.

It is instructive to calculate the uncertainty relationship here, just as in the case of the square well. We shall perform the calculations for the ground state.

$$<x> = \int_{-\infty}^{\infty} x\psi_0{}^2 \, dx = 0$$

The integral vanishes because $x\psi_0{}^2$ is an odd function, integrated over an even interval.

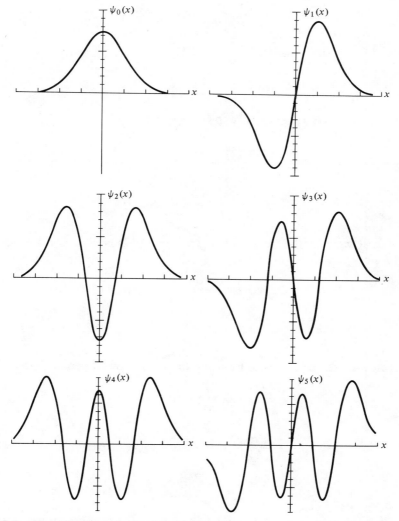

Figure 7-7   Wave functions for the harmonic oscillator.

TABLE 7-1 WAVE FUNCTIONS FOR THE HARMONIC OSCILLATOR

$$\psi_0 = \left(\frac{\alpha}{\sqrt{\pi}}\right)^{1/2} \exp\left(-\frac{\alpha^2 x^2}{2}\right)$$

$$\psi_1 = \left(\frac{\alpha}{2\sqrt{\pi}}\right)^{1/2} 2\alpha x \exp\left(-\frac{\alpha^2 x^2}{2}\right)$$

$$\psi_2 = \left(\frac{\alpha}{8\sqrt{\pi}}\right)^{1/2} (4\alpha^2 x^2 - 2) \exp\left(-\frac{\alpha^2 x^2}{2}\right)$$

$$\psi_n = \left(\frac{\alpha}{2^n n!\sqrt{\pi}}\right)^{1/2} H_n(\alpha x) \exp\left(-\frac{\alpha^2 x^2}{2}\right)$$

$$\alpha = \left(\frac{mk_s}{\hbar^2}\right)^{1/4} = \left(\frac{m\omega}{\hbar}\right)^{1/2}$$

$$y = \alpha x$$

$$H_n(y) = (-1)^n \exp(y^2) \frac{d^n}{dy^n} \exp(-y^2)$$

$$\mathcal{E}_n = (n + \tfrac{1}{2})\hbar\omega$$

$$<p> = \frac{\hbar}{i} \int_{-\infty}^{\infty} \psi_0 \frac{d\psi_0}{dx}\, dx = 0$$

This integral vanishes because ψ_0 is even and $d\psi_0/dx$ is odd. Their product is an odd function.

$$<x^2> = \int_{-\infty}^{\infty} x^2 \psi_0{}^2\, dx = \frac{\alpha}{\sqrt{\pi}} \int_{-\infty}^{\infty} x^2 \exp(-\alpha^2 x^2)\, dx$$

$$= \frac{1}{2\alpha^2}$$

We have evaluated the integral by using the results of Appendix V-5.

$$<p^2> = -\hbar^2 \int_{-\infty}^{\infty} \psi_0 \frac{d^2}{dx^2} \psi_0\, dx$$

$$= -\frac{\hbar^2\alpha}{\sqrt{\pi}} \int_{-\infty}^{\infty} (\alpha^4 x^2 - \alpha^2) \exp(-\alpha^2 x^2)\, dx$$

$$= -\frac{\hbar^2\alpha^3}{\sqrt{\pi}} \left(\alpha^2 \frac{\sqrt{\pi}}{2} \alpha^{-3} - \sqrt{\pi}\, \alpha^{-1}\right) = \tfrac{1}{2}\hbar^2\alpha^2$$

We recall that $(\Delta x)^2 = <x^2> - <x>^2$ and $(\Delta p)^2 = <p^2> - <p>^2$. Then the uncertainty relation yields

$$\Delta x \, \Delta p = \tfrac{1}{2}\hbar. \tag{7-68}$$

We notice that the uncertainty product for the ground state of the oscillator is as small as it can ever be.

7-7 The Square Well in Three Dimensions

So far in our discussion of quantum mechanics we have considered only one-dimensional problems. The reason for this restriction has been that we wished to keep to a minimum of algebraic complexity while introducing the new physical and mathematical ideas that are necessary. In order to consider problems that resemble more closely genuine physical systems it is necessary to use a three-dimensional formalism. As an illustration of the technique we shall use the problem of the three-dimensional infinite square well.

Our physical system corresponds to a particle of mass m which is confined in a rectangular solid defined by

$$0 < x < a$$

$$0 < y < b$$

$$0 < z < c$$

The potential energy inside this region is zero. Outside it is infinite. The Schroedinger equation becomes

$$T\psi = \mathcal{E}\psi \tag{7-69}$$

where T is the kinetic energy operator. In three dimensions

$$T = \frac{\mathbf{p} \cdot \mathbf{p}}{2m} = \frac{p_x{}^2 + p_y{}^2 + p_z{}^2}{2m}$$

$$= -\frac{\hbar^2}{2m}\left(\frac{\partial^2}{\partial x^2} + \frac{\partial^2}{\partial y^2} + \frac{\partial^2}{\partial z^2}\right) = -\frac{\hbar^2}{2m}\nabla^2$$

where ∇^2 is called the Laplacian operator.

The Schroedinger equation, which can now be written in the form

$$\frac{\partial^2\psi}{\partial x^2} + \frac{\partial^2\psi}{\partial y^2} + \frac{\partial^2\psi}{\partial z^2} + \frac{2m\mathcal{E}}{\hbar^2}\psi = 0 \tag{7-70}$$

is a partial differential equation. It is readily solved by the technique of separation of variables. This method is of great use in solving partial differential equations which arise elsewhere in physics. The procedure is to assume that the function ψ can be written as a product of three functions, X, Y, and Z, each of which depends on only one of the coordinates:

$$\psi = X(x)\, Y(y)\, Z(z) \tag{7-71}$$

We substitute this product in Equation (7-70), using primes to denote differentiation with respect to the appropriate variables—that is,

$$X' = \frac{\partial X}{\partial x} = \frac{dX}{dx} \qquad Y' = \frac{dY}{dy} \qquad \text{etc.}$$

Then the equation reads

$$X''YZ + XY''Z + XYZ'' + \frac{2m\mathcal{E}}{\hbar^2} XYZ = 0$$

Next we divide the equation by XYZ; this operation is permissible because the function $\psi = XYZ$ cannot be identically zero and still obey

$$\int_0^a \int_0^b \int_0^c |\psi|^2 \, dz \, dy \, dx = 1 \qquad (7\text{-}72)$$

the required normalization condition. After dividing we obtain

$$\frac{X''}{X} + \frac{Y''}{Y} + \frac{Z''}{Z} + \frac{2m\mathcal{E}}{\hbar^2} = 0 \qquad (7\text{-}73)$$

We notice that each term of this equation depends on a different variable and that the three variables are independent. The last term, of course, is constant. The only way for the equation to remain valid for all values of x, y, and z in our interval is for *each* term of Equation (7-73) to be constant. Therefore

$$\frac{X''}{X} = -\alpha^2 \qquad \frac{Y''}{Y} = -\beta^2 \qquad \frac{Z''}{Z} = -\gamma^2$$

where α, β, and γ are constants. The first of these equations can be rewritten as

$$X'' + \alpha^2 X = 0 \qquad (7\text{-}74)$$

From our experience in Section 7-2 we can write immediately the solution to Equation (7-74) which will go to zero at $x = 0$ and at $x = a$.

$$X = \sqrt{\frac{2}{a}} \sin \frac{n_x \pi x}{a} \qquad n_x = 1, 2, 3, \ldots \qquad (7\text{-}75a)$$

Similar results occur for Y and Z:

$$Y = \sqrt{\frac{2}{b}} \sin \frac{n_y \pi y}{b} \qquad n_y = 1, 2, 3, \ldots \qquad (7\text{-}75b)$$

$$Z = \sqrt{\frac{2}{c}} \sin \frac{n_z \pi z}{c} \qquad n_z = 1, 2, 3, \ldots \qquad (7\text{-}75c)$$

The product of these functions is

$$\psi = \left(\frac{8}{abc}\right)^{1/2} \sin \frac{n_x \pi x}{a} \sin \frac{n_y \pi y}{b} \sin \frac{n_z \pi z}{c} \qquad (7\text{-}76)$$

If we now substitute the functions in Equation (7-75) back into Equation (7-73), we obtain

$$-\frac{n_x{}^2\pi^2}{a^2} - \frac{n_y{}^2\pi^2}{b^2} - \frac{n_z{}^2\pi^2}{c^2} + \frac{2m\mathcal{E}}{\hbar^2} = 0$$

This result can be solved for $\mathcal{E}$.

$$\mathcal{E} = \frac{\pi^2\hbar^2(n_x{}^2/a^2 + n_y{}^2/b^2 + n_z{}^2/c^2)}{2m} \tag{7-77}$$

We have completed the problem, since we now know ψ for all allowed values of the integers n_x, n_y, and n_z. It remains to provide an interpretation of these results.

We notice first of all that there are three quantum numbers that are necessary to describe each state. This fact is a general property of three-dimensional systems. In this problem, as in the one-dimensional square well, none of the quantum numbers may be zero. The lowest energy, the ground state, occurs when $n_x = n_y = n_z = 1$. The set of quantum numbers which defines the first excited state will depend on the relative sizes of a, b, and c. As an instructive example let us examine the case in which the particle is confined in a cubical region. Then $a = b = c$ and the energy levels of Equation (7-77) become

$$\mathcal{E}_{\text{cube}} = \frac{\pi^2\hbar^2(n_x{}^2 + n_y{}^2 + n_z{}^2)}{2ma^2}$$

The ground state still has $n_x = n_y = n_z = 1$, but there are three possibilities for the first excited state:

$$n_x = 2 \qquad n_y = n_z = 1 \qquad (1)$$
$$n_y = 2 \qquad n_x = n_z = 1 \qquad (2)$$
$$n_z = 2 \qquad n_x = n_y = 1 \qquad (3)$$

It is easy to see that many of the higher energy states have this same situation; that is, several distinct quantum states possess the same energy. This property is called *degeneracy;* it is another way in which a multidimensional system differs from a one-dimensional system. In Chapter 8 we shall deal with the hydrogen atom, a three-dimensional quantum system. It shares with the cubical box the possession of three quantum numbers and degenerate states.

7-8   Selection Rules

In our consideration of quantum mechanics we have been assuming that our systems are constant with respect to time. A more general treatment would require us to abandon this restriction, and we would need to solve such problems as describing the transition of a system from one quantum state to another; for instance, a particle in an infinite square well might be in the first excited state. From our work so far we could expect it simply to remain in the first excited state. Nevertheless it is possible for the particle to fall into the ground state, emitting energy perhaps in the form of a photon. In keeping with the probabilistic interpretation of quantum mechanics, it is not possible to predict exactly how long the particle will remain in the first excited state. It *is* possible to predict *on the average* how long the particle will stay in the upper state. The details of the calculation are too lengthy to be presented here. The interested reader will find them under the heading of time-dependent perturbation theory in the references listed at the end of this chapter. We shall simply state the result.

If a system is initially in a state described by a wave function ψ_i and then the system is influenced by a force described by a potential energy V, the probability of making a transition to a final state (described by ψ_f) is proportional to

$$P_{if} \sim \left| \int_\tau \psi_f^* V \psi_i \, d\tau \right|^2 \tag{7-78}$$

where the region τ includes all of space. The integral is usually called the *matrix element* of the potential V between the initial and final states. It is evident that the transition from initial to final state cannot occur if the matrix element vanishes. For a given system it is usually possible to find a set of rules that will determine whether or not the matrix element is zero; these are called *selection rules*.

The most interesting and useful selection rules are those that govern emission and absorption of photons. For concreteness we shall discuss a one-dimensional particle with charge q that absorbs a photon. The electric field E associated with the photon will displace the particle somewhat, doing work in the process. This work is just the interaction potential needed to calculate matrix elements for the transition of the particle from one state to another. If the original expected value of the position of the particle is $<x>$, then $x - <x>$ is a measure of the displacement; since qE is the force exerted by the electric field, then

$$V = qE(x - <x>) \tag{7-79}$$

is the interaction potential. The matrix element is given by

$$qE \int_{-\infty}^\infty \psi_f^* (x - <x>)\psi_i \, dx \tag{7-80}$$

Selection rules based on the value (zero or nonzero) of this matrix element are called *electric dipole selection* rules in recognition of the fact that $q(x - <x>)$ is simply the electric dipole moment of the system. It is a fact that this matrix element also describes emission of a photon as well as absorption.

To illustrate the use of the ideas presented on selection rules, let us calculate dipole matrix elements for a particle in an infinite square well, as in Section

7-2. We are interested in determining the rules that govern the change of the particle from the initial state (n_i) to the state (n_f). The matrix element is

$$M_{if} = \frac{2}{a} \int_0^a \sin \frac{n_f \pi x}{a} \left(x - \frac{a}{2} \right) \sin \frac{n_i \pi x}{a} \, dx \qquad (7\text{-}81)$$

It is assumed that $n_i \neq n_f$. Then the term

$$\frac{2}{a} \cdot \frac{a}{2} \int_0^a \sin \frac{n_f \pi x}{a} \sin \frac{n_i \pi x}{a} \, dx$$

vanishes. The remaining term yields

$$M_{if} = \frac{2}{a} \left\{ x \left[\frac{\sin (n_i - n_f)\pi x/a}{2(n_i - n_f)^2 \pi^2 / a^2} - \frac{\sin (n_i + n_f)\pi x/a}{2(n_i + n_f)^2 \pi^2 / a^2} \right] \right.$$

$$\left. + \frac{\cos (n_i - n_f)\pi x/a}{2(n_i - n_f)^2 \pi^2 / a^2} - \frac{\cos (n_i + n_f)\pi x/a}{2(n_i + n_f)^2 \pi^2 / a^2} \right\}_0^a$$

$$= \frac{a}{\pi^2} \left[\frac{(-1)^{n_i - n_f} - 1}{(n_i - n_f)^2} - \frac{(-1)^{n_i + n_f} - 1}{(n_i + n_f)^2} \right] \qquad (7\text{-}82)$$

If n_i and n_f are both even integers or if they are both odd integers, then $M_{if} = 0$. If one is even and the other odd, then $M_{if} \neq 0$. We conclude from this calculation that transitions of the particle are governed by the selection rule that the *parity* (evenness or oddness) of its quantum number must change in order for a photon to be emitted or absorbed. Figure 7-8 illustrates allowed transitions for this problem.

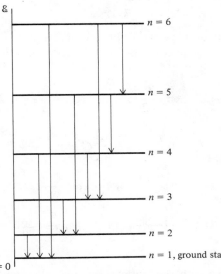

Figure 7-8 Energy-level diagram for the one-dimensional infinite square well showing allowed transitions that connect states with $n \leq 6$.

It is important to realize that the dipole selection rules are not absolute. By this statement we mean that Equation (7-78) is only an approximation, so that the vanishing of a matrix element does not mean that the transition never occurs. It usually means, however, that the probability of its occurrence is very small.

7-9 Penetration of a Potential Barrier

In this section we return to the topic of collisions and consider a problem that is similar to the one discussed in Section 7-5. In this instance we consider the case of a wave incident onto a potential barrier such as the one shown in Figure 7-9. We suppose that the wave represents a particle with energy $\mathcal{E}$. Then from previous work we can write the wave numbers for the problem. In regions I and III we have

$$k_1{}^2 = \frac{2m\mathcal{E}}{\hbar^2} \tag{7-83}$$

and in region II

$$k_2{}^2 - \frac{2m(\mathcal{E} - V_0)}{\hbar^2} \tag{7-84}$$

We can also write the wave function in the three regions

$$\psi_1 = A \exp(ik_1x) + B \exp(-ik_1x)$$
$$\psi_2 = C \exp(ik_2x) + D \exp(-ik_2x) \tag{7-85}$$
$$\psi_3 = E \exp(ik_1x)$$

As in our previous examples, we require that ψ and its first derivative be continuous at the points $x = a/2$ and $x = -a/2$. These conditions give rise to four equations:

$$A \exp\left(\frac{-ik_1a}{2}\right) + B \exp\left(\frac{ik_1a}{2}\right) = C \exp\left(\frac{-ik_2a}{2}\right) + D \exp\left(\frac{ik_2a}{2}\right)$$

$$A \exp\left(\frac{-ik_1a}{2}\right) - B \exp\left(\frac{ik_1a}{2}\right) = \frac{k_2}{k_1}\left[C \exp\left(\frac{-ik_2a}{2}\right) - D \exp\left(\frac{ik_2a}{2}\right)\right]$$

$$C \exp\left(\frac{ik_2a}{2}\right) + D \exp\left(\frac{-ik_2a}{2}\right) = E \exp\left(\frac{ik_1a}{2}\right) \tag{7-86}$$

$$C \exp\left(\frac{ik_2a}{2}\right) - D \exp\left(\frac{-ik_2a}{2}\right) = \left(\frac{k_1}{k_2}\right) E \exp\left(\frac{ik_1a}{2}\right)$$

These four equations with five unknowns are quite analogous to Equations (7-48) and (7-49) for the step potential. The procedure to be followed is to

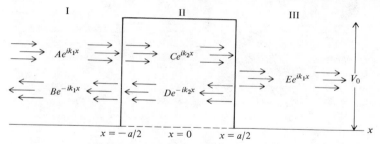

Figure 7-9   A rectangular one-dimensional potential barrier showing the traveling waves that exist in each region of space.

consider A (the incident amplitude) as capable of being adjusted experimentally and then to solve for the other four variables as ratios. The most interesting of these is the ratio E/A, which yields

$$\frac{E}{A} = \frac{\exp\,(-ik_1a)}{\cos k_2a - \frac{1}{2}i(k_1/k_2 + k_2/k_1)\sin k_2a} \tag{7-87}$$

This expression is most readily used when the incident energy $\mathcal{E}$ is greater than the potential V_0 of the barrier. In that case k_2 is a real number. Classically, the particle can climb over the barrier and continue into region III. In the quantum treatment, since $|E/A|^2$ differs from unity, there will be some reflection, although the effect will be small if $\mathcal{E} \gg V_0$.

If $\mathcal{E} < V_0$, then k_2 is an imaginary number, and the circular trigonometric functions in Equation (7-87) can be replaced by hyperbolic functions. If we use the definition

$$K = ik_2 \tag{7-88}$$

the transmitted amplitude becomes

$$\frac{E}{A} = \frac{\exp\,(-ik_1a)}{\cosh Ka + \frac{1}{2}i(K/k_1 - k_1/K)\sinh Ka} \tag{7-89}$$

Again the quantum prediction contradicts classical physics. Even though the energy is not sufficient for a particle to clear the barrier, some of its wave function can extend into region III, meaning that there is a nonzero probability of penetration. This transmission probability is

$$T = \left|\frac{E}{A}\right|^2 = \frac{1}{\cosh^2 Ka + \frac{1}{4}(K/k_1 - k_1/K)^2 \sinh^2 Ka} \tag{7-90}$$

Figure 7-10 shows the transmission probability as a function of energy for several different rectangular barriers. The qualitative features are exactly what one might expect. At low energy the transmission probability is low and at high energy it approaches unity. The classical result is

$$\begin{aligned} T_{\text{classical}} &= 0 \qquad \mathcal{E} < V_0 \\ &= 1 \qquad \mathcal{E} > V_0 \end{aligned} \tag{7-91}$$

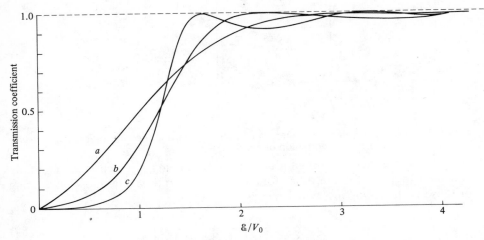

Figure 7-10 Transmission coefficient as a function of incident energy for three different rectangular potential barriers: (a) low barrier, (b) intermediate, (c) high barrier.

The curve corresponding to the greatest barrier height is the one that most nearly approximates the classical result.

An interesting effect occurs whenever the De Broglie wavelength of the particle in region II is an integral submultiple of twice the length of the barrier. The transmission probability becomes exactly equal to unity and the result is perfect transmission. In effect, the two sources of reflected wave (the two interfaces) produce two waves that have equal amplitude and opposite phase, yielding no reflection at all. The situation is similar to the physics of lenses that are coated to reduce reflection.

An interesting and important special case of barrier transmission is that for which the barrier height is high and the length a to be traversed is great. In this limiting case the barrier is nearly opaque and $T \ll 1$. In Equation (7-90) we can replace the hyperbolic functions with an exponential, since

$$\sinh x \approx \cosh x \approx \tfrac{1}{2} \exp x, \qquad x \gg 1 \qquad (7\text{-}92)$$

After some algebra we obtain

$$T \approx \frac{16k_1{}^2 K^2}{(k_1{}^2 + K^2)^2} \exp(-2Ka) \qquad (7\text{-}93)$$

The importance of this approximate form for T is that it can be used to arrive at an estimate of the transmission through a barrier of arbitrary shape. The estimation consists of noticing that T is proportional to $\exp(-2Ka)$, and that

the product Ka is effectively an area that can be represented by $\int_{x1}^{x2} K(x)\, dx$,

since K varies with position because the potential energy is no longer held constant. If we recall the definition of K (and that of k_2) and denote the potential energy by $V(x)$, we obtain

$$T \approx \exp\left[\frac{2}{\hbar}\int_{x1}^{x2}\sqrt{2m(V-\mathcal{E})}\,dx\right] \qquad (7\text{-}94)$$

The integration proceeds over that part of the barrier for which $V > \mathcal{E}$. This approximate form will prove useful in Section 14-11 in the discussion of alpha decay of nuclei.

An interesting example of tunneling from classical physics is the phenomenon of the total internal reflection of light. Figure 7-11a shows light incident on a right-angled prism with acute angles of 45 degrees. If the index of refraction of the glass is about 1.5, the internal angle of incidence on the hypotenuse will be greater than the critical angle, and the light will be totally reflected. If an identical prism is placed in contact with the first one, as shown in Figure 7-11b, the rays of light will no longer be totally reflected but will pass straight through the two surfaces. The two diagonal surfaces do not need to be in actual contact but may be separated a distance of a few wavelengths of light and still have some light go through this barrier. This can be shown with the arrangement sketched in Figure 7-11c in which the diagonal surface of the second prism has been polished around the edges, making it a lens-shaped surface with a large radius of curvature. The rays of light will pass completely through the circle of contact, will be totally internally reflected when the distance between the surfaces is greater than a few wavelengths of light, and will be partially reflected in a zone surrounding the center when the distance of separation is of the order of a few wavelengths of light. If white light is incident on this system, the pattern observed by the eye will consist of a central bright circle surrounded by a narrow band of reddish light because the longer wavelengths travel farther than the shorter ones in the space between the surfaces.

In terms of the modern theory of light, the rays are analogous to the passage of photons through the system. The electromagnetic theory of light predicts that the wave will actually penetrate into the space beyond the totally reflecting surface for a distance of a few wavelengths. This wave turns out to be a longitudinal wave and carries no energy. Bringing a second surface near the first one will produce a penetration of the barrier between the two surfaces, with a transverse wave going through the second prism. The phenomenon of

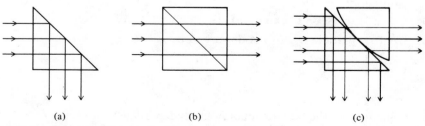

(a)   (b)   (c)

Figure 7-11   Frustrated total reflection.

the tunneling of a light wave through the barrier is sometimes called "frustrated total internal reflection." It can be used as a valve for controlling the transmission of light through a system.

An important quantum mechanical phenomenon is that known as the *Josephson effect* in which superconducting electrons pass through a thin nonconducting layer between two superconductors. This will be discussed in Section 11-8.

Problems

7-1. Prove that the Schroedinger equation (7-11) is linear. In other words, show that if ψ_i and ψ_j are both solutions, then $(A\psi_i + B\psi_j)$ is also a solution.

7-2. Starting from Equation (7-21), show that $n/2$ de Broglie waves fit into an infinite square well.

7-3. Starting from Equation (7-20), show that $n/2$ de Broglie waves fit into an infinite square well.

7-4. (a) For a particle (mass m, quantum number n) in an infinite square well of width a, what is the probability of finding the particle within a distance ϵ of the left-hand edge? (b) Find the limit as $n \to \infty$. (c) For finite n, find the limit as $\epsilon/a \ll 1$.

7-5. (a) Set up the Schroedinger equation for a particle in an infinite square well where

$$V = 0 \qquad -a < x < +a$$
$$V = \infty \qquad \text{elsewhere}$$

(b) Solve the equation to find $\mathcal{E}$ and ψ.

7-6. Show that $<x^2> = a^2/3 - a^2/(2n^2\pi^2)$ for the particle in an infinite square well with limits $(0, a)$.

7-7. Prove that if ψ is real (that is, $\psi^* = \psi$) and $\psi \neq 0$ only in the interval (a, b), then $<p> = 0$.

7-8. Verify by direct substitution that Equation (7-15) is a solution of Equation (7-14).

7-9. Verify that $\psi = C \exp (\pm ikx)$ is a solution to Equation (7-36).

7-10. A free particle $(V = 0)$ is described by $\psi = C \exp (ikx)$ and

$$\int_{-\infty}^{\infty} \psi^* \psi \, dx = 1$$

Show that the kinetic energy of the particle is exactly $\hbar^2 k^2/2m$ by calculating $<T>$ and $<T^2> - <T>^2$.

7-11.   For a plane wave incident on a step potential (Fig. 7-5) calculate the reflection coefficient R for the case in which $\mathcal{E} < V_0$.

7-12.   What ratio of $\mathcal{E}/V_0$ is necessary for scattering from a one-dimensional step potential so that the transmission probability will be 50 percent?

7-13.   In the text we defined the new variable y by the equation

$$y = (mk_s/\hbar^2)^{1/4}x$$

Prove that y is dimensionless.

7-14.   By using the integrals in Appendix V-5 prove that the first three wave functions of the harmonic oscillator (Table 7-1) satisfy $\displaystyle\int_{-\infty}^{\infty} \psi_n^2\, dx = 1$.

7-15.   A pendulum with a length of one meter has a bob with a mass of 0.1 kg. What is its zero-point energy?

7-16.   The pendulum in Problem 7-15 is displaced 5 cm from its equilibrium position and then released. What is the quantum number associated with its wave function?

7-17.   Calculate $<x>$ and $<x^2>$ for the first excited state of a harmonic oscillator.

7-18.   Calculate $<p>$ and $<p^2>$ for the first excited state of a harmonic oscillator.

7-19.   From Problems 7-17 and 7-18 calculate $\Delta x\,\Delta p$ for the first excited state of a harmonic oscillator.

7-20.   Draw an energy-level diagram for a particle of mass m in a cubical box of length a, including all states with energy up to $15\pi^2 n^2/(ma^2)$.

7-21.   A particle is trapped in the region $0 < x < a$ but the potential is not zero. Instead, it is such that the wave function is $\psi = Ax(a - x)$. (a) Is this the ground state or some excited state? (b) Find A. (c) Find $<x>$. (d) Find $<x^2>$. (e) Is this particle more localized or less localized than one that is in a square well of length a?

7-22.   (a) For the classical harmonic oscillator show that the time-average of the kinetic energy is equal to the time-average of the potential energy. (b) For the ground state of the quantum mechanical harmonic oscillator show that the space-averages of the kinetic and potential energies are equal.

7-23.   Show that for a very large quantum number the probability of finding a particle in any small interval in a one-dimensional infinite square well is independent of the position within the well.

7-24.   A stream of electrons, each having an energy $\mathcal{E} = 4$ eV is directed toward a potential barrier of height $V_0 = 5$ eV. The width of the barrier is

2×10^{-7} cm. Calculate the percentage transmission of this beam through this barrier.

7-25. Determine the eigenvalues for an alpha particle inside the nucleus of a uranium atom by assuming a square-well potential with a height $V_0 = 29$ MeV and a width $a = 10^{-12}$ cm. Compare these eigenvalues with the energies of the alpha particles emitted by uranium.

7-26. Assume that an alpha particle has an energy of 10 MeV and approaches a potential barrier of height equal to 30 MeV. Determine the width of the potential barrier if the transmission coefficient is 0.002.

7-27. Derive the condition for perfect transmission through a rectangular potential barrier.

7-28. Obtain the amplitude for the wave reflected from a rectangular potential barrier and show directly that probability is conserved in the collision process.

7-29. Show that Equation (7-93) is the limit of T as Ka becomes very large.

7-30. An object in one dimension is described by a wave function

$$\psi = x\sqrt{3} \qquad 0 < x < 1$$
$$\psi = 0 \qquad \text{elsewhere}$$

(a) What is the probability of finding the object within the interval $(0, 0.5)$? (b) What is the average position of the object?

7-31. The three lowest states of a certain system are described by

$$\psi_a = \exp\left(\frac{-x^2}{2}\right)$$

$$\psi_b = x \exp\left(\frac{-x^2}{2}\right)$$

$$\psi_c = (2x^2 - 1) \exp\left(\frac{-x^2}{2}\right)$$

Determine which transitions are allowed among these states and which ones are forbidden.

7-32. Two different particles are confined in the same one-dimensional box. If particle 1 were alone, its wave function would be $\psi_1(x_1)$. If particle 2 were alone, its wave function would be $\psi_2(x_2)$. Assume that the particles do not interact. (a) Write the Schroedinger equation for this two-body system. (b) Investigate which of the following wave functions will describe the system: $(\psi_1 + \psi_2)$, $(\psi_1 - \psi_2)$, $\psi_1\psi_2$, or ψ_1/ψ_2? (c) If $\mathscr{E}_1$ and $\mathscr{E}_2$ are the single-particle energies, which of the following will give the two-particle energy: $(\mathscr{E}_1 + \mathscr{E}_2)$, $(\mathscr{E}_1 - \mathscr{E}_2)$, $\mathscr{E}_1\mathscr{E}_2$, or $\mathscr{E}_1/\mathscr{E}_2$?

7-33. A one-dimensional particle has a wave function given by

$$\psi = (2a)^{-1/2} \exp i(kx - \omega t)$$

within the interval $(-a, a)$. (a) What is the average position of the particle? (b) What is its average momentum? (c) An inspired theorist guesses that $A(\partial/\partial t)$ might be the form of the energy operator, where A is a constant. Show that this form is reasonable and evaluate A.

7-34. (a) How many nodes, maxima, minima, and points of inflection are there in each wave function for a particle in a one-dimensional square well? (b) How many of these same special points are there for harmonic oscillator wave functions?

References

Beard, D. B., and G. B. Beard, *Quantum Mechanics with Applications*. Boston: Allyn and Bacon, 1970.

Bohm, D., *Quantum Theory*. Englewood Cliffs, N. J.: Prentice-Hall, 1951.

Feynman, R. P., R. B. Leighton, and M. Sands, *The Feynman Lectures on Physics,* Vol. III. Reading, Mass.: Addison-Wesley Publishing Company, 1965.

Pauling, L., and E. B. Wilson, *Introduction to Quantum Mechanics*. New York: McGraw-Hill Book Company, 1935.

Powell, J. L., and B. Crasemann, *Quantum Mechanics*. Reading, Mass.: Addison-Wesley Publishing Company, 1961.

Schiff, L. I., *Quantum Mechanics*. New York: McGraw-Hill Book Company, 1968 (3rd ed.).

Sherwin, C. W., *Quantum Mechanics*. New York: Holt, Rinehart and Winston, 1959.

PART TWO

The Extranuclear Structure of the Atom

8 | The Hydrogen Atom

8-1 Historical Survey

Hydrogen, the simplest of all the elements, has been investigated most extensively both experimentally and theoretically. The knowledge obtained from this study has acted as a guide to the study of more complex elements. In this chapter we limit our considerations to hydrogen; in the next two chapters we discuss more complicated elements.

In the latter part of the nineteenth century much work was done with spectroscopes to obtain both the emission and absorption spectra of various elements. As a result of this work, certain spectral lines in specific elements were found to be related in a simple manner by means of an equation suggested by Rydberg (1889) in which the reciprocal of the wavelength (that is, the wave number) of a spectral line is given as the difference between two numbers or two *terms*.

As long ago as 1885 Balmer succeeded in obtaining a simple relationship among the wave numbers of the lines in the visible region of the hydrogen spectrum. Balmer's equation expressed in modern notation is

$$\frac{1}{\lambda} = \bar{\nu} = R_\mathrm{H}\left(\frac{1}{2^2} - \frac{1}{n^2}\right) \qquad n = 3, 4, 5, \ldots \tag{8-1}$$

where λ is the wavelength, $\bar{\nu}$ is the wave number of the spectral line, R_H is a constant known as Rydberg's constant for hydrogen, and n is an integer greater than 2. The empirical value of Rydberg's constant for hydrogen is

$$R_\mathrm{H} = 109,677.581 \text{ cm}^{-1}$$

By substituting for n in Equation (8-1) the successive values 3,4,5,6, . . . , we obtain the wave numbers of lines in the Balmer series. Figure 8-1 is a photograph of these lines, and Figure 8-2 shows the positions on a scale that in-

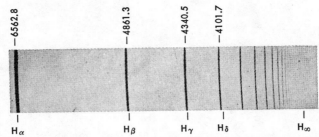

Figure 8-1 Photograph of the emission spectrum of hydrogen showing the Balmer series lines in the visible and near ultraviolet region. H_∞ shows the theoretical position of the series limit. (Reprinted by permission from *Atomic Spectra and Atomic Structure*, by G. Herzberg, Dover Publications.)

creases uniformly in wave number. It is clear from Balmer's equation that as *n* approaches infinity the lines crowd together and approach the *series limit*. The value of this limit for the Balmer series is $R_H/2^2 = 27,419.40$ cm^{-1}.

The first quantitatively correct derivation of the Balmer formula on the basis of an atomic model was given by Bohr (1913). Beginning with Rutherford's nuclear atom, Bohr constructed a planetary model which we shall consider more fully in the next section. In order to obtain agreement with the spectroscopic data summarized in Balmer's equation, it was necessary to depart from classical physics in three respects:

1. Charged electrons circling around a charged nucleus experience an acceleration and, according to Maxwell's formulation of classical electromagnetism, should emit light; Bohr was forced to assume that they do not radiate.

2. Bohr assumed that atomic electrons can move only in orbits for which the angular momentum is an integer multiple of $\hbar$.

3. Bohr also assumed that atoms emit light only when an electron jumps from one permissible orbit to another; such a jump is accompanied by emission of a single photon whose energy is the difference in the energies of the initial and final orbits.

In spite of the quantum assumptions which Bohr included in his theory, the basic idea is still classical. The position and momentum of an atomic electron are considered to be (in principle) measurable to arbitrary accuracy, in violation of the uncertainty principle (which was not yet known in 1913). Therefore it is not surprising that the Bohr theory is incorrect in detail. It is nevertheless essential for the serious student to master at least the rudiments of the Bohr theory, since some of the fundamental physical quantities that arise therein also play a part in the more nearly correct theories. The simple Bohr model also remains the least toilsome method of "deriving" the Balmer formula, being a good example of an heuristic proof—that is, a quick, incorrect method for getting a correct result.

The Bohr theory in its simplest form uses circles as the orbits of the electrons. An obvious refinement in the theory was made by Sommerfeld (1916) by considering elliptical orbits. Sommerfeld's treatment, although interesting and informative, is primarily of historical interest and is not needed for an understanding of the quantum solution of the hydrogen problem. For this reason our discussion of elliptic orbits is deferred to Section 8-9, which can be read or omitted without affecting the continuity of the discussion. Besides allowing for elliptic orbits, Sommerfeld expanded the Bohr model to even greater complexity by including the corrections due to special relativity. This inclusion increases the mathematical complexity of the problem without touching on the basic shortcomings of the Bohr theory.

As already stated, the Bohr model incorporates only a few quantum features, and so we know that it must somewhere disagree with experimental observation. Two very serious difficulties exist. First, the Bohr model predicts the correct spectrum only for hydrogenic atoms (atomic systems with only one electron H, He^+, Li^{++}, etc.). The neutral helium atom, with two electrons that interact, has energy levels at values different from those predicted by the Bohr theory. Similar discrepancies occur for all multielectron atoms. A second difficulty with the Bohr theory is its inability to predict transition probabilities. For a given element, not all spectral lines possess the same intensity, and it would be hoped that a correct theory could enable us to calculate the relative strengths of lines.

The Schroedinger theory, introduced in Chapter 7, was successful in eliminating both fundamental problems. In Section 8-4 we shall consider this method as applied to hydrogen. Further refinements in the quantum mechanical description have been made. The details of these additional theories are beyond the level of this book, but we mention their nature to complete the history of the problem. Dirac (1928) succeeded in formulating a relativistic quantum theory of the electron, a theory that can be applied to the hydrogen problem. The results of the Dirac theory agree with the nonrelativistic calculation, except for a few details that are difficult to test experimentally. In a series of papers beginning in 1947 Lamb and Retherford published experimental results on fine structure in the hydrogen spectrum which could not be accounted for by using any of the theories mentioned so far. However, calcu-

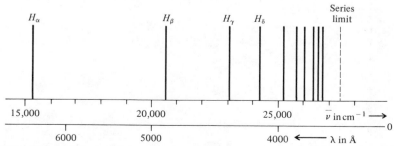

Figure 8-2 Graph of the positions of the Balmer series lines on wave number and wavelength scales.

lations based on relativistic quantum field theory have been made by a number of people to explain the Lamb shift (see Section 9-20).

8-2 Bohr's Theory of the Hydrogen Atom

We restate Bohr's postulates as listed in the preceding section and use them to develop a quantitative description.

I. *An electron in a circular orbit can retain its motion under the influence of the Coulomb force without radiating.* We consider a stationary nucleus with a charge equal to Ze and an orbiting electron with charge $-e$. The Coulomb force is thus given by

$$F = -k\frac{Ze^2}{r^2} \qquad (8\text{-}2)$$

where r is the radius of the circular orbit and k is determined by the choice of electrical units; if e is in electrostatic units and r in centimeters, then $k=1$ and F is in dynes; if e is in coulombs and r is in meters, then $k = 1/4\pi\epsilon_0$ and F is in newtons. From Newton's second law of motion, the force on the electron is equal to its mass m multiplied by its centripetal acceleration.

$$F = -\frac{mv^2}{r} \qquad (8\text{-}3)$$

where v is the speed of the electron (constant for a given circular orbit). Equating these two expressions for the force yields

$$mv^2 = \frac{kZe^2}{r} \qquad (8\text{-}4)$$

It is a standard procedure in electrostatics to use Equation (8-2) to obtain the potential energy of the electron:

$$V = -\frac{kZe^2}{r} \qquad (8\text{-}5)$$

The negative sign is present because the force is attractive rather than repulsive. The kinetic energy is $\frac{1}{2}mv^2$. Therefore the total energy $\mathcal{E}$, being the sum of the potential and kinetic energies, is

$$\mathcal{E} = \frac{1}{2}mv^2 - \frac{kZe^2}{r} \qquad (8\text{-}6)$$

which, on the substitution of Equation (8-5), becomes

$$\mathcal{E} = -\frac{kZe^2}{2r} \qquad (8\text{-}7)$$

In this form the negative sign refers to the fact that we are discussing a bound state of the electron. The zero point of energy corresponds to an atom that is just barely ionized; that is, $\mathcal{E} = 0$ when $r = \infty$.

II. *The angular momentum in a stable orbit is given by an integer multiple of $\hbar$.* In symbols,

$$mvr = n\hbar \qquad (8\text{-}8)$$

where $n = 1, 2, 3, \ldots$ Using Equations (8-4) and (8-8), we can eliminate v to obtain

$$r = \frac{n^2\hbar^2}{mkZe^2} \qquad (8\text{-}9)$$

The radius of the smallest possible orbit is obtained by setting $n = 1$ and using the known values of the other quantities. For hydrogen $Z = 1$, and the smallest orbit is

$$r_1 = \frac{\hbar}{mke^2} = 0.529 \times 10^{-10} \text{ m} = 0.529 \text{ Å} \qquad (8\text{-}10)$$

This distance is called the *Bohr radius*. It is of the same order of magnitude as the radius of an atom as determined by the kinetic theory of gases.

The value of the radius from Equation (8-9) can be substituted into Equation (8-7) to get an expression for the energy that depends on known quantities.

$$\mathcal{E} = -\frac{mk^2Z^2e^4}{2n^2\hbar^2} \qquad (8\text{-}11)$$

The ground-state energy for hydrogen is obtained by setting $n = Z = 1$ in this equation.

$$\mathcal{E}_1 = -13.58 \text{ eV} \qquad (8\text{-}12)$$

This energy is just the amount needed to ionize a hydrogen atom and is sometimes denoted by $\mathcal{E}_I$ for ionization energy. Then

$$\mathcal{E}_n = \frac{\mathcal{E}_I}{n^2} \qquad (8\text{-}13)$$

is the expression for the energy of the nth level.

III. *When an electron changes from one orbit to another, a photon is emitted whose energy is equal to the difference in energy between the two orbits.* In symbols

$$h\nu = \mathcal{E}_i - \mathcal{E}_f \qquad (8\text{-}14)$$

where ν is the frequency of the emitted photon. Then

$$\nu = \frac{\mathcal{E}_I}{h}\left(\frac{1}{n_f^2} - \frac{1}{n_i^2}\right) \qquad (8\text{-}15)$$

which exhibits the same general form as Balmer's formula. To compare this result with spectroscopic data it is more convenient to write the equation using wave numbers where

$$\bar\nu = \frac{1}{\lambda} = \frac{\nu}{c} \tag{8-16}$$

to yield

$$\bar\nu = \frac{\mathcal{E}_I}{2\pi\hbar c}\left(\frac{1}{n_f^2} - \frac{1}{n_i^2}\right) \tag{8-17}$$

The student is invited to use $\mathcal{E}_I = 13.58$ eV and to find $\hbar c$ in Appendix I and compare the numerical value so obtained with the value of R_H given in the preceding section. The agreement is sufficient to lend considerable impetus to the Bohr theory.

In this simple model of the hydrogen atom the nucleus is at the center of the atom, whereas the electron may be in any one of the circular orbits characterized by the quantum numbers $n = 1, 2, 3, 4, \ldots$. As long as the electron re-

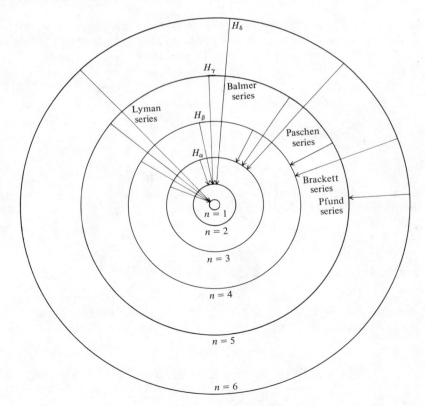

Figure 8-3 Quantum jumps giving rise to the different spectral series of hydrogen.

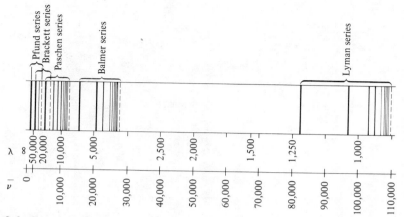

Figure 8-4 Relative positions of the lines of the different spectral series of hydrogen.

mains in its orbit no energy is radiated, but whenever an electron jumps from an outer orbit to an inner orbit energy is radiated in the form of light. A line of the Balmer series corresponds to a jump of the electron from an initial orbit of quantum number n greater than 2 to the final orbit for which $n = 2$ (see Fig. 8-3). The red line or H_α line of the Balmer series corresponds to a transition from orbit of quantum number $n_i = 3$ to orbit of quantum number $n_f = 2$. Any one atom at any instant can emit only one photon of frequency ν, but, since there are always many atoms in any quantity of hydrogen that is examined spectroscopically, there are always other atoms that emit photons of different frequencies. The result is the series of lines actually observed. The relative number of atoms in which the electrons go from a given initial state to a given final state determines the relative intensity of the particular spectral line corresponding to this electron transition.

In addition to the Balmer series, other groups or series of spectral lines of hydrogen have been discovered. The Lyman series lies entirely in the ultraviolet region; the wave numbers of the lines of the Lyman series are given by

$$\bar{\nu} = \frac{\mathcal{E}_I}{2\pi\hbar c}\left(\frac{1}{1^2} - \frac{1}{n_i^2}\right) \qquad n_i = 2, 3, 4, \ldots. \qquad (8\text{-}18)$$

that is, an electronic jump from any outer orbit to the innermost orbit ($n_f = 1$) gives rise to a line of the Lyman series.

Three other series of lines are known for hydrogen; these series are all in the infrared region and are known as the Paschen series for which $n_f = 3$, the Brackett series for which $n_f = 4$, and the Pfund series for which $n_f = 5$. The relative positions of these spectral series are shown in Figure 8-4. A different way of representing the transitions that lead to the spectral series is shown in Figure 8-5, an energy-level diagram similar to those developed in the wave mechanics of Chapter 7.

Transitions from higher to lower energy levels can occur spontaneously.

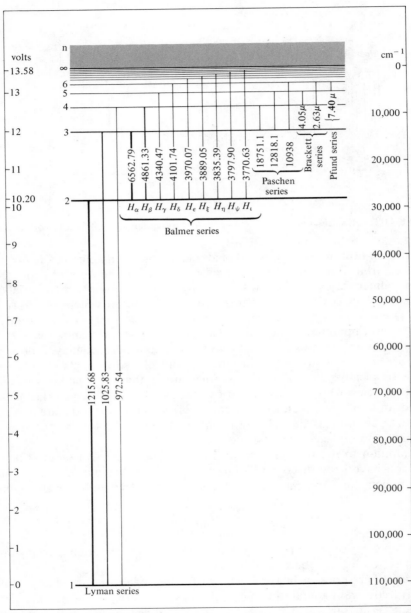

Figure 8-5 Energy-level diagram for hydrogen. Wavelengths of the lines of the Lyman, Balmer, and Paschen series are in angstroms.

In order to get into one of these higher energy states, the atom must be *excited* by some external agency. Such excitation is accomplished by collisions of hydrogen atoms with electrons, photons, or with other hydrogen atoms. It is essential that the bombarding projectile have enough kinetic energy to cause the transition with some energy to spare to ensure conservation of momentum. From Figure 8-5 it is clear that a projectile with an energy of less than 10.2 eV can have only elastic collisions with hydrogen atoms.

A simplified experimental arrangement for producing collisions between electrons and hydrogen atoms is shown in Figure 8-6. The tube contains hydrogen at low pressure. Electrons liberated from the hot filament F are accelerated toward the grid G by a difference of potential V. When they enter the region between the grid G and the plate P, these electrons will have an amount of energy equal to $Ve = \frac{1}{2}mv^2$. If these electrons suffer no energy loss on collision, they will travel to the plate and be registered by the galvanometer. As the voltage is increased, the current to the plate will be increased until the energy of the electrons is just the right amount to raise the hydrogen atom from the normal state to one of its excited states. When this value is reached, there will be a drop in the current to the plate, indicating that many electrons have given up their energy to hydrogen atoms to raise them to an excited state. These low-energy electrons are prevented from reaching the plate P by a small difference of potential of about 0.5 volt maintained between the grid and plate, represented schematically by the battery B_1.

The first decrease in current occurs when the difference of potential between F and P is about 10.2 volts; atoms of hydrogen are thus raised from the normal state $n = 1$ to the first excited state $n = 2$. This value of the potential difference is sometimes called the first *resonance potential*. Atoms in the first excited state for which $n = 2$ can now return to the normal state $n = 1$ with the emission of radiation of wavelength $\lambda = 1215.7$ Å, the most intense line of the Lyman series. This line is in the far ultraviolet region and can be observed only if the discharge tube is made of quartz or has a quartz window or a window of

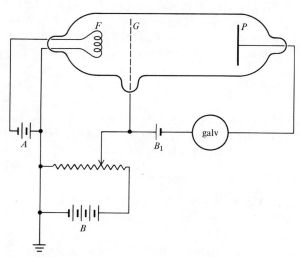

Figure 8-6 Method of determining the ionization potential of hydrogen.

some other substance that will transmit ultraviolet light of this wavelength. A second drop in current, or a second resonance potential, will be observed when the potential difference between F and P is increased to about 12 volts. At this value of the voltage atoms of hydrogen will be raised from the normal state $n = 1$ to the level $n = 3$. These atoms may now return to the normal state, either directly with the emission of ultraviolet light of wavelength 1025.8 Å or by first going from $n = 3$ to $n = 2$ with the emission of light of wavelength 6562.8 Å, the red line of the Balmer series, followed by the transition from $n = 2$ to $n = 1$ with the emission of the Lyman series line. The red line of the Balmer series can be observed coming from the gas in the tube. The higher resonance potentials may also be determined in this manner. As the voltage is increased still further, a point will be reached at which ionization of the gas will set in. The result is a very large increase in the current through the tube. At this stage the light coming from the gas will consist of lines of all the spectral series. At this value of the voltage, known as the *ionization potential,* the electrons from the filament have just sufficient energy to ionize the hydrogen atom—that is, to remove the electron from the lowest orbit of hydrogen $n = 1$ to $n = \infty$ (outside the atom). The numerical value of this ionization potential can be computed from the values of the two energy levels $\mathcal{E}_\infty$ and $\mathcal{E}_1$ and is

$$V = 13.58 \text{ volts} \tag{8-19}$$

in good agreement with the Bohr theory.

The experiment just described is very similar to one performed by Franck and Hertz (1914), who used mercury instead of hydrogen. They were not aware of Bohr's work when they performed their experiments, and their results were widely recognized as independent evidence that electrons in atoms can move only in discrete energy states.

Although De Broglie's hypothesis about the wave nature of the electron came earlier in our discussion (Chapter 6), it followed Bohr's theory historically. The two ideas can be reconciled in an interesting fashion. From De Broglie's theory the wavelength of an electron is given by

$$\lambda = \frac{h}{mv} \tag{8-20}$$

An electron in orbit is to be considered as a standing wave. Such a wave must close on itself, requiring that the circumference of the orbit contain an integral number of wavelengths. For an orbit of radius r

$$2\pi r = n\lambda \tag{8-21}$$

Eliminating λ from Equations (8-20) and (8-21) yields

$$mvr = n\hbar \tag{8-22}$$

which is identical to the content of Bohr's second postulate.

8-3 Motion of the Hydrogen Nucleus

It was mentioned in Section 8-1 that a number of refinements of the semi-classical Bohr theory are possible. Most of them, however, do not aid materially in an understanding of the quantum treatment, but one classical refinement is necessary in this regard. The theory, as presented in the preceding section, assumes that the electron moves in a circle about a fixed nucleus; this assumption cannot be correct unless the nucleus is infinitely massive. For a nucleus of mass M both the nucleus and the electron will rotate about the center of mass of the two-body system. They will have a common angular velocity ω. In Figure 8-7 the electron and the nucleus are shown, along with O, the position of the center of mass; M and m rotate about an axis through O, perpendicular to the paper. If a is the distance of the electron from the axis of rotation and A the distance of the nucleus from the same axis, then from the definition of the center of mass

$$am = AM \tag{8-23}$$

If r is the distance between the electron and the nucleus, then

$$r = a + A \tag{8-24}$$

If we eliminate A from these two equations, we get

$$a = r\frac{M}{M + m} \tag{8-25}$$

If we eliminate a, we get

$$A = r\frac{m}{M + m} \tag{8-26}$$

Further, if v and V are the speeds of the electron and nucleus, respectively, then

$$\begin{aligned} v &= a\omega \\ V &= A\omega \end{aligned} \tag{8-27}$$

The kinetic energy of the system is the sum of the kinetic energies of the two particles separately:

$$\begin{aligned} \mathscr{E}_k &= \tfrac{1}{2}MV^2 + \tfrac{1}{2}mv^2 \\ &= \tfrac{1}{2}MA^2\omega^2 + \tfrac{1}{2}ma^2\omega^2 \end{aligned} \tag{8-28}$$

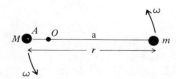

Figure 8-7 Rotation of the nucleus and electron about a common axis through O.

Substituting the values of a and A from Equations (8-25) and (8-26), we can reduce the expression for the kinetic energy of the system to

$$\mathscr{E}_k = \frac{1}{2} \frac{mM}{M+m} r^2 \omega^2 \qquad (8\text{-}29)$$

It is customary to introduce the symbol μ for the product of the two masses divided by their sum and to call μ the *reduced mass:*

$$\mu = \frac{mM}{M+m} = \frac{m}{1+m/M} \qquad (8\text{-}30)$$

Then the kinetic energy becomes

$$\mathscr{E}_k = \tfrac{1}{2} \mu r^2 \omega^2 \qquad (8\text{-}31)$$

a form that differs from the corresponding expression for infinite M in that m is replaced by μ. Examination of Equation (8-30) shows that

$$\lim_{M \to \infty} \mu = m \qquad (8\text{-}32)$$

indicating that Equation (8-31) has the correct limiting behavior.

In a similar manner it can be shown that the potential energy is

$$\mathscr{E}_p = -\frac{kZe^2}{r} = -\mu\omega^2 r^2 \qquad (8\text{-}33)$$

leading to a total energy

$$\mathscr{E} = -\tfrac{1}{2}\mu\omega^2 r^2 = -\frac{kZe^2}{2r} \qquad (8\text{-}34)$$

identical in form with the preceding treatment.

Bohr's Postulate II that the angular momentum is quantized now reads

$$MA^2\omega + ma^2\omega = n\hbar \qquad (8\text{-}35)$$

Again we use Equations (8-25) and (8-26) to obtain

$$\mu r^2 \omega = n\hbar \qquad (8\text{-}36)$$

The algebra now proceeds as in the case of infinite nuclear mass to yield

$$r = \frac{n^2\hbar^2}{\mu kZe^2} \qquad (8\text{-}37)$$

and

$$\mathscr{E} = -\frac{\mu k^2 Z^2 e^4}{2n^2\hbar^2} \qquad (8\text{-}38)$$

It is clear from this last equation that the reduced-mass correction introduces a small numerical change in the value of $\mathscr{E}_I$, the ionization energy of hydrogen.

The change is not important for slide-rule accuracy, but it is not outside the limits of spectroscopic measurement. We can use Bohr's third postulate to obtain

$$\bar{\nu} = \frac{\mu k^2 Z^2 e^4}{2ch\hbar^2}\left(\frac{1}{n_f^2} - \frac{1}{n_i^2}\right)$$

$$= Z^2 R\left(\frac{1}{n_f^2} - \frac{1}{n_i^2}\right) \tag{8-39}$$

leading to the notion that the Rydberg constant also has a small dependence on the nuclear mass. It is customary to denote by R_∞ the value of R when μ is replaced by m:

$$R = R_\infty \frac{1}{1 + m/M}$$

and

$$R_\infty = \frac{mk^2e^4}{2ch\hbar^2}$$

An examination of Equation (8-39) leads to three very interesting conclusions. In the first place, spectral series similar to those of hydrogen should exist for ions which have a hydrogenlike structure—that is, a nucleus of charge $+Ze$ and a single external electron; for example, singly ionized helium He$^+$ of nuclear charge $+2e$ should give a series of spectral lines whose wave numbers are given by the equation

$$\bar{\nu} = 4R\left(\frac{1}{n_f^2} - \frac{1}{n_i^2}\right) \tag{8-40}$$

Except for the small change in the value of the Rydberg constant, due to the difference in the nuclear masses, these wave numbers are four times as large as the wave numbers of the lines in the corresponding series of hydrogen. Such lines have actually been observed in the spectrum of helium.

Other hydrogenlike spectra have been observed for doubly ionized lithium, triply ionized beryllium, on up to multiply ionized oxygen. Table 8-1 lists the values of the Rydberg constant for hydrogen and hydrogenlike ions and shows its dependence on the mass of the nucleus.

A second interesting conclusion that follows from Equation (8-39) is that a knowledge of the Rydberg constant for hydrogen and ionized helium can be used to calculate the ratio of the mass of the proton to the mass of the electron. Using the subscripts H and He for the quantities characteristic of hydrogen and ionized helium, we get

$$\frac{R_H}{R_{He}} = \frac{1 + m/M_{He}}{1 + m/M_H} \tag{8-41}$$

TABLE 8-1   DEPENDENCE OF THE RYDBERG CONSTANT ON THE MASS OF THE
NUCLEUS

Nucleus	Rydberg Constant in cm⁻¹	Nucleus	Rydberg Constant in cm⁻¹
$^{1}_{1}\mathrm{H}$	109,677.581	$^{7}_{3}\mathrm{Li}$	109,728.723
$^{2}_{1}\mathrm{H}$	109,707.419	$^{9}_{4}\mathrm{Be}$	109,730.623
$^{3}_{1}\mathrm{H}$	109,717.348	$^{11}_{5}\mathrm{B}$	109,731.835
$^{3}_{2}\mathrm{He}$	109,717.344	$^{12}_{6}\mathrm{C}$	109,732.286
$^{4}_{2}\mathrm{He}$	109,722.264	$^{14}_{7}\mathrm{N}$	109,733.004
$^{6}_{3}\mathrm{Li}$	109,727.295	$^{16}_{8}\mathrm{O}$	109,733.539

Values taken from *Atomic Energy States*, edited by Charlotte E. Moore, Bureau of Standards, Circular 467.

By substituting the value

$$M_{\mathrm{He}} = 3.9717 M_{\mathrm{H}}$$

obtained from mass spectroscopic data, as well as the values of R_{H} and R_{He}, into the above equation we get

$$\frac{M_{\mathrm{H}}}{m} = 1837$$

in excellent agreement with values determined by other methods.

A third and extremely important conclusion that can be drawn from Equation (8-39) is that even for the same value of Z—that is, for the same type of atom—there should be lines of slightly different wave numbers for the nuclei of different masses. This has led directly to the discovery of the hydrogen isotope of mass number 2, now called *deuterium*. The history of the discovery of deuterium is very interesting. As a result of Aston's very precise measurements of the masses of many isotopes, the relative chemical atomic weights could be computed, taking into consideration the fact that oxygen consisted not only of the isotope of mass number 16 but also of small quantities of mass numbers 17 and 18. The relative chemical atomic weights of hydrogen and oxygen computed by Aston differed by about two parts in 10,000 from that determined by direct physical and chemical methods. Birge and Menzel (1931) suggested that this discrepancy could be explained by assuming the existence of two isotopes of hydrogen, $^{1}\mathrm{H}$ and $^{2}\mathrm{H}$, in the ratio of 4500:1.

Urey, Murphy, and Brickwedde (1932) performed a series of experiments on the spectrum of hydrogen to find the isotope $^{2}\mathrm{H}$. They used a 21-ft concave diffraction grating and photographed the lines of the Balmer series. The dispersion of the instrument was 1.3 Å/mm. They first used ordinary tank hydrogen in the discharge tube and obtained a faint trace of a line slightly displaced from the regular H_{β} line. On the assumption that this faint line was due to the presence of a small quantity of deuterium in the hydrogen, they decided to prepare samples of hydrogen containing larger concentrations of deuterium and thus increase the relative intensity of this line. To accomplish this they took liquid hydrogen, allowed most of it to evaporate, and used the small part that remained. In the process of evaporation the lighter constituent evaporates at a greater rate, leaving the residue with a greater concentration of the heavier constituent. Two different samples were used: (a) the part that remained after the liquid hydrogen evaporated at atmospheric pressure and (b) the part that remained after the liquid hydrogen evaporated at a pressure slightly higher than the triple point pressure. With these samples the intensity of the displaced line was greatly enhanced, showing that they were now much richer in the isotope ^{2}H. The results of their experiment on four lines of the Balmer series are given in Table 8-2.

The discovery of deuterium led rapidly to methods for isolating it in comparatively large quantities, enabling scientists to use it in many different fields of investigation in chemistry and biology as well as in physics. In physics, deuterium and the ionized atom known as the *deuteron* have been of inestimable value in the study of atomic nuclei.

8-4   The Schroedinger Solution of the Hydrogen Atom

The greatest initial success of the Schroedinger theory was the solution of the hydrogen problem without the use of the special postulates introduced by Bohr. Instead, the basic postulates of quantum mechanics described in Section 7-1 are used.

We can write the total kinetic energy for the electron and the nucleus in the form

$$\mathcal{E}_k = \frac{P^2}{2M} + \frac{p^2}{2m} \tag{8-42}$$

TABLE 8-2   SEPARATION OF THE BALMER LINES DUE TO THE TWO ISOTOPES OF HYDROGEN

Spectrum Lines	$^1H_\alpha - {}^2H_\alpha$	$^1H_\beta - {}^2H_\beta$	$^1H_\gamma - {}^2H_\gamma$	$^1H_\delta - {}^2H_\delta$
Calculated	1.793 Å	1.326 Å	1.185 Å	1.119 Å
Observed using ordinary H		1.346	1.206	1.145
Observed using evaporated H(a)		1.330	1.199	1.103
Observed using evaporated H(b)	1.791	1.313	1.176	1.088

where **P** and **p** are the momenta of the nucleus and the electron, respectively. If we denote the respective Cartesian coordinates by (X, Y, Z) and (x, y, z), we see immediately that the Schroedinger equation becomes a partial differential equation in six variables. The method of separation of variables used in Section 7-7 will not work for this choice of coordinates; so we use the concept of reduced mass from the preceding section, replacing the mass of the electron by μ where

$$\mu = \frac{mM}{M + m} \tag{8-43}$$

The Cartesian coordinates of the electron are replaced by spherical polar coordinates (r, θ, ϕ), where r is the distance between the electron and the nucleus and θ, ϕ are angles as defined in Figure 8-8. The Cartesian coordinates of the nucleus are replaced by the Cartesian coordinates of the center of mass, for which the quantum mechanical description is that of a free particle, already solved in Section 7-4. For the rest of this discussion we shall ignore the translatory motion of the center of mass and concentrate on the internal motion of the atom. The kinetic energy in the new coordinate system is

$$\mathcal{E}_k = \frac{p^2}{2\mu} = -\frac{\hbar^2}{2\mu}\left(\frac{\partial^2}{\partial x^2} + \frac{\partial^2}{\partial y^2} + \frac{\partial^2}{\partial z^2}\right) \tag{8-44}$$

$$= -\frac{\hbar^2}{2\mu}\nabla^2$$

It is necessary to write the Laplacian operator ∇^2 in spherical coordinates so that the Schroedinger equation will become separable. The coordinates transform according to the equations

$$x = r \sin\theta \cos\phi$$
$$y = r \sin\theta \sin\phi$$
$$z = r \cos\theta \tag{8-45}$$

as can be seen by inspecting Figure 8-8. The transformation of ∇^2 into spherical coordinates is a long and tedious but not difficult operation; the details can be found in books on partial differential equations or on electromagnetic theory. The result is

$$\nabla^2 = \frac{1}{r^2}\frac{\partial}{\partial r}\left(r^2\frac{\partial}{\partial r}\right) + \frac{1}{r^2\sin\theta}\frac{\partial}{\partial\theta}\left(\sin\theta\frac{\partial}{\partial\theta}\right) + \frac{1}{r^2\sin^2\theta}\frac{\partial^2}{\partial\phi^2} \tag{8-46}$$

The potential energy is given by

$$V = -k\frac{Ze^2}{r} \tag{8-47}$$

and the Schroedinger equation is

$$\nabla^2\psi = -\frac{2\mu}{\hbar^2}\left(\mathcal{E} - V\right)\psi \tag{8-48}$$

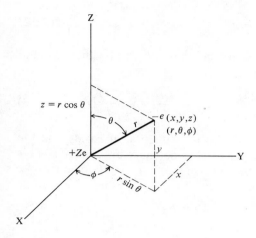

Figure 8-8

where $\mathcal{E}$ is the total energy of the system. Since V is a function of r only, it is possible to use the technique of separation of variables in which it is assumed that $\psi(r,\ \theta,\ \phi)$ is a product of three functions: $R(r)$, a function of r only, $\Theta(\theta)$, a function of θ only, and $\Phi(\phi)$, a function of ϕ only.

$$\psi(r,\ \theta,\ \phi) = R(r)\ \Theta(\theta)\ \Phi(\phi) \tag{8-49}$$

When this product form is substituted into Equation (8-48), the latter can be broken up into three ordinary differential equations. We perform this substitution and divide the equation by ψ to get

$$\frac{1}{r^2 R}\frac{d}{dr}\left(r^2\frac{dR}{dr}\right) + \frac{1}{\Theta r^2 \sin\theta}\frac{d}{d\theta}\left(\sin\theta\frac{d\Theta}{d\theta}\right) + \frac{1}{\Phi r^2 \sin^2\theta}\frac{d^2\Phi}{d\phi^2} = -\frac{2\mu}{\hbar^2}(\mathcal{E}-V)$$
$$\tag{8-50}$$

where we have used ordinary derivatives instead of partial derivatives because each function depends on only one variable.

If we multiply this equation by $r^2 \sin^2\theta$, the third term on the left-hand side will become a function of ϕ only; that is, it becomes

$$\frac{1}{\Phi}\frac{d^2\Phi}{d\phi^2}$$

Transferring this term to the right-hand side and all others to the left, we get

$$\frac{\sin^2\theta}{R}\frac{d}{dr}\left(r^2\frac{dR}{dr}\right) + \frac{\sin\theta}{\Theta}\frac{d}{d\theta}\left(\sin\theta\frac{d\Theta}{d\theta}\right) + \frac{2\mu}{\hbar^2}r^2\sin^2\theta(\mathcal{E}-V) = -\frac{1}{\Phi}\frac{d^2\Phi}{d\phi^2}$$
$$\tag{8-51}$$

Since the left-hand side of this equation depends on r and θ, and the right-hand side is a function of ϕ only, this equation can be satisfied only if each side is equal to a constant; let us call this constant m_l^2. We now have two equations, each one equal to m_l^2. One of these is

$$\frac{1}{\Phi}\frac{d^2\Phi}{d\phi^2} = -m_l{}^2 \tag{8-52}$$

and the other, rearranged so that the terms containing r are on the left-hand side and those containing θ are on the right, is

$$\frac{1}{R}\frac{d}{dr}\left(r^2\frac{dR}{dr}\right) + \frac{2\mu}{\hbar^2}(\mathscr{E} - V) = \frac{m_l{}^2}{\sin^2\theta} - \frac{1}{\Theta\,\sin\,\theta}\frac{d}{d\theta}\left(\sin\,\theta\,\frac{d\Theta}{d\theta}\right) \tag{8-53}$$

Since the left-hand side of Equation (8-53) is a function of r only and the right-hand side does not contain r, again each side must be constant. This time we denote the constant by b. We thus obtain the following two equations:

$$\frac{1}{R}\frac{d}{dr}r^2\frac{dR}{dr} + \frac{2\mu r^2}{\hbar^2}(\mathscr{E} - V) = b \tag{8-54}$$

and

$$\frac{m_l{}^2}{\sin^2\theta} - \frac{1}{\Theta\,\sin\,\theta}\frac{d}{d\theta}\left(\sin\,\theta\,\frac{d\Theta}{d\theta}\right) = b \tag{8-55}$$

The original partial differential equation (8-48) has now been broken up into three ordinary differential equations (8-52), (8-54), and (8-55), one equation for each of the three coordinates. Each equation will result in the introduction of its own quantum number, just as in Section 7-7 in which a three-dimensional problem was considered.

Of these three ordinary differential equations the one involving ϕ, Equation (8-52), is easiest to solve. We write it in the form

$$\frac{d^2\Phi}{d\phi^2} + m_l{}^2\Phi = 0 \tag{8-56}$$

from which it is apparent that a solution is

$$\Phi = \exp\,(im_l\phi) \tag{8-57}$$

where $i = \sqrt{-1}$. Postulate III of Section 7-1 requires that the wave function must be continuous. In order for this condition to be met, the function Φ must be periodic in the angle ϕ with period equal to 2π; otherwise there would be a jump discontinuity at some value of ϕ. The periodic condition is

$$\Phi(\phi + 2\pi) = \Phi(\phi) \tag{8-58}$$

Substituting Equation (8-57) into this condition, we get

$$\exp\,[im_l\,(\phi + 2\pi)] = \exp\,(im_l\phi) \tag{8-59}$$

Since $\exp\,(ix)$ never vanishes for real x, we may divide Equation (8-59) by $\exp\,(im_l\phi)$ to obtain

$$\exp\,(2\pi im_l) = 1 \tag{8-60}$$

Recalling Euler's identity,

$$\cos 2\pi m_l + i \sin 2\pi m_l = 1 \tag{8-61}$$

The real and imaginary parts of this equation must be equalities taken separately. Taking just the real part,

$$\cos 2\pi m_l = 1 \tag{8-62}$$

and therefore

$$m_l = 0 \pm 1, \pm 2, \pm 3, \ldots \tag{8-63}$$

The quantum number m_l has arisen as a natural consequence of the Schroedinger equation instead of being introduced as one of the initial hypotheses.

The solutions of the equations for θ and r are more involved, and will not be solved here in detail. We shall discuss just a few features of their solutions. The θ equation can be arranged to take the form of Legendre's differential equation, which is solved in several of the references listed at the end of this chapter. There exists a continuous infinity of solutions of Legendre's equation, most of which become infinite for $\theta = 0°$ or $\theta = 180°$. The only solutions that remain finite are those for which the separation constant b assumes the form

$$b = l(l + 1) \tag{8-64}$$

where $l \geq 0$ and also $l > |m_l|$.

Equation (8-54) for r is the only one that contains the energy $\mathcal{E}$. Again there is a continuous infinity of solutions. This time most of the solutions become infinite as r approaches infinity. The only ones that remain finite are those for which

$$\mathcal{E} = -\frac{\mu k^2 Z^2 e^4}{2\hbar^2} \cdot \frac{1}{n^2} \tag{8-65}$$

where n is an integer (a quantum number) which must be greater than l. We see that this condition agrees with the result of the Bohr theory in Equation (8-38).

Although we do not propose to solve Equation (8-54) here, we may profitably examine its solution for a special case, namely, the ground state of hydrogen. For the ground state the quantum numbers assume their smallest possible values, which are $m_l = 0$, $l = 0$, and $n = 1$. Since $l = 0$, it follows that $b = 0$, and Equation (8-54) becomes (after multiplying by R)

$$r^2 \frac{d^2 R}{dr^2} + 2r \frac{dR}{dr} - \frac{2\mu r^2}{\hbar^2} (\mathcal{E} - V) = 0 \tag{8-66}$$

We can guess as a solution the form

$$R = \exp\left(-\frac{r}{a}\right) \tag{8-67}$$

When we perform the requisite differentiations and substitute into the differential equation, also including the known form of $V(r)$ from Equation (8-47), we get

$$\frac{r^2}{a^2} - 2\frac{r}{a} + \frac{2\mu r^2}{\hbar^2}\mathcal{E} + \frac{2\mu r}{\hbar^2} \cdot kZe^2 = 0 \tag{8-68}$$

This algebraic equation is really two equations, for, if $\exp(-r/a)$ is to be a solution, the expression (8-68) must not depend on r; this implies that the coefficients of r and of r^2 must vanish separately.

$$\frac{1}{a^2} + \frac{2\mu}{\hbar^2}\mathcal{E} = 0 \tag{8-69a}$$

$$-\frac{2}{a} + \frac{2\mu}{\hbar^2} \cdot kZe^2 = 0 \tag{8-69b}$$

TABLE 8-3   SOME EIGENFUNCTIONS OF HYDROGEN

$$\psi_{100} = \frac{1}{\sqrt{\pi}}\left(\frac{Z}{a_0}\right)^{3/2} \exp\left(\frac{-Zr}{a_0}\right)$$

$$\psi_{200} = \frac{1}{4\sqrt{2\pi}}\left(\frac{Z}{a_0}\right)^{3/2}\left(2 - \frac{Zr}{a_0}\right)\exp\left(\frac{-Zr}{2a_0}\right)$$

$$\psi_{210} = \frac{1}{4\sqrt{2\pi}}\left(\frac{Z}{a_0}\right)^{3/2}\frac{Zr}{a_0}\exp\left(\frac{-Zr}{2a_0}\right)\cos\theta$$

$$\psi_{21\pm1} = \frac{1}{8\sqrt{\pi}}\left(\frac{Z}{a_0}\right)^{3/2}\frac{Zr}{a_0}\exp\left(\frac{-Zr}{2a_0}\right)\sin\theta\exp(\pm i\phi)$$

$$\psi_{300} = \frac{1}{81\sqrt{3\pi}}\left(\frac{Z}{a_0}\right)^{3/2}\left(27 - 18\frac{Zr}{a_0} + 2\frac{Z^2r^2}{a_0^2}\right)\exp\left(\frac{-Zr}{3a_0}\right)$$

$$\psi_{310} = \frac{\sqrt{2}}{81\sqrt{\pi}}\left(\frac{Z}{a_0}\right)^{3/2}\left(6 - \frac{Zr}{a_0}\right)\frac{Zr}{a_0}\exp\left(\frac{-Zr}{3a_0}\right)\cos\theta$$

$$\psi_{31\pm1} = \frac{1}{81\sqrt{\pi}}\left(\frac{Z}{a_0}\right)^{3/2}\left(6 - \frac{Zr}{a_0}\right)\frac{Zr}{a_0}\exp\left(\frac{-Zr}{3a_0}\right)\sin\theta\exp(\pm i\phi)$$

$$\psi_{320} = \frac{1}{81\sqrt{6\pi}}\left(\frac{Z}{a_0}\right)^{3/2}\frac{Z^2r^2}{a_0^2}\exp\left(\frac{-Zr}{3a_0}\right)(3\cos^2\theta - 1)$$

$$\psi_{32\pm1} = \frac{1}{81\sqrt{\pi}}\left(\frac{Z}{a_0}\right)^{3/2}\frac{Z^2r^2}{a_0^2}\exp\left(\frac{-Zr}{3a_0}\right)\sin\theta\cos\theta\exp(\pm i\phi)$$

$$\psi_{32\pm2} = \frac{1}{162\sqrt{\pi}}\left(\frac{Z}{a_0}\right)^{3/2}\frac{Z^2r^2}{a_0}\exp\left(\frac{-Zr}{3a_0}\right)\sin^2\theta\exp(\pm 2i\phi)$$

$$a_0 = \frac{\hbar^2}{\mu ke^2}$$

Solving Equation (8-69b) for a, we get

$$a = \frac{\hbar^2}{\mu k Z e^2} \tag{8-70}$$

a result that agrees with the first Bohr radius, as in Equation (8-37). If this value of a is substituted into Equation (8-69a), the energy of the ground state is obtained in agreement with previous results. We see that the Bohr radius has reappeared in the Schroedinger formalism, but with a difference in meaning. In the Bohr theory the electron was always separated from the nucleus by the same radius. In the quantum-mechanical theory the electron is by no means so localized, but the Bohr radius is a measure of how far from the nucleus the electron will be on the average. The value of this radius for $Z = 1$, hydrogen, is often denoted by a_0:

$$a_o = \frac{\hbar^2}{\mu k e^2} = 0.529 \times 10^{-10}\, \text{m} \tag{8-71}$$

The solutions of the equations for r, θ, and ϕ, when calculated in detail, can be multiplied to obtain the solution of the original partial differential equation

$$\psi_{nlm_l} = N_{nlm_l}\, R(r)\, \Theta(\theta)\, \Phi(\phi) \tag{8-72}$$

The quantity N is a normalization factor, different for each set of quantum numbers, but constant in r, θ, and ϕ. It is evaluated by using the condition

$$\int\int\int \psi^*_{nlm_l}\, \psi_{nlm_l}\, dv = 1 \tag{8-73}$$

in agreement with Postulate IV of Chapter 7. The triple integral is taken over all space by using the element of volume in spherical coordinates

$$dv = r^2 \sin\theta\, dr\, d\theta\, d\phi \tag{8-74}$$

with the limits of ϕ from 0 to 2π, those for θ from 0 to π, and those for r from 0 to ∞. Table 8-3 is a list containing some of the wave functions for low-lying states of the hydrogen atom, including the normalization factors.

8-5 Interpretation of the Schroedinger Solution

The outstanding feature of the Schroedinger solution of the hydrogen atom is the appearance of the three quantum numbers n, l, and m_l. They must satisfy the rules we mentioned in the preceding section and which we summarize here, each quantum number being accompanied by the variable that led to its appearance,

$$\begin{array}{ll} r: & n = 1, 2, 3, 4, 5, \ldots \\ \theta: & l = 0, 1, 2, \ldots, (n-1) \\ \phi: & m_l = 0, \pm 1, \pm 2, \ldots, \pm l \end{array} \tag{8-75}$$

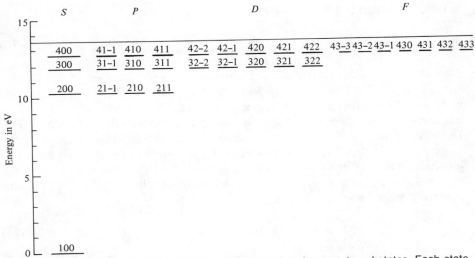

Figure 8-9 Energy-level diagram for hydrogen showing degenerate substates. Each state is denoted by the values of n, l, m_l.

As we noted in the Section 8-4, the equations for θ and ϕ do not contain the energy $\mathcal{E}$, and we saw that $\mathcal{E}$ is determined entirely by the value of n, the *principal quantum number*. For each value of n that is larger than unity there are several permissible sets of values of l and m_l, each set defining a different quantum state, for all of which the energy is the same. The energy-level diagram for hydrogen is therefore more complicated than shown in Figure 8-5. In Figure 8-9 we show a diagram that includes these *degenerate* states—that is, states with different values of the quantum numbers but with the same energy. It should be noted that the appearance of degeneracy in two out of three quantum numbers for a problem in spherical coordinates is unique to the hydrogen problem, being a peculiarity of the $1/r$ potential law.

The interpretation of the quantum numbers l and m_l is provided by the theory of angular momentum in quantum mechanics. This theory is one of the most fruitful and highly developed theories in the field of modern physics; we cannot present a full treatment of the theory here, but we shall present enough to provide an understanding of the connection between angular momentum and the numbers l, m_l.

Without any knowledge of quantized angular momentum, we can already see that it is reasonable to consider l to be the magnitude of a vector whose component along a fixed axis is m_l. This concept is suggested by the fact that l cannot have negative values and m_l takes on values between $-l$ and $+l$. This vector interpretation can be verified by looking at expectation values of angular momentum operators.

In classical mechanics the definition of the angular momentum $\mathbf{L}$ of a particle is the vector product of its displacement $\mathbf{r}$ from an axis and its linear momentum $\mathbf{p} = m\mathbf{v}$; that is,

$$\mathbf{L} = \mathbf{r} \times \mathbf{p} \tag{8-76}$$

The vector $\mathbf{r}$ has components x, y, z which are represented by themselves as quantum-mechanical operators (see Postulate II of Section 7-1). The momentum $\mathbf{p}$ has components p_x, p_y, p_z which have the operator forms $(\hbar/i)(\partial/\partial x)$. $(\hbar/i)(\partial/\partial y)$, $(\hbar/i)(\partial/\partial z)$, respectively. The vector $\mathbf{L}$ has components that can be written as follows:

$$\begin{aligned}
L_x &= yp_z - zp_y \\
L_y &= zp_x - xp_z \\
L_z &= xp_y - yp_x
\end{aligned} \tag{8-77}$$

In operator form these equations become

$$L_{x\ \text{op}} = \frac{\hbar}{i}\left(y\frac{\partial}{\partial z} - z\frac{\partial}{\partial y}\right)$$

$$L_{y\ \text{op}} = \frac{\hbar}{i}\left(z\frac{\partial}{\partial x} - x\frac{\partial}{\partial z}\right) \tag{8-78}$$

$$L_{z\ \text{op}} = \frac{\hbar}{i}\left(x\frac{\partial}{\partial y} - y\frac{\partial}{\partial x}\right)$$

These operators can be applied twice to a function and the results can be added to yield

$$L_{\text{op}}^2 = L_{x\ \text{op}}^2 + L_{y\ \text{op}}^2 + L_{z\ \text{op}}^2 \tag{8-79}$$

In order to calculate the expectation value of L_{op}^2 we need to evaluate the integral

$$<L^2> = \iiint \psi_{nlm_l}^* L_{\text{op}}^2 \psi_{nlm_l} \, dv \tag{8-80}$$

There is a choice of coordinate systems to use in this integration. The algebra is somewhat easier in spherical coordinates, but there is a conceptual difficulty involved because the element of volume dv depends on r and θ and it is not obvious how to handle the differentiation implied in the operator. This difficulty is removed in Cartesian coordinates, where $dv = dx\,dy\,dz$, a constant. The additional algebra in Cartesian coordinates consists in the necessity of writing the wave function in terms of x, y, and z, in which form the differentiation is lengthier. We shall not perform the integration here, but we state the result, which the student is urged to check for specific choices of n, l, and m_l.

$$<L^2> = l(l+1)\,\hbar^2 \tag{8-81}$$

This result is somewhat surprising in view of the vector interpretation of l we suggested. We might have expected $l^2\hbar^2$ as the result of this calculation; the fact that we did not get the expected answer is a peculiarity of quantum mechanics and cannot be interpreted in any classical way.

It is an easier job to calculate $<L_z>$ than it was to calculate $<L^2>$; the reason is that $L_{z\ \mathrm{op}}$ has a very simple form in spherical coordinates.

$$L_{z\ \mathrm{op}} = \frac{\hbar}{i} \frac{\partial}{\partial \phi} \tag{8-82}$$

Therefore the element of volume $r^2 \sin \theta \, dr \, d\theta \, d\phi$ is constant with respect to this operator and the algebra is rather simple. The result is

$$<L_z> = m_l \hbar \tag{8-83}$$

We see from these two results [Equations (8-81) and (8-83)] that the identification of l and m_l with a vector and its component (along the z axis) is reasonable as long as we keep in mind that quantum mechanics leads to an unusual value for the length of the vector. As a consequence of this identification, the quantum number l is called the *orbital angular momentum* quantum number; m_l is variously referred to as the *azimuthal* quantum number, the *magnetic* quantum number, or the *component of orbital angular momentum*.

We conclude this section by pointing out what is perhaps the most drastic physical difference between the Bohr theory and the Schroedinger theory. In the Bohr theory the principle quantum number n determines both the energy and the orbital angular momentum. The ground state in this theory has angular momentum equal to $\hbar$. In the Schroedinger theory the energy and angular momentum are associated with separate quantum numbers and the angular momentum vanishes for the ground state (and for many of the excited states as well). From the classical point of view it is strange that the angular momentum of the electron can be zero; however, from the idea that the electron is a sort of probability cloud it seems more reasonable. We shall discuss this idea in more detail in Section 8-7.

8-6 Selection Rules for Hydrogen

Whenever an electron is in one of the excited states of an atom there is always the possibility that it can undergo a change of its state into one with lower energy. This transition is usually accompanied by the emission of a photon. As discussed in Section 7-8, the probability of such transitions is governed by selection rules that can be calculated. In this case the appropriate dipole operator is proportional to the product $r\Theta_1\Phi$, or r multiplied by the angular part of any hydrogen wave function for which $l = 1$. We shall call this operator Y_1. From Table 8-3 we see that the choices are

$$Y_1 = \begin{cases} r \cos \theta \\ r \sin \theta \exp{(\pm i\phi)} \end{cases} \tag{8-84}$$

The matrix element for a transition from an initial state with quantum numbers n', l', m_l' to a final state n, l, m_l is given by

$$M_{if} = \int\int\int \psi_{n'l'm'_l} Y_1 \psi_{nlm_l} \, dv$$

If $M_{if} = 0$ for all three choices of Y_1, the transition is *forbidden*. If any choice of Y_1 yields $M_{if} \neq 0$, the transition is *allowed*. A general calculation is tedious, but it has been performed, and the result is the set of dipole selection rules

$$\Delta l = \pm 1 \tag{8-85a}$$

$$\Delta m_l = 0 \quad \text{or} \quad \pm 1 \tag{8-85b}$$

with no restrictions on the changes in the principal quantum number n. These selection rules do not imply that they represent the only transitions permitted; they are merely the most favored ones. Changes in the orbital angular momentum such as $\Delta l = \pm 2$ are possible but much less likely to occur than those for which $\Delta l = \pm 1$.

8-7   Electronic Orbits of Hydrogen

In addition to any analytical treatment of a problem such as that of the hydrogen atom, it is always useful and very satisfying to have either a picture or a mechanical model as an aid to understanding the problem. Such a picture or model is also very helpful in discussing more complex problems that do not lend themselves readily to analytical treatment; in this case, for example, to atoms with more than one electron. The concept of electronic orbits introduced by Bohr is still an exceedingly useful one, even though the results of the Schroedinger solution of the problem lead to a hazier or cloudier picture of such orbits; for example we shall sometimes talk about the Bohr orbit of a muonic atom, a pionic atom, or a kaonic atom, where the negatively charged particle that is moving around the nucleus of an atom is one that was not known in the days of Bohr and Schroedinger (see Section 18-3).

Instead of assuming some particular form for the orbit of an electron of a hydrogen atom, the usual procedure is to determine the probability $P(r)\,dr$ of an electron having a radial coordinate whose value lies between r and $r + dr$. As shown in Chapter 7, using Postulate IV and Equation (7-4), the probability of finding an electron in the volume element dv is

$$\text{prob} = \psi^*_{nlm_l} \psi_{nlm_l} \, dv$$

with

$$dv = r^2 \sin\theta \, dr \, d\theta \, d\phi \quad \text{and} \quad \psi_{nlm_l} = R \cdot \Theta \cdot \Phi \tag{8-86}$$

The integrals over θ and ϕ are normalized to unity so that the radial probability $P(r)\,dr$ becomes

$$P(r)\,dr = r^2 R^* R \, dr \tag{8-87}$$

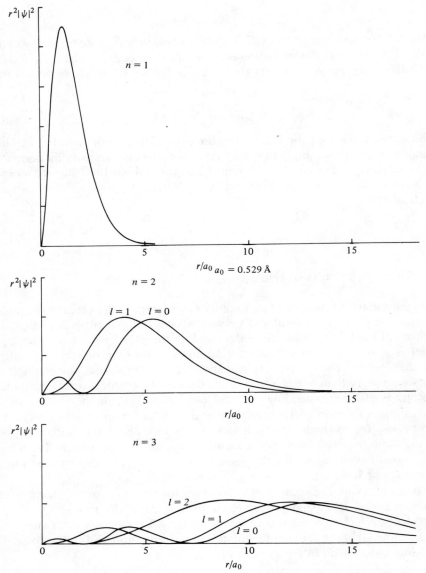

Figure 8-10 Curves showing the probability of finding the electron in the hydrogen atom within a spherical shell of radius r and thickness dr for some values of n and l.

Figure 8-10 is a graph of

$$P_{nl}(r) = r^2 \, |R|^2$$

the radial probability density as a function of the radial distance from the nucleus in units of a_0 for a few values of n and l.

A glance at the graph of Figure 8-10 shows, for example, that when $n = 1$ and $l = 0$ the electron may be at the nucleus or anywhere with a radial distance of $5a_0$ from it, with the maximum probability of being found at a_0, the radius of the first Bohr orbit. For $n = 2$ and $l = 1$, a set of quantum numbers corresponding to the second circular Bohr orbit, the probability density has a maxi-

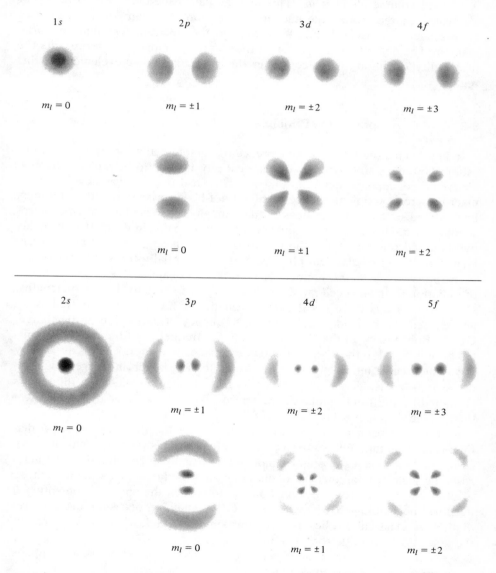

Figure 8-11   Pictorial representation of density distribution of electronic charge, or electronic cloud, for selected states of the hydrogen atom. (Adapted from R. Leighton, *Principles of Modern Physics*, McGraw-Hill, New York, 1959.)

mum at $4a_0$ with a wide range of values of r extending out to about $14a_0$. For $n = 2$ and $l = 0$ the curve shows two maxima with a zero probability at a distance of $2a_0$.

If the value for the probability is multiplied by the electronic charge, the quantity $e\psi^*\psi \, dv$ will represent the probability of finding the electronic charge in a small volume element dv. This may be called the density distribution of electronic charge around the nucleus of the atom. We can construct a set of figures, such as those in Figure 8-11, to show the electronic cloud in a pictorial manner. The density at any region in the electronic cloud then represents the probability of finding the electron in the particular volume element in this region.

8-8   The Correspondence Principle

It has been noted in this and preceding chapters that some of the well-established ideas and concepts of classical physics had to be replaced by new ideas and concepts of quantum physics. Classical physics was developed for macroscopic phenomena; hence it should not be surprising to find that it must be modified severely by the ideas of quantum physics in dealing with the microscopic world. However, there must be a realm or region in which the two fields overlap, and for this region the results of quantum physics must be in agreement with those of classical physics. This is essentially the statement of the *correspondence principle* enunciated by Bohr; for example, the classical theory of radiation from an accelerated charge is known to hold for long wavelengths. One result of this theory is that the frequency of the radiation emitted by an accelerated charge is identical with the frequency of oscillation of the charge. On the Bohr theory of the hydrogen atom, the frequency of the radiation emitted by it is equal to the energy difference between two energy levels divided by the Planck constant h. However, as we go toward larger and larger quantum numbers—that is, as $n \to \infty$ —or as we go to longer wavelengths, the results predicted by the Bohr theory should approach those predicted by the classical theory (see Problem 8-1).

It will have been noted that the appearance of the Planck constant $\hbar$ distinguishes quantum physics from classical physics. This is a very small quantity compared with values of angular momentum usually encountered on the macroscopic scale. The predictions of quantum physics should be the same as those of classical physics for those phenomena for which the angular momentum $\hbar$ becomes negligible—that is, when $\hbar \to 0$. We shall, on occasion, use the predictions of classical physics as first approximations and, when known, present the appropriate results from quantum theory.

We include here a few examples of the correspondence principle as it applies to systems we have already studied.

In the Schroedinger treatment of hydrogen we have seen that the magnitude squared of the angular momentum is $l(l+1)\hbar^2$. If we consider very large values

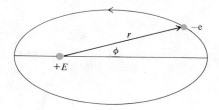

Figure 8-12 Elliptic orbit of the electron in the hydrogen atom.

of l, we see that $l^2 \gg l$ so that the correct form becomes indistinguishable from the semiclassical form for the angular momentum.

For a particle in an infinite square well, the probability density is sinusoidal (see Section 7-2). As the quantum number becomes larger, the probability density behaves more and more like a uniform distribution, which would agree with the classical idea that a particle moving in a box could be found anywhere in the box with equal probability (see Problem 7-23).

As a final example, we recall the problem of barrier penetration from Section 7-9. In the quantum solution an incident wave is partially reflected and partially transmitted. In classical mechanics the probability of transmission is either zero or one, depending on the energy of the particle. But these are just the limits of the transmission probability for low energy or high energy, respectively, in the quantum case.

The student is encouraged to think of other examples to which the correspondence principle applies. It is also fruitful to compare it with the special theory of relativity for which the formulas must agree with the Newtonian forms when $v \ll c$.

8-9 Elliptic Orbits for Hydrogen

Bohr's original theory, which dealt only with circular orbits, was extended by Sommerfeld to include elliptic orbits. To accomplish this, Sommerfeld generalized Bohr's first postulate for the determination of the permissible orbits to read

$$\oint p_i \, dq_i = n_i h \tag{8-88}$$

where q_i is a coordinate which varies periodically, p_i is the corresponding value of the momentum, and n_i is an integer. The symbol $\oint$ means that the integration is taken over a whole period of motion. In the case of circular orbits, there is only one coordinate that varies periodically, namely, the angle ϕ that the radius vector makes with the x axis. In the case of elliptic motion, not only does the angle ϕ vary but the length of the radius vector r also varies periodically, as shown in Figure 8-12. The elliptic orbits will therefore be determined by the two quantum conditions

$$\oint p_\phi \, d\phi = n_\phi h \tag{8-89}$$

$$\oint p_r \, dr = n_r h \tag{8-90}$$

where n_ϕ is called the *angular* or *azimuthal* quantum number and n_r is called the *radial* quantum number. Let the origin of coordinates be taken at the nucleus, which will be considered fixed, and let the mass of the electron be constant; that is, neglect the relativity variation of mass with velocity. The first integral can be evaluated very easily, since the momentum p_ϕ corresponding to the coordinate ϕ is merely the angular momentum p of the electron in the elliptic orbit, and this, from Kepler's law, is a constant [see Appendix V-3, Equation (6)]. Integrating Equation (8-89) over one period, from 0 to 2π, yields

$$\oint_0^{2\pi} p_\phi \, d\phi = n_\phi h$$

or

$$p_\phi = p = \frac{n_\phi h}{2\pi} \tag{8-91}$$

that is, the angular momentum is always an integral multiple of $\hbar$.

The second integral, when evaluated (see Appendix V-7), yields the equation

$$n_r h = \frac{2\pi p}{(1 - \epsilon^2)^{1/2}} - 2\pi p \tag{8-92}$$

where ϵ is the eccentricity of the ellipse. Substituting the value for p from Equation (8-91) yields

$$n_r h = \frac{n_\phi h}{(1 - \epsilon^2)^{1/2}} - n_\phi h$$

or

$$n_r + n_\phi = \frac{n_\phi}{(1 - \epsilon^2)^{1/2}}$$

If we set

$$n = n_r + n_\phi \tag{8-93}$$

then

$$1 - \epsilon^2 = \frac{n_\phi^2}{n^2} \tag{8-94}$$

n is called the *principal* quantum number. The total energy of the electron in the elliptic orbit depends only on the length of its semimajor axis (see Appendix V-7) and is given by

$$\mathcal{E} = -\frac{kZe^2}{2a} \tag{8-95}$$

The total energy can also be expressed in terms of the eccentricity (see Appendix V-7), as follows:

$$\mathcal{E} = -\frac{mk^2Z^2e^4(1 - \epsilon^2)}{2p^2} \tag{8-96}$$

Substituting the values for ϵ and p from Equations (8-94) and (8-91), respectively, we get

$$\mathcal{E} = -\frac{2\pi^2mke^4Z^2}{n^2h^2} \tag{8-97}$$

which is identical with the expression for the energy of the electron in a circular orbit of quantum number n. The introduction of elliptic orbits does not result in the production of new energy terms; hence no new spectral lines are to be expected because of this *multiplicity* of orbits. However, since the spectral lines of hydrogen do show fine structure when examined with instruments of high resolving power, its explanation must lie elsewhere. It was not satisfactorily explained until the introduction of the concept of *electron spin* in 1925 (see Section 9-4).

It is interesting to determine the possible electronic orbits for any given principal quantum number n. The length of the semimajor axis a is obtained from Equations (8-95) and (8-97):

$$a = n^2 \frac{h^2}{4\pi^2mke^2Z} = n^2\frac{a_0}{Z} \tag{8-98}$$

whereas the semiminor axis is given by

$$b = a(1 - \epsilon^2)^{1/2} \tag{8-99}$$

so that

$$b = nn_\phi\frac{A_0}{Z} \tag{8-100}$$

where

$$a_0 = \frac{h^2}{4\pi^2mke^2} = 0.529 \times 10^{-8} \text{ cm}$$

is the radius of the first Bohr orbit. These equations show that the length of the semimajor axis is determined solely by the principal quantum number n, whereas the length of the semiminor axis depends on the azimuthal quantum number n_ϕ as well as n. For the first orbit corresponding to the lowest energy

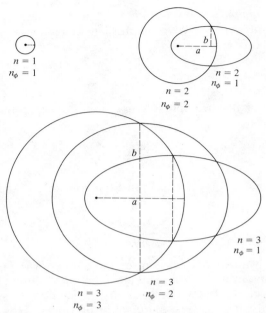

$n = 1$
$n_\phi = 1$

$n = 2$
$n_\phi = 1$

$n = 2$
$n_\phi = 2$

$n = 3$
$n_\phi = 1$

$n = 3$
$n_\phi = 2$

$n = 3$
$n_\phi = 3$

Figure 8-13 Possible electronic orbits for a given total quantum number n.

level or the normal state of hydrogen, the principal quantum number $n = 1$. Since the sum of n_r and n_ϕ must be unity and each must be an integer, when $n_r = 0$, $n_\phi = 1$, and when $n_r = 1$, $n_\phi = 0$. On the basis of this theory of the structure of the atom it was decided that n_ϕ can never be zero, since that would mean that the ellipse would be reduced to a straight line and that the electron would have to pass through the nucleus twice during every period. The smallest possible value for n_ϕ is thus always unity. With $n = n_\phi = 1$, the first orbit is a circle identical with the first Bohr orbit. With $n = 2$, n_ϕ may have the values 1 or 2, so that there are two possible orbits for $n = 2$, a circle and an ellipse. Similarly there are three possible orbits for $n = 3$, a circle and two ellipses, as shown in Figure 8-13. For ionized helium, $Z = 2$, the orbits are similar but the radius of the first orbit is $a_0/2$. The orbits for the other hydrogenlike atoms can be constructed in the same manner.

It may at first sight appear strange that with the introduction of two quantizing conditions instead of one no new energy levels and no new spectral lines are predicted. An examination of these two conditions shows, however, that both have exactly the same periodicity—that is, as the angle ϕ goes from 0 to 2π, the radius vector r goes from its maximum value through the minimum value and back to its maximum value. A mathematical examination of multiply periodic systems shows that whenever the ratio of two of the periods of such a system is a rational number, the two quantum conditions degenerate into a single quantum condition. But if the ratio of the two periods is an irrational number—that is, if the two periods are incommensurable—then there will be

two independent quantum conditions. In general, there will be as many independent quantum conditions of the form

$$\oint p_i \, dq_i = n_i h \qquad\qquad (8\text{-}88)$$

as there are incommensurable periods in the motion. In such cases the system
is referred to as a *nondegenerate system*. One method for removing this
degeneracy, in the case of the elliptic motion of the electron in hydrogen,
is to take into consideration the relativity change of momentum as the velocity
of the electron in its orbit changes. Sommerfeld has carried out this calculation
and has shown that the path of an electron is a rosette, as shown in Figure 8-14,
which may be considered as an ellipse whose major axis precesses slowly
in the plane of the ellipse about an axis through one of the foci. The equation
of the path of the electron is

$$\frac{1}{r} = \frac{1 + \epsilon \cos \psi\phi}{a(1 - \epsilon^2)}$$

which differs from that of an ellipse in that the angle ϕ is replaced by the
angle $\psi\phi$, where ψ is a number less than unity and is given by

$$\psi^2 = 1 - \frac{k^2 Z^2 e^4}{c^2 p^2}$$

When the angle ϕ is increased by 2π, r does not return to its original value

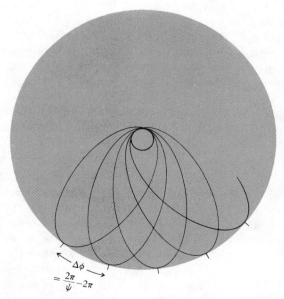

Figure 8-14   Rosette figure of the path of an electron in hydrogen when the relativity correction is taken into consideration.

but reaches it only after the angle $\psi\phi$ has been increased by 2π or when ϕ has been increased by the angle $2\pi/\psi$. Hence the radius vector r returns to its original value only after the axis of the ellipse has precessed through an angle

$$\Delta\phi = \frac{2\pi}{\psi} - 2\pi$$

The effect of the relativity correction on the expression for the total energy of the orbit is to introduce an additional term in Equation (8-97). This additional term is

$$\Delta\mathcal{E} = -\frac{2\pi^2 mk^2 e^4}{h^2} Z^4 \alpha^2 \left(\frac{n}{n_\phi} - \frac{3}{4}\right)\frac{1}{n^4} \tag{8-101}$$

where

$$\alpha = \frac{2\pi ke^2}{ch} = 7.297 \times 10^{-3} \doteq \frac{1}{137} \tag{8-102}$$

is known as the Sommerfeld *fine structure constant*. This term shows that the energy does depend on the azimuthal quantum number n_ϕ which has the effect of splitting the energy level into n terms very close together. The energy of the first orbit, $n = 1$, can have only one possible value since $n_\phi = 1$, or there is only one energy level for $n = 1$. For principal quantum number $n = 2$ the energy can have two possible values corresponding to the two values of n_ϕ, 1 and 2—that is, there are two energy levels for $n = 2$. Similarly there are three possible energy values or energy levels for $n = 3$, and so on. For the first line of the Balmer series, corresponding to the change in the principal quantum numbers $n_i = 3$ to $n_f = 2$, there are six possible transitions for the different values of n_ϕ. This means that, with a spectroscope of very high resolving power, the H_α line should appear to consist of six lines very close together. Actually the H_α line has fewer components. To make experiment and theory agree, some of the transitions have to be ruled out by some *selection* principle. The selection principle chosen is that the azimuthal quantum number n_ϕ can change only by $+1$ or -1 or, expressed in mathematical form,

$$\Delta n_\phi = \pm 1 \tag{8-103}$$

The application of this selection rule to the Balmer series shows that each line should consist of three components; similarly, each line of the Paschen series should consist of five components, whereas the lines of the Lyman series should all be single lines. The fine structure of the Balmer lines and some of the lines of ionized helium have been carefully studied but the agreement with the theoretical predictions is not very good. Most of these discrepancies were later removed by the introduction of the hypothesis of electron spin (see Chapter 9). Further discussion of the fine structure of spectral lines will be postponed until the subject of electron spin has been considered.

Problems

8-1.   Discuss the relationship between the frequency of rotation of the electron in the circular Bohr orbits and the frequency of the radiation emitted when the quantum number changes by unity. Show that for very large quantum numbers the frequency of rotation of the electron and the frequency of the radiation emitted when the quantum number changes by unity approaches the value $\nu = 2cR/n^3$.

8-2.   Using the data from Table 8-1, plot a curve showing the dependence of Rydberg's constant on the nuclear mass. Determine the value of R_∞ from this curve.

8-3.   Hydrogen of mass number 3 sufficient for spectroscopic examination is put into a tube containing ordinary hydrogen. Determine the separation of the H_α lines that should be observed.

8-4.   (a) Draw an energy level diagram for the energy levels characterized by the principal quantum numbers $n = 2$ and $n = 3$, taking the relativity corrections into consideration. Show the transitions permitted by the selection rule. (b) Draw a diagram showing the positions of these lines relative to the position of the line predicted without relativity correction.

8-5.   Ultraviolet light of 800 Å is incident on hydrogen in a quartz tube. Calculate the kinetic energies with which electrons will be ejected from the hydrogen atoms. Express the results in electron volts.

8-6.   Calculate the limit of the Paschen series in wave numbers, angstrom units, and electron volts.

8-7.   In an electron tube containing hydrogen, such as that shown in Figure 8-6, the maximum kinetic energy of the electron is 13.0 eV. Determine which lines of the hydrogen spectrum will be emitted by the gas.

8-8.   Show that, in the nonrelativistic case, the electronic orbits are circular if $n = n_\phi$.

8-9.   Determine the wavelengths of the first two spectral lines of singly ionized helium that correspond, or are the analogs, of the first two lines (a) of the Lyman series and (b) the Balmer series. In what region of the spectrum, visible or ultraviolet, will these lines be found?

8-10.   Determine the ratio of the electrostatic force of attraction to the gravitational force of attraction between the proton and electron of hydrogen. The value of the universal constant of gravitation is $G = 6.68 \times 10^{-8}$ dyne cm²/gm².

8-11. Determine (a) the first resonance potential and (b) the ionization potential of singly ionized helium.

8-12. Suppose a new kind of hydrogen were formed in which the electron is replaced by an object just like it except for its mass, which is about 212 times greater than the mass of the electron. What would be the value of the first Bohr radius for such an atom?

8-13. (a) Calculate v/c numerically for the first Bohr orbit of hydrogen. (b) Calculate its angular velocity ω. (c) Compare the frequency of this circular motion with the frequency of the Balmer line for which $n = 1$.

8-14. Calculate the numerical value of the ratio $2\pi r_1/\lambda_c$, where λ_c is the Compton wavelength of the electron.

8-15. Calculate the De Broglie wavelength of an electron which is just able to cause a transition from a state with $n = 2$ to a state with $n = 3$ in hydrogen.

8-16. (a) Write the Schroedinger equation for a particle of mass m which is confined in a spherical container of radius R. (b) Show that the ground-state wave function is $\psi = (1/r) \sin kr$. (c) What is k?

8-17. Draw an energy-level diagram for the hydrogen atom, including all states with $n \leqq 5$. Include the substates in l and m_l. Show all the allowed transitions that connect pairs of states in your diagram.

8-18. Choose several hydrogenic wave functions and determine their behavior under parity $(x, y, z \rightarrow -x, -y, -z)$. Your results should show that the parity operator is equivalent to multiplication by $(-1)^l$.

8-19. Calculate $<L^2>$ and $<L_z>$ for the state ψ_{322}.

8-20. Calculate $<L^2>$ and $<L_z>$ for the state ψ_{200}.

8-21. Show that the transition $\psi_{210} \rightarrow \psi_{100}$ is allowed.

8-22. Show that the transition $\psi_{320} \rightarrow \psi_{100}$ is forbidden.

8-23. Show that the transition $\psi_{200} \rightarrow \psi_{100}$ is forbidden.

8-24. Calculate $<r>$ for the ground state of the hydrogen atom.

8-25. Assuming that an accelerated charge e radiates energy at the rate R given by

$$R = \frac{2}{3}\frac{ke^2a^2}{c^3} = \frac{d\mathcal{E}}{dt}$$

where a is its acceleration, (a) show that the electron of a hydrogen atom rotating with an angular velocity ω given by

$$\omega^2 = \frac{ke^2}{mr^3}$$

will spiral in toward the nucleus. (b) Show that the rate at which the radius decreases with time is given by

$$\frac{dr}{dt} = -\frac{4}{3}\frac{k^2e^4}{c^3m^2} \cdot \frac{1}{r^2}$$

(c) Calculate the time taken for the electron to spiral into the nucleus, starting from an orbit for which $r = 2 \times 10^{-8}$ cm.

References

Beard, D. B., and G. B. Beard, *Quantum Mechanics with Applications*. Boston: Allyn and Bacon, 1970, Chapter VIII.

Feynman, R. P., R. B. Leighton, and M. Sands, *The Feynman Lectures on Physics*, Vol. III. Reading, Mass.: Addison-Wesley Publishing Company, 1965, Chapter 19.

Livesey, D. K., *Atomic and Nuclear Physics*. Waltham, Mass.: Blaisdell Publishing Company, 1966, Chapter 7.

Pauling, L., and E. B. Wilson, *Introduction to Quantum Mechanics*. New York: McGraw-Hill Book Company, 1935.

Series, G. W., *The Spectrum of Atomic Hydrogen*. London: Oxford University Press, 1957.

Taylor, B. N., W. H. Parker, and D. N. Langenburg, *Rev. Mod. Phys.*, **41**, 375 (1969)

Tipler, P. A., *Foundations of Modern Physics*. New York: Worth Publishers, 1969, Chapter 7.

Tralli, N., and F. R. Pomilla, *Atomic Theory, An Introduction to Wave Mechanics*. New York: McGraw-Hill Book Company, 1969, Chapter 5.

9 | Optical Spectra and Electronic Structure

9-1 Introduction

The study of atomic spectra has yielded invaluable information concerning the arrangement and distribution of the electrons within the atom. Most of the principles and rules used in spectroscopy have been obtained empirically, but with the development of wave mechanics many of them have been placed on a good theoretical foundation. One of the most important of these principles is Bohr's frequency condition, which states that the frequency of any line of the spectrum is proportional to the difference between the values of the energies of two states of the atom emitting the radiation; that is,

$$\nu = \frac{\mathcal{E}_i - \mathcal{E}_f}{h} \tag{9-1}$$

where ν is the frequency of the emitted radiation, $\mathcal{E}_i$ is the energy of the initial state of the atom, $\mathcal{E}_f$ is the energy of the final state of the atom, and h is the Planck constant. Expressed in terms of the corresponding wave number $\bar{\nu}$, this equation becomes

$$\bar{\nu} = \frac{\mathcal{E}_i}{ch} - \frac{\mathcal{E}_f}{ch} \tag{9-2}$$

where c is the velocity of light. Equation (9-2) shows that the wave number of any spectral line can be expressed as the difference between two terms:

$$\bar{\nu} = T_i - T_f \tag{9-3}$$

where each term T, expressed in wave numbers, represents an atomic energy state or energy level.

Atomic spectra can be grouped in two general classifications: (1) optical spectra and (2) x-ray spectra. For any given element, the wave numbers of the lines in the x-ray spectra are much greater than those in the optical spectra. From this it can be inferred that the difference in energies between two states of an atom which emits an x-ray spectral line is very large and that the energy values of these atomic states are also very large. The wave numbers of the lines of the optical spectra are comparatively small. It will be shown not only that the difference between two atomic energy states is small but that the energies of the atomic states giving rise to these optical lines are also small. In general, the optical spectrum of any given element is much more complex than the x-ray spectrum of the same element. In this chapter a few typical optical spectra will be considered, together with the relationship of these spectra to the extranuclear structure of the atom. X-ray spectra will be discussed in the next chapter.

9-2 Optical Spectral Series

A great deal of work had been done in analyzing optical spectra in the century preceding the publication of Bohr's theory of hydrogen. The spectral lines of an element had been arranged in several *series,* and as aids in selecting lines belonging to the same series, various types of evidence were used such as (a) the physical appearance of the lines, whether "sharp" or "diffuse," (b) the methods used in producing the spectra, whether with the aid of an arc or a spark, and (c) the behavior of the lines when the emitting atoms were subjected to external electric and magnetic fields—for example, the Zeeman effect.

Rydberg (1889) suggested that the optical series then known could be arranged in such a way that the wave number of any line in the series would be given by the difference between two terms as follows:

$$\bar{\nu} = \bar{\nu}_{\infty} - \frac{RZ^2}{(n + \phi)^2} \tag{9-4}$$

in which R is Rydberg's constant, n is an integer, and ϕ is a fraction less than unity which is practically constant for all lines of the series. The series approaches a limit for very large values of n; the term $\bar{\nu}_{\infty}$ is the wave number approached by the series in the limit as n approaches infinity. Z has the value unity for series due to neutral atoms, the value two for singly charged ions, three for doubly charged ions, and so on. The similarity between Rydberg's formula and Bohr's frequency condition is obvious. In each case the wave number of a line of a spectral series is given as the difference between a fixed term and a variable term. The fixed term is the wave number of the series limit represented by either ν_{∞} or T_l. The variable term is a wave number associated with an atomic state described by a specific value of the integer n and the constant ϕ.

Of the several series of spectral lines from any one element, the most intense are the principal series, the sharp series, the diffuse series, and the fundamental or Bergmann series. In terms of Rydberg's formula these series are represented by the following equations:

Principal series

$$\bar{\nu} = P_\infty - \frac{R}{(n+P)^2} \qquad (n = 2, 3, 4, \ldots)$$

Sharp series

$$\bar{\nu} = S_\infty - \frac{R}{(n+S)^2} \qquad (n = 2, 3, 4, \ldots)$$

Diffuse series

$$\bar{\nu} = D_\infty - \frac{R}{(n+D)^2} \qquad (n = 3, 4, 5, \ldots)$$

Fundamental series

$$\bar{\nu} = F_\infty - \frac{R}{(n+F)^2} \qquad (n = 4, 5, 6, \ldots)$$

(9-5)

The fixed term $\bar{\nu}_\infty$ is replaced by P_∞, S_∞, D_∞, or F_∞, and the constant ϕ in the variable term is replaced by P, S, D, or F in Rydberg's formula. The constants P, S, D, and F all have different values. It has been found empirically that the fixed terms have the following values:

$$P_\infty = \frac{R}{(1+S)^2}$$

$$S_\infty = \frac{R}{(2+P)^2}$$

$$D_\infty = \frac{R}{(2+P)^2}$$

and

$$F_\infty = \frac{R}{(3+D)^2}$$

It will be noticed that the sharp and diffuse series both have the same series limit. A shorthand notation is frequently used in writing the equations for the different series. This is done by using the letters that appear in the denominator of the particular term to represent the term. Thus nP is written as an abbreviation of the term $R/(n+P)^2$, nS for $R/(n+S)^2$, nD for $R/(n+D)^2$, and so on. In this notation the lines of the different series are written as follows:

Principal series

$$\bar{\nu} = 1S - nP \qquad (n = 2, 3, 4, 5, \ldots)$$

Sharp series

$$\bar{\nu} = 2P - nS \qquad (n = 2, 3, 4, 5, \ldots)$$

Diffuse series

$$\bar{\nu} = 2P - nD \qquad (n = 3, 4, 5, 6, \ldots)$$

Fundamental series

$$\bar{\nu} = 3D - nF \qquad (n = 4, 5, 6, 7, \ldots)$$

(9-6)

From the analyses of their spectra, term values have been computed for many of the energy states of all the elements. Although the wave number of any line can be expressed as the difference between two terms, the converse is not always true; that is, not all the differences that can be formed between the term values of an atom represent spectral lines. In order to account for the nonappearance of certain lines, selection rules are necessary. Such selection rules, originally formulated empirically, can be derived by means of wave mechanics. They are best stated in terms of possible changes in quantum numbers intimately related to the structure of the atom. These changes in quantum numbers can be followed on an energy level diagram, which will make it clear what will be the energy of a photon emitted in a transition. This was done for the hydrogen atom in Chapter 8. It is extremely difficult to use this method for more complex atoms, since it involves the solution of Schroedinger's equation for systems of three or more particles. Approximate methods have been developed for many of these problems, the approximations used generally determined by knowledge of the specific properties of the system.

There is a comparatively simple method of arriving at many of the important properties of atoms by the use of a model, the so-called *vector model* of the atom. One reason for using this model is that angular momentum, which is a vector quantity and is quantized, plays an extremely important role in determining the behavior of the atomic particles. The vector model has also been extended to nuclear particles and has been very useful in explaining many properties of nuclei. We shall describe the vector model in the following sections and use it extensively in the rest of the book.

9-3   Vector Model of the Atom: Orbital Angular Momentum

In our discussion of the hydrogen atom it was shown that the angular momentum of an electron in an orbit can be represented by a vector **L** drawn perpendicular to the plane of the orbit. The unit of angular momentum used in

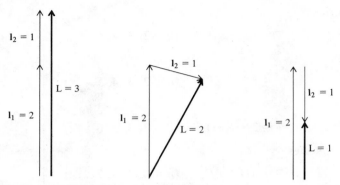

Figure 9-1 Method of addition of angular momentum vectors.

this discussion is $\hbar = h/2\pi$, a unit first introduced by Bohr in his theory of the hydrogen atom. In Bohr's theory the orbital angular momentum was given by

$$n_\phi \hbar$$

where n_ϕ is the azimuthal quantum number. In the Schroedinger theory the orbital angular momentum is given by

$$\sqrt{l(l+1)}\,\hbar$$

where l is the orbital angular momentum quantum number which is restricted to values given by

$$l = 0, 1, 2, \ldots, n-1$$

where n is the principal quantum number.

In dealing with atoms having more than one electron we will use the vector **l** to represent the orbital angular momentum vector for a single electron and the vector **L** for the total angular momentum vector of all the electrons in the atom. In the vector model of the atom the numerical value assigned to **l** is the value of the angular momentum quantum number l. Thus, when $l = 2$, we shall write **l** = 2 with the understanding that the angular momentum has the value $\sqrt{l(l+1)}\,\hbar$, which in this case is $\sqrt{6}\,\hbar$.

The total angular momentum vector **L** of an atom containing more than one electron is the vector sum of the orbital angular momentum vectors of the individual electrons; thus

$$\mathbf{L} = \mathbf{l}_1 + \mathbf{l}_2 + \mathbf{l}_3 + \cdots \tag{9-7}$$

with the restriction that the vector **L** is restricted to integral values; for example, for the case of two electrons, one of which has $\mathbf{l}_1 = 2$ and the other $\mathbf{l}_2 = 1$, the vector $\mathbf{L} = \mathbf{l}_1 + \mathbf{l}_2$ can have any one of three values, 3, 2, or 1. The method of adding the vectors $\mathbf{l}_1$ and $\mathbf{l}_2$ is shown in Figure 9-1. The total angular momentum of the two electrons when $\mathbf{L} = 3$ is

$$\sqrt{L(L+1)}\,\hbar = \sqrt{12}\,\hbar$$

9-4 Electron Spin

In order to account for the fine structure of the lines in the spectral series of some of the elements and also to account for the anomalous Zeeman effect, Uhlenbeck and Goudsmit (1925) introduced the hypothesis that the electron rotates or spins about an axis just like a top. The angular momentum of the electron due to its spin $\mathbf{p}_s$ is assigned the value

$$\mathbf{p}_s = s\,\hbar \tag{9-8}$$

where s has the value $\frac{1}{2}$. Vectorially this can be represented by s of length $\frac{1}{2}$ in units of $\hbar$. Again it must be noted that, according to wave mechanics, the magnitude of the spin angular momentum is $\sqrt{s(s+1)}\,\hbar$—that is, $\frac{1}{2}\sqrt{3}\,\hbar$.

In the vector treatment of the atom the *spin quantum number s* will be used, but in actual calculation its wave mechanics value will be substituted.

The vector sum **S** of the spin angular momenta of several electrons is subject to the following restrictions: for an odd number of electrons **S** must be an odd multiple of $\frac{1}{2}$; for an even number of electrons **S** must be an integer. This means that the vectors representing the spin must always be parallel or antiparallel—that is, oppositely directed. This is shown in two typical cases in Figure 9-2, one for three electrons for which **S** can have the values $\frac{1}{2}$ or $\frac{3}{2}$, the other for four electrons for which **S** can have the values 0, 1, or 2.

We have already mentioned (Section 4-11) that the distribution of electrons in solids is given by a statistical theory developed by Fermi and Dirac. The distinguishing feature of particles whose energy distribution is given by the Fermi-Dirac statistics is that they possess an intrinsic spin or angular momentum of $\frac{1}{2}\hbar$; such particles are called *fermions*. The electron, proton, and neutron are fermions. Particles whose intrinsic spins or angular momenta are integral multiples of $\hbar$ or zero are called *bosons;* their energy distribution is given by a statistical theory developed by Bose and Einstein. The alpha particle of spin zero and the deuteron (the nucleus of deuterium) of spin 1 are examples of bosons.

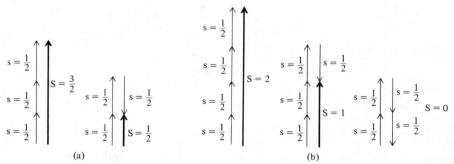

Figure 9-2 Addition of electron spin vectors (a) for three electrons, (b) for four electrons.

9-5 Total Angular Momentum Vector

In many cases, for example, in the alkali elements, the changes in the atomic configurations giving rise to the optical spectrum are produced by a change in state of a single electron. The total angular momentum of a single electron is the vector sum of the orbital and spin angular momenta of the single electron. The total angular momentum is given by $j\hbar$, where j is the total angular momentum quantum number. The vector $\mathbf{j}$, representing the total angular momentum, is defined by the equation.

$$\mathbf{j} = \mathbf{l} + \mathbf{s} \tag{9-9}$$

with the restriction that the vector sum must always be an odd multiple of $\frac{1}{2}$. Since s is always equal to $\frac{1}{2}$, j can have only two values for a given value of l, namely, $l + \frac{1}{2}$ and $l - \frac{1}{2}$, except when $l = 0$, in which case j can have the value $\frac{1}{2}$ only. Thus for $l = 2$ and $s = \frac{1}{2}$—see Figure 9-3a—j can have the values $\frac{5}{2}$ and $\frac{3}{2}$.

Again it must be remarked that, from wave mechanical considerations, the magnitude of the total angular momentum is $\sqrt{j(j+1)}\,\hbar$. In the addition of vectors $\mathbf{l}$ and $\mathbf{s}$ to form the vector $\mathbf{j}$, the magnitude of $\mathbf{l}$ is taken as $\sqrt{l(l+1)}$ and that of $\mathbf{s}$ is taken as $\sqrt{s(s+1)}$. The numerical values of l, s, and j, which are needed for determining the wave mechanical values of the corresponding vectors, are the values obtained from the vector model of the atom. The angle between the vectors $\mathbf{l}$ and $\mathbf{s}$ in Figure 9-3b can be obtained from the figure with the aid of the cosine law, yielding

$$\cos\,(s,\,l) = \frac{j(j+1) - l(l+1) - s(s+1)}{2\sqrt{s(s+1)}\,\sqrt{l(l+1)}} \tag{9-10}$$

If the changes in atomic states are produced by the action of two or more

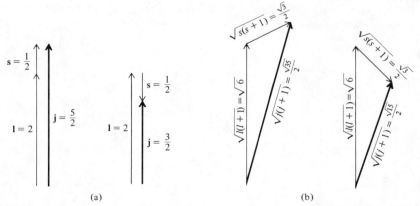

(a)   (b)

Figure 9-3 (a) Addition of the vectors **l** and **s** to form **j** according to the vector model. (b) Addition of the vectors **l** and **s** to form **j** according to wave mechanics.

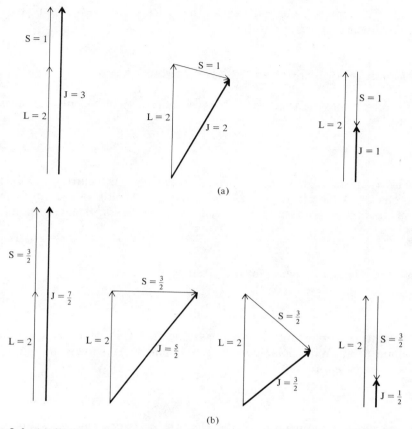

Figure 9-4 (a) Addition of the vectors **L** and **S** to yield integral values of **J**. (b) Addition of the vectors **L** and **S** to yield odd half-integral values of **J**.

electrons, the value of the total angular momentum of these electrons, denoted by $J\hbar$, will depend on the interaction or the coupling between the orbital and the spin angular momenta. Experience has shown that the type of coupling which occurs most frequently is the Russell-Saunders or **L-S** type of coupling. In this type of coupling all the orbital angular momentum vectors of the electrons combine to form a resultant **L** and, independently, all their spin angular momentum vectors combine to form a resultant **S**. The total angular momentum vector of the atom is then given by the relation

$$\mathbf{J} = \mathbf{L} + \mathbf{S} \qquad (9\text{-}11)$$

with the restriction that **J** must be an integer if **S** is an integer, and **J** must be an odd multiple of $\frac{1}{2}$ if **S** is an odd multiple of $\frac{1}{2}$. This type of coupling is illustrated in Figure 9-4 for $L = 2$, $S = 1$, and $L = 2$, $S = \frac{3}{2}$. It can be seen from the figure that the number of possible values of **J**, for $L > S$, is $2S + 1$. The reader can construct similar figures for other values of **L** and **S** and show that, when $L < S$, **J** can have $2L + 1$ values. In particular, if $L = 0$, **J** can have only one value, namely **J** = **S**.

9-6 Magnetic Moment of an Orbital Electron

An electron moving in a plane orbit of area A is equivalent to a current i given by

$$i = \frac{e}{cT} \qquad (9\text{-}12)$$

where T is the period of the electron in its orbit, i, the current in em units, e, the charge of the electron in es units, and c, the ratio of the em to the es unit of charge. A plane circuit carrying current has a magnetic moment μ given by

$$\mu = iA$$

so that the magnetic moment of the orbital electron is

$$\mu = \frac{eA}{cT} \qquad (9\text{-}13)$$

To evaluate this magnetic moment, assume that the electron is moving in an elliptic orbit. With polar coordinates r and ϕ, as shown in Figure 9-5, the area can be expressed as

$$A = \tfrac{1}{2} \int_0^{2\pi} r^2 \, d\phi$$

Now the angular momentum of the electron p_ϕ is constant and can be expressed as

$$p_\phi = mr^2 \frac{d\phi}{dt}$$

The elimination of r^2 between the last two equations yields

$$A = \tfrac{1}{2} \int_0^T \frac{p_\phi}{m} \, dt = \tfrac{1}{2} \frac{p_\phi}{m} T \qquad (9\text{-}14)$$

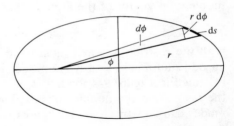

Figure 9-5 Elliptic orbit of an electron.

The magnetic moment μ is therefore

$$\mu = \frac{e}{2mc}\, p_\phi \tag{9-15}$$

The angular momentum $\mathbf{p}_\phi$ can also be expressed in terms of the orbital quantum number $\mathbf{l}$ as

$$\mathbf{p}_\phi = \mathbf{l}\hbar$$

from which

$$\mu = \mathbf{l}\frac{e\hbar}{2mc} \tag{9-16}$$

The magnetic moment of the orbital electron is therefore an integral multiple of the quantity $e\hbar/2mc$. This quantity is known as the magnetic moment of a *Bohr magneton* and is represented by the symbol M_B. Substitution of the numerical values for the constants yields

$$M_\mathrm{B} = \frac{e\hbar}{2mc} = 9.2741 \times 10^{-21} \text{ erg gauss}^{-1}$$

Since the electronic charge is negative, the magnetic moment due to its orbital motion

$$\mu = -\mathbf{l}M_\mathrm{B} \tag{9-17}$$

can be represented by a vector opposite to that of $\mathbf{l}$.

9-7 Magnetic Moment due to Spin

An electron spinning about its axis should also behave as a tiny magnet and possess a magnetic moment due to this spin. However, nothing is known about the shape of an electron or the manner in which its charge is distributed. Lacking this information, it is impossible to calculate its spin magnetic moment in a manner analogous to that used for the orbital motion. In order to obtain agreement with experimental results, it is necessary to assign to the spin magnetic moment the value

$$\mu_s = 2 \cdot \frac{e}{2mc}\, p_s \tag{9-18}$$

where

$$p_s = s\hbar$$

so that

$$\mu_s = 2s\frac{e\hbar}{2mc} \tag{9-19}$$

On the basis of the vector model of the atom, s is always $\frac{1}{2}$, so that the magnetic moment due to spin would have the value of one Bohr magneton. According to wave mechanics this value of s should be replaced by $\sqrt{s(s+1)} = \sqrt{3}/2$. In this case

$$\mu_s = \sqrt{3}\,\frac{e\hbar}{2mc}$$

$$= 1.62 \times 10^{-20} \text{ erg gauss}^{-1}$$

9-8 Magnetic Quantum Numbers

When the atoms of an element are placed in a very strong magnetic field of flux density **B,** the electrons, because of their magnetic moments, will experience torques tending to orient them. One of the effects of the introduction of this external magnetic field is that there now exists a definite direction in space to which the vector quantities may be referred. If the magnetic field is strong enough to break down the coupling between the electrons so that

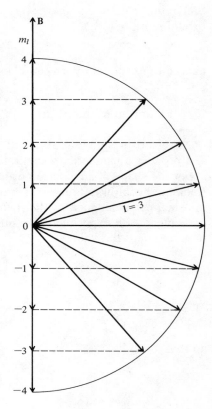

Figure 9-6 Projection of **l** in the direction of the magnetic field determines the magnetic orbital quantum number m_l.

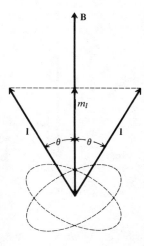

Figure 9-7 Precession of the angular momentum vector **l** about **B** as an axis.

each electron acts independently, then the spin and angular momentum vectors will take up definite angles in space with respect to the magnetic field. These vectors, however, may be oriented only in certain definite directions with respect to the magnetic field. According to wave mechanics, the directions that the vector **l** may assume are such that its projection in the direction of the magnetic field must always have an integral value.

The projection of **l** on the magnetic field direction is denoted by m_l and is called the *magnetic orbital quantum number*. The possible values of m_l are $l, l - 1, l - 2, \ldots, 0, \ldots, -l$—that is, there are $2l + 1$ possible values of m_l. This is illustrated in Figure 9-6 for $l = 3$. The angle θ between **l** and **B** is given by

$$m_l = l \cos \theta \qquad (9\text{-}20)$$

and, as shown in Section 8-9, this relationship becomes

$$\cos \theta = \frac{m_l}{\sqrt{l(l + 1)}}$$

in terms of the magnitudes of the vectors.

The torque due to the magnetic field causes the angular momentum vector **l** to precess about the direction of the magnetic field as an axis, always maintaining the same inclination θ, as shown in Figure 9-7. The additional energy $\Delta\mathscr{E}$ due to the action of the magnetic field is given by

$$\Delta\mathscr{E} = \mu B \cos \theta \qquad (9\text{-}21)$$

Substitution of the values of μ and $\cos \theta$ from Equations (9-16) and (9-20) yields

$$\Delta\mathscr{E} = \frac{e\hbar}{2mc} B m_l \qquad (9\text{-}22)$$

This equation will be useful in the discussion of the Zeeman effect.

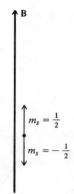

Figure 9-8   Projection of the spin vector **s** in the direction of the magnetic field showing the two possible values of m_s.

The vector **s** representing the spin angular momentum can assume only two possible positions with respect to the magnetic field: it may be parallel to it or antiparallel; that is, oppositely directed to it. Its projection along the direction of the magnetic field is denoted by m_s, which is called the magnetic spin quantum number, and can have only two values, $+\frac{1}{2}$ or $-\frac{1}{2}$, as illustrated in Figure 9-8.

There are similar restrictions on the positions that the total angular momentum vector **j** can assume in the presence of a magnetic field. Since we are dealing with only a single electron, **j** can have only odd half-integral values; the restriction on the positions of **j** is that m_j, the projection of **j** on the direction of the magnetic field, must have odd half-integral values. Figure 9-9(a) shows the possible values of m_j for $\mathbf{j} = \frac{3}{2}$.

On the basis of wave mechanics, m_j remains a half integer for the corre-

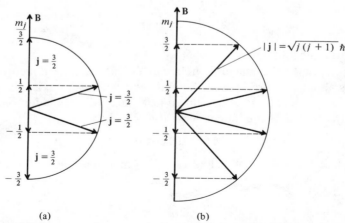

(a)   (b)

Figure 9-9   (a) Projection of **j** in the direction of the magnetic field **B** showing the possible values of m_j according to the vector model of the atom. (b) Projection of the vector **j** to give the same values of m_j as in (a) according to wave mechanics.

sponding values of $\mathbf{j}$ even though the magnitude of $\mathbf{j}$ is $\sqrt{j(j+1)}$. The orientations of $\mathbf{j}$ with respect to the magnetic field are, however, slightly different, as illustrated in Figure 9-9(b). It will be noted that there are $2j + 1$ values for m_j in both methods of projection. The term *space quantization* is usually applied to the above restrictions imposed on the orientation of the vectors $\mathbf{l}$, $\mathbf{s}$, and $\mathbf{j}$ in the presence of a magnetic field.

TABLE 9-1 POSSIBLE NUMBER OF ELECTRONS IN A GIVEN GROUP

n	l $0, 1, 2 \cdots (n-1)$	m_l $l, l-1 \cdots 0, \cdots l$	m_s $\pm\frac{1}{2}$	Number of Electrons in Subgroup	Number of Electrons in Completed Group
1	0	0	$+\frac{1}{2}$		
1	0	0	$-\frac{1}{2}$	2	2
2	0	0	$+\frac{1}{2}$		
2	0	0	$-\frac{1}{2}$	2	
2	1	-1	$+\frac{1}{2}$		
2	1	-1	$-\frac{1}{2}$		8
2	1	0	$+\frac{1}{2}$		
2	1	0	$-\frac{1}{2}$	6	
2	1	1	$+\frac{1}{2}$		
2	1	1	$-\frac{1}{2}$		
3	0	0	$+\frac{1}{2}$		
3	0	0	$-\frac{1}{2}$	2	
3	1	-1	$+\frac{1}{2}$		
3	1	-1	$-\frac{1}{2}$		
3	1	0	$+\frac{1}{2}$		
3	1	0	$-\frac{1}{2}$	6	
3	1	1	$+\frac{1}{2}$		
3	1	1	$-\frac{1}{2}$		
3	2	-2	$+\frac{1}{2}$		18
3	2	-2	$-\frac{1}{2}$		
3	2	-1	$+\frac{1}{2}$		
3	2	-1	$-\frac{1}{2}$		
3	2	0	$+\frac{1}{2}$		
3	2	0	$-\frac{1}{2}$	10	
3	2	1	$+\frac{1}{2}$		
3	2	1	$-\frac{1}{2}$		
3	2	2	$+\frac{1}{2}$		
3	2	2	$-\frac{1}{2}$		

9-9 Pauli's Exclusion Principle

All these quantum numbers having been introduced, the problem now is to assign the appropriate set of quantum numbers to each electron in the atom in order to specify the *state* of that electron. Wave mechanics does not offer any guiding principle for the assignment of these quantum numbers. However, Pauli (1925) introduced a principle, known as *Pauli's exclusion principle,* for the assignment of quantum numbers to the electrons in the atom. *Pauli's exclusion principle states that no two electrons in an atom can exist in the same state.* Now the state of any one electron can be completely specified by a group of four quantum numbers, such as n, l, m_l, m_s or n, l, j, m_j. Hence Pauli's exclusion principle states that the group of values assigned to the four quantum numbers must be different for different electrons.

Electrons having the same value for the principal quantum number n form a definite *group, shell,* or *energy level.* Electrons with the same n are further subdivided according to the value of the orbital angular momentum l. Differences in l values, for the same value of n, denote comparatively smaller energy differences than equal values of l and different values of n. Those electrons that possess the same value of l for a given n are said to be in the same subgroup, or subshell, or sublevel. The possible number of electrons in a subgroup depends on the possible orientations of the vectors $\mathbf{l}$ and $\mathbf{s}$—that is, on the possible values of m_l and m_s or on the possible values of $\mathbf{j}$ and m_j. Table 9-1 shows the maximum possible number of electrons in a given group for $n = 1$, 2, and 3.

Several interesting facts can be obtained from a study of Table 9-1. The maximum number of subgroups for a given value of n is n. The maximum possible number of electrons in a given subgroup is $2(2l + 1)$. A subgroup is filled or completed when the sum of the m_l is zero. Also the sum of the m_s is zero for a completed subgroup. From wave mechanical considerations it can be shown that *for a closed shell* $\mathbf{S} = 0$, $\mathbf{L} = 0$, *and* $\mathbf{J} = 0$; that is, the contributions of the electrons in a closed shell to the total angular momentum of an atom are zero. Hence, in order to determine the angular momentum of an atom, only those electrons that are outside closed shells need be considered.

9-10 Distribution of Electrons in an Atom

In assigning electrons to the different groups and subgroups in an atom, we must have recourse not only to optical and x-ray spectra but also to other phenomena, such as the magnetic and chemical behavior of the element. The normal state of an atom is one in which all the electrons are in the lowest possible energy levels. In the atom of the simplest element, hydrogen, $Z = 1$, the normal state is characterized by the quantum numbers $n = 1$, $l = 0$; m_l is, of course, zero, and m_s may be either $+\frac{1}{2}$ or $-\frac{1}{2}$. The atom of the next element, helium, $Z = 2$, has both its electrons in the shell $n = 1$, $l = 0$; m_s is $+\frac{1}{2}$ for

TABLE 9-2 DISTRIBUTION OF ELECTRONS IN THE ATOMS

X-Ray Notation	K	L		M			N				Element	Atomic number Z	First ionization potential in volts
Values of n, l	1,0	2,0	2,1	3,0	3,1	3,2	4,0	4,1	4,2	4,3			
Spectral Notation	$1s$	$2s$	$2p$	$3s$	$3p$	$3d$	$4s$	$4p$	$4d$	$4f$			Lowest spectral term

Element	Atomic number Z	First ionization potential in volts	$1s$	$2s$	$2p$	$3s$	$3p$	$3d$	$4s$	$4p$	$4d$	$4f$	Lowest spectral term
H	1	13.598	1										$^2S_{1/2}$
He	2	24.587	2										$^1S_0$
Li	3	5.392	2	1									$^2S_{1/2}$
Be	4	9.322	2	2									$^1S_0$
B	5	8.298	2	2	1								$^2P_{1/2}$
C	6	11.260	2	2	2								$^3P_0$
N	7	14.534	2	2	3								$^4S_{3/2}$
O	8	13.618	2	2	4								$^3P_2$
F	9	17.422	2	2	5								$^2P_{3/2}$
Ne	10	21.564	2	2	6								$^1S_0$
Na	11	5.139		Neon configuration 10 electron core		1							$^2S_{1/2}$
Mg	12	7.646				2							$^1S_0$
Al	13	5.986				2	1						$^2P_{1/2}$
Si	14	8.151				2	2						$^3P_0$
P	15	10.486				2	3						$^4S_{3/2}$
S	16	10.360				2	4						$^3P_2$
Cl	17	12.967				2	5						$^2P_{3/2}$
A	18	15.759				2	6						$^1S_0$
K	19	4.341		Argon configuration 18 electron core					1				$^2S_{1/2}$
Ca	20	6.113							2				$^1S_0$
Sc	21	6.54						1	2				$^2D_{3/2}$
Ti	22	6.82						2	2				$^3F_2$
V	23	6.74						3	2				$^4F_{3/2}$
Cr	24	6.766						5	1				$^7S_3$
Mn	25	7.435						5	2				$^6S_{5/2}$
Fe	26	7.870						6	2				$^5D_4$
Co	27	7.86						7	2				$^4F_{9/2}$
Ni	28	7.635						8	2				$^3F_4$
Cu	29	7.726						10	1				$^2S_{1/2}$
Zn	30	9.394						10	2				$^1S_0$
Ga	31	5.999						10	2	1			$^2P_{1/2}$
Ge	32	7.899						10	2	2			$^3P_0$
As	33	9.81						10	2	3			$^4S_{3/2}$
Se	34	9.752						10	2	4			$^3P_2$
Br	35	11.814						10	2	5			$^2P_{3/2}$
Kr	36	13.999						10	2	6			$^1S_0$

TABLE 9-2 DISTRIBUTION OF ELECTRONS IN THE ATOMS (Continued)

X-ray notation			K	L	M	N				O					P						Lowest spectral term
Values of n,l			1,0	2,0,1	3,0,1,2	4,0	4,1	4,2	4,3	5,0	5,1	5,2	5,3	5,4	6,0	6,1	6,2	6,3	6,4	6,5	
Spectral notation			$1s$	s,p	s,p,d	$4s$	$4p$	$4d$	$4f$	$5s$	$5p$	$5d$	$5f$	$5g$	$6s$	$6p$	$6d$	$6f$	$6g$	$6h$	
Element	Atomic number Z	First ionization potential in volts																			
Rb	37	4.177	Krypton configuration 36 electron core							1											$^2S_{1/2}$
Sr	38	5.695								2											$^1S_0$
Y	39	6.38						1		2											$^2D_{3/2}$
Zr	40	6.84						2		2											$^3F_2$
Nb	41	6.88						4		1											$^6D_{1/2}$
Mo	42	7.099						5		1											$^7S_3$
Tc	43	7.28						6		1											$^6S_{5/2}$
Ru	44	7.37						7		1											$^5F_5$
Rh	45	7.46						8		1											$^4F_{9/2}$
Pd	46	8.34						10													$^1S_0$
Ag	47	7.576	Palladium configuration 46 electron core							1											$^2S_{1/2}$
Cd	48	8.993								2											$^1S_0$
In	49	5.786								2	1										$^2P_{1/2}$
Sn	50	7.344								2	2										$^3P_0$
Sb	51	8.641								2	3										$^4S_{3/2}$
Te	52	9.009								2	4										$^3P_2$
I	53	10.451								2	5										$^2P_{3/2}$
Xe	54	12.130								2	6										$^1S_0$
Cs	55	3.894	Xenon configuration 54 electron core												1						$^2S_{1/2}$
Ba	56	5.212													2						$^1S_0$
La	57	5.577	Shells $1s$ to $4d$ contain 46 electrons							2	6	1			2						$^2D_{3/2}$
Ce	58	5.47							1	2	6	1			2						
Pr	59	5.42							2	2	6	1			2						
Nd	60	5.49							3	2	6	1			2						$^5I_4$
Pm	61	5.55							4	2	6	1			2						
Sm	62	5.63							5	2	6	1			2						$^7F_0$
Eu	63	5.67							6	2	6	1			2						$^8S_{7/2}$
Gd	64	6.14							7	2	6	1			2						$^9D_2$
Tb	65	5.85							8	2	6	1			2						
Dy	66	5.93							9	2	6	1			2						
Ho	67	6.02							10	2	6	1			2						
Er	68	6.10							11	2	6	1			2						
Tm	69	6.18							13	2	6	0			2						$^2F_{7/2}$
Yb	70	6.254							14	2	6	0			2						$^1S_0$
Lu	71	5.426							14	2	6	1			2						$^3D_{5/2}$

TABLE 9-2 DISTRIBUTION OF ELECTRONS IN THE ATOMS (Concluded)

Elements	Atomic number Z	First ionization potential in volts	K 1	L 2	M 3	N 4	5s (5,0)	5p (5,1)	5d (5,2)	5f (5,3)	5g (5,4)	6s (6,0)	6p (6,1)	6d (6,2)	6f (6,3)	6g (6,4)	6h (6,5)	7s (7,0)	7p (7,1)	Lowest spectral term
Hf	72	7.0							2			2								$^3F_2$
Ta	73	7.89							3			2								$^4F_{3/2}$
W	74	7.98	Shells 1s to 5p contain 68 electrons						4			2								$^5D_0$
Re	75	7.88							5			2								$^6S_{5/2}$
Os	76	8.7							6			2								$^5D_4$
Ir	77	9.1							7			2								$^4F_{9/2}$
Pt	78	9.0							9			1								$^3D_3$
Au	79	9.225							10			1								$^2S_{1/2}$
Hg	80	10.437										2								$^1S_0$
Tl	81	6.108	Shells 1s to 5d contain 78 electrons									2	1							$^2P_{1/2}$
Pb	82	7.416										2	2							$^3P_0$
Bi	83	7.287										2	3							$^4S_{3/2}$
Po	84	8.42										2	4							
At	85	—										2	5							
Rn	86	10.748										2	6							$^1S_0$
Fr	87	—	Radon configuration 86 electron core															1		
Ra	88	5.279																2		$^1S_0$
Ac	89	6.9										2	6	1				2		$^2D_{3/2}$
Th	90									1		2	6	1				2		$^3F_2$
Pa	91									2		2	6	1				2		
U	92									3		2	6	1				2		$^5L_6$
Np	93									4		2	6	1				2		
Pu	94	5.8								5		2	6	1				2		
Am	95	6.0								6		2	6	1				2		
Cm	96									7		2	6	1				2		
Bk	97									8		2	6	1				2		
Cf	98									9		2	6	1				2		
E	99									10		2	6	1				2		
Fm	100									11		2	6	1				2		
Mv	101									12		2	6	1				2		
No	102									13		2	6	1				2		
Lr	103									14		2	6	1				2		
	104																			
Ha	105																			

Note: The lowest spectral term for each element obtained from Circular 467, Atomic Energy Levels, National Bureau of Standards; the first ionization potentials from NSRDS-NBS-34, Ionization Potentials and Ionization Limits Derived from Analyses of Optical Spectra by C. E. Moore, 1970.

one electron and $-\frac{1}{2}$ for the second electron. This shell is now completed or closed. It will be recalled that helium is one of the inert gases; therefore it may be expected to have a very stable electron configuration. This should also be true of all the other inert gases.

In the atom of the next element, lithium, $Z = 3$, two electrons can be put in the shell $n = 1$, $l = 0$, but the third electron must be put into a new shell $n = 2$, $l = 0$. Lithium is one of the alkali elements and has a valence of unity. This means that a single electron, in shell $n = 2$, can be detached easily from the atom to form the lithium ion, Li^+. This is indicated by the fact that its ionization potential is only 5.39 volts, whereas for He it is 24.58 volts (see Table 9-2). Another interesting point is that the lithium ion, Li^+, has the same configuration as neutral helium. We may expect the atoms of the other alkali elements—sodium, potassium, rubidium, and cesium—to be built up in a similar manner—that is, a single valence electron starting a new shell outside a closed configuration typical of an inert gas. This is shown in Table 9-2, which gives the distribution of electrons in the atoms of the elements.

It is convenient at this point to introduce the x-ray notation for the different groups. The group or shell for which $n = 1$ is called the K shell, $n = 2$, the L shell, $n = 3$, the M shell, and so on. Beryllium, for example, with $Z = 4$, has two electrons in the completed K shell and two additional electrons in the L shell, thus completing the first subgroup in this shell. Beryllium is one of the alkaline earth elements with a valence of 2. The atoms of the other elements of this group—magnesium, calcium, strontium, barium, and radium—should have similar structures—that is, two electrons outside an inert gas or closed shell configuration. This can be verified from Table 9-2.

Boron, $Z = 5$, has two electrons in the completed K shell, three in the L shell, and two in the completed subgroup $n = 2$, $l = 0$, the third electron starting the new subgroup $n = 2$, $l = 1$. The atoms of the other elements in this group —aluminum, gallium, indium, and thallium—similarly have three electrons outside a closed shell, two in a completed subgroup $l = 0$ and one in the next subgroup $l = 1$.

This process of atom building can be continued by the addition of an electron to the L-shell subgroup $l = 1$, since the element of atomic number $Z + 1$ is formed from element of atomic number Z. In each case the positive charge of the nucleus must be increased by one. The L shell will be completed with the element neon, $Z = 10$, with two electrons in the K shell and eight electrons in the L shell. Neon is one of the inert gases and has a very stable configuration. Fluorine, $Z = 9$, has two electrons in the K shell and seven electrons in the L shell. In chemical action it is found that fluorine has a valence of -1, indicating that it very easily forms an ion F^- by adding an electron to the L shell to form a stable configuration similar to that of neon.

The next eight elements, from sodium, $Z = 11$, to argon, $Z = 18$, are formed by adding the additional electrons to the M shell for which $n = 3$. Sodium has an electron ($n = 3$, $l = 0$) outside a closed shell; magnesium has two electrons outside this closed shell, both in the subgroup $l = 0$, thus completing it.

The next subgroup with $l = 1$ is begun with aluminum, $Z = 13$, and completed with argon, $Z = 18$. It may be remarked here that the chemical properties of an element are determined mostly by the electrons in the outer shell of the atom.

Potassium, $Z = 19$, retains the argon configuration of the first eighteen electrons, but the nineteenth electron starts a new group, $n = 4$, belonging to the N shell. Calcium, $Z = 20$, has two electrons in the N shell $n = 4$, $l = 0$. It might have been expected that these electrons would have been placed in the still incomplete M shell $n = 3$, $l = 2$, but spectroscopic evidence is against this. However, the next group of atoms from scandium, $Z = 21$, to copper, $Z = 29$, have their additional electrons placed in the M shell $n = 3$, $l = 2$, which is then completed. From gallium, $Z = 31$, to krypton, $Z = 36$, an inert gas, the additional electrons are added to the N shell $n = 4$, $l = 1$. By examining Table 9-2 the reader will find the order in which electrons have been assigned to the various groups and subgroups. It will be of interest to check this assignment with the chemical properties of the elements, remembering that these properties are controlled essentially by the outer electrons.

9-11   Spectral Notation

In the course of the development of spectroscopy several types of notation have been used. The following is the modern notation. In describing the electron configuration small letters are used to represent the values of l as follows:

$$l = 0, 1, 2, 3, 4, 5, \ldots$$

$$s, p, d, f, g, h, \ldots$$

that is, if an electron is in a shell for which $l = 0$, it is called an s electron, for $l = 1$, a p electron, and so on. The value of the principal quantum number n is written as a prefix to the letter representing its l value. The number of electrons having the same n and l values is indicated by an index written at the upper right of the letter representing their l value. Thus the 11 electrons of sodium in the normal state are designated as follows:

$$1s^2\ 2s^2\ 2p^6\ 3s$$

that is, there are two $1s$ electrons, two $2s$ electrons, six $2p$ electrons, and one $3s$ electron. One must be careful not to confuse the symbol s written for $l = 0$ with the same symbol used for the spin quantum number.

Capital letters are used to represent the total orbital angular momentum of an atom according to the following scheme:

$$L = 0, 1, 2, 3, 4, 5, \ldots$$

$$S, P, D, F, G, H, \ldots$$

The value of the total angular momentum of the atom J is written as a subscript at the lower right of the letter representing the particular L value of the atomic state. The multiplicity of the total spin is written as a superscript at the upper left of the letter representing the L value. If S is the total spin, the multiplicity is equal to $2S + 1$; for example, a state with $L = 1$, $S = \frac{1}{2}$, and $J = \frac{3}{2}$ would be written $^2P_{3/2}$ and read "doublet P three-halves"; a state with $L = 2$, $S = 1$, and $J = 2$ would be written $^3D_2$ and read "triplet D two."

9-12   Spectrum of Sodium

The optical spectrum of sodium is typical of the spectra of all the alkali atoms. In its normal state the sodium atom consists of a closed core of 10 electrons and one additional electron in the $3s$ state. Since the closed core contributes nothing to the angular momentum of the atom, only the states of this eleventh or optical electron need be considered in discussing the spectrum of neutral sodium.

The atoms of sodium can be raised from the normal state to higher energy states by bombarding them with electrons, by subjecting them to high temperatures in a flame or in an electric arc, or by allowing them to absorb radiant energy from an external source. The atom in one of the higher energy states is said to be in an *excited* state. When the atom returns to a state of lower energy, radiation is emitted in the form of a photon of very definite frequency given by Bohr's frequency condition. The spectrum of sodium, as shown in Figure 9-10, consists of several series of spectral lines, some of which were mentioned in Section 9-2. When these spectral lines are examined with instruments of high resolving power, it is found that many of the lines consist of doublets—that is, two lines very close together. Such lines are said to exhibit *fine structure;* for example, the well-known yellow line of sodium, frequently referred to as the sodium D line, consists of two lines close together of wavelengths 5889.96 and 5895.93 Å—that is, they are separated by about 6 Å.

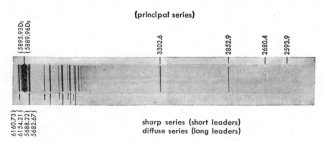

Figure 9-10   The emission spectrum of sodium showing lines of three series. The upper numbers are wavelengths, in Å, of the lines of the principal series. The short leaders below the spectrum indicate the lines of the sharp series, while the long leaders indicate the lines of the diffuse series. (Reprinted by permission from *Atomic Spectra and Atomic Structure*, by G. Herzberg, Dover Publications.)

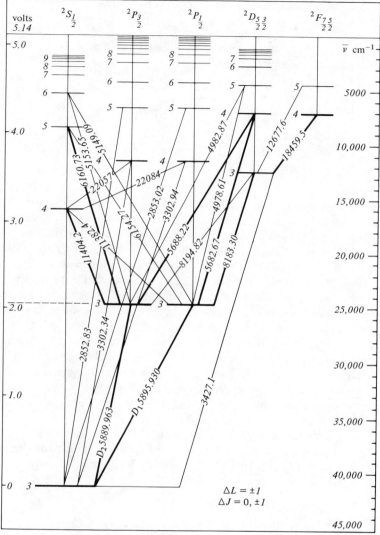

Figure 9-11 Energy-level diagram of sodium. The numbers along the lines are the wavelengths, in angstroms, emitted during the indicated transitions.

These lines form one of the doublets of the principal series. The other lines of the principal series are in the ultraviolet region. Lines of the principal series are due to transitions from a P state to the lowest S state. Since the smallest value of the principal quantum number for the optical electron of sodium is $n = 3$, the lowest state is designated as a $3S$ state. Since $l = 0$, the value of j for this state is $j = s = \frac{1}{2}$. For the P state $L = l = 1$ and, since $J = L + \frac{1}{2}$ and $L - \frac{1}{2}$, the total angular momentum of the P state is $\frac{1}{2}$ or $\frac{3}{2}$. Since there are two values of J, the P state is a doublet state and is designated as

$$^2P_{1/2}, \, ^2P_{3/2}$$

Similarly for the D terms for which $L = l = 2$, $J = j = \frac{3}{2}$ or $\frac{5}{2}$, so that the D term is a doublet and is designated as

$$^2D_{3/2}, \, ^2D_{5/2}$$

and the F term $L = l = 3$, $J = j = \frac{5}{2}$ or $\frac{7}{2}$, is designated as

$$^2F_{5/2}, \, ^2F_{7/2}$$

The S state is always a single state with $J = S = \frac{1}{2}$, since $L = 0$, but since $2S + 1 = 2$ for this state, it is written as

$$^2S_{1/2}$$

The energy-level diagram of sodium (Fig. 9-11) shows the relative positions of these energy levels, drawn approximately to scale. The $^2P_{3/2}$ level is actually slightly above the $^2P_{1/2}$ level, but the separation is too small to be shown in the figure; for example, the separation of the ^2P levels giving rise to the yellow lines of sodium is only 17 cm^{-1}, whereas the term value is about 25,000 cm^{-1}. Similarly the ^2D and ^2F levels are drawn as single levels. The principal quantum number n is written for each term in the figure. Notice that the large wave numbers are associated with the low-energy terms. This is due to the fact that the zero level of energy is taken as the energy of the ionized atom; the energy values are all negative, but the minus signs have been omitted as a matter of convenience. The lowest energy level is the $3^2S_{1/2}$ level and its numerical value is 41,449.0 cm^{-1}. This is equivalent to 5.14 eV and is the energy that must be supplied to remove the electron from the $3^2S_{1/2}$ level to infinity. For this reason the higher voltages coincide with the higher energy levels and the ionization potential, 5.14 volts, is placed at $n = \infty$.

The principal series of sodium is produced by transitions from the ^2P states to the lowest state, $3^2S_{1/2}$. These lines are all doublets, since they originate in the $^2P_{1/2, 3/2}$ levels and end in the $^2S_{1/2}$ level. The yellow lines of sodium are due to the transitions

$$\lambda = 5895.93 \text{ Å}, \, 3^2S_{1/2} - 3^2P_{1/2} \qquad (D_1 \text{ line})$$

$$\lambda = 5889.96 \text{ Å}, \, 3 \, S_{1/2} - 3^2P_{3/2} \qquad (D_2 \text{ line})$$

The wave number of any line of the principal series is given by

$$\bar{\nu} = 3^2S_{1/2} - n \, ^2P_{1/2} \qquad (n = 3, 4, 5, \cdots)$$

or

$$\bar{\nu} = 3^2S_{1/2} - n \, ^2P_{3/2} \qquad (n = 3, 4, 5, \cdots)$$

The lines of the sharp series are due to transitions from the higher $^2S_{1/2}$ levels to the $3^2P_{1/2, 3/2}$ levels and their wave numbers are given by

$$\bar{\nu} = 3^2P_{1/2} - n \, ^2S_{1/2} \qquad (n = 4, 5, 6, \ldots)$$

$$\bar{\nu} = 3^2P_{3/2} - n \, ^2S_{1/2} \qquad (n = 4, 5, 6, \ldots)$$

Transitions from the ^2D levels to the 3^2P levels give rise to the diffuse series and those from the ^2F levels to the 3^2D levels give rise to the fundamental series.

Transitions can take place between S and P states, P and D states, and D and F states, but under normal conditions no transitions can take place between S and D states, S and F states, or P and F states. The transitions that can take place are given by the following selection rules for the vectors **L** and **J**:

$$\Delta\mathbf{L} = \pm 1 \qquad\qquad\qquad (9\text{-}23a)$$

$$\Delta\mathbf{J} = \pm 1 \text{ or } 0 \qquad\qquad (9\text{-}23b)$$

The selection rule for **J** prohibits transitions between some of the doublet levels, even though they are not ruled out by the selection rule for **L**; for example, in the diffuse series the transition $^2P_{1/2} - {}^2D_{5/2}$ is forbidden, whereas the other three transitions are permitted.

Although the selection rules had empirical origins, they can be derived, at least for the simplest systems, by means of wave mechanics or quantum electrodynamics. The necessity for the selection rules can be inferred from the fact that only a single photon is emitted in a transition between two energy states; the photon carries away an amount of angular momentum equal to $\hbar$. Since the total angular momentum of the atom is J, the emission of a photon would lead to the selection rule that $\Delta J = \pm 1$. That $\Delta J = 0$ is also permissible can be understood from the fact that the direction of spin of an electron in an orbit can have only one of two possible values, $\pm\frac{1}{2}$, so that a change in S of ± 1, when added vectorially to the change in orbital angular momentum, $\Delta L = \pm 1$, can lead to $\Delta J = 0$.

The doublet character of the energy levels is typical not only of sodium and the other alkali elements but also of hydrogen and the singly ionized alkaline earth elements such as Be^+, Mg^+, and Ca^+. A glance at Table 9-2 will show that the singly ionized atoms of the alkaline earths have exactly the same electronic structure as the neutral alkali atoms, that is, a single electron outside a closed core typical of the configuration of the atoms of the inert elements. It should be emphasized that the doublet character of the energy levels is satisfactorily accounted for by the hypothesis that the electron possesses a spin.

9-13 Absorption of Energy

If white light is sent through sodium vapor and then examined with a reflection grating, it is found that those wavelengths that correspond to the lines of the principal series of sodium are missing. Such a spectrum is called an *absorption* spectrum. R. W. Wood and his collaborators performed a series of experiments on the absorption spectrum of sodium. In one such experiment the vapor was obtained by heating metallic sodium in a steel tube faced with quartz windows. It was necessary to use quartz windows in this experiment,

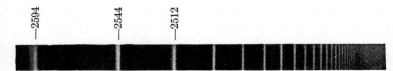

Figure 9-12 Photograph of the absorption spectrum of sodium showing some of the lines in the ultraviolet region. The numbers are the wavelengths of the lines in angstroms. (Reprinted by permission from *Atomic Spectra and Atomic Structure*, by G. Herzberg, Dover Publications.)

since most of the lines of the principal series of sodium are in the ultraviolet region. As many as 60 lines were observed in this absorption spectrum. That only lines of the principal series appear in the absorption spectrum is due to the fact that most of the atoms in the tube are in the lowest state, $3^2S_{1/2}$. A photograph of the absorption spectrum of sodium is shown in Figure 9-12. In emission the lines of the principal series correspond to transitions from the $^2P_{1/2,3/2}$ levels to the ground state $3^2S_{1/2}$; in absorption the transitions are from the lowest state $3^2S_{1/2}$ to the $^2P_{1/2,3/2}$ levels.

An interesting experiment would be to send monochromatic light of wavelength equal to that of the sodium D lines into a tube containing sodium vapor. R. W. Wood performed such an experiment, using the yellow light from an oxyhydrogen flame containing sodium. The yellow light was focused on the axis of an evacuated test tube containing sodium vapor. On looking down into the test tube it was found that the sodium vapor, near the wall of the tube where the incident beam entered, emitted yellow fluorescent radiation. Other investigations showed that the fluorescent radiation consisted only of the yellow lines of sodium. By referring to the energy level diagram it can be seen that the atoms in the normal state, $3^2S_{1/2}$, were raised to the next higher states, $3^2P_{1/2,3/2}$, by the absorption of the yellow D lines. On returning to the normal state these atoms emitted radiation of the same wavelength. This type of fluorescent radiation is called *resonance radiation.*

Another method of raising the atoms of sodium from their normal to their excited states is to bombard them with electrons of appropriate energies. Figure 9-13 is a schematic diagram of the apparatus that can be used to accomplish this. This is essentially the same type of experiment as that used in raising hydrogen atoms from their normal to their excited states (see Section 8-2). The major difference is that this tube now contains some sodium vapor at a low pressure; also, since the lower lying optical levels will yield radiation in the visible region, the light will readily pass through the walls of the glass tube. It can then be analyzed with a spectroscope.

As the voltage between the filament F and the grid G is increased, the current to the plate P is increased until the voltage reaches the value of 2.1 volts, when there is a pronounced decrease in current through the galvanometer G. Furthermore, the sodium vapor in the region between G and P is observed to emit yellow light. An examination of this light with a spectrograph shows that

it consists of the sodium resonance lines only. The interpretation of this phenomenon is that an electron, on colliding with a sodium atom, loses an amount
of energy, equivale.it to 2.1 eV, to the sodium atom, thereby raising it from the
normal to the next higher state, $3^2P_{1/2, 3/2}$. On returning to its normal state, the
sodium atom then emits the resonance radiation. This means that the energy
of the incident electron must be at least equal to the quantum of energy corresponding to the sodium D lines. This can be checked by substituting the appropriate values in the formula

$$Ve = h\nu = \frac{hc}{\lambda}$$

and calculating λ. This yields $\lambda = 5898$ Å, in good agreement (within the limits
of experimental error) with the wavelengths of the D lines of sodium. This
potential, at which the resonance lines appear, is called the *resonance potential.*

When the voltage between the filament and grid in the above apparatus is
increased to about 4 volts, the color of the light emitted by the sodium vapor
changes, indicating that additional spectral lines are being emitted. At this
voltage the spectrogram shows, in addition to the D lines, the presence of the
doublet $3^2S_{1/2} - 4^2P_{1/2, 3/2}$ of wavelengths 3302 and 3303 Å of the principal
series and the doublet $3^2P_{1/2, 3/2} - 5^2S_{1/2}$ of wavelengths 6154 and 6161 Å of
the sharp series. Other lines appear at 4.4 and 4.6 volts. At 5.14 volts ionization of the sodium vapor occurs, as indicated by the very large increase in
current from the filament to the plate, and, at the same time, the spectrograph
records the appearance of the entire optical spectrum of sodium.

The emission of the entire optical spectrum when the voltage reaches the
value of the ionization potential, 5.14 volts, can readily be explained by the
fact that the electrons from the filament which have an amount of energy equal
to 5.14 eV are capable of ionizing the sodium atoms. In this process electrons

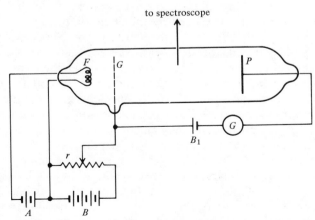

Figure 9-13 A diagram of the apparatus for determining the resonance and ionization
potentials of sodium.

are removed from the normal state $3^2S_{1/2}$ of the sodium atoms. An electron returning to an ionized atom may enter any one of the excited states and finally reach the normal state by a series of successive quantum jumps. Corresponding to each quantum jump there is an emission of radiation of appropriate frequency, giving rise to the lines observed in the optical spectrum. The intensity of a spectral line is determined by the number of atoms in which identical transitions take place simultaneously. The transitions giving rise to the intense spectral lines must have a greater probability of occurrence than those producing the weaker lines. The most probable transitions are those permitted by the selection rules. The probability that transitions not permitted by the selection rules will occur is vanishingly small under ordinary conditions.

9-14   The Zeeman Effect

More than a century ago Faraday placed a sodium flame in a strong magnetic field to determine whether any change would be produced in the wavelength or frequency of its spectral lines. He was unable to observe any effect, probably because the spectroscope did not have sufficient resolving power. In 1896 P. Zeeman, using apparatus of greater resolving power, was able to show that a magnetic field acting on a source of light did produce a distinct change in the character of the spectral line. Using a theory developed by H. A. Lorentz, he was able to determine the nature of the charge emitting the light and to measure its charge-to-mass ratio. The Lorentz explanation of the Zeeman effect was based on the idea that monochromatic light was emitted by an electron revolving in a circular orbit at an angular velocity ω, which is related to the frequency ν of the light by the well-known equation

$$\omega = 2\pi\nu$$

and that the radiation was produced by the acceleration of the charge. Although this explanation has been superseded by the quantum theory, it will be worthwhile to present a simplified treatment of the Lorentz theory of the Zeeman effect. One reason is that it does lead to the correct equation for determining the value of e/m from experiments involving the *singlet* lines which are emitted by atoms in states of spin angular momentum $S = 0$, so that $\mathbf{L} = \mathbf{J}$. Atoms of elements in the second group of the periodic table, for example, Mg or Ca, possess such states (see Fig. 9-23). Another reason is that considerable insight into the physical process can be gained by considering both the classical and quantum mechanical approaches to this phenomenon.

To study the normal Zeeman effect let us consider a calcium arc placed between the pole pieces of a strong electromagnet, one capable of producing a flux density of about 30,000 gauss. The light may be viewed in a direction perpendicular to the direction of the magnetic field or parallel to the magnetic field; to make the latter possible a small hole may be drilled along the axis of one of the pole pieces of the magnet and the light coming through it sent through a spectroscope, or a small mirror may be inserted near one of the pole pieces to reflect such light into a spectroscope. Let us confine our attention to one specific spectral line of calcium of wavelength $\lambda = 4226.7$ Å, which is an intense line of the singlet series (see Fig. 9-23) and exhibits the normal Zeeman effect.

When there is no magnetic field present, the spectral line can be sharply focused and will appear at a certain position in the spectroscope. Let us call the frequency of this line ν_0 when no magnetic field is present. Suppose now that a strong magnetic field of flux density **B** is applied to the source of light and that the light emitted in a direction transverse to the magnetic field is viewed in the spectroscope. The single line will be found to be split into three lines or three components, as shown in Figure 9-14. One of these components will be in the same position as the original line and will thus have the same frequency ν_0; the other two components of frequencies ν_1 and ν_2 will be seen on either side of the original line and equally displaced from the central component ν_0. When this experiment is repeated with the light that is emitted parallel to the direction of **B**, it will be observed that the original line splits into two components only and that these two components have the frequencies ν_1 and ν_2, respectively; this time there is no central or undeviated component.

Although the light emitted by the source is unpolarized, the Zeeman components of the spectral line are polarized. The three components seen when the light is viewed perpendicular to the magnetic field are linearly polarized. By means of an analyzer such as a sheet of Polaroid or a Nicol prism it can be shown that the outer components are polarized at right angles to the undisplaced components. The direction of vibration of the electric vector of the electromagnetic wave of the outer components is perpendicular to the magnetic flux density **B** and the direction of vibration of the electric vector of the undisplaced component is parallel to **B**.

The two components of the spectral line that are produced when the light is emitted parallel to the direction of the magnetic field are found to be circu-

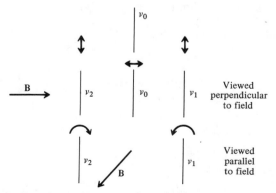

Figure 9-14 The normal Zeeman effect showing the splitting of a spectral line when the source of light is in a magnetic field.

larly polarized in opposite directions. The directions of these polarizations with respect to the direction of **B** are shown in Figure 9-14; these directions can be determined with the aid of a suitable quarter-wave plate as described in Section 4-4.

9-15   Explanations of the Normal Zeeman Effect

The classical explanation of the normal Zeeman effect is based on Lorentz's electron theory. Assume that an electron in the atom is moving in a circular orbit of radius r under the action of some central force F_0, as shown in Figure 9-15. Then from Newton's second law of motion

$$F_0 = \frac{mv_0^2}{r} = m\omega_0^2 r \tag{9-24}$$

where v_0 is its linear velocity in the orbit and ω_0 is its angular velocity. If an external magnetic field is applied perpendicular to the plane of the orbit of the electron, two effects will be produced. During the time that the magnetic field is being established, there will be an electric field tangent to the orbit because of the emf produced by the changing magnetic flux through it. At the same time there will be an additional force on the electron that will be perpendicular to the direction of the magnetic field and to the velocity of the electron—that is, the force will be radial. A simple analysis shows that if the tangential electric field is such that it causes an increase in the velocity of the electron, the radial force will be directed toward the center, thus providing the additional force needed to keep it moving in the same orbit with this higher velocity. Conversely, if the tangential electric field decreases the velocity of the electron, the radial force due to the magnetic field will be directed away from the center, thus decreasing the centripetal force to the amount needed to keep the electron moving in the same circular orbit at the smaller velocity. The simplified analysis given above is a special case of a very famous theorem due to Larmor which will be discussed in greater detail later.

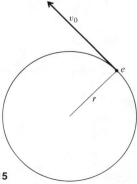

Figure 9-15

Suppose that the velocity of the electron has been increased to v_1 by the application of the magnetic field of flux density B; then the force F_B due to the magnetic field is

$$F_B = Bev_1 = Be\omega_1 r \tag{9-25}$$

where

$$v_1 = \omega_1 r$$

Equation (9-25) can be used with mks units or with cgs electromagnetic units. In the latter case B is in gauss and e, in electromagnetic units of charge. If e is in statcoulombs and B in gauss, the conversion factor c must be used and the equation will read

$$F_B = B\frac{e}{c}v_1 = B\frac{e}{c}\omega_1 r \tag{9-26}$$

We shall use Equation (9-25) in the following development.

Since the force on the electron is directed toward the center, the total force acting radially is

$$F_0 + F_B = m\omega_1{}^2 r$$

Substituting the values for F_0 and F_B, we get

$$m\omega_0{}^2 r + Be\omega_1 r = m\omega_1{}^2 r \tag{9-27}$$

Solving Equation (9-27) for ω_1, we get

$$\omega_1 = \frac{eB/m \pm \sqrt{(eB/m)^2 + 4\omega_0{}^2}}{2}$$

It can be shown that

$$\left(\frac{eB}{m}\right)^2 \ll 4\omega_0{}^2$$

therefore we can write

$$\omega_1 = \omega_0 + \frac{eB}{2m} \tag{9-28}$$

Only the positive sign is retained, since the effect of the magnetic field is small and can produce only a slight change in the magnitude of the angular velocity. If the charge should be rotating in the opposite direction, its angular velocity will be decreased by the amount $eB/2m$, so that, in general, its angular velocity will be

$$\omega = \omega_0 \pm \frac{eB}{2m} \tag{9-29}$$

This equation may be put in terms of the frequency of rotation with the aid of the equations

$$\omega = 2\pi\nu$$

$$\omega_0 = 2\pi\nu_0$$

where ν is the frequency corresponding to the angular velocity ω. Equation (9-29) then becomes

$$\nu = \nu_0 \pm \frac{eB}{4\pi m} \tag{9-30}$$

The quantity $eB/4\pi m$, where e is in electromagnetic units, is called the normal Zeeman separation in a magnetic field of flux density B. The quantity e/m can thus be determined from a measurement of the normal Zeeman separation of a single spectrum line. Determinations of e/m from measurements of the Zeeman effect yield

$$\frac{e}{m} = 1.759 \times 10^7 \text{ emu/gm}$$

which is almost identical with the value of e/m obtained for electrons by means of electric and magnetic deflection experiments.

To compare prediction with experimental observation consider the direction of the magnetic flux density **B** as that of the positive x axis; the current in the electromagnet producing this field can be represented as circular in the Y-Z plane, as shown in Figure 9-16. In the actual source of light the atoms will have all possible orientations. Since light is a transverse wave motion, only those components of the acceleration of the electron that are perpendicular to the line of sight will be effective in sending radiation in this direction. Those electrons moving in orbits parallel to the Y-Z plane will have their frequencies increased or decreased by an amount $eB/4\pi m$. Any uniform circular motion can be resolved into two simple harmonic vibrations at right angles to one another and differing in time phase by a quarter of a period. If the light coming

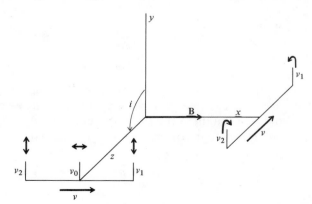

Figure 9-16 The direction of vibration of the components of a spectral line exhibiting the normal Zeeman effect in relation to the direction of the current producing the magnetic field.

from these electrons is viewed along the z axis, only the y component of the vibration will be observed and the frequencies will be

$$\nu_1 = \nu_0 + \frac{eB}{4\pi m} \tag{9-31}$$

and

$$\nu_2 = \nu_0 - \frac{eB}{4\pi m} \tag{9-32}$$

Since only the y component of the acceleration is effective in sending light in this direction, these components will be linearly polarized with the direction of vibration perpendicular to the magnetic field. If the light is viewed along the x axis, both the y and z vibrations will be effective in sending out radiations and the two component lines will be circularly polarized. Experiment shows that the higher frequency component is circularly polarized in the same direction as that of the current in the electromagnet. An analysis of the direction of the force due to the magnetic field shows that only a negative charge rotating in the same direction as that of the polarization can have its frequency increased by the magnetic field (see Fig. 9-16). The component of lower frequency is circularly polarized in the opposite direction and again must be due to a negative charge rotating in this direction. Thus it appears that the light emitted by an atom originates in negatively charged particles for which the value of e/m is identical with that observed for electrons.

To explain the presence of the undeviated component of frequency ν_0 consider those vibrations that are parallel to the x axis. Since the motion of the electron is parallel to the direction of the magnetic field, there will be no additional force acting on it and its frequency will be unchanged. This component will therefore be observed when the light is viewed perpendicular to the magnetic field; the light of frequency ν_0 will be linearly polarized with the direction of vibration parallel to the direction of the magnetic field. This component will not be observed when the source is viewed parallel to the magnetic field, since no light can be emitted parallel to the direction of vibration of the charge.

It was remarked earlier that one of the reasons for the introduction of an electron spin was to explain the anomalous Zeeman effect. If the spin of the electron is left out of consideration, the only angular momentum possessed by the electron is that due to its orbital motion of amount $l\hbar$. Consideration of this orbital motion provides the quantum-mechanical explanation of the normal Zeeman effect. In the presence of a magnetic field of flux density $\mathbf{B}$ the vector $\mathbf{l}$ precesses around the direction of the magnetic field as axis. The angular velocity of precession may be obtained by direct calculation or from a famous theorem due to Larmor, which states that the effect of a magnetic field on an electron moving in an orbit is to superimpose on the orbital motion a precessional motion of the entire orbit about the direction of the magnetic field with angular velocity Ω (omega) given by

$$\Omega = \frac{e}{2mc} B \qquad (9\text{-}33)$$

in which e is in statcoulombs. Figure 9-7 shows two positions of the vector **l** as it precesses about the magnetic field at constant inclination and the corresponding positions of the electronic orbit. The additional energy of the electron due to this precessional motion is given by

$$\Delta \mathcal{E} = \mu B \cos \theta \qquad (9\text{-}21)$$

$$\Delta \mathcal{E} = m_l \frac{e\hbar}{2mc} B \qquad (9\text{-}22)$$

where m_l is the projection of **l** on B. In terms of the Larmor precession, the expression for the additional energy can be written as

$$\Delta \mathcal{E} = m_l \Omega \hbar \qquad (9\text{-}34)$$

Since m_l is restricted to the $(2l+1)$ integral values $l, l-1, \cdots, 0, \ldots, -l$, the effect of the magnetic field is to split up each energy level into $2l+1$ components spaced an amount $(e\hbar/2mc)B$ apart. This is illustrated in Figure 9-17

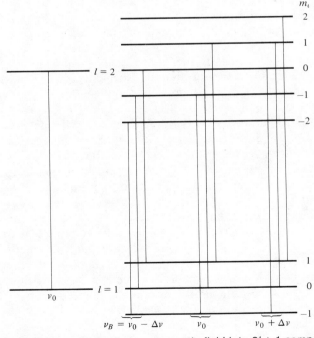

Figure 9-17 Splitting of energy levels in a magnetic field into $2l + 1$ components. Section on the left represents the single transition in the absence of a magnetic field, while that at the right represents the splitting of these two energy levels in a magnetic field and the possible transitions yielding the normal Zeeman effect. Selection rule: $\Delta m_l = 0, \pm 1$.

for two energy levels, one for which $l = 2$, the other, $l = 1$. If $\mathcal{E}_0'$ represents the energy of the level $l = 1$ in the absence of a magnetic field and $\mathcal{E}_B'$ represents the energy of this level in the presence of the magnetic field of flux density B, then

$$\mathcal{E}_B' = \mathcal{E}_0' + \Delta\mathcal{E}' = \mathcal{E}_0' + m_l' \frac{e\hbar}{2mc} B$$

Similarly, if $\mathcal{E}_0''$ and $\mathcal{E}_B''$ represent the energies of the level $l = 2$ without and with the magnetic field respectively, then

$$\mathcal{E}_B'' = \mathcal{E}_0'' + \Delta\mathcal{E}'' = \mathcal{E}_0'' + m_l'' \frac{e\hbar}{2mc} B$$

The quantity of energy radiated in the presence of the magnetic field is given by

$$h\nu_B = \mathcal{E}_B'' - \mathcal{E}_B' = \mathcal{E}_0'' - \mathcal{E}_0' + (m_l'' - m_l') \frac{e\hbar}{2mc} B$$

$$= h\nu_0 + \Delta m_l \frac{e\hbar}{2mc} B \qquad (9\text{-}35)$$

from which

$$\nu_B = \nu_0 + \Delta m_l \frac{eB}{4\pi mc} \qquad (9\text{-}36)$$

where ν_B is the frequency of the radiation emitted with the magnetic field present and ν_0 is the frequency of the radiation in the absence of the magnetic field. The restrictions imposed on the changes in the magnetic quantum number m_l are given by the selection rule

$$\Delta m_l = 0 \qquad \text{or} \pm 1 \qquad (9\text{-}37)$$

Application of this selection rule to Equation (9-37) yields

$$\nu_B = \nu_0 \quad \text{for} \quad \Delta m_l = 0 \qquad (9\text{-}38a)$$

and

$$\nu_B = \nu_0 \pm \frac{eB}{4\pi mc} \quad \text{for} \quad \Delta m_l = \pm 1 \qquad (9\text{-}38b)$$

These frequencies are identical with those obtained on the classical theory for the normal Zeeman effect. Although there are nine possible transitions for the energy levels shown in Figure 9-17, they are grouped into only three different frequency components as indicated by Equations (9-38a) and (9-38b). It is thus evident that the orbital angular momentum alone is not sufficient to account for the anomalous Zeeman effect (see Section 9-17), although adequate for the normal Zeeman effect.

9-16 The Landé g Factor

With the introduction of electron spin, the total angular momentum of the atom **J** becomes the vector sum of the orbital and spin angular momenta **L** and **S**, all in units of $\hbar$; thus

$$\mathbf{J} = \mathbf{L} + \mathbf{S} \tag{9-11}$$

Because of the interaction between these two angular momenta, the vectors **L** and **S**, although maintaining their relative orientations, precess about their resultant **J**. The magnetic moment due to the orbital motion $\boldsymbol{\mu}_L$ is given by

$$\boldsymbol{\mu}_L = \mathbf{L}\frac{e\hbar}{2mc} \tag{9-16}$$

Because of the negative charge, $\boldsymbol{\mu}_L$ is oppositely directed to **L**. The magnetic moment due to the spin of the electron is given by

$$\boldsymbol{\mu}_S = 2\mathbf{S}\frac{e\hbar}{2mc} \tag{9-19}$$

Again $\boldsymbol{\mu}_S$ is oppositely directed to **S** because of the negative charge of the electron. The relationships between the magnetic moments and the angular momenta are shown in Figure 9-18. In the scale chosen $\boldsymbol{\mu}_L$ is drawn twice the length of **L**; hence $\boldsymbol{\mu}_S$ must be drawn four times the length of **S**. The resultant magnetic moment $\boldsymbol{\mu}$ is therefore *not* along **J**. Because the vectors **L** and **S** precess about **J**, $\boldsymbol{\mu}_L$ and $\boldsymbol{\mu}_S$ must also precess about **J**. If each of these vectors is resolved into two components, one parallel to **J** and the other perpendicular to it, the value of the perpendicular component of each vector, averaged over a period of the motion, will be zero, since it is constantly changing direction. The effective magnetic moment of the atom will therefore be $\boldsymbol{\mu}_J$, the sum of the components of $\boldsymbol{\mu}_L$ and $\boldsymbol{\mu}_S$ along **J**. This is given by

$$\mu_J = \mu_L \cos(L, J) + \mu_S \cos(S, J) \tag{9-39}$$

where $\cos(L, J)$ represents the cosine of the angle between **L** and **J** and similarly $\cos(S, J)$ represents the cosine of the angle between **S** and **J**. By applying the cosine law to the triangle formed by **L**, **S**, **J** in a manner analogous to that used in deriving Equation (9-10), we get

$$\cos(L, J) = \frac{L(L+1) + J(J+1) - S(S+1)}{2\sqrt{L(L+1)}\sqrt{J(J+1)}}$$

and

$$\cos(S, J) = \frac{S(S+1) + J(J+1) - L(L+1)}{2\sqrt{S(S+1)}\sqrt{J(J+1)}}$$

Substituting these values in Equation (9-39) as well as the values of $\boldsymbol{\mu}_L$ and $\boldsymbol{\mu}_S$ from Equations (9-16) and (9-19), respectively, we get

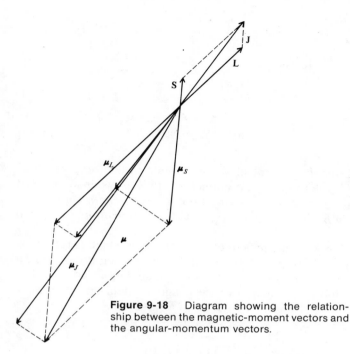

Figure 9-18 Diagram showing the relationship between the magnetic-moment vectors and the angular-momentum vectors.

$$\mu_J = \frac{e\hbar}{2mc} \frac{3J(J+1) + S(S+1) - L(L+1)}{2\sqrt{J(J+1)}}$$

Multiplying the numerator and denominator of the above equation by $\sqrt{J(J+1)}$, we get

$$\mu_J = \frac{e\hbar}{2mc} \sqrt{J(J+1)} \left[1 + \frac{J(J+1) + S(S+1) - L(L+1)}{2J(J+1)} \right]$$

or

$$\mu_J = \frac{e\hbar}{2mc} g\sqrt{J(J+1)} \tag{9-40}$$

where

$$g = 1 + \frac{J(J+1) + S(S+1) - L(L+1)}{2J(J+1)} \tag{9-41}$$

The quantity g is called the Landé *g factor*. It determines the splitting of the energy levels in the presence of a weak external magnetic field and shows that this splitting is determined by the values of L, S, and J. For levels in which the total spin S is zero, $\boldsymbol{\mu}_J$ will be opposite in direction to **L**, and the energy levels will split up in a magnetic field in a manner identical with that shown for the normal Zeeman effect.

Equation (9-40) is the defining equation for the Landé *g* factor. Remembering that

$$M_B = \frac{e\hbar}{2mc}$$

we can rewrite Equation (9-40) as

$$\mu_J = M_B g J$$

so that

$$g = \frac{\mu_J/M_B}{J} \tag{9-42}$$

Now μ_J/M_B is the magnetic moment of the electronic system expressed in Bohr magnetons and J is the total angular momentum of this system expressed in units of $\hbar$. *Thus the Landé g factor is the ratio of the magnetic moment of the system, expressed in Bohr magnetons to its angular momentum, expressed in units of $\hbar$.*

If the atom is placed in a magnetic field of flux density B which is relatively weak so that the coupling between **L** and **S** will not be broken down, the resultant **J** will precess about the direction of the magnetic field as an axis. The additional energy $\Delta\mathcal{E}$ due to the action of the magnetic field on this atomic magnet will be

$$\Delta\mathcal{E} = \mu_J B \cos (J, B)$$
$$= g \frac{e\hbar}{2mc} B \sqrt{J(J+1)} \cos (J, B)$$

But $\sqrt{J(J+1)} \cos (J, B)$ is the projection of the vector **J** on the direction of the magnetic field and is given by the magnetic quantum number m_J; so that

$$\Delta\mathcal{E} = \frac{e\hbar}{2mc} B g m_J \tag{9-43}$$

The quantity

$$\frac{e\hbar}{2mc} B$$

is called a *Lorentz unit;* it is a unit of energy used for expressing the splitting of the energy levels in a magnetic field.

9-17 Anomalous Zeeman Effect

When the light from a sodium flame or arc, which has been placed in a magnetic field of about 30,000 gauss, is examined with the aid of a spectroscope of high resolving power, it is found that each of the lines of the principal series exhibits the following anomalous Zeeman pattern: the longer wavelength component, $3^2S_{1/2} - {}^2P_{1/2}$, splits into four lines, whereas the shorter wavelength component splits into six lines. The splitting of the energy levels giving

TABLE 9-3

State	L	S	J	g	m_J	gm_J
$3\,^2S_{1/2}$	0	$\frac{1}{2}$	$\frac{1}{2}$	2	$\frac{1}{2}, -\frac{1}{2}$	$1, -1$
$3\,^2P_{1/2}$	1	$\frac{1}{2}$	$\frac{1}{2}$	$\frac{2}{3}$	$\frac{1}{2}, -\frac{1}{2}$	$\frac{1}{3}, -\frac{1}{3}$
$3\,^2P_{3/2}$	1	$\frac{1}{2}$	$\frac{3}{2}$	$\frac{4}{3}$	$\frac{3}{2}, \frac{1}{2}, -\frac{1}{2}, -\frac{3}{2}$	$2, \frac{2}{3}, -\frac{2}{3}, -2$

rise to these lines can be determined with the aid of Equation (9-43). This will be done for the sodium D lines as typical of the lines of the principal series. For the $3\,^2S_{1/2}$ energy level, $L = 0$, $S = \frac{1}{2}$, $J = \frac{1}{2}$; hence from Equation (9-41)

$$g = 1 + \frac{\frac{1}{2} \cdot \frac{3}{2} + \frac{1}{2} \cdot \frac{3}{2}}{2 \cdot \frac{1}{2} \cdot \frac{3}{2}} = 2$$

Since m_J can have the values $\frac{1}{2}$ and $-\frac{1}{2}$, gm_J can have the values $+1$ and -1. Table 9-3 gives the values for the quantum numbers necessary for the determination of the splitting factor gm_J for each of the energy levels of the sodium D lines.

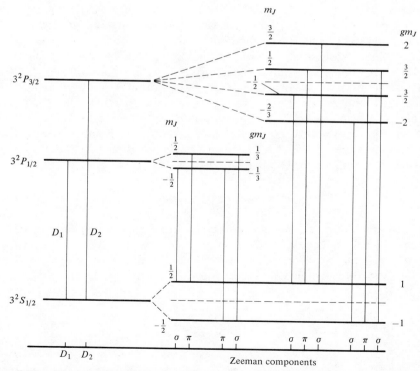

Figure 9-19   Diagram showing the splitting of the energy levels of the sodium D lines in a weak magnetic field, and the transitions giving rise to the anomalous Zeeman pattern according to the selection rule $\Delta m_J = 0, \pm 1$.

Figure 9-19 shows the Zeeman components that appear when the light from the source is viewed perpendicular to the direction of the magnetic field. The lines designated by σ are polarized with the electric vector perpendicular to the direction of the magnetic field, whereas the lines designated by π are polarized parallel to the magnetic field. The polarization of these lines can be predicted from wave mechanical considerations which lead to the result that transitions for which $\Delta m_J = 0$ give rise to the π components and the transitions for which $\Delta m_J = \pm 1$ give rise to the σ components. When the light is viewed parallel to the magnetic field, only the lines for which $\Delta m_J = \pm 1$ appear and these are now circularly polarized. The direction of circular polarization for $\Delta m_J = +1$ is opposite to that for which $\Delta m_J = -1$. Thus the introduction of electron spin has led to complete agreement between the experimental results and the theory of the anomalous Zeeman effect.

9-18   The Stern-Gerlach Experiment and Electron Spin

A direct experimental demonstration of the existence of the magnetic moment of an electron, particularly that due to its spin, is given in an experiment first performed by Stern and Gerlach (1921), using neutral silver atoms. Similar experiments were performed later with other atoms such as hydrogen, lithium, sodium, potassium, copper, and gold. A glance at Table 9-2 will show that the normal state of each of these atoms is a $^2S_{1/2}$ state, for which $L = 0$, $J = S = \frac{1}{2}$; that is, in the normal state of a silver atom its entire magnetic moment is due to the spin of one electron. It has been shown that when such atoms are placed in a magnetic field of induction B they become oriented in such directions that m_J, the projection of J in the direction of the magnetic field, can have the two values $+\frac{1}{2}$ and $-\frac{1}{2}$ only. The additional energy of the atom due to its position in the magnetic field is given by

$$\Delta \mathcal{E} = \frac{e\hbar}{2mc} B g m_J \tag{9-43}$$

If these small atomic magnets are placed in a uniform magnetic field, they will experience torques which will orient them with respect to the magnetic field. If the magnetic field is inhomogeneous, each atomic magnet will also experience a force which will accelerate it. The magnitude of this force on

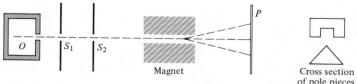

Figure 9-20   Arrangement of apparatus in the Stern-Gerlach experiment.

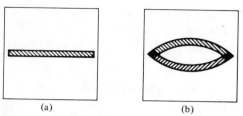

(a)   (b)

Figure 9-21 Type of pattern on photographic plate made by a beam of silver atoms (a) without magnetic field on, (b) with magnetic field on.

each magnet can be determined by differentiating Equation (9-43) with respect to the space coordinate, say x, which yields

$$F = \frac{e\hbar}{2mc}\frac{\partial B}{\partial x} \cdot gm_J \tag{9-44}$$

where $\partial B / \partial x$ determines the inhomogeneity of the magnetic field.

In the Stern-Gerlach experiment a narrow beam of silver atoms coming from an oven O, after passing through the defining slits S_1 and S_2, was allowed to pass through an inhomogeneous magnetic field and recorded on plate P, as shown in Figure 9-20. The entire apparatus was in an evacuated chamber. The inhomogeneous magnetic field was produced by an electromagnet with specially designed pole pieces. One pole piece was in the form of a knife-edge, whereas the other had a channel cut in it parallel to the knife edge. Each silver atom could assume only one of two possible directions in the magnetic field, given by $m_J = +\frac{1}{2}$ or $-\frac{1}{2}$. It has been shown that for the $^2S_{1/2}$ state, $g = 2$, and $gm_J = +1$ or -1, so that the force experienced by each atom due to the inhomogeneity of the field is

$$F = \pm \frac{e\hbar}{2mc} \cdot \frac{\partial B}{\partial x} = \pm M_B \frac{\partial B}{\partial x} \tag{9-45}$$

where

$$M_B = \frac{e\hbar}{2mc} = 0.927 \times 10^{-20} \text{ erg gauss}^{-1}$$

In terms of the vector model, those atoms with electron spins directed parallel to the magnetic field will experience a force in one direction, whereas those with oppositely directed spins will experience a force in the opposite direction. According to this, the beam of atoms should split into two beams in its passage through the inhomogeneous magnetic field. This splitting of the beam into two parts of approximately equal intensity was actually observed in these experiments. Figure 9-21 shows the type of pattern observed in these experiments. From the amount of the separation of the two beams and the degree of inhomogeneity of the magnetic field, it was shown that the component of the magnetic moment of the atom in the direction of the field was equal to one Bohr magneton, M_B.

The results of the Stern-Gerlach experiment, together with the explanation of the multiplicity of atomic energy levels and the anomalous Zeeman effect, strongly support the hypothesis of the existence of an electron spin.

9-19 Fine Structure of the Hydrogen Energy Levels

Hydrogen atoms and hydrogenlike ions, such as He^+, Li^{2+}, and Be^{3+}, have similar electronic structures: a single electron revolving around a nucleus. Hence their energy levels can be designated in a manner similar to that of sodium. Figure 9-22 shows the energy levels of hydrogen for $n = 1, 2, 3$, and 4 in the spectroscopic notation. Because hydrogen atoms have the simplest structure, they have been the subject and object of all theories concerning the structure of the atom, atomic processes, and interactions of atoms and radiation. As mentioned in the preceding chapter, the Schroedinger theory is not capable of predicting these energy levels, since it does not take the spin of the electron into account. The Dirac theory, which does take spin into consideration, predicts the fine structure of the energy levels. There were some slight differences between the predictions of this theory and experimental results; these differences were removed by the application of quantum electrodynamics, which considered, among other things, the interaction between the

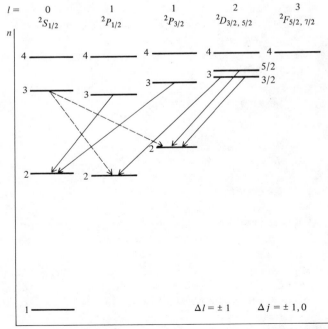

Figure 9-22   Some energy levels of hydrogen. Only transitions corresponding to the H_α line are shown.

electron and the radiation field. Experiments performed by Lamb and Rether-ford, using radiofrequency techniques, verified these theoretical predictions.

Among the important results obtained by Lamb and Retherford are (a) that the $2S$ level is shifted upward so that the $2^2S_{1/2}$ level lies above the $2^2P_{1/2}$ level by 0.0353 cm^{-1} and (b) that the separation between the $2^2P_{3/2}$ and $2^2P_{1/2}$ levels is 0.3652 cm^{-1}. In the energy-level diagram of Figure 9-22 the transi-tions are shown only for the H_α line; these are the transitions allowed by the selection rules. The intensities of the seven components vary considerably. The most intense components are those corresponding to the transitions $^2D_{5/2} \rightarrow {}^2P_{3/2}$ and $^2D_{3/2} \rightarrow {}^2P_{1/2}$ in the approximate ratio of 2:1.

9-20 Spectra of Two-Electron Atoms

Atoms of the alkaline earth elements, beryllium, magnesium, calcium, strontium, barium, and radium, have two electrons outside a closed con-figuration typical of the inert elements. The total angular momentum of any one of these atoms is merely the sum of the angular momenta of the two elec-trons outside the closed core. When the atom is in the normal state, both electrons have the same principal quantum number n and are in the completed subgroup for which $l = 0$. The application of Pauli's principle leads to the conclusion that the two electrons must have their spins in opposite directions, so that the total spin quantum number $S = s_1 + s_2 = \frac{1}{2} - \frac{1}{2} = 0$. The total angular momentum $J = L + S = 0$, so that the normal state is a singlet state and is designated by the symbol $^1S_0$.

The excited states of a two-electron system can arise from a variety of con-figurations of the two electrons consistent with the Pauli exclusion principle. The P states, for example, are due to those combinations of the angular momenta l_1 and l_2 whose vector sum $L = l_1 + l_2$ is unity. For the D states $L = l_1 + l_2 = 2$ and so on. The spin vector S can have only two values $S = \frac{1}{2} + \frac{1}{2} = 1$ or $S = \frac{1}{2} - \frac{1}{2} = 0$. If $S = 0$, then $J = L$, and the state is a singlet state such as $^1P_1$ for $L = 1$, $^1D_2$ for $L = 2$, and so on. If $S = 1$, then J can have the three values $L + 1$, L, $L - 1$, yielding a triplet state such as $^3P_2$, $^3P_1$, $^3P_0$, for $L = 1$, $^3D_3$, $^3D_2$, $^3D_1$ for $L = 2$, and so on. A two-electron system therefore has two distinct sets of energy levels—the singlet set arising from configurations in which the electron spins are in opposite directions, $S = 0$, and the triplet set arising from configurations in which the electron spins are in the same direction, $S = 1$. In the latter set must be included those states for which $L = 0$ and $S = 1$, even though J has only the single value unity. These states are designated as $^3S_1$ states and are customarily called triplet states.

The energy-level diagram of calcium showing some of the singlet and triplet states is given in Figure 9-23. These energy levels are for those con-figurations in which one electron always remains in the $4s$ state ($n_1 = 4$, $l_1 = 0$), whereas the second electron changes its state. Other energy levels, in which neither electron is in the $4s$ level, are known to exist, but they will not be

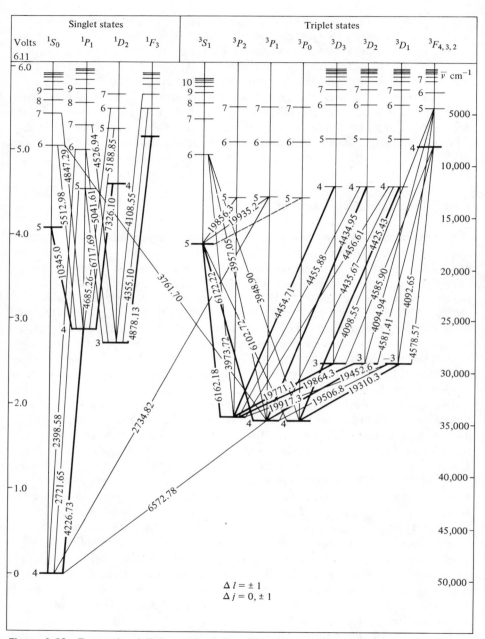

Figure 9-23 Energy-level diagram of calcium. Wavelengths are in angstroms.

considered here. Some of the transitions giving rise to spectral lines are indicated in the figure. It is found that, as a general rule, singlet terms combine with other singlet terms and triplet terms combine with other triplet terms.

Intercombination lines due to transitions between triplet and singlet states have been found but are few in number and are usually less intense than the nonintercombination lines. Several intercombination lines are shown in the figure.

When the changes in the atomic states are produced by jumps of a single electron, only those spectral lines will be observed that are permitted by the selection rule $\Delta l = \pm 1$. In such cases the selection rule for L is also $\Delta L = \pm 1$. The selection rule for J is $\Delta J = 0$ or ± 1, with the exception that the transition from $J = 0$ to $J = 0$ is forbidden. Using the principles of Section 7-8, it is possible to understand why $J = 0$ to $J = 0$ transitions are forbidden. The wave function for a state with $J = 0$ does not depend on the angles θ and ϕ of a spherical coordinate system. The dipole operator behaves like a vector and is a quantity that is proportional to $\cos \theta$. The matrix element for the transition is then proportional to $\int_0^\pi \cos \theta \sin \theta \, d\theta$, a quantity that vanishes. Since the matrix element vanishes, the transition is forbidden.

Transitions between singlet states produce spectral series consisting of singlet lines. Transitions between triplet states produce spectral series consisting of lines that exhibit fine structure when examined with instruments of high resolving power. The lines of the principal series, due to the transitions $5^3S_1 - n^3P_{2,1,0}$, never have more than three components. Similarly the lines of the sharp series, due to the transitions $4^3P_{2,1,0} - n^3S_1$, may have three components. The lines of the diffuse series $4^3P_{2,1,0} - n^3D_{3,2,1}$ and of the fundamental series $3^3D_{3,2,1} - n^3F_{4,3,2}$ may have as many as six components.

If a calcium atom should find itself in a 4^3P_2 state or a 4^3P_0 state, then, according to the selection rules, it will not be able to return to the normal state with the emission of radiation. Such states are called *metastable* states. The atom may go from the metastable state to the normal state if it gives up the appropriate amount of energy to another atom during a collision process. The atom may absorb radiation that will raise it from the metastable state to a higher energy state, from which, selection rules permitting, it may return to the normal state with the emission of radiation.

It is interesting to note that the lines of the singlet series exhibit the normal Zeeman effect, whereas the lines of the triplet series exhibit the anomalous Zeeman effect. This is in agreement with deductions from the Landé g formula. For the singlet series $S = 0$, $J = L$, and therefore $g = 1$, so that the splitting up of the energy levels will be in whole multiples of the Lorentz unit and the permitted transitions will produce the normal Zeeman pattern.

In addition to the alkaline earth elements, there are other elements such as zinc, cadmium, and mercury which possess singlet and triplet sets of atomic energy levels determined from spectroscopic analysis. Examination of Table 9-2 shows that these atoms contain two electrons outside closed configurations. Helium, which possesses only two electrons, must also be included in the above group. An interesting point about helium is that no intercombination lines due to transitions between its triplet and singlet states have ever been found. Because of this fact it was at one time supposed that there were two different kinds of helium: parhelium, possessing singlet states, and ortho-helium, possessing triplet states. The difference between the two sets of states is, of course, due to the two possible orientations of the electron spin axes.

9-21 The Laser

One of the most interesting and useful devices developed in the last decade is the *laser,* a source of *coherent* optical radiation. The term laser is an acronym of the words *l*ight *a*mplification by the *s*timulated *e*mission of *r*adiation. The laser is an outgrowth of the earlier development of the maser (1954) by C. H. Townes of the United States and N. Basov and A. M. Prokhorov of the U.S.S.R; the term maser is an acronym for the words *m*icrowave *a*mplification by the *s*timulated *e*mission of *r*adiation. In 1958 A. L. Shawlow and C. H. Townes conceived the idea of extending the principle of the maser to the optical region, and the first optical maser or lasers was produced by T. H. Maimon in 1960. There are now many different types of laser available commercially and used for a variety of scientific and industrial purposes. There is also an extensive literature on the subject; a few references to it will be found at the end of this chapter. Here we simply present the principles at the basis of the operation of a laser that utilizes gases, such as the helium-neon laser.

Two basic ideas are involved in the operation of a laser; one is that of the stimulated emission of radiation, first introduced by Einstein (1917) and used in the development of the Planck radiation equation (Section 4-7), and the other is that of the amplification of a particular frequency of the radiation emitted by a system. Let us consider two energy levels of an atom, $\mathcal{E}_2$ and $\mathcal{E}_1$, with $\mathcal{E}_2 > \mathcal{E}_1$. If the atom is in state $\mathcal{E}_2$ and the transition from $\mathcal{E}_2$ to $\mathcal{E}_1$ is an allowed transition, the atom will go spontaneously from $\mathcal{E}_2$ to $\mathcal{E}_1$ in a comparatively short time of the order of $10^{-7} - 10^{-8}$ sec with the emission of radiation of frequency ν given by

$$\mathcal{E}_2 - \mathcal{E}_1 = h\nu \tag{9-46}$$

In a gas containing a large number of such atoms radiation of this frequency will be emitted by many of them in a short time interval with no definite phase

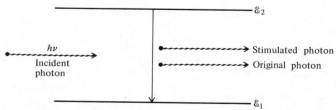

Figure 9-24 Atom in state $\mathcal{E}_2$ stimulated to emit photon by incident radiation.

relationships between them; the radiation emitted is said to be *incoherent* radiation and is the type usually emitted by a typical source.

Another method by which the atom can go from state $\mathscr{E}_2$ to state $\mathscr{E}_1$ is by the action of an external radiation field of frequency ν and density $\rho(\nu)$ (see Fig. 9-24). If an atom in state $\mathscr{E}_2$ is stimulated to emit a photon of frequency ν by a photon of the external radiation field, these two photons, the incident photon and the emitted photon, will travel in the same direction in the same phase. Because of this definite phase relationship the radiation is said to be coherent. If we can increase the number of atoms in state $\mathscr{E}_2$ appreciably, the stimulated emission from one atom can become the external radiation field to stimulate the emission from other atoms and thus build up the intensity of the coherent radiation, all of frequency ν, to a very large value. This cannot be done with only two energy levels; at least three energy levels are needed for the success of this process.

In any large assemblage of atoms of a substance at temperature T, the number of atoms n_i in energy state $\mathscr{E}_i$ is given by the Boltzmann distribution law

$$n_i = N \exp\left(-\frac{\mathscr{E}_i}{kT}\right) \tag{9-47}$$

where k is the Boltzmann constant. Thus most of the atoms are in the lowest or normal energy state which we shall designate as $\mathscr{E}_0$. In order to increase the number of atoms in an excited state $\mathscr{E}_2$ above the amount normally in this state, it is necessary to supply energy from an outside source. Several such methods are available; for example, light of frequency $\nu_{2,0} = (\mathscr{E}_2 - \mathscr{E}_0)/h$ can be incident on the material. This is called optical pumping and has been known for many years. Another is to bombard the atoms with electrons of the appropriate energy $(\mathscr{E}_2 - \mathscr{E}_0)$. A third method involves collisions between (1) atoms of a different element that have been raised to an energy state higher than $\mathscr{E}_2$ and (2) atoms of the substance in state $\mathscr{E}_0$.

Another condition necessary for the operation of a laser is that the state $\mathscr{E}_2$ be a metastable state, that is, one in which the lifetime is slightly larger than a lower state $\mathscr{E}_1$ so that the population n_2 of state $\mathscr{E}_2$ can be built up to a larger value than n_1 of state $\mathscr{E}_1$. The lifetime of $\mathscr{E}_2$ can be of the order of microseconds to milliseconds if that of $\mathscr{E}_1$ of the order of 10^{-7}–10^{-8} sec. If $n_2 > n_1$, stimulated emission from $\mathscr{E}_2$ to $\mathscr{E}_1$ can result in an intense beam, that is, in amplification, of the radiation emitted in the process. It was shown in Chapter 4 that the probability of emission of stimulated radiation B_{21} is proportional to the density of the radiation $\rho(\nu_{21})$ in the region containing the atoms in state $\mathscr{E}_2$. The density $\rho(\nu_{21})$ can be increased by putting the substance in a *resonant cavity*. If the substance is in the form of a gas, it can be put into a tube of length L whose ends are flat mirrors, M_1 and M_2 as illustrated in Figure 9-25. The distance between the mirrors is chosen so that the radiation traveling between them will be reinforced by constructive interference and the radiation field will be coherent. Another method commonly used in gas lasers is to place the mirrors outside the tube both to increase the path length and for

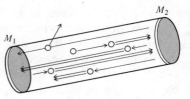

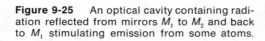

Figure 9-25 An optical cavity containing radiation reflected from mirrors M_1 to M_2 and back to M_1 stimulating emission from some atoms.

easier adjustment. Frequently these mirrors are curved, with their principle axes coinciding with the axis of the tube. The mirrors are usually made of several thin coatings of a dielectric material with a reflection coefficient of 99 percent for the particular wavelength incident on it.

Let us assume that the gas in the tube shown in Figure 9-25 has three energy levels $\mathcal{E}_2$, $\mathcal{E}_1$, and $\mathcal{E}_0$, with $\mathcal{E}_2 > \mathcal{E}_1 > \mathcal{E}_0$. Suppose that an atom in state $\mathcal{E}_2$ emits a photon of frequency ν_{21} spontaneously. If the direction of emission is at some angle other than zero to the axis of the tube, the photon will leave through the wall. If the direction of emission is parallel to the axis, it may go toward the mirror M_1 and be reflected toward M_2 and back again. Some time during this motion it may encounter another atom in state $\mathcal{E}_2$ and stimulate it to emit a photon of frequency ν_{21}. The action of the electromagnetic field of the incident photon is such that it will cause the two photons, the incident and emitted ones, to travel in the same direction in the same phase. These two photons, with their electromagnetic fields in phase, will now travel between the two mirrors and can stimulate two other atoms in state $\mathcal{E}_2$ to emit radiation in phase with them. If, by any one of the methods mentioned earlier, the number of atoms n_2 in state $\mathcal{E}_2$ has been raised to a value greater than n_1, the number of atoms in state $\mathcal{E}_1$, the coherent radiation of frequency ν_{21} can have its intensity built up to a very large value. Most of the atoms in state $\mathcal{E}_1$ will go spontaneously to the state $\mathcal{E}_0$ in a comparatively short time with the emission of radiation of frequency $\nu_{1,0}$. This radiation is incoherent and will leave through the walls of the tube. Very few atoms in state $\mathcal{E}_1$ will be stimulated to absorb radiation of frequency ν_{21} by the radiation field.

The coherent radiation leaves the tube through a small aperture in one of the mirrors. It may do so continuously, or some type of shutter may be used to allow the intensity of the radiation in the tube to be built up to a high value and then released as a short pulse of radiation.

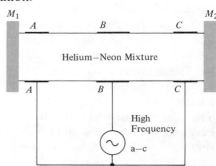

Figure 9-26 Helium-neon laser with external electrodes A, B, and C connected to source of high frequency alternating current. M_1 and M_2 are mirrors.

One of the most common gas lasers is the helium-neon laser because one of the spectral lines emitted by this laser is in the visible region; it is a red line of wavelength 6328 Å. Figure 9-26, a simplified diagram of such a laser, illustrates its method of operation. In one type the tube contains a mixture of helium and neon in the ratio of about 5:1 at a pressure of about 0.6 mm of mercury; that is, there are about 5 atoms of helium to 1 atom of neon. The atoms that are stimulated to emit the coherent radiation are those of neon. Atoms of helium and neon are bombarded by electrons and some are raised to excited states. The electrons are produced by means of an external high-frequency electromagnetic field whose electrodes are outside the tube. Electrons of sufficient energy raise some helium atoms to an energy state designated as $\mathcal{E}_3$; this state is a metastable state (see Fig. 9-27). Some of these excited helium atoms give up their energy in radiationless collisions with neon atoms, raising them to state $\mathcal{E}_2$; other neon atoms may also be raised to state $\mathcal{E}_2$ by electron bombardment. Initially a few atoms of neon in state $\mathcal{E}_2$ emit photons of wavelength 6328 Å in going to energy state $\mathcal{E}_1$. Some of these photons that are emitted parallel to the axis of the tube travel back and forth between mirrors M_1 and M_2 and constitute the radiation field $\rho(\nu_{21})$ that can now stimulate other neon atoms in state $\mathcal{E}_2$ to emit photons of the same frequency. The energy state $\mathcal{E}_1$ has a comparatively short lifetime of the order of 10^{-8} sec and goes to state $\mathcal{E}_0'$ with the emission of incoherent photons of wavelength 6680 Å. This particular state is itself metastable and is depopulated

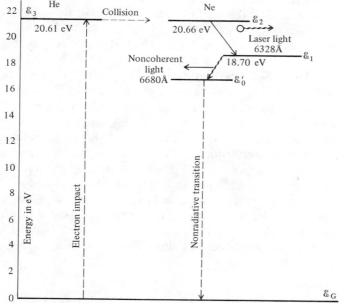

Figure 9-27 Helium-neon energy levels involved in the production of red laser light of 6328 Å wavelength.

by collisions with the walls of the tube; during these collisions, the excited atoms give up their energy in radiationless transitions to the ground state $\mathcal{E}_G$. The ratio of the length of the tube to its diameter is chosen so that these collisions occur often enough.

The operation of the helium neon laser for the production of the coherent radiation of wavelength 6328 Å involves four energy levels of neon as shown in Figure 9-27. The helium atoms are essential for inverting the population of the excited states of neon so that $n_2 > n_1$. The light is emitted by a helium-neon laser as a parallel beam through an aperture in one of the mirrors. There is a slight angular width θ of the beam which is of the order of magnitude of the diffraction angle, that is, λ/d, where d is the diameter of the beam of light. The radiation field builds up to density $\rho(\nu_{21})$, which is called a saturation value when the rate at which energy leaves the tube is equal to the energy produced in stimulated emission.

The technology and applications of lasers have progressed greatly in recent years. Lasers have been made of a variety of materials. We have already mentioned gases. Some liquids and many solids are useful for this purpose. Many different frequencies from microwaves through ultraviolet are now available as laser light. Some lasers, such as the helium-neon gas laser, can run continuously; others are used in a pulsed mode with the light emitted in short bursts. Many types of application are still in the early stages of development; it is to be expected that existing uses will expand and new uses will be invented. Laser light is superior to light emitted from usual sources in three ways: (a) the intensity can be made much higher; (b) the light can be monochromatic to better accuracy; and (c) the light is coherent. Each of these three properties is valuable in applications.

The high intensity is used in cases in which it is desired to concentrate a great deal of energy in a small region. In this way a lens is used to focus laser light onto materials for cutting metal, for drilling small holes, or for reattaching a detached retina in a human eye. High intensity of light implies a large electric field, which can be used in research to study nonlinear optical properties of materials.

Properties (b) and (c) are not really independent. Taken together, they make it possible to perform interferometry over great distances, permitting greatly increased precision in alignment and in measurement of distances. The Stanford Linear Accelerator (two miles in length; see Section 12-5) is aligned by a laser. The property of coherence has led to the science of holography, which is photography with laser light. The photographic film from such photography looks strange when viewed in ordinary light, but when viewed using laser light the original object appears to be three dimensional.

Problems

9-1. Using vector diagrams, determine the different values for the total orbital angular momentum of a two-electron system for which $l_1 = 3$ and $l_2 = 2$.

9-2. Using vector diagrams, determine the possible values of the total angular momentum of an f electron (a) according to the vector model, (b) according to wave mechanics. (c) Determine the angle between the vectors s and l in part (b).

9-3. Using vector diagrams, determine the possible values of the total angular momentum of an electron system for which (a) $L = 2$, $S = 3$, (b) $L = 3$, $S = 2$.

9-4. Using vector diagrams, determine the possible values of the total angular momentum of an electron system for which (a) $L = 3$, $S = \frac{5}{2}$, (b) $L = 2$, $S = \frac{5}{2}$.

9-5. Determine the electron configuration of (a) barium in the normal state and (b) mercury in the normal state.

9-6. Determine the angular velocity of the precessional motion of an electron orbit when a source of light is placed in a magnetic field of 30,000 gauss. Compare this precessional velocity of the orbit with the change in the angular velocity of the electron on the basis of the classical Lorentz theory.

9-7. Determine the maximum change in the energy of a p electron due to the precessional motion of its orbit in a magnetic field of 30,000 gauss.

9-8. Draw a diagram to show the relative separations of the sodium D lines and their Zeeman components produced by a magnetic field of 30,000 gauss. Use a wave-number scale.

9-9. (a) Using the wavelengths given in the energy level diagram of calcium, determine the values of the two lowest resonance potentials. (b) If electrons of 2.8 eV energy are used to excite the calcium atoms, which spectral lines will be emitted by calcium?

9-10. Two resonance potentials have been observed with mercury vapor, 4.86 and 6.67 volts. The ionization potential of mercury is 10.43 volts. Compute (a) the wavelengths of the mercury resonance radiation, (b) the wave number of the lowest state of mercury. Check your results by reference to standard tables or energy-level diagrams.

9-11. The wavelengths of the lines obtained on a spectrogram were measured and classified into three series as follows:

Principal series	Sharp series	Diffuse series
6707.85 Å	8126.5 Å	6103.5 Å
3232.6	4971.9	4603.0
2741.3	4273.3	4132.3
2562.5	3985.8	3915.0
2475.3	3838.2	3794.7

From the above data the series limit, expressed in wave numbers, was determined as $43,486 \text{ cm}^{-1}$ for the principal series and $28,582 \text{ cm}^{-1}$ for the sharp and diffuse series.

(a) Convert the wavelengths given above to wave numbers.
(b) Construct an energy-level diagram using a wave-number scale.
(c) Determine the ionization potential of this element and then place a voltage scale on the energy-level diagram.
(d) Identify the element.
(e) Determine the first resonance potential of this element.
(f) Determine the principal quantum numbers for each energy level.

9-12. A Stern-Gerlach type of experiment is performed with a narrow beam of free electrons emitted by a hot filament. The electrons are accelerated to a velocity of 10^9 cm/sec and then enter a long inhomogeneous magnetic field where the gradient is 500 gauss/cm. (a) What is the Landé g value for free electrons? (b) Determine the force on these electrons in the magnetic field. (c) What should the length of path be to produce a separation of 2 cm between electrons of opposite spins when they emerge from the magnetic field?

9-13. Discuss the feasibility of performing a Stern-Gerlach type of experiment with a narrow beam of protons coming from a hydrogen discharge tube operated at 50,000 volts. If the magnetic-field gradient was the same as that in Problem 9-12, (a) what force would the protons experience and (b) how long would the path have to be to produce a separation of 2 cm between oppositely oriented protons? (c) What value of the magnetic-field gradient would be needed to produce a separation of 0.4 cm in a path length of 2000 cm of the oppositely oriented proton spins?

9-14. (a) Sketch an energy-level diagram for helium, including states with $n \leqq 4$. (b) On your diagram indicate all of the allowed transitions.

9-15. Two electrons have orbital angular momenta $l_1 = 1$ and $l_2 = 3$. (a) What are the possible values of L? (b) What are the possible values of S? (c) What are the possible values of J? (d) Write the spectroscopic notation for all the states available to these two electrons.

9-16. What is the maximum number of electrons in an atom that can have the same value n for the principal quantum number?

9-17. (a) What are L, S, and J for the ground state of the carbon atom? (b) What other terms are possible for excited states of carbon which have the same configuration $(1s^2\ 2s^2\ 2p^2)$ as the ground state?

Problems for Students Who Know Some Matrix Algebra

The quantum mechanics of electrons (or any particles with spin $\frac{1}{2}$) can be expressed in a matrix form. Wave functions are two-component column vectors.

Wherever ψ^* was used before, we now need $\psi^\dagger$, where the dagger means transpose to a row vector and perform complex conjugation. The components of the spin angular momentum are 2×2 matrices:

$$S_x = \frac{1}{2}\begin{pmatrix} 0 & 1 \\ 1 & 0 \end{pmatrix} \qquad S_y = \frac{1}{2}\begin{pmatrix} 0 & -i \\ i & 0 \end{pmatrix} \quad \text{and} \quad S_z = \frac{1}{2}\begin{pmatrix} 1 & 0 \\ 0 & -1 \end{pmatrix}$$

9-18. Show that $S_x{}^2 + S_y{}^2 + S_z{}^2$ is equal to the identity matrix multiplied by $S(S + 1)$.

9-19. For a spin state $\begin{pmatrix} 1 \\ 0 \end{pmatrix}$ find $<S_x>$, $<S_y>$ and $<S_z>$.

9-20. Repeat Problem 9-19 for the state $\begin{pmatrix} 0 \\ 1 \end{pmatrix}$.

9-21. Using expectation values, show that the $\dfrac{1}{\sqrt{2}}\begin{pmatrix} 1 \\ 1 \end{pmatrix}$ corresponds to a state with the spin in the $+x$ direction.

9-22. Show that $2S_x$ is the flip operator (the operator that reverses the z component of the spin).

9-23. Show that $S_xS_y - S_yS_x = iS_z$. This is an example of the noncommutative property of matrix algebra. We often denote $S_xS_y - S_yS_x$ by $[S_x, S_y]$

9-24. Evaluate $[S_y, S_z]$.

9-25. Evaluate $[S_z, S_x]$.

9-26. Evaluate $[S_x, S^2]$ and similarly for S_y and S_z.

References

Born, M., *Atomic Physics*. New York: Hafner Publishing Company, 1959, Chapter VI.

Candler, C., *Atomic Spectra and the Vector Model*. Princeton: D. Van Nostrand Company, 1964.

Heavens, O. S., *Optical Masers*. New York: John Wiley & Sons, 1964.

Herzberg, G., *Atomic Spectra and Atomic Structure*. New York: Dover Publications, 1944, Chapters I and II.

Leighton, R. B., *Principles of Modern Physics*. New York: McGraw-Hill Book Company, 1959, Chapters 5-8.

Levine, A. K. (Ed.), *Lasers*. New York: Marcel Dekker, 1966.

Marshall, S. L. (Ed.), *Laser Technology and Applications*. New York: McGraw-Hill Book Company, 1968, Chapters 1-6.

Melia, T. B., *An Introduction to Masers and Lasers*. London: Chapman and Hall, 1967.

Moore, Charlotte E., *Atomic Energy Levels*. Washington, D. C.: National Bureau of Standards, Circular 467, Vol. I, 1949; Vol. II, 1952; Vol. III, 1958.

Sawyer, R. A., *Experimental Spectroscopy*. Englewood Cliffs, N. J.: Prentice-Hall, 1951.

Steele, Earl L., *Optical Lasers in Electronics*. New York: John Wiley & Sons, 1968.

White, H. E., *Introduction to Atomic Spectra*. New York: McGraw-Hill Book Company, 1934.

Wood, R. W., Physical Optics. New York: Macmillan Company, 1934, Chapters V, XVII, and XXI.

10 | X-Ray Spectra

10-1 Characteristic X-Ray Spectra

Moseley (1913) made a systematic investigation of the characteristic x-ray spectra of the elements. The elements investigated were used as targets in x-ray tubes and the radiation from each target was analyzed with the aid of a single crystal spectrometer. A potassium ferrocyanide crystal was mounted on the spectrometer table and the spectrum was recorded on a photographic plate. The spectrometer and the photographic plate were placed in an evacuated chamber to avoid absorption of the long wavelength x-rays in the air. The spectral lines observed were grouped into two series, a short wavelength group known as the K series and a comparatively long wavelength group known as the L series. The wide separation of these two series of lines is illustrated in Figure 10-1 for the case of silver in which the K series wavelengths extend from 0.486 to 0.563 Å and the L series lines are in the wavelength range 3.3 to 4.7 Å. Other investigators have found two other series of lines of still longer wavelengths in the heavier elements, $Z > 66$, classified as M series and N series.

Moseley found that the character of a given series was almost the same for all the elements studied and that the frequency of a particular line of a series

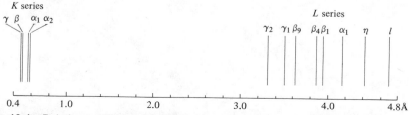

Figure 10-1 Relative positions of the K and L x-ray series spectral lines of silver.

309

varied in a regular manner from element to element in the periodic table. By plotting the square root of the frequency of one of the lines, say the K_α line (the most intense line of the K series), against the atomic number of the element emitting this line, Moseley obtained a straight line. Figure 10-2 shows such a graph, now known as a Moseley diagram. The K_α line is actually a doublet; in Moseley's work this doublet was not resolved but appeared as a single line. The equation of any one of the lines on a Moseley diagram, to a good approximation, is given by

$$\nu = C(Z - a)^2 \tag{10-1}$$

where C and a are constants. For the K_α line C was found to be equal to $\frac{3}{4}Rc$, where R is the Rydberg constant, c is the speed of light, and a was found to be nearly 1. The equation for the frequency of the K_α line of any element can therefore be written as

$$\nu_{K_\alpha} = \tfrac{3}{4}Rc(Z - 1)^2 \tag{10-2}$$

It must be remembered that Moseley did this work more than a half century ago. Some of the elements now known were then unknown. The atomic number of an element merely represented its position in the arrangement of the elements according to their atomic weights. In order to obtain a straight line for the curve, $\sqrt{\nu}$ against Z, Moseley had to rearrange the orders of nickel and cobalt, assigning the lower atomic number to the element of higher atomic weight. Furthermore, a gap had to be left at $Z = 43$ to show the existence of an element, technetium, then unknown.

Moseley's work followed closely on the introduction of Rutherford's nuclear

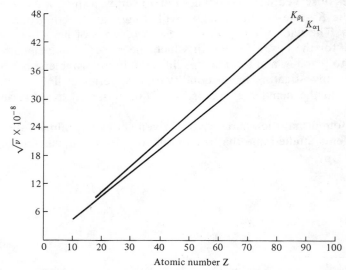

Figure 10-2 Moseley diagram in which the square root of the frequency is plotted against the atomic number of the emitting element for two lines of the K series.

theory of the atom and Bohr's theory of hydrogen. The relationship between Bohr's theory and Moseley's work can best be shown by rewriting Equation (10-2) for the frequency of the K_α line to read

$$\nu_{K_\alpha} = cR(Z - 1)^2\left(\frac{1}{1^2} - \frac{1}{2^2}\right) \tag{10-3}$$

The interpretation of this equation on the Bohr theory is that the K_α line is emitted when an electron goes from the orbit of principal quantum number $n = 2$ to the orbit of principal quantum number $n = 1$. The appearance of the factor $(Z - 1)$ rather than Z in Equation (10-3) can be explained by assuming that the electron which goes from orbit $n = 2$ to $n = 1$ is "screened" from the total nuclear charge Z by the negative charge of a single electron. This explanation can best be understood by considering the manner in which x-rays are produced. The element in the target consists of neutral atoms in which the first shell, $n = 1$, contains two electrons and, according to Pauli's principle, no more electrons can get into this K shell. The only time an electron can go from the L shell, $n = 2$, to the K shell is when one of the electrons is missing from the K shell. The obvious inference is that during the operation of the x-ray tube a cathode ray knocks out an electron from the K shell of an atom. Since most of the other shells have their full quota of electrons, this K electron will have to go either to one of the unoccupied outer levels or completely outside the atom, depending on the amount of energy transferred to the atom by the incident cathode ray. As a result of this process the K shell will now have only one electron in it. If an electron from the L shell should go into the K shell, it will do so with the emission of a quantum of radiation whose frequency is that of the K_α line. The electron that goes from the L shell to the K shell moves in an electric field that is essentially that of the positive nuclear charge and the negative charge of the single electron still remaining in the K shell. This electric field is therefore equivalent to that of a positive charge of magnitude $(Z - 1)e$. The effect of the outer electrons on this electric field can be shown to be very small.

10-2 X-Ray Energy-Level Diagram

A simplified energy-level diagram can be used to show the changes in atomic configuration that give rise to the K and L series of x-ray spectral lines. In this diagram (Fig. 10-3) the zero energy level is taken as that of the ground state of the neutral atom. This differs from the optical case in which the zero level of energy is that of the ionized atom. Let us assume that the electron that is incident on an atom has sufficient energy to remove one electron from the K shell to the outside of the atom. If $\mathcal{E}_K$ represents the work done in removing this K electron, the energy of the system can be represented at a level $\mathcal{E}_K$ above the zero level. This atom is now ionized with one electron missing from the K shell.

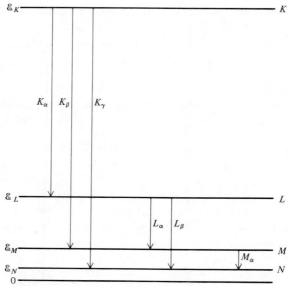

Figure 10-3 Simplified x-ray energy-level diagram.

Let us consider the same neutral atom once more and suppose that the impinging electron lacks sufficient energy to remove an electron from the K shell, but has sufficient energy to remove one from the L shell to the outside. Then, if the atom is ionized by the removal of one electron from the L shell, the energy of the system will be $\mathcal{E}_L$ and can be represented by a line at the proper height above the zero level. Similarly, $\mathcal{E}_M$ represents the work done in ionizing a neutral atom by removing an electron from the M shell, and $\mathcal{E}_N$ represents the work done in removing an electron from the N shell of a neutral atom.

Suppose that the atom is now in the energy state $\mathcal{E}_K$; that is, one electron is missing from the K shell. If an electron goes from the L shell to the K shell, the atom will then be in the energy state represented by $\mathcal{E}_L$; that is, one electron will now be missing from the L shell. The frequency of the spectral line radiated when an electron goes from the L to the K shell, or when the energy state of the atom is changed from $\mathcal{E}_K$ to $\mathcal{E}_L$, is given by Bohr's frequency condition:

$$\nu_{K_\alpha} = \frac{\mathcal{E}_K - \mathcal{E}_L}{h} \tag{10-4}$$

There is also a definite probability that an electron might go from the M shell directly to the K shell, leaving the atom in the energy state $\mathcal{E}_M$. The line emitted in this transition is the K_β line; its frequency is given by

$$\nu_{K_\beta} = \frac{\mathcal{E}_K - \mathcal{E}_M}{h} \tag{10-5}$$

Or an electron may go from the N shell directly to the K shell with the emission of the K_γ spectral line of frequency

$$\nu_{K_\gamma} = \frac{\mathcal{E}_K - \mathcal{E}_N}{h} \tag{10-6}$$

Similar analyses can be used for the transitions producing the L and M series of spectral lines; for example, if the atom is in the energy state $\mathcal{E}_L$, an electron may go from the M shell to the L shell with the emission of the L_α line of frequency

$$\nu_{L_\alpha} = \frac{\mathcal{E}_L - \mathcal{E}_M}{h} \tag{10-7}$$

When the voltage across the x-ray tube is sufficiently high, a very large number of atoms in the target of the tube will be raised to the energy state $\mathcal{E}_K$, others to the energy state $\mathcal{E}_L$, and so on, by the action of the electrons incident on the target. The K series of spectral lines will be emitted by those atoms in which the electrons go directly from the L, M, or N shells to the K shell; the L series of spectral lines will be emitted by those atoms in which the electrons go directly from the M or N shells to the L shell. The intensity of a spectral line will be proportional to the number of atoms in which the appropriate transitions take place. The K_α line, for example, is the most intense line of the K series, whereas the K_γ line is the faintest one. In most of the atoms in the energy state $\mathcal{E}_K$, therefore, electrons go from the L shells

TABLE 10-1   WAVELENGTHS OF THE K SERIES LINES AND THE K ABSORPTION
LIMIT FOR SOME ELEMENTS
(Wavelengths in Å*)

Line	$K_{\alpha2}$	$K_{\alpha1}$	$K_{\beta2}$	$K_{\beta1}$	K_γ	K Absorption Limit
Transitions	KL_{II}	KL_{III}	KM_{II}	KM_{III}	$KN_{\text{II,III}}$	K_∞
Z Element						
24 Cr	2.293606	2.2897		2.0849		2.0702
29 Cu	1.54439	1.540562		1.3922		1.3806
42 Mo	0.71359	0.709300	0.63287	0.63229	0.6210	0.6198
46 Pd	0.58982	0.58545	0.52112	0.52052	0.51023	0.5092
47 Ag	0.56380	0.5594075	0.49769	0.49707	0.48703	0.4859
50 Sn	0.49505	0.49060	0.45518	0.45455	0.4450	0.4247
74 W	0.21383	0.2090100*	0.18518	0.18437	0.1796,4	0.1784
78 Pt	0.19038	0.18551	0.16450	0.16368	0.1594,2	0.1582
79 Au	0.18508	0.18020	0.15981	0.15898	0.1548,6	0.1536
82 Pb	0.17029	0.16538	0.14681	0.14597	0.1421,19	0.1409
92 U	0.13097	0.12595	0.11230	0.11139	0.1084,2	0.1072

*Wavelengths based on $WK_{\alpha1} = 0.2090100$ Å as standard.
Values obtained from Tables V and VI in J. A. Bearden's paper, "X-ray Wavelengths," *Rev. Mod. Phys.*, **39**, 78 (1967).

to the K shells. Stated in a different manner, the probability that an electron will go from the L shell to the K shell is much greater than the probability that an electron will go from the M shell directly to the K shell. The probability that an electron will go from the N shell to the K shell is very small.

With the precision and resolving power available in modern x-ray spectroscopy, many of the lines have been resolved into two or more components. The K_α line, for example, has been resolved into two components, $K_{\alpha 1}$ and $K_{\alpha 2}$, for all elements of atomic number greater than 11. The K_β line has been resolved into two components for most of the elements of atomic number greater than 29. The transitions for the α_1, α_2 lines and the β_1, β_2 lines are shown in Table 10-1. Another K_β line, designated in the literature as β_5, is due to the transition $KM_{\mathrm{IV\,V}}$ and appears in elements beginning with $Z = 19$. The K_γ line has also been resolved into several components. The fine structure observed in x-ray spectral lines must obviously be due to the multiplicity of some of the energy levels. One might determine this multiplicity from an analysis of the emission spectra. It will be more instructive, however, to show how this multiplicity of the energy levels can be determined by more direct experiments in which electrons from the inner shells of atoms are removed from them by the action of x-rays from an external source. This is an extension of the photoelectric effect to the region of x-ray wavelengths. There are two general experimental methods for investigating this phenomenon. One is to study the absorption spectra of the x-rays; the other is to determine the energies of the electrons ejected from the atoms.

10-3 X-Ray Absorption Spectra

A method for studying the x-ray absorption spectrum of an element is illustrated in Figure 10-4. The continuous spectrum from some suitable target T is used in this experiment. A narrow beam of x-rays coming through the slits S_1 and S_2 is incident on the absorbing material A containing the element under investigation. The transmitted beam is then analyzed by the crystal C of the x-ray spectrometer, and the intensity of each wavelength λ is measured by the ionization it produces in the ionization chamber I. A photographic plate may be used in place of the ionization chamber; the intensity will then be determined by the blackening on the photographic plate.

Each particular setting of the crystal corresponds to a definite wavelength given by Bragg's law $n\lambda = 2d \sin \theta$. The usual procedure is to measure the intensity I_0 of a given wavelength with the absorbing material removed from the path of the x-rays, then to insert the absorbing material in the path of the x-rays, and measure the new intensity I for the same wavelength. This procedure is repeated over a wide range of wavelengths. For each particular wavelength

$$I = I_0 \exp \left(-\mu x \right) \qquad (10\text{-}8) \quad [\text{Chap. 5, Eq. (5-18)}]$$

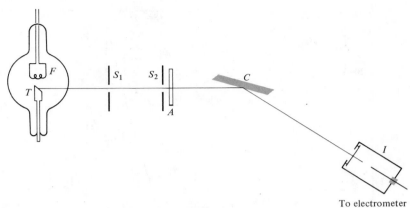

Figure 10-4   Diagram showing the arrangement of apparatus for measuring the absorption of x-rays.

where μ is the absorption coefficient for the wavelength used and x is the thickness of the absorbing material.

The results of a typical experiment, using a thin foil of silver as the absorbing material, are shown in Figure 10-5 where the mass absorption coefficient μ/ρ (see Section 5-13) is plotted against the wavelength λ. It is found that the mass absorption coefficient increases with the wavelength, approximately as λ^3, until a particular wavelength λ_K, at the position K, is reached, at which wavelength μ/ρ drops suddenly to a lower value. In the wavelength region between K and L_I the mass absorption coefficient again increases as λ^3; at L_I it drops in value. There are three such breaks in the curve close together marked L_I, L_{II}, and L_{III}. The wavelength at which the first break occurs is called the wavelength of the K absorption limit. The other breaks occur at the L_I, L_{II}, and L_{III} absorption limits.

In the case of silver the wavelength of the K absorption limit is $\lambda_K = 0.4859$ Å. This is slightly less than the shortest wavelength which occurs in the K series lines, the K_γ line, for which $\lambda = 0.4870$ Å (see Table 10-1). In the production of the lines of the K series it was found that an amount of energy, $\mathcal{E}_K$, had first to be supplied to the atom to remove an electron from the K shell, after which an electron from some outer shell, in going into the K shell, would emit a quantum of radiation. The energy of this quantum is, of course, always less than $\mathcal{E}_K$, as shown by Equations (10-4), (10-5), and (10-6). When the element forms the target of an x-ray tube, this energy, $\mathcal{E}_K$, comes from the kinetic energy of the incident cathode rays or electrons. When the element, however, is used as an absorber of x-rays, this energy must come from the incident x-rays. If the energy of an incident quantum or photon, $h\nu$, is greater than $\mathcal{E}_K$, the photon will be able to knock an electron out of the K shell, thus raising the atom to the energy state $\mathcal{E}_K$. The smallest value of the energy of a photon which will be able to remove an electron from the K shell is the following:

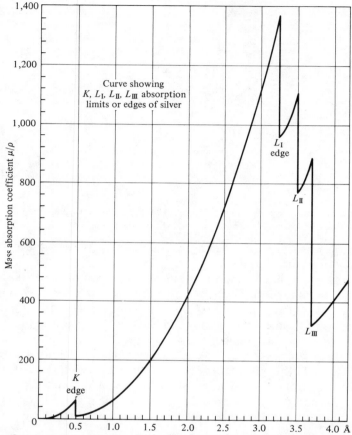

Figure 10-5   Graph showing the K and L x-ray absorption limits of silver.

$$h\nu_K = \mathcal{E}_K = \frac{hc}{\lambda_K} \qquad (10\text{-}9)$$

where λ_K and ν_K represent the wavelength and frequency, respectively, of the K absorption limit. If the energy of the incident photon $h\nu$ is less than $\mathcal{E}_K$, the photon will not be able to remove an electron from the K shell but it may have enough energy to be able to remove an electron from one of the higher levels, L, M, or N.

The fact that there are three absorption limits, L_I, L_{II}, and L_{III}, very close together and in the range of wavelengths of the L series indicates that the L shell probably consists of three energy levels. The wavelength of the L_I absorption limit is smaller than that of any line of the L series. The energy of each one of the states L_I, L_{II}, and L_{III} can be obtained from the graph with the aid of equations of the type of Equation (10-9); thus, for example,

$$\mathcal{E}_{L_I} = \frac{hc}{\lambda_{L_I}} \tag{10-10}$$

It is difficult to obtain the wavelengths of the M absorption limits by this method because of the absorption of these long wavelength x-rays by the element under investigation. It will be recalled that the absorption coefficient varies as the cube of the wavelength. In the case of silver the wavelength of the M limit is about 10 times that of the L limit so that the absorption would be approximately 1000 times as great. However, the wavelengths of the M absorption limits have been determined in this manner for some of the heavier elements. In each case five absorption limits were found in the M region, indicating that the M shell consists of five energy levels close together.

10-4   X-Ray Critical Voltages

The correctness of the above interpretation of the absorption limits can be demonstrated in a fairly direct manner by using the element under investigation as the target in an x-ray tube and controlling carefully the voltage across the tube. The maximum kinetic energy which the electrons incident on the target possess is Ve, where V is the voltage across the tube. The energy required to remove an electron from the K shell of an atom can be expressed as

$$\mathcal{E}_K = h\nu_K = \frac{hc}{\lambda_K} = eV_K \tag{10-11}$$

The critical voltage V_K is the minimum voltage which, when applied between the anode and the cathode of the x-ray tube, will give the impinging electron enough energy to remove electrons from the K shells. If the voltage across the tube is less than V_K, the atoms in the target cannot be raised to the energy state $\mathcal{E}_K$; hence the lines of the K series will not appear in the spectrum. When the voltage across the tube is increased until the value V_K is reached, all the lines of the K series appear simultaneously. This indicates that many of the atoms were raised to the energy state $\mathcal{E}_K$, making possible transitions to lower energy states with the radiation of quanta corresponding to the lines of the K series.

The critical voltage for the production of the K lines of silver, for example, can be determined from Equation (10-11) by substituting for λ_K its value 0.4859 Å and the accepted values for e, h, and c, yielding $V_K = 25.6\,\mathrm{kV}$. When the voltage across a silver target x-ray tube is, say, 20 kV, the spectrum is found to consist of the L series lines superimposed on the continuous radiation. When the voltage across the tube is increased, the intensities of these two spectra are increased, but no new lines appear until the voltage across the tube reaches the value 25.6 kV, when all the lines of the K series appear simultaneously. Further increase in voltage produces an increase in the intensities of the characteristic spectra relative to the continuous spectrum.

10-5 Magnetic Spectrograph

The x-ray energy levels of an atom can be determined in an independent way by utilizing the photoelectric effect with x-rays of known wavelength. If an x-ray quantum of energy $h\nu$ ejects an electron from an inner shell of the atom, thereby raising the atom to a higher energy state, the kinetic energy of the ejected electron, in the nonrelativistic range, will be given by

$$\tfrac{1}{2}mv^2 = h\nu - \mathcal{E}_{K,L}\cdots \qquad (10\text{-}12)$$

where $\mathcal{E}_{K,L}\cdots$ represents the appropriate final energy state of the atom. By measuring the kinetic energy of the ejected electron it is possible to determine the particular energy level from which it was ejected.

H. R. Robinson developed the magnetic spectrograph for determining the velocities of the ejected electrons. A diagram of the apparatus is shown in Figure 10-6. The element to be investigated, usually in the form of a very thin foil, is placed on a holder at C. A narrow beam of monochromatic x-rays, usually the intense K_α line from a known target material, enters through the thin window W in the evacuated box and strikes the element on C. The entire apparatus is in a uniform magnetic field of flux density B perpendicular to the plane of the figure. The electrons which are ejected from the element at C move in circular paths under the influence of the magnetic field. Only those electrons that pass through the slit S will be able to strike the photographic plate P. Because of the geometry of the apparatus, all electrons with the same velocity will be focused at the same distance from the source C. The radius of the electronic path is half the distance from C to the point on the photographic plate which has been blackened by the electrons.

The velocities of the electrons ejected from C can be found with the aid of the well-known equation

$$Bev = \frac{mv^2}{r}$$

or

$$v = Br \cdot \frac{e}{m} \qquad (10\text{-}13)$$

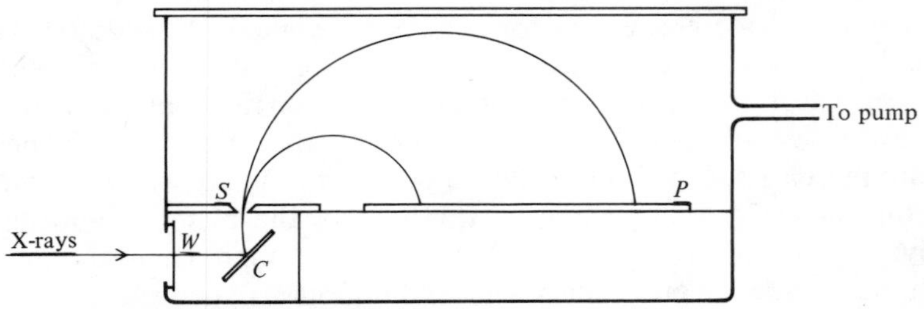

Figure 10-6 Robinson's magnetic spectograph.

where r is the radius of the path of an electron. In any one experiment, the electron ejected from the innermost shell of the atom will have the smallest velocity and its path will have the smallest radius r. Electrons ejected from the outer shells, such as the M and N shells, will have greater energy and be focused farther out on the photographic plate. From the positions of the lines on the photographic plate the energies of the ejected electrons can be computed. By substituting the value of the electronic energy in Equation (10-12) the energy required to remove the electron from its particular shell can be determined.

The values of the energy levels of a large number of elements were determined with the magnetic spectrographs by Robinson. More recently K. Siegbahn and his co-workers, using a magnetic spectrograph of very high precision, determined the energy levels of many atoms. References to these determinations are given at the end of this chapter.

10-6 X-Ray Terms and Selection Rules

The x-ray term structure or energy level diagram can be built up from analyses of the emission and absorption spectra of the elements and from the determination of the energy levels by the magnetic analysis of the energies of electrons ejected from these levels. The complete energy level diagram of uranium, $Z = 92$, is shown in Figure 10-7, with the x-ray notation for these levels and some of the transitions giving rise to the x-ray spectral lines. It must be remembered that each x-ray energy level represents a state of an atom which has one electron missing from a closed shell. Pauli pointed out that in a configuration in which an electron is missing from a completed subgroup the spectral term is the same as if that one electron alone occupied the subgroup. Normally there are two $1s$ electrons in the K shell; if one electron is removed, the energy state of the atom will be that of a $1s$ electron, namely, $^2S_{1/2}$, just as in the case of an alkali atom. The L shell consists of two subgroups, one for which $l = 0$ and the other $l = 1$. There are normally two electrons in the subgroup $n = 2$, $l = 0$; if one $2s$ electron is removed, the energy state of the atom will be that of a single electron for which $n = 2$, $l = 0$, namely, a $2s$ $^2S_{1/2}$ configuration. This is the energy level L_I. There are normally six $2p$ electrons in the completed subgroup $n = 2$, $l = 1$. If one $2p$ electron is removed, the energy state of the atom will be a P state; this state will be a doublet state corresponding to the two possible values of j; that is, $j = \frac{1}{2}$, or $j = \frac{3}{2}$. The two energy states L_{II} and L_{III} therefore correspond to the doublet terms $^2P_{1/2}$ and $^2P_{3/2}$, respectively, again similar to the terms of the alkali atoms. This procedure can be carried out for each of the subgroups; for example, the removal of an electron from a subgroup for which $l = 2$ gives rise to the $^2D_{3/2}$ and $^2D_{5/2}$ terms. The optical notations and the (n, l, j) values for each energy level are shown in the figure. One difference between x-ray energy levels and alkali terms should be noted: the higher x-ray energy states are designated by the smaller quantum numbers.

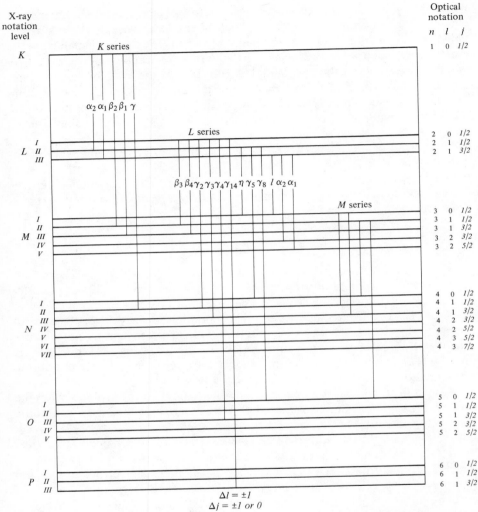

Figure 10-7 X-ray energy-level diagram for a heavy element such as uranium showing some of the transitions giving rise to spectral lines. (Not drawn to scale.)

The selection rules for the permitted transitions are the same as those for the alkali atoms; namely,

$$\Delta L = \pm 1$$

$$\Delta J = 0 \text{ or } \pm 1$$

Although no restrictions are placed on changes in the principal quantum number n, transitions for which $\Delta n = 0$ are very rare and the lines produced by such transitions are of very small intensity. The above selection rules account for the transitions that give rise to the more intense lines of the x-ray spectral series. Some fainter lines, forbidden by the above selection rules, have also been observed, as well as some lines which cannot be accounted for by any transitions between states represented in the energy-level diagram.

10-7   Radiationless Transitions—Auger Effect

When an electron is ejected from the K level of an atom by an x-ray photon, there is a probability that an electron from the L level of the same atom will go to the K level, with the consequent emission of a K_α x-ray photon. Transitions of this type give rise to the fluorescent spectra of the atom. The electron ejected from the K level is a photoelectron. Auger (1925), in a cloud chamber investigation of the photoelectrons ejected from atoms of argon, observed many instances in which the K photoelectron track had associated with it a smaller electron track. The smaller track represented an electron with much less energy than the K photoelectron. Furthermore, the two tracks had a common origin. Auger interpreted this effect as due to a radiationless transition; that is, instead of a K_α x-ray photon being emitted when an L electron went into the K level, the energy was used to eject a second electron from the L level, leaving it doubly ionized. This process might be imagined as one in which a K_α photon is formed when an electron goes from the L to the K level and that this photon then ejects an electron from the already ionized L level, leaving it doubly ionized. This would correspond to an *internal conversion* in which the K_α photon ejects a photoelectron from the ionized L level of the same atom. Since no radiation gets outside the atom, this is actually a radiationless transition. This effect is called the *Auger effect* and the electron emitted in this radiationless transition is called an *Auger electron*. For a number of years this two-step process was believed to account for internal conversion. More modern analysis referring to nuclear work (see Section 14-14) has shown that this effect takes place as a single-step phenomenon.

We can apply the photoelectric equation to the Auger effect. Before the emission of the Auger electron the atom was in the K state; after the ejection of the Auger electron the atom has two electrons missing from the L level; we shall designate this as the LL state. The kinetic energy of the Auger electron will be the difference between the energy of the K state, $\mathcal{E}_K$, and the energy of the LL state, $\mathcal{E}_{LL}$—that is,

$$\tfrac{1}{2}mv^2 = \mathcal{E}_K - \mathcal{E}_{LL} \tag{10-14}$$

There are other possibilities besides the K-LL transition; for example, after an L electron has gone to the K level, an Auger electron may be ejected from the M level, so that the final state of the atom is one in which L and M levels are each singly ionized. This state can be designated as an LM state and the radiationless transition designated as a K-LM transition. Other possible Auger transitions can readily be inferred.

Although the Auger effect was first discovered in connection with the x-ray photoelectric effect, any other method of ionizing an atom by the removal of an electron from an inner shell may lead to an Auger transition. Such effects have been observed in some optical spectra and, of greater interest, in readjustments of the electronic levels following nuclear disintegrations, particularly K-electron capture, and internal conversion following gamma-ray emission (see Section 14-14).

10-8 Production of Characteristic X-Ray Spectra

Although characteristic x-ray spectra are most commonly produced by bombarding elements with electrons, other methods are also used. Any method that causes the removal of an electron from one of the inner levels will result in the production of characteristic x-ray spectra. When photons of sufficient energy bombard a substance, as, for example, in the photoelectric effect with x-rays, the subsequent readjustment of the electrons in the atoms results in the emission of the characteristic x-ray spectra. Such spectra are frequently termed *x-ray fluorescent spectra*. The x-ray fluorescent spectra are usually of low intensity but frequently are the best way of analyzing the nature of the element in an unknown sample.

Characteristic x-ray spectra can also be produced by bombarding a target with positive ions, such as high-energy protons and alpha particles.

Characteristic x-ray spectra whose origins can be traced to readjustments in the electronic configurations of atoms following nuclear disintegrations have also been observed. In one type of nuclear disintegration known as *electron capture*, in which the nucleus of an atom captures one of the electrons from the same atom, the strongest evidence for this process is the emission of the characteristic x-ray spectrum. The origin of this spectrum is traced by comparing it with the known x-ray spectra of neighboring elements. If the atomic number of the original element is Z, the x-ray spectrum observed is always found to be the characteristic spectrum of the element of atomic number $Z - 1$; for example, it was inferred that one of the radioactive isotopes of cadmium, $Z = 48$, $A = 107$, disintegrates by electron capture to form silver,

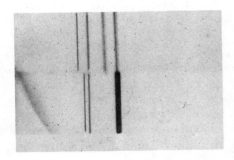

Figure 10-8 Photograph obtained with curved crystal x-ray spectrometer. Upper half shows the K_α, K_β, and K_γ lines of Ag emitted as a result of the decay of Cd^{107} by electron capture. Lower half shows K_α lines of Ag and Cd used for calibration; these are followed by the K_β lines. (Photograph by J. E. Edwards, M. L. Pool, and F. C. Blake.)

$Z = 47$, $A = 107$. To check this inference Edwards, Pool, and Blake (1945) examined the x-rays emitted by a small sample of this isotope of cadmium, using a curved crystal x-ray spectrograph with a mica crystal. A copy of the photograph obtained by them is shown in Figure 10-8. The upper half shows the K_α, K_β, and K_γ lines of silver emitted by the radioactive source; the lower half is a calibration photograph showing the K_α lines of Cd and Ag. This photograph shows that the x-rays are emitted by the silver formed in the radioactive decay (electron capture) of cadmium.

Problems

10-1. After studying the x-ray spectrum of platinum determine the minimum voltage that must be used across the x-ray tube to produce the K series lines.

10-2. Using the data in Table 10-1, calculate the energies, in electron volts, of the L_{II} and L_{III} energy levels for the elements listed in the table. Plot the square roots of these energies against atomic number and discuss any regularities observed in these curves. How do these curves compare with the Moseley diagram?

10-3. In an experiment with the magnetic spectrograph the $K_{\alpha 1}$ line from a silver target x-ray tube is incident on a thin copper foil inside the spectrograph. If the intensity of the magnetic field is 100 gauss, calculate the radius of the smallest electron path that will be observed. What is the origin of these electrons?

10-4. The K_α x-ray photons of molybdenum are incident on a thin foil of tungsten in a magnetic spectrograph in which the field intensity is 150 gauss. The L_I, L_{II}, L_{III} absorption limits of tungsten are 1.025, 1.075, and 1.216 Å, respectively. (a) Determine the energies of the electrons ejected from these levels. (b) Determine the radii of their paths in the magnetic field. (c) Calculate the separation of the lines of this magnetic spectrum.

10-5. The $K_{\alpha 1}$ x-ray line of uranium is used to investigate the energy levels of silver. A silver foil is placed inside a magnetic spectrograph and bombarded with uranium $K_{\alpha 1}$ x-ray photons. The spectrograph is in a magnetic field of 300 gauss. (a) Determine the kinetic energies of the electrons ejected from the K and L levels of silver. (b) Determine the radii of their paths in the magnetic field. Solve this problem by using the relativistic equations for energy and momentum. (c) Determine the velocity of the electrons ejected from the K level of silver.

10-6. The L_I absorption limit of bismuth is 0.757 Å. Determine the minimum energy of a beam of x-ray photons which can produce the L and M fluorescent spectral series of bismuth.

10-7. (a) Assuming that the energy required to remove a second electron from the L level is approximately equal to that required to remove the first electron, show that the kinetic energy of an Auger electron from an atom which is left doubly ionized in the L state is given by

$$\tfrac{1}{2}mv^2 = \mathcal{E}_K - 2\mathcal{E}_L$$

(b) Calculate the kinetic energy of such an Auger electron emitted by a copper atom if the L_{I} absorption limit of copper is 13 Å.

10-8. An Auger electron is ejected from a tungsten atom, leaving the latter ionized in the L and M states. Calculate the kinetic energy of this electron if the L_{I} absorption limit of tungsten is 1.025 Å and the M_{I} absorption limit is 4.41 Å.

References

Bearden, J. A., "X-ray Wavelengths," *Rev. Mod. Phys.* **39**, 78 (1967).

Bearden, J. A., and A. F. Burr, "Re-evaluation of X-ray Atomic Energy Levels," *Rev. Mod. Phys.* **39**, 125 (1967).

Compton, A. H., and S. K. Allison, *X-rays in Theory and Experiment.* New York: D. Van Nostrand Company, 1935, Chapters I, VII, and VIII.

Leighton, R. B., *Principles of Modern Physics.* New York: McGraw-Hill Publishing Company, 1959, Chapter 12.

Richtmyer, F. K., E. H. Kennard, and J. N. Cooper, *Introduction to Modern Physics,* sixth ed. New York: McGraw-Hill Book Company, 1969.

Siegbahn, K., Ed., *Alpha-, Beta-, and Gamma-Ray Spectroscopy.* Amsterdam, The Netherlands: North Holland Publishing Company, 1965, Chapter III.

Siegbahn, M., *The Spectroscopy of X-Rays.* New York: Oxford University Press, 1925, Chapters IV-VI, VIII.

11 | Selected Applications of Quantum Physics

11-1 Introduction

In the earlier chapters of this part of the book we set forth the basic ideas of quantum mechanics and used them to interpret the structure of the atom. In this chapter we shall consider further applications of these principles in the areas of molecular structure, statistical mechanics, and solid-state physics. In a certain sense these subjects represent a digression from atomic and nuclear physics, since all three topics deal with collections of atoms. Yet it is conventional to include them in treatments of atomic physics and there are good reasons for doing so. Historically, quantum principles were applied to systems with many atoms at an early stage in the development of the theory. Understanding of atomic physics made progress in parallel with these other fields and in many cases contributions to several areas were made by the same individual. In recent years each of these subjects has grown into a full-scale field of physical research and to deal adequately with any one area would require an entire book. Therefore we shall consider here only certain applications, chosen according to criteria of simplicity, historical significance, or practical utility.

11-2 Molecular Physics

The field of molecular physics could be interpreted to include much that is traditionally called chemistry. Our purpose here is not to examine the more complicated molecules but rather to keep the discussion at a low level of difficulty. Many principles, when understood for a simple system, will serve well to interpret more complex problems. The simplest molecules, such as molecules of noble gases, are, of course, monatomic. The structure they possess is actually atomic structure and has already been considered.

The next most simple type of molecule is diatomic; we shall restrict our

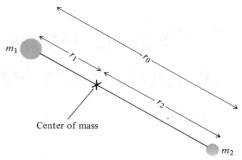

Figure 11-1 The rigid rotator as a model for a diatomic molecule.

discussion to this type. A wide variety of compounds such as CO, CN, H_2, N_2, O_2, NO, Cl_2, and HCl, has just two atoms, so that this restricted treatment can be expected to yield results that may be compared with experiment. In order to make such a comparison, the theory begins with a model for describing a diatomic molecule and proceeds to a quantum description. From our experience with atomic structure we expect that the result of the quantum calculation will consist of a number of wave functions, each describing a possible state of the molecule. We expect also that there will be discrete energy levels available for the molecule and that a transition from one level to another should be accompanied by emission or absorption of a photon. These expectations are realized in practice; the energies involved are often somewhat less than those of atomic physics, and the photons are typically in the infrared or the microwave portion of the electromagnetic spectrum.

As a simple model we begin with the rigid rotator in which the two atoms are considered to be at a constant distance r_0 from each other, rotating about the center of mass of the two-body system. The arrangement is shown in Figure 11-1. If r_1 and r_2 are the distances from the atoms to the center of mass and m_1 and m_2 are the masses of the atoms, then the position of the center of mass is defined by

$$m_1 r_1 = m_2 r_2 \qquad (11\text{-}1)$$

As in Chapter 8, we define the reduced mass

$$\mu = \frac{m_1 m_2}{m_1 + m_2} \qquad (11\text{-}2)$$

and observe that if the molecule rotates with an angular frequency of ω radians per second the kinetic energy is given by

$$\mathcal{E}_K = \tfrac{1}{2}\mu\, r_0{}^2 \omega^2 \qquad (11\text{-}3)$$

Typically the molecule is free to rotate without hindrance, so that the potential energy can be taken to be zero. Therefore we may drop the subscript to $\mathcal{E}$ and recognize it as the total energy.

As a matter of convenience, we define the moment of inertia I of the molecule about its center of mass:

$$I \equiv m_1 r_1{}^2 + m_2 r_2{}^2 \tag{11-4}$$

It can be shown (see Problems) that

$$I = \mu r_0{}^2 \tag{11-5}$$

and we can write the energy in the form

$$\mathcal{E} = \tfrac{1}{2} I\, \omega^2 \tag{11-6}$$

In Chapter 7 we observed that momentum is a more useful quantity than velocity in the quantum description of a system. In this instance the angular momentum is the important quantity defined by

$$P_r = m_1 r_1 v_1 + m_2 r_2 v_2 = m_1 r_1{}^2 \omega + m_2 r_2{}^2 \omega \tag{11-7}$$

Here we have used explicitly the fact that both atoms in the molecule rotate with the same angular velocity. From the definition of I we obtain

$$P_r = I\omega \tag{11-8}$$

Squaring Equation (11-8) and inserting it into Equation (11-6), we get

$$\mathcal{E} = \frac{P_r{}^2}{2I} \tag{11-9}$$

This form is similar to the expression $p^2/(2m)$ which was used for kinetic energy in Chapter 7. The Schroedinger equation for the rigid rotator is seen to be

$$\frac{1}{2I}\,(P_r)^2_{\mathrm{op}}\,\psi = \mathcal{E}\,\psi \tag{11-10}$$

where the subscript refers to the fact that we need the operator form of the angular momentum. The exact form of this operator and the details of solution depend on the type of coordinates used. The most felicitous choice is spherical polar coordinates, using a polar angle θ and an aximuthal angle ϕ to designate the orientation of the axis of the molecule with respect to an axis (called the z direction) fixed in space. In this coordinate system the Schroedinger equation looks similar to the one used in Chapter 8 to describe the hydrogen atom. In the problem at hand there is no potential energy, no radial dependence (because the separation of the two atoms is constant), and therefore no derivatives with respect to r. The appropriate form is

$$-\frac{\hbar^2}{2I}\left(\frac{\partial^2 \psi}{\partial \theta^2} + \cot\theta\,\frac{\partial \psi}{\partial \theta} + \frac{1}{\sin^2\theta}\frac{\partial^2 \psi}{\partial \phi^2}\right) = \mathcal{E}\psi \tag{11-11}$$

In this equation, just as in the problem of the hydrogen atom, the correct approach is to separate the variables, assuming that ψ can be written as the product of two functions, one of which depends only on θ and the other only on ϕ. Since Equation (11-11) is the same as the angular part of the hydrogen atom, the solutions take the same form. Two angular momentum quantum numbers appear in the solution, analagous to l and m for hydrogen. We shall refer to them as J and M. The number J can take on any of the values 0, 1, 2, 3, . . . , and M can be any value such that $-J \leq M \leq J$.

For the hydrogen atom the calculation of the energy levels was complicated because of the existence of a radial equation. Here the situation is simpler, and we can use the result that for a state with quantum number J

$$<P_r{}^2> = J(J + 1)\hbar^2 \tag{11-12}$$

This fact can be inserted into Equation (11-9) to yield

$$\mathcal{E} = \frac{J(J + 1)\hbar^2}{2I} \tag{11-13}$$

Figure 11-2 shows the resulting energy-level diagram. Each level actually consists of $2J + 1$ sublevels, each with the same energy, corresponding to the possible values of M.

The dipole selection rules for transitions between pairs of states are the same

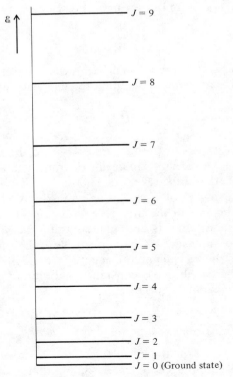

Figure 11-2 Energy levels for the rigid rotator.

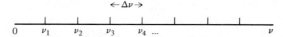

Figure 11-3   Frequency spectrum for the rigid rotator. The Jth frequency is given by $\nu_J = J\hbar/(2\pi I)$ and $\Delta\nu = \hbar/(2\pi I)$.

for the rigid rotator as for atomic electrons. This result follows from the fact that the wave functions have the same form for the two problems. For the rotator we have

$$
\begin{aligned}
|\Delta J| &= 1 \\
|\Delta M| &= 0, 1
\end{aligned}
\tag{11-14}
$$

Since the energies are determined by J, we can use the first of these rules to predict the spectrum of light emitted or absorbed by the rotator. The energy of a photon absorbed in a transition from a state of angular momentum J to the next higher state $(J + 1)$ will be just the difference in the energies of the two states.

$$
\begin{aligned}
\Delta\mathcal{E} &= \frac{(J + 1)(J + 2)\hbar^2}{2I} - \frac{J(J + 1)\hbar^2}{2I} \\
&= \frac{\hbar^2}{I}(J + 1)
\end{aligned}
\tag{11-15}
$$

but the frequency of the light absorbed is related in the usual way to this energy difference:

$$
\Delta\mathcal{E} = h\nu
\tag{11-16}
$$

Therefore we find that

$$
\nu = \frac{\hbar}{2\pi I}(J + 1),
\tag{11-17}
$$

or the observed frequencies are equally spaced, as shown in Figure 11-3. It is to be emphasized that the energy levels are not equally spaced (see Fig. 11-2) but rather the *differences* between successive energy levels are equally spaced.

Although the rigid rotator model is an oversimplification of the diatomic molecule, it leads to results that can be compared with experiment. The series of equally spaced frequencies it predicts can be readily observed in the infrared as rather closely spaced lines, often referred to as rotational bands. Measurement of the spacing $\Delta\nu$ can be performed, yielding I, since from Equation (11-17)

$$
I = \frac{\hbar}{2\pi\Delta\nu}
\tag{11-18}
$$

From Equation (11-5) the moment of inertia is just $\mu r_0{}^2$; the reduced mass, μ, is known because the atomic masses are known. Therefore a measurement of the frequencies of two adjacent rotational lines provides a measure of the interatomic spacing r_0 for the molecule.

In order to increase our understanding of the diatomic molecule, it is necessary to improve the model. We do this by recognizing that the interatomic distance is not really constant—that the atoms can oscillate back and forth along the line which joins them in such a way that the center of mass remains fixed. If we assume that the restorative force is proportional to the amount by which the interatomic distance r deviates from r_0, the Schroedinger equation for the radial part of the motion becomes

$$-\frac{\hbar^2}{2\mu}\frac{d^2\psi}{dr^2} + \tfrac{1}{2}k_s\,(r-r_0)^2\psi = \mathcal{E}\psi \tag{11-19}$$

where k_s is the effective spring constant for the system. If we define $x \equiv r - r_0$, this equation becomes

$$\frac{d^2\psi}{dx^2} + \frac{2\mu}{\hbar^2}\,(\mathcal{E} - \tfrac{1}{2}k_s x^2)\psi = 0 \tag{11-20}$$

This equation was considered in Section 7-6 and solved in Appendix V-6. The quantum number which must be introduced in its solution we shall designate here as v. The energy levels are equally spaced and are given by the expression

$$\mathcal{E} = (v + \tfrac{1}{2})\,\hbar\omega_0 \tag{11-21}$$

where $\omega_0 = \sqrt{k_s/\mu}$, the angular frequency the oscillator would possess classically. The dipole selection rule for emission or absorption of a photon by the oscillator is

$$|\Delta v| = 1 \tag{11-22}$$

so that $\Delta\mathcal{E}$ is the same for all allowed transitions. Therefore a frequency spectrum for the oscillator will consist of a line at just one frequency, $\nu = 2\pi\omega_0$. This frequency is typically in the infra-red, as was the case for the rigid rotator. However, there is a tendency for the vibrational frequencies of molecules to be higher than the rotational frequencies.

In order to improve the model to achieve still better agreement with experiment, it is possible to couple the rotation and vibration of the molecule to solve a three-dimensional Schroedinger equation. Such a calculation requires algebraic complexity beyond the scope of this treatment, but we shall indicate the qualitative features of the result. A functional form for the potential energy as a function of interatomic separation was introduced by P. Morse (1929):

$$V(r) = D\{1 - \exp\,[-a\,(r - r_0)]\}^2 \tag{11-23}$$

where D, a, and r_0 are constants; D is the dissociation energy of the molecule, the value that $V(r)$ approaches as r, the interatomic separation, becomes infinitely large. The quantity a is an inverse length whose value determines the width of the potential well; r_0 is the equilibrium separation of the two atoms. Figure 11-4 portrays the Morse potential. Of course, the form (11-23) is not the

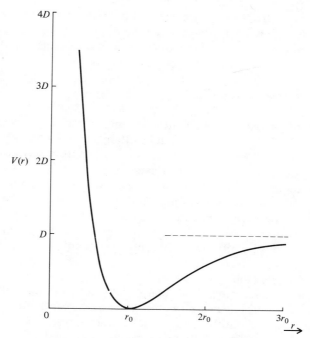

Figure 11-4 The Morse potential for a diatomic molecule.

correct form for the potential energy of any real molecule. However, it is an excellent approximation, and it has the advantage of being expressible analytically. The energy levels obtained from solving the Schroedinger equation with the Morse potential are similar to those obtained from separate treatment of vibration and rotation. We can understand the differences qualitatively by considering how rotation affects vibration and vice versa.

An energy-level diagram for pure vibration consists of equally spaced levels, based on the assumption of a purely parabolic potential

$$V(x) = \tfrac{1}{2}k_s x^2 \tag{11-24}$$

The Morse potential spreads out more than a parabola at distances far from the minimum. In so doing it resembles the shape of the Coulomb potential of the hydrogen atom (where the levels get closer together as the energy increases) rather than the shape of a square well (where the levels get farther apart with increasing energy). Near the equilibrium radius the shape of the Morse potential is close to being parabolic. As a result of this paragraph, we conclude that the energy levels from the Morse potential are approximately equally spaced but that as the energy increases spacing becomes closer.

To infer the effect of vibration on the rotational levels we shall consider a classical model with two rotating masses joined by a spring. As the rotational energy increases, the spring will tend to stretch, thus causing an increase in the moment of inertia I. Since I appears in the denominator of Equation (11-13), it is clear that the higher energy states will be somewhat closer together than would be predicted by the model in which the rotator is considered to be rigid.

11-3 Statistical Mechanics

The subject of statistical mechanics originated many years earlier than quantum mechanics. It can be studied at length by using Newtonian mechanics as a basis without the introduction of ideas from quantum physics. The introduction of quantum physics led to considerable progress in statistical mechanics, and in the latter sections of this chapter we shall examine some of the results of this additional understanding. In this section we shall introduce the basic ideas, including their application to a simple classical system, the ideal gas. We choose here to recognize the profound influence of quantum mechanics on this field of physics by introducing directly the quantum form of the theory, recovering the classical result as a limiting case, as an illustration of the correspondence principle (see Section 8-8).

The motivation for learning about statistical mechanics extends beyond its use for the specific topics discussed in this chapter. Several of the applications considered here come from the area of solid-state physics, and it should not seem surprising that in that area there are many other applications of statistical methods. In the theory of plasmas (ionized gases) statistical mechanics is of utmost importance. In brief, a statistical approach is required for the study of any system composed of many bodies, since the detailed calculations needed for nearly any other treatment would be prohibitively difficult. In areas of physics in which the statistical approach is not directly used, some of the language of statistical mechanics is employed. Finally, statistical mechanics is a theory of great elegance and beauty, deserving of study for its own intrinsic worth.

The quantum discussion of a system containing many identical particles requires the consideration of the wave function

$$\psi(\mathbf{r}_1, \mathbf{r}_2, \mathbf{r}_3, \ldots)$$

of a many-body state. The vector $\mathbf{r}_1$ is the position vector of the first particle. A basic assumption is that if the particles are truly identical the interchange of any pair of them will leave the wave function unchanged except for the sign of ψ, which may or may not change:

$$\psi(\mathbf{r}_2, \mathbf{r}_1, \ldots) = \pm\psi(\mathbf{r}_1, \mathbf{r}_2, \ldots) \tag{11-25}$$

The sign in this equation depends on the type of particles that make up the collection. Particles that have integer spins (in units of $\hbar$) have wave functions for which the $+$ sign holds. Such particles are said to obey *Bose-Einstein statistics* and are called *bosons*. Other particles with spins that are odd half-integers ($\frac{1}{2}, \frac{3}{2}, \frac{5}{2}, \ldots$) obey *Fermi-Dirac statistics* and are called *fermions*.

Wave functions are symmetric with respect to interchange of two bosons, antisymmetric with respect to interchange of two fermions. As a consequence of the antisymmetry of fermion wave functions, no two fermions may possess identical sets of quantum numbers.

We can use a simple example to show why an antisymmetric wave function implies that the quantum numbers must be different for two identical fermions. Consider a system with one particle for which the wave function is $\psi(\mathbf{r})$. If two different noninteracting particles (with equal masses) are placed in this system, the wave function will be a product $\psi_a(\mathbf{r}_1)\,\psi_b(\mathbf{r}_2)$, where a and b refer to the sets of quantum numbers of the two particles respectively (see Problem 7-32). Next, suppose that the particles are identical. Then the product wave function must be rewritten to take into account symmetry (or antisymmetry) under interchange. For two bosons the correct form is the symmetric function

$$\psi_S = \frac{\psi_a(\mathbf{r}_1)\,\psi_b(\mathbf{r}_2) + \psi_a(\mathbf{r}_2)\,\psi_b(\mathbf{r}_1)}{\sqrt{2}}$$

For two fermions we get a change in sign.

$$\psi_A = \frac{\psi_a(\mathbf{r}_1)\,\psi_b(\mathbf{r}_2) - \psi_a(\mathbf{r}_2)\,\psi_b(\mathbf{r}_1)}{\sqrt{2}}$$

In both cases the factor of $\sqrt{2}$ ensures that the wave function is normalized to unity. For ψ_A it is clear that if the set of quantum numbers a is the same as the set b, then ψ_A will vanish and will thereby be unacceptable as a wave function. Therefore no two identical fermions can have the same set of quantum numbers. Since ψ_S does not vanish when $a = b$, it is permissible to have any number of bosons in the same quantum state.

Electrons are fermions (since they have spin $\frac{1}{2}$), and we have already seen that no two of them may occupy the same quantum state, a result known as Pauli's exclusion principle (Section 9-9). At the present time all known particles seem capable of description within the framework presented here. It is possible to construct mathematical theories in which particles are neither bosons nor fermions and in which the symmetry is more complicated. Such theories are called *parastatistics*. So far no examples have been found of systems that obey parastatistics.

Our statistical treatment of a collection of particles begins by recognizing that the energies of any two of these particles are likely to be different. We shall denote by $n(\mathcal{E})\,d\mathcal{E}$ the number of particles whose energies are between $\mathcal{E}$ and $(\mathcal{E} + d\mathcal{E})$. We shall further assume that this number can be written in the form

$$n(\mathcal{E})\,d\mathcal{E} = f(\mathcal{E})\,g(\mathcal{E})\,d\mathcal{E} \qquad (11\text{-}26)$$

The quantity $f(\mathcal{E})$ is the expected density of particles with energy at $\mathcal{E}$, assuming that there is only one state available with energy $\mathcal{E}$; $g(\mathcal{E})\,d\mathcal{E}$ is the number of such states. We shall examine f and g in more detail.

The form of the function $f(\mathcal{E})$ depends on whether the particles in the collection are bosons or fermions. In Appendix V-8 we show that for Bose-Einstein statistics

$$f_{BE}(\mathcal{E}) = \frac{1}{\exp\,[(\mathcal{E} - \mu)/(kT)] - 1} \tag{11-27}$$

and that for Fermi-Dirac statistics

$$f_{FD}(\mathcal{E}) = \frac{1}{\exp\,[(\mathcal{E} - \mu)/(kT)] + 1} \tag{11-28}$$

The difference, of course, is the sign that appears in the denominator. This apparently trivial difference will be seen by subsequent examples to lead to profound differences in the behavior of the two general types of particle. The quantity μ is called the chemical potential. Its value and interpretation will be discussed in the examples; T is the temperature in kelvins (or degrees kelvin). The quantity k is Boltzmann's constant

$$k = 1.38062 \times 10^{-23}\,\text{joule K}^{-1} \tag{11-29}$$

In numerical calculations it is usually handier to use

$$k = \frac{1\;\text{eV}}{1.1605 \times 10^4 \text{K}} \tag{11-30}$$

and for rapid informal calculation from memory

$$kT = \tfrac{1}{40}\;\text{eV at 300K, room temperature} \tag{11-31}$$

If the energy per particle is sufficiently large, both forms of f become rather small, implying that the density of particles will decrease and that the distance of interparticle separation will increase. Under these circumstances quantum effects are of lesser importance and the collection of particles will behave as predicted by classical physics. In this same limit $\mathcal{E} \gg kT$ both forms of f approach the same value

$$f_{MB}\,(\mathcal{E}) = C \exp\left(-\frac{\mathcal{E}}{kT}\right) \tag{11-32}$$

where the subscript refers to the Maxwell-Boltzmann statistics obeyed by classical particles; C is a constant. A comparison of these three forms for f is presented in Table 11-1.

The quantity $g(\mathcal{E})$ refers to the number of states available at the energy $\mathcal{E}$. For a quantum system this number could be counted precisely if a complete energy-level diagram were available for the system. In practice, such information is not available, and, further, the number of states is very large. The procedure is to assume that there are infinitely many of these states and that instead of adding them up we perform an integral over an interval of energy. Then $g(\mathcal{E})$ is referred to as the *density of states*. Another way to understand what g means—a method that prefigures the way it is used in practice—is to visualize a three-dimensional momentum space in which the axes measure the

TABLE 11-1   COMPARISON OF BOSE-EINSTEIN, FERMI-DIRAC, and MAXWELL-BOLTZMANN STATISTICS

Bose-Einstein	*Fermi-Dirac*	*Maxwell-Boltzmann*
Bosons Spin = 0, 1, 2, . . . ψ is even under interchange	Fermions Spin = $\frac{1}{2}, \frac{3}{2}, \frac{5}{2}, \ldots$ ψ is odd under interchange	Classical, distinguishable particles
$f_{BE} = \dfrac{1}{\exp\left[(\mathcal{E} - \mu)/kT\right] - 1}$	$f_{FD} = \dfrac{1}{\exp\left[(\mathcal{E} - \mu)/kT\right] + 1}$	$f_{MB} = C \exp\left(-\dfrac{\mathcal{E}}{kT}\right)$
No upper limit to the number of particles per state	Only one particle per state	

Cartesian components of the momentum vector **p** of a particle. If the magnitude of the momentum can vary between p and $p + dp$, there will be an infinite number of states (points in momentum space) on which the end of the vector **p** may lie. If that number were finite, it could be established by counting. Since it is infinite, we refer to $G(p)\, dp$ as the *measure of the set of points available to the tip of the momentum vector.* Using the ideas discussed in Appendix V-9, we set

$$g(\mathcal{E})\, d\mathcal{E} = G(p)\, dp \qquad (11\text{-}33)$$

Having discussed the two factors that form $n(\mathcal{E})$, we next consider how to use it in calculations. Since $n(\mathcal{E})\, d\mathcal{E}$ is the number of particles in a small energy interval, the total number N, of particles in the collection will be the sum over all such energy intervals; in other words, an integral over all values of energy.

$$N = \int_0^\infty n(\mathcal{E})\, d\mathcal{E} \qquad (11\text{-}34)$$

If Q is a physical quantity that depends on the energy, we can recognize that $Q(\mathcal{E})\, n(\mathcal{E})\, d\mathcal{E}$ will be the amount of that quantity possessed by those particles whose energy is near $\mathcal{E}$. The total amount of Q for the entire system is

$$Q_{total} = \int_0^\infty Q(\mathcal{E})\, n(\mathcal{E})\, d\mathcal{E} \qquad (11\text{-}35)$$

The amount of Q possessed by a single molecule is, on the average,

$$<Q> = \frac{1}{N} \int_0^\infty Q(\mathcal{E})\, n(\mathcal{E})\, d\mathcal{E} \qquad (11\text{-}36)$$

The student is urged to notice the similarity between the use of $n(\mathcal{E})$ and the use of $\psi^*\psi$ in the quantum mechanics of a single particle (see Chapter 7). The differences between the two are not to be ignored, since $\psi^*\psi$ is normalized to unity and $n(\mathcal{E})$ is normalized to N, the total number of particles. Further, $\psi^*\psi$ is used with operators, whereas Q is always a multiplicative quantity in an averaging process.

In the remainder of this section we shall apply the abstract ideas of statistical mechanics to a concrete system: a classical ideal gas in a container. The gas is visualized as consisting of many molecules of mass m, each in motion with some energy $\mathcal{E}$. The assumption of classical molecules implies that the energy is distributed according to Maxwell-Boltzmann statistics, [Equation (11-32)]. The ideal-gas assumption means that the particles do not interact and that the total energy of a molecule is just its kinetic energy,

$$\mathcal{E} = \frac{p^2}{2m} \quad \text{or} \quad p = (2m\mathcal{E})^{1/2} \tag{11-37}$$

The first task is to evaluate $G(p)$. Figure 11-5 depicts a momentum vector $\mathbf{p}$ in momentum space. The volume available to its tip, assuming its magnitude to be between p and $p + dp$, is a spherical shell of radius p and thickness dp. Only part of the entire shell is shown in the diagram. The volume of the entire shell is just $G(p)\, dp$. Clearly

$$G(p)\, dp = 4\pi p^2\, dp \tag{11-38}$$

Then, since $g(\mathcal{E})\, d\mathcal{E} = G(p)\, dp$, and from differentiating Equation (11-37),

$$g(\mathcal{E}) = G(p) \frac{dp}{d\mathcal{E}} = 4\pi p^2 \cdot \tfrac{1}{2} (2m\mathcal{E})^{-1/2} \cdot 2m \tag{11-39}$$

$$= 4\pi (2m^3\mathcal{E})^{1/2}$$

Since the Maxwell-Boltzmann distribution already has an arbitrary multiplicative constant, we do not need to carry along the constants contained in g, and we can write

$$n(\mathcal{E})\, d\mathcal{E} = C\mathcal{E}^{1/2} \exp\left(-\frac{\mathcal{E}}{kT}\right) d\mathcal{E} \tag{11-40}$$

We can evaluate the constant C by recalling the normalization requirement [Equation (11-34)]. We perform the required integration by changing the variable from $\mathcal{E}$ to $y = \mathcal{E}^{1/2}$ and then using the results of Appendix V-5.

$$\int_0^\infty n(\mathcal{E})\, d\mathcal{E} = C \int_0^\infty \mathcal{E}^{1/2} \exp\left(-\frac{\mathcal{E}}{kT}\right) d\mathcal{E}$$

$$= 2C \int_0^\infty y^2 \exp\left(-\frac{y^2}{kT}\right) dy$$

$$= 2C \cdot \tfrac{1}{4}\sqrt{\pi}\, (kT)^{3/2} = N \tag{11-41}$$

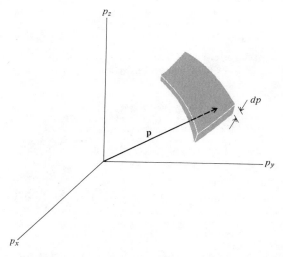

Figure 11-5 The measure of the set of points available to the tip of the momentum vector. The region shown is part of a spherical shell of thickness dp and radius p.

From this result we can obtain C and write

$$n(\mathscr{E})\,d\mathscr{E} = \left(\frac{2}{\sqrt{\pi}}\right) N(kT)^{-3/2}\,\mathscr{E}^{1/2}\exp\left(-\frac{\mathscr{E}}{kT}\right)d\mathscr{E} \qquad (11\text{-}42)$$

As an illustration of the use of $n(\mathscr{E})$, let us use it to calculate the total energy of the gas (often called the internal energy).

$$\mathscr{E}_{\text{total}} = \int_0^\infty \mathscr{E}\,n(\mathscr{E})\,d\mathscr{E}$$

$$= C\int_0^\infty \mathscr{E}^{3/2}\exp\left(-\frac{\mathscr{E}}{kT}\right)d\mathscr{E} \qquad (11\text{-}43)$$

This integral can be performed by parts, differentiating $\mathscr{E}^{3/2}$ and integrating the exponential function. The integrated term vanishes, leaving

$$\mathscr{E}_{\text{total}} = C\,\tfrac{3}{2}kT\int_0^\infty \mathscr{E}^{1/2}\exp\left(-\frac{\mathscr{E}}{kT}\right)d\mathscr{E} \qquad (11\text{-}44)$$

In this expression we can recognize all the ingredients of Equation (11-41), so that

$$\mathscr{E}_{\text{total}} = \tfrac{3}{2}NkT \qquad (11\text{-}45)$$

Thus for a monatomic ideal gas the total energy increases linearly with temperature. The average energy per molecule is

$$<\mathcal{E}> = \frac{1}{N}\, \mathcal{E}_{\text{total}} = \tfrac{3}{2} kT \qquad (11\text{-}46)$$

The specific heats of an ideal monatomic gas can be calculated from the total energy by using some simple results from thermodynamics. The specific heat at constant volume is

$$C_v = \frac{1}{n}\frac{d}{dt}\, \mathcal{E}_{\text{total}} = \frac{3}{2}\frac{Nk}{n} = \tfrac{3}{2} R \qquad (11\text{-}47)$$

where n is the number of moles of gas and R is the familiar gas constant, the product of Boltzmann's constant with Avogadro's number.

For systems that are more complicated than a monatomic ideal gas the methods of statistical mechanics can be used to prove the theorem of *equipartition of energy*, which we state without proof in a somewhat nonrigorous form. The theorem states that the average energy per molecule is $\tfrac{1}{2} kT$ times the number of quadratic degrees of freedom per molecule. To determine this number it is necessary to write the classical expression for the energy of the molecule and then count the number of quadratic terms. For the example we have just considered the energy was assumed in Equation (11-37) to be

$$\mathcal{E} = \frac{p^2}{2m} = \frac{1}{2m}(p_x{}^2 + p_y{}^2 + p_z{}^2) \qquad (11\text{-}48)$$

The three quadratic terms imply that the average energy per molecule should be $\tfrac{3}{2} kT$ in agreement with what we have just calculated.

An important application of the theorem of equipartition of energy is provided by the ideal diatomic gas. Assume that the total mass of a molecule is M, the reduced mass is μ, and the classical spring constant for vibration is k_s. The moment of inertia appropriate for rotation around the line joining the two atoms (the long axis) we call I_3; the moment of inertia about any axis perpendicular to the long axis we call I. Then the classical expression for the total energy of a diatomic molecule is

$$\mathcal{E} = \frac{1}{2M}(p_x{}^2 + p_y{}^2 + p_z{}^2) + \frac{P_{r1}{}^2}{2I} + \frac{P_{r2}{}^2}{2I} + \frac{P_{r3}{}^2}{2I_3} + \frac{p_{\text{osc}}{}^2}{2\mu} + \tfrac{1}{2} k_s\, s^2 \qquad (11\text{-}49)$$

where p_{osc} is the effective oscillator momentum and s is the deviation from the equilibrium separation of the two atoms in the molecule. Since there are eight quadratic terms in this expression, the equipartition theorem leads us to expect $4kT$ for the average energy per molecule and $4R$ as the specific heat at constant volume. This prediction does not agree with experiment; the discrepancy can be explained by using quantum ideas. In quantum mechanics all rotational and vibrational energy levels are quantized for a diatomic molecule, as we saw in Section 11-2. To have any rotation at all a molecule must have P_r of the order of $\hbar$; to have any vibration the energy must be of the order of $\tfrac{1}{2}\hbar\omega_0$ (where ω_0 is the classical angular frequency). At low temperatures, where kT is less than $P_r{}^2/(2I)$ and much less than $\tfrac{1}{2}\hbar\omega_0$, only the first three terms in Equation (11-49) are effective—the same three translational terms that appear for a monatomic ideal gas. The specific heat at constant volume is $\frac{3}{2}R$. At somewhat higher temperatures the two rotational terms involving I can contribute and C_v increases to $\frac{5}{2}R$. At still higher temperatures the two vibrational terms enter and C_v becomes $\frac{7}{2}R$. The rotational term involving I_3 requires a very high temperature before it can contribute because I_3 is very small and it appears in the denominator.

11-4 Blackbody Radiation

In Section 4-6 the topic of blackbody radiation was discussed, including a presentation of the experimental results. Since it was this topic that led to the introduction of the idea of a quantum of energy, it was fitting to introduce it early in the study of quantum ideas. In this section we shall re-examine blackbody radiation from the standpoint of statistical mechanics.

We shall assume that the radiation in a cavity can be considered as a collection of many photons or a "photon gas." Photons have unit spin and are therefore bosons, so we shall use the Bose-Einstein distribution to describe them. For simplicity we shall assume that the photons are contained in a cubical box of length L and volume $V = L^3$. Our first task is to evaluate $G(p)\,dp$, the density-

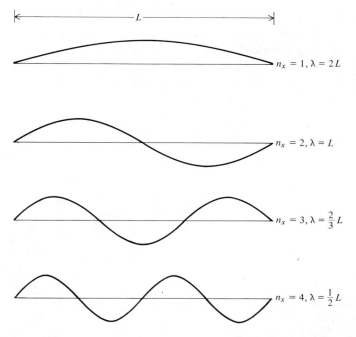

Figure 11-6 Standing waves along the x direction, confined to a region of length L. By inspection, $\lambda = 2\,L/n_x$.

of-states factor. We must recognize that since the photons are confined to a container they must behave like standing waves. From Figure 11-6 we can see that there will be a countably infinite number of standing waves as projected onto the x axis. If n_x is the number of the particular mode, the expression

$$\lambda = \frac{2L}{n_x} \tag{11-50}$$

provides the wavelength. For a photon corresponding to light at wavelength λ, the momentum in the x direction will be

$$p_x = \frac{h}{\lambda} = \frac{n_x h}{2L} \tag{11-51}$$

In momentum space, along the x direction, the possible states are thus uniformly spaced with an amount of momentum equal to $h/(2L)$ separating two adjacent points. The same spacing will hold in the y and z directions, since we chose a cubical box to contain the photons. The situation is shown in Figure 11-7, where a subspace of momentum space, the plane $p_z = 0$, is depicted. For given values of p and dp the dots inside the spherical shell are the allowed states. For each dot there are two possible states corresponding to left and right circular polarization and contributing a factor of two to $G(p)\,dp$. Then

$$G(p)\,dp = 2 \cdot \frac{1}{8} \cdot \frac{4\pi p^2}{(h/2L)^3}\,dp \tag{11-52}$$

The factor of $\frac{1}{8}$ comes from the fact that for our count of standing waves we need only positive values of n_x and therefore only positive values of p_x; hence we need only one octant of momentum space. We recognize $4\pi p^2\,dp$ as the total volume of the spherical shell, and $(h/2L)^3$ as the volume of momentum space per state. Thus

$$G(p)\,dp = \frac{8\pi V p^2\,dp}{h^3} \tag{11-53}$$

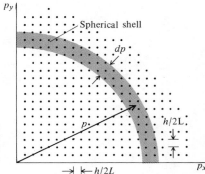

Figure 11-7 A two-dimensional view of momentum space. Each dot represents a possible end point for a momentum vector. Only those dots inside the spherical shell contribute to the quantity $G(p)\,dp$.

The transformation to energy is simple, since for photons

$$\mathcal{E} = pc, \qquad d\mathcal{E} = c\,dp \tag{11-54}$$

Therefore, using the method of Appendix V-9,

$$g(\mathcal{E})\,d\mathcal{E} = \frac{8\pi V \mathcal{E}^2\,d\mathcal{E}}{h^3 c^3} \tag{11-55}$$

This result was used as Equation (4-22). Next we multiply $g(\mathcal{E})$ with $f(\mathcal{E})$, using Equation (11-27) for f. We set μ equal to zero because the physical meaning of μ in this case is the energy needed to divide a single photon into two photons of lower frequency; the processes of photon division and recombination are assumed to occur at the walls of the cavity without any loss or gain of energy. Then we obtain

$$n(\mathcal{E})\,d\mathcal{E} = \frac{8\pi V \mathcal{E}^2\,d\mathcal{E}}{h^3 c^3 \{\exp\,[\mathcal{E}/(kT)] - 1\}} \tag{11-56}$$

The result just obtained is the number of photons at a certain energy, but in optical experiments it is more usual to measure the energy per unit volume possessed by those photons at a certain frequency. This quantity is given by

$$\rho_\nu\,d\nu = \frac{1}{V}\,h\nu\,n(\mathcal{E})\,d\mathcal{E} \tag{11-57}$$

Once again we need to transform a probability density function, this time using the equations

$$\mathcal{E} = h\nu, \qquad d\mathcal{E} = h\,d\nu \tag{11-58}$$

to define the change of variable. We obtain

$$\rho_\nu(\nu)\,d\nu = \frac{8\pi h}{c^3}\,\frac{\nu^3}{\exp\,[h\nu/(kT)] - 1}\,d\nu \tag{11-59}$$

For comparison with certain experimental results it is convenient to use the wavelength as the variable.

$$\nu = \frac{c}{\lambda}, \qquad d\nu = (-)\frac{c}{\lambda^2}\,d\lambda$$

$$\rho_\lambda(\lambda)\,d\lambda = \frac{8\pi hc}{\lambda^5}\,\frac{d\lambda}{\exp\,[hc/(\lambda kT)] - 1} \tag{11-60}$$

These results have appeared previously as Equations (4-24) and (4-25). Notice that in the expressions for ρ_ν and ρ_λ the volume of the cavity no longer appears. We assert that the results are valid for a cavity of arbitrary size and shape.

Wien's law, Equation (4-14), can be derived from Equation (11-60) by calculating $\partial \rho_\lambda/\partial \lambda$ and setting the result equal to zero. The Stefan-Boltzmann law,

Equation (4-13), can be obtained from Equation (11-59) by calculating $\int_0^\infty \rho_\nu \, d\nu$.

11-5 Specific Heat of Solids

The earliest generalization concerning the specific heat of various solids was that of Dulong and Petit who noticed that for a wide variety of substances the molar specific heat is $3R$, or about 6 cal/deg. We can understand this result on the basis of equipartition of energy. Each atom of a solid is held in place in such a way that it experiences a restorative force if displaced from its equilibrium position. For a small displacement $\mathbf{r}$ the force will be approximately harmonic, and the classical expression for the energy of a molecule of mass m will be

$$\mathcal{E} = \frac{1}{2m} \mathbf{p} \cdot \mathbf{p} + \tfrac{1}{2} k_s \mathbf{r} \cdot \mathbf{r} \tag{11-61}$$

Each dot product consists of three terms, so that every atom will possess six quadratic degrees of freedom. Then the average energy per atom will be $3kT$ at a temperature T. The total energy of the solid is

$$\mathcal{E}_{\text{total}} = 3NkT \tag{11-62}$$

where N is the number of atoms. Calculating as in Equation (11-47) we obtain for the specific heat

$$C_v = 3R \tag{11-63}$$

If measurements are made at room temperature, the law of Dulong and Petit agrees well with experiment for some solids but not for others. For all substances the specific heat goes to zero as the temperature approaches absolute zero. Since C_v does not depend on temperature in the approximation of Dulong and Petit, we see that a better theory is needed for the description at low temperature. An improvement was made by Einstein (1907) who reasoned by analogy with blackbody radiation that the oscillations of the atoms should be quantized and that the average energy per atom should be

$$<\mathcal{E}> = 3kT \frac{h\nu/(kT)}{\exp\left[h\nu/(kT)\right] - 1} \tag{11-64}$$

instead of $3kT$ as predicted by classical equipartition. In this equation ν is the frequency of oscillation of an atom in the crystal. This form came much closer to describing the experimental data than did the law of Dulong and Petit, but enough deviation existed so that a better theory was desired.

The next stage of improvement was provided by P. Debye (1912). To understand Debye's theory we recognize that in a solid there is a regular crystalline

structure with the atoms spaced on a three-dimensional lattice. One atom cannot vibrate without influencing its neighbors, and the result of any such collective motion is a wave disturbance in the crystal. This wave propagates at
the speed of sound, since it is a form of sound. Further, the frequencies of
standing sound waves in a crystal are quantized, analogous to standing light
waves in a cavity. A quantum of sound energy is called a *phonon*, a further
analogy with electromagnetic waves. Thermal motion of atoms in a crystal,
then, is conceived of as a random mixture of quantized sound waves or a
"phonon gas." The density of states for a photon gas was shown in Equation
(11-55) to be

$$g(\mathcal{E}) \, d\mathcal{E} = \frac{8\pi V \mathcal{E}^2 \, d\mathcal{E}}{h^3 c^3}$$

We can use this result here if we modify two quantities in the equation. Equation (11-55) contains a factor of two which was introduced because photons
have two independent states of polarization (or equivalently, spin orientation)
available to them. Phonons have spin zero, so we must divide by two. The other
change involves replacing c with the speed of sound, a replacement that is
complicated by the fact that sound waves in a crystal can be transverse as well
as longitudinal and the two types of wave do not propagate at the same speed.
We state without proof that if v_t and v_l are the velocities for transverse and
longitudinal sound propagation, respectively, an effective sound velocity v_s
can be defined by

$$\frac{1}{v_s{}^3} = \frac{2}{v_t{}^3} + \frac{1}{v_l{}^3} \tag{11-65}$$

This form incorporates the idea that there exist two independent transverse
modes but only one longitudinal mode.

The resulting density of states is

$$g(\mathcal{E}) \, d\mathcal{E} = \frac{4\pi V \mathcal{E}^2 \, d\mathcal{E}}{h^3 v_s{}^3} \tag{11-66}$$

This form can be transformed (see Appendix V-9) into a function of phonon
frequency:

$$\mathcal{E} = h\nu \qquad d\mathcal{E} = h \, d\nu$$

$$g(\nu) \, d\nu = \frac{4\pi V \nu^2 \, d\nu}{v_s{}^3} \tag{11-67}$$

The next (and the most crucial) step is to recognize that the number of states
of collective oscillation in a crystal is finite. The truth of this statement becomes
apparent when one considers that for high frequencies the wavelength of a
sound wave would become so small that it would require several oscillations
in the empty space between atoms, where no matter exists to oscillate. We
can infer the number of possible frequency states available by considering the

normal modes of oscillation of an N-body system. For a classical system of N bodies we can in principle obtain a complete description of the system by writing Newton's second law for each of the bodies. Since the second law is a vector equation, we have three scalar equations for each body, or $3N$ equations in all. They will be second-order differential equations, coupled together, and tedious to solve if N is very large. The result of such a solution is that there will be $3N$ independent states of motion. Three of them correspond to pure translation of the system and several more refer to pure rotation about the center of mass. The rest correspond to vibrational modes, and for N of the order of 10^{23} we make essentially no error if we use $3N$ as the total number of independent vibrational states. Each of these states has a definite natural frequency of oscillation; many frequencies occur, and two separate modes may have the same frequency or they may have different frequencies, yet all frequencies are finite. The spectrum of normal frequencies is discrete, but for large values of N they are spaced so closely that we can approximate the spectrum by a continuous function. The highest of the $3N$ frequencies is called the *Debye frequency* and is denoted by ν_D. The $3N$ vibrational states can be counted by adding up the allowed states or approximated by integrating over the density of states

$$\int_0^{\nu_\mathrm{D}} g(\nu)\, d\nu = 3N \tag{11-68}$$

Figure 11-8 shows the density of states as a function of frequency and illustrates how the Debye frequency serves as a cut-off, chosen to make the total number of states equal to $3N$. We can insert Equation (11-67) into Equation (11-68) and perform the integration (assuming that the speed of sound does not depend on the frequency). We obtain

$$\nu_\mathrm{D}{}^3 = \frac{9N v_s{}^3}{4\pi V} \tag{11-69}$$

We next rewrite Equation (11-67) so that it depends explicitly on the Debye frequency.

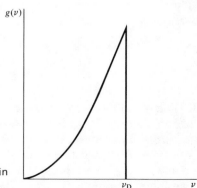

Figure 11-8   Density of states for phonons in a crystal.

$$g(\nu)\ d\nu = \frac{9N}{\nu_D{}^3}\ \nu^2\ d\nu$$

Then the number of phonons near the frequency ν is

$$n(\nu)\ d\nu = \frac{9N}{\nu_D{}^3}\ \frac{\nu^2\ d\nu}{\exp\ [h\nu/(kT)] - 1} \tag{11-70}$$

The total energy of the crystal can now be calculated by multiplying n by the energy per phonon and summing (integrating).

$$\mathcal{E}_{\text{total}} = \int_0^{\nu_D} n(\nu)\ h\nu\ d\nu = \frac{9Nh}{\nu_D{}^3} \int_0^{\nu_D} \frac{\nu^3\ d\nu}{\exp\ [h\nu/(kT)] - 1} \tag{11-71}$$

It is convenient to change the variable of integration to $x = h\nu/(kT)$, using x_m to denote $h\nu_D/(kT)$. It is also customary to recognize that since $h\nu$ and kT both have the same dimensions (i.e., energy) we can define Debye temperature by

$$kT_D \equiv h\nu_D \tag{11-72}$$

After making these substitutions, we obtain

$$\mathcal{E}_{\text{total}} = 9N\left(\frac{T}{T_D}\right)^3 kT \int_0^{x_m} \frac{x^3\ dx}{\exp x - 1} \tag{11-73}$$

To get the specific heat we need to calculate

$$C_v = \frac{1}{n}\frac{d}{dT}\mathcal{E}_{\text{total}} \tag{11-74}$$

where n is the number of moles. The differentiation is nontrivial, since both the integrand and the upper limit of the integral depend on temperature. It is rather more instructive to find approximations for $\mathcal{E}_{\text{total}}$ and C_v at extremes of the temperature scale.

At temperatures high enough that $kT \gg h\nu_D$, the quantity x will be small enough that we can approximate $\exp x$ by $1 + x$, the first two terms of the Maclaurin series. Then

$$\mathcal{E}_{\text{total}} \approx 9N\left(\frac{T}{T_D}\right)^3 kT \int_0^{x_m} \frac{x^3\ dx}{x} = 3NkT \tag{11-75}$$

Then in this limit C_v becomes equal to $3R$, proving that the high-temperature limit of the Debye theory is just the generalization of Dulong and Petit.

For the low-temperature approximation we recognize that $x \gg 1$ and that x_m must be very large indeed, since it is the largest value that x can assume. The integral in Equation (11-73) with an infinite upper limit is available in various collections of mathematical formulas:

$$\int_0^{\infty} \frac{x^3\ dx}{\exp x - 1} = \frac{\pi^4}{15} \tag{11-76}$$

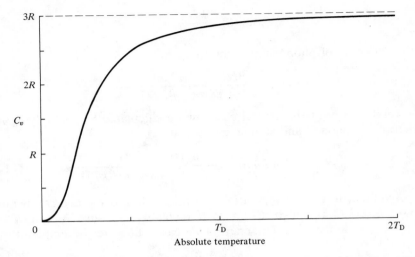

Figure 11-9 Specific heat of a solid as a function of temperature. The line $C_v = 3R$ is the high-temperature limit and is the result of Dulong and Petit. Near $T = 0$, $c_v \sim T^3$.

Then the total energy becomes

$$\mathcal{E}_{\text{total}} = \frac{3\pi^4 N k}{5 T_D^3} T^4 \tag{11-77}$$

a result that is easy to differentiate. We see that near absolute zero the specific heat of a solid increases like T^3. The resulting shape for the specific heat of a solid is shown in Figure 11-9. We see that if the temperature of the solid is any larger than the Debye temperature the generalization of Dulong and Petit will be a reasonable approximation. As a rule of thumb, the harder a substance, the higher its Debye temperature; diamond has T_D equal to 1860K, whereas most metals have T_D of a few hundred kelvins. The principle is that hardness in a solid is associated with large values of the effective spring constant; hence a high value for ν_D and a high value for T_D.

Further refinement of the theory of the specific heats of solids was carried out by Born and von Kármán, who took into account the exact type of crystal lattice. Their result agrees better with data than that of Debye but the qualitative features are similar.

11-6 Free Electrons in Metals

In our treatment of the specific heat of solids we discussed the motion of the atoms that make up a crystal. In effect, we were including only the central core of each atom, since the valence electrons are often free to move. We calculated the specific heat without any reference to the possibility of motion on the part of these electrons, obtaining results that agree well with experiment. In this section we shall examine the motion of electrons in a metal from the point of view of quantum statistics.

We shall assume that the nuclei and inner electrons of a metal are fixed at their lattice sites in a crystal and that the outer (valence) electrons are free to move through the crystal. As is usual in our quantum calculations, this model is just an approximation and does not describe quantitatively any real system. The qualitative features, however, are similar to those present in a more complicated model. In our approximation the electrons do not interact with each other or with the nuclei at the lattice sites. In effect, we describe an ideal gas consisting of electrons contained in a crystal of volume V. Since the spin of the electron is $\frac{1}{2}$, we shall require Fermi-Dirac statistics. Further, each electron has two possible spin orientations, and therefore the measure of the set of points (in momentum space) which are available to the momentum vector of an electron with momentum p will be

$$G(p)\ dp = \frac{8\pi V p^2\ dp}{h^3} \tag{11-78}$$

the same result used in Section 11-4 to describe a photon gas. The energy of an electron of mass m is given by

$$\mathcal{E} = \frac{p^2}{2m} \tag{11-79}$$

since we are ignoring any interactions. Using the familiar technique for transforming a probability density function, we find that

$$g(\mathcal{E})\ d\mathcal{E} = \frac{8\sqrt{2}\,\pi V m^{3/2}}{h^3}\,\mathcal{E}^{1/2}\ d\mathcal{E} \tag{11-80}$$

It will be convenient to denote by C the fraction in this equation, since it consists entirely of constants. The number of electrons that have energies near the value $\mathcal{E}$ will be

$$n(\mathcal{E})\ d\mathcal{E} = \frac{C\mathcal{E}^{1/2}\ d\mathcal{E}}{\exp\ [(\mathcal{E} - \mathcal{E}_F)/(kT)] + 1} \tag{11-81}$$

We have used $\mathcal{E}_F$ instead of μ in the Fermi-Dirac distribution function. Its meaning becomes apparent if we consider the shape of the function $n(\mathcal{E})$ when the temperature is at absolute zero. If $\mathcal{E}$ is larger than $\mathcal{E}_F$, the argument of the exponential is positive and infinitely large, implying that $n(\mathcal{E})$ is zero. If $\mathcal{E}$ is less than $\mathcal{E}_F$, the same argument is negative and infinitely large; in this case the exponential vanishes and $n(\mathcal{E})$ simplifies. The results are

$$\text{at } T = 0, \qquad n(\mathcal{E})\ d\mathcal{E} = C\mathcal{E}^{1/2}\ d\mathcal{E} \qquad \mathcal{E} < \mathcal{E}_F$$

$$= 0 \qquad \mathcal{E} > \mathcal{E}_F \tag{11-82}$$

This function was plotted in Figure 4-16, and its physical significance was discussed there. The conclusion is that the electrons fill all the available energy levels up to the value $\mathcal{E}_F$, which is called the Fermi energy. If the temperature of the crystal is raised to some value above absolute zero, the majority of the electrons are not disturbed at all; only a few of those near the Fermi level are elevated to a higher energy. We have therefore attained the reason why the Debye theory for specific heats can provide reasonably good agreement with experiment, even though it ignores the motion of the electrons. Only those electrons with energies close to $\mathcal{E}_F$ can contribute to the specific heat and their number is small in comparison with the number of atoms in the crystal.

The behavior of an electron gas is often compared with that of molecules of water in the liquid state. Water is essentially incompressible, and in a large body of water the most violent motion occurs only in the form of surface phenomena. The electron gas is likewise incompressible because of the Pauli exclusion principle, and the system is often referred to as the "Fermi sea." The electrons near the Fermi energy are analogous to water molecules near the surface of the ocean. The same ideas and terminology are also used to describe nuclear matter; since protons and neutrons have spin $\frac{1}{2}$, they are fermions, and a heavy nucleus with many protons and neutrons similarly has its lowest energy states completely filled.

It is possible to use the normalization condition, Equation (11-34), to obtain an expression for the Fermi energy. This equation must be true at any temperature, but we shall evaluate it for $T = 0$ because the function $n(\mathcal{E})$ takes on a very simple form at absolute zero, the form obtained in Equation (11-82). Since $n(\mathcal{E})$ vanishes above the Fermi energy when $T = 0$, we need to integrate only from $\mathcal{E} = 0$ to $\mathcal{E} = \mathcal{E}_F$.

$$N = C \int_0^{\mathcal{E}_F} \mathcal{E}^{1/2} \, d\mathcal{E} = \tfrac{2}{3} C \mathcal{E}_F^{3/2} \tag{11-83}$$

From Equation (11-80) we recall the form of the quantity C and then solve for $\mathcal{E}_F$:

$$\mathcal{E}_F = \frac{h^2}{2m} \left(\frac{3N}{8\pi V}\right)^{2/3} \tag{11-84}$$

This result was stated earlier without proof as Equation (4-41). In using this result it is necessary to remember that N is the total number of valence electrons in the crystal.

11-7 The Band Theory of Conduction-Electrons

In the preceding section we considered the behavior of electrons in a crystal, using as a model the ideal Fermi gas. This model, of course, is too simple to be correct for any substance that actually exists. The most serious oversimplification was the assumption that the electrons do not interact. In fact they

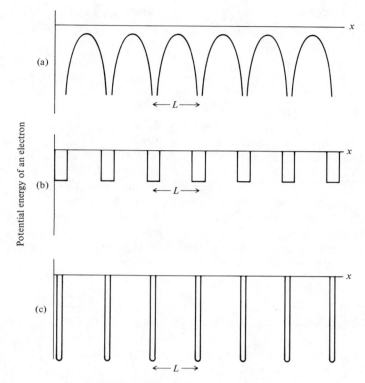

Figure 11-10   Potential energy of an electron in a one-dimensional crystal: (a) typical crystal, (b) Kronig-Penney model, (c) Kronig-Penney model simplified by making the potential wells very narrow and very deep.

interact with each other and with the positive charges associated with nuclei which are essentially at fixed points in the crystal lattice. In this section we wish to consider the nature of the changes in the free-electron model which occur when we consider the interaction between a single electron and the ion at a lattice site.

The important thing about this interaction is that the force involved is periodic and so is the potential energy of the electron. In Figure 11-10 this periodic potential is shown for a typical crystal, considering the problem as one-dimensional. Even with the simplification to one dimension the problem at hand is difficult. It is necessary to solve the Schroedinger equation with the complicated mathematical form for the periodic potential. The essential features of the problem are retained in a further simplification which is due to Kronig and Penney (1930); the potential is represented as being a series of square wells, recurring periodically. Unfortunately, even this periodic square-well problem requires a tedious amount of algebra for its solution. We choose to simplify the problem still further by assuming that an electron in the crystal is free (i.e., that the potential energy V is constant) except when it is very near to an ion and that the force for close approach is infinitely large. This assumption corresponds to the potential shown in Figure 11-10c. Our problem is to solve the Schroedinger equation

$$-\frac{\hbar^2}{2m}\frac{d^2\psi}{dx^2} + V\psi = \mathcal{E}\psi \qquad (11\text{-}85)$$

for this periodic potential.

It is clear that we need to examine carefully the conditions on the continuity of ψ and its derivatives. Inspection of the Schroedinger equation shows that since V is infinite at each lattice site, at least one other term must be infinite there, as well. The only possible choice is the second derivative. It follows that the first derivative of ψ will have a jump discontinuity; ψ itself is continuous. We shall assume that the discontinuity in $d\psi/dx$ is proportional to the value of ψ at the singular point; the proof of this idea is left as a problem, but we note here that the result is reasonable, since the effect of the lattice center on an electron should be related to some measure of the probability that the electron is present. In symbols we write

$$\Delta\left(\frac{d\psi}{dx}\right) = -K\psi \qquad (11\text{-}86)$$

where K is a constant. The minus sign is not essential to the treatment; it is present because we assumed attractive wells with $V < 0$ at the singularities.

Because of a theorem by Bloch, which we shall not prove here, it is possible to assert that for any periodic potential the wave function can be written in the form

$$\psi(x) = u(x) \exp(ikx) \qquad (11\text{-}87)$$

where u is periodic with the same characteristic length as that of the lattice itself:

$$u(x - L) = u(x) \qquad (11\text{-}88)$$

The quantity k is the effective wave number of the electron as it propagates through the crystal. From Equations (11-87) and (11-88) it is readily proved that

$$\psi(x) = \exp(ikL)\,\psi(x - L) \qquad (11\text{-}89)$$

In using this result, we must recognize that the points x and $(x - L)$ lie on different sides of one of the singular points in the potential. We shall have to use a functional form for $\psi(x)$ different from that for $\psi(x - L)$ because of the intervening singularity which led to discontinuous derivatives. We shall use ψ_0 and ψ_1 to refer to the wave functions in two adjoining regions, and we rewrite the equation in the form

$$\psi_1(x) = \exp(ikL)\,\psi_0(x - L) \qquad (11\text{-}90)$$

The continuity of ψ and the discontinuity of ψ' can now be rewritten.

$$\psi_1(L) = \psi_0(L)$$

$$\psi_1'(L) - \psi_0'(L) = -K \,\psi_0(L) \qquad (11\text{-}91)$$

Since in the regions between lattice points $V = 0$, we recall that the wave functions can be written in the form

$$\psi_0(x) = A \exp(i\alpha x) + B \exp(-i\alpha x) \qquad (11\text{-}92)$$

$$\alpha = \frac{\sqrt{2m\mathscr{E}}}{\hbar}$$

Here m is the mass of an electron. Using Equation (11-90), we obtain for ψ_1

$$\psi_1(x) = \exp(ikL)\,\{A \exp[i\alpha(x - L)] + B \exp[-i\alpha(x - L)]\} \quad (11\text{-}93)$$

The conditions (11-91) can now be written explicitly:

$$\exp(ikL)\,[A + B] = A \exp(i\alpha L) + B \exp(-i\alpha L)$$

$$i\alpha \exp(ikL)\,[A - B] - i\alpha[A \exp(i\alpha L) - B \exp(-i\alpha L)]$$

$$= -K[A \exp(i\alpha L) + B \exp(-i\alpha L)] \quad (11\text{-}94)$$

These two equations can be rearranged and placed in the form

$$rA + sB = 0$$

$$tA + uB = 0 \qquad (11\text{-}95)$$

where r, s, t, and u are all sums of exponentials. This set of linear equations will have consistent solutions for A and B if and only if the determinant $(ru - st)$ vanishes. After carrying out the algebra, we obtain (using $\exp ix = \cos x + i \sin x$) the result

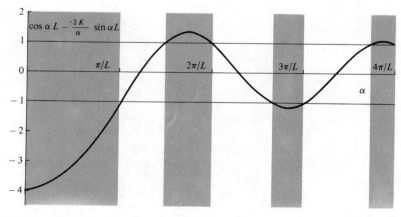

Figure 11-11   Curve showing the origin of energy bands for propagation of a wave in a simple periodic structure. Forbidden values of α are shaded. Allowed values are defined by $|\cos \alpha L - 2(K/\alpha) \sin \alpha L| \leq 1$. The curve shown corresponds to $2KL = 5$.

$$\cos kL = \cos \alpha L - \frac{2K}{\alpha} \sin \alpha L \qquad (11\text{-}96)$$

In this equation k is real (corresponding to a wave that propagates rather than attenuates) and so the magnitude of the left-hand side never exceeds unity. Thus it is essential that the right-hand side should likewise remain below unity in absolute value. The right-hand side is graphed as a function of α in Figure 11-11 for a particular choice of K. We see that the condition implied here is not met for certain values of α, called *forbidden bands,* shaded in the diagram. The *allowed bands* are unshaded and are defined by the fact that the condition on the right-hand side of Equation (11-96) is satisfied.

From Equation (11-92) we recall that α is determined by the energy of the electron. Since α is allowed to take on only certain values, it is apparent that $\mathscr{E}$ is likewise restricted. We assert that the existence of allowed and forbidden energy bands is characteristic of the periodic structure of the potential and that we may expect the bands to appear in more complicated and more realistic models than the simple one for which we have shown the calculation.

Applying this band theory to valence electrons in a crystal, we conclude that the electrons do not constitute a uniform "Fermi sea" as predicted in the preceding section. Rather, the sea contains layers, corresponding to allowed bands. The layer that contains the Fermi energy is called the *conduction band.* The electrical conductivity of the crystal depends greatly on whether the Fermi energy is near the top or the bottom of the conduction band. In Figure 11-12

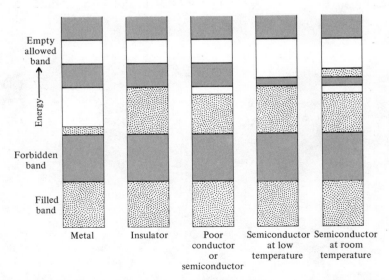

Figure 11-12   Schematic representation of energy bands for materials with different electrical conductivity. Dotted areas are filled with electrons. Shaded areas are forbidden bands.

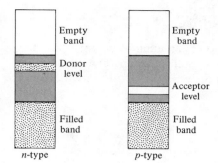

Figure 11-13 Band structure for impurity semi-conductors.

we illustrate how this can occur. For metals only a few electrons are in the conduction band and many states are available, permitting these electrons to move easily. The electrical conductivity is large. The opposite extreme occurs for insulators, in which the band that would have been the conduction band is completely filled, and a wide forbidden band intervenes, making it very unlikely that an electron can get enough energy to rise to the next allowed band. An intrinsic semiconductor is often a substance in which the highest occupied band is filled, but the next allowed band is separated by an energy of about 1 eV, making it possible for some of the electrons to gain sufficient thermal energy to get into the higher band.

Even for metals, the electrical conductivity is finite at room temperature. The reason is that there are other effects that have not been considered in our model. At room temperature, for most crystals, the outstanding effect is that of electron-phonon collisions which impede the motion of the electrons. As the temperature is lowered, fewer phonons are present, and other effects are more important. Typically these are irregularities in the crystal with which electrons can collide. Such will be the case for impurities whose atoms do not fit into the crystal lattice, hence destroying the perfection of the periodic structure. Even for pure materials, however, imperfections of various types can appear because of strains in the crystal.

Impurities do not always decrease the conductivity of a substance. For some insulators the addition of the proper kind of impurity can cause the crystal to behave like a semiconductor. This effect comes about because the impurity can add levels to the band structure, creating allowed states in a forbidden band. Two elements of great practical importance in this regard are silicon and germanium. Both are capable of having their conductivity improved by the addition of trivalent impurities. Two cases exist. Impurities with three *P*-wave electrons (P, As, or Sb) are called donor impurities. They contribute extra levels near the top of the forbidden band, as shown in Figure 11-13. These extra levels are filled with electrons, which easily gain the energy required to reach an empty band, where they become mobile. The result is called an *n*-type semiconductor. Another possibility is to use as an impurity a substance with only one *P*-wave electron (B, Al, Ga, or In) to provide an empty level at the lower part of a forbidden band, giving rise to a p-type semiconductor. These impurities are called acceptors.

By inspecting Figure 11-13 it is possible to gain a qualitative understanding of the behavior of an n-p junction. It is clear that electrons can move more easily from n to p than the reverse. This property has led to the widespread use of crystal diodes as rectifiers.

11-8   Josephson Effect

A very interesting example of tunneling and one that has already had many important applications was first described by Brian D. Josephson (1962). If two superconducting materials are separated by a very thin layer of non-superconducting material, there will be a tunneling of superconducting electrons through the barrier. Figure 11-14 shows a *Josephson junction*—two superconductors separated by a very thin nonsuperconducting layer—with a d-c source of potential across the superconductors. A current will be produced in this circuit which will consist of two parts: (a) there will be a direct component or direct supercurrent when the potential difference across the junction is zero and (b) there will be superposed a high frequency alternating supercurrent.

In the theory of superconductivity developed by J. Bardeen, L. N. Cooper, and J. R. Schrieffer, pairs of electrons with their spins oppositely directed are assumed to be the carriers of the electric charges and thus carry a charge $2e$; they are sometimes referred to as *Cooper pairs*. If a Cooper pair in superconductor I is at a potential V_1 and another pair in superconductor II is at a potential V_2, the transfer of a pair of electrons through the barrier will involve the emission or absorption of energy equal to

$$2e(V_1 - V_2) = 2eV \qquad (11\text{-}97)$$

This emission or absorption of energy may be in the form of radiation of frequency ν given by

$$h\nu = \Delta\mathcal{E} = 2eV \qquad (11\text{-}98)$$

from which

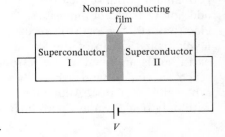

Figure 11-14   Josephson junction.

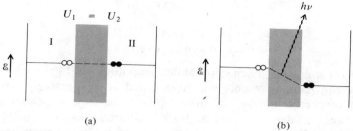

(a)   (b)

Figure 11-15 (a) Electrochemical potentials $U_1 = U_2$, direct Josephson supercurrent; (b) $h\nu = 2U$.

$$\nu = \frac{2eV}{h} \tag{11-99}$$

We can refer to an energy-level diagram for the Josephson junction such as that shown in Figure 11-15 to account for the two types of current—the direct and the alternating currents. In Figure 11-15a the energy levels of an electron pair in the two superconductors are equal, which means that there is zero potential difference between them. In this case the electrochemical potentials of the two superconductors are equal and electron pairs can tunnel through the barrier at zero potential difference, giving rise to the direct Josephson supercurrent. In Figure 11-15b the superconductors have chemical potentials $+U$ and $-U$ so that the energy difference is $2U$; hence the tunneling of an electron pair through the barrier must be accompanied by the emission of energy, in this case a photon of energy

$$h\nu = 2U = 2eV \tag{11-100}$$

For the process to be reversed, energy of the same frequency would have to be absorbed from an external radiation source.

Equation (11-100) shows that a straightforward method exists for measuring the ratio of the fundamental constants e/h by measuring the chemical potential with a potentiometer calibrated with a standard cell and determining the frequency of the radiation emitted or absorbed with modern high-frequency techniques. A simple calculation shows that if V is of the order of microvolts, ν is of the order of gigahertz.

In the quantum-mechanical treatment of the Josephson effect the wave function $\Psi(\mathbf{r}, t)$ is not limited to a single Cooper pair but describes a very large number of them. The Cooper pairs are formed by the interaction between the electrons and the ions of the crystal lattice giving rise to a net attractive force between two electrons, thus reducing their energy relative to free electrons. The wave function describing these pairs extends over a much larger distance than one describing individual electrons. These wave functions must all have the same phase in the superconductor; that is, there is phase coherence among these waves. In a sense this is an extension of quantum dynamics to macroscopic phenomena.

The wave functions for the two superconductors may be written as

$$\Psi_1(\mathbf{r}, t) = \psi_1(\mathbf{r}) \exp(i\theta_1) \tag{11-101a}$$

and

$$\Psi_2(\mathbf{r}, t) = \psi_2(\mathbf{r}) \exp(i\theta_2) \tag{11-101b}$$

in which $\psi_1(\mathbf{r})$ and $\psi_2(\mathbf{r})$ are the probability amplitudes of the charge densities of the electron pairs in superconductors I and II, respectively, and θ_1 and θ_2 are the respective phases of these waves; the time dependence of the wave functions resides in θ_1 and θ_2. By substituting these equations into the time-dependent Schroedinger equations and solving them with appropriate boundary conditions, we can obtain equations for the superconducting current J_s and for the phases θ_1 and θ_2 as a function of time. For the superconducting current at zero potential difference we have

$$J_s = J_0 \sin (\theta_1 - \theta_2) \tag{11-102}$$

Thus the direct supercurrent depends on the sine of the differences in the phases of the waves on the two sides of the barrier.

When a difference of potential V is placed across the superconductors, the phases will change with time as follows:

$$\frac{d}{dt} (\theta_1 - \theta_2) = \frac{2eV}{h} \tag{11-103}$$

so that after a time t

$$\theta_1(t) - \theta_2(t) = \theta_1(0) - \theta_2(0) + \frac{2eV}{\hbar} t \tag{11-104}$$

where $\theta_1(0)$ and $\theta_2(0)$ are constants. Thus the supercurrent will oscillate with an angular frequency

$$\omega = \frac{2eV}{\hbar} = 2\pi\nu \tag{11-105}$$

If electromagnetic radiation from an outside source with a frequency ν_0, which is a whole number n times the Josephson frequency ν, is incident on the Josephson junction, beats will be produced between the two whenever V is a whole multiple of $h\nu/2e$ or when

$$nh\nu = 2eV$$

$$V = \tfrac{1}{2}n\frac{h}{e}\nu \tag{11-106}$$

These values of V will appear as steps in the current-voltage curve for the Josephson tunnel junction. These steps were observed by S. Shapiro (1963) who used a microwave field of 9.75 GHz on a niobium-oxide-lead junction at 4.2K, as shown in Figure 11-16.

As mentioned previously (Section 4-9), the Josephson effect was used by Parker, Langenberg, Denenstein, and Taylor (1969) to make a precise de-

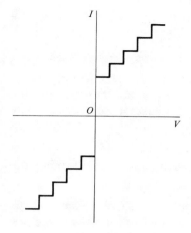

Figure 11-16 Current I plotted as a function of the voltage V for a Josephson junction in a microwave field. [After S. Shapiro, *Phys. Rev. Letters* **11**, 80 (1963).]

termination of the value of e/h. They used a Josephson tunnel junction of tin-tin oxide-lead and point contact junctions with tantalum or lead maintained at sufficiently low temperatures down to about 1.2K and irradiated the junctions with microwaves with a frequency of 10 GHz ($\lambda = 3$ cm). The values determined by them are given as

$$\frac{2e}{h} = 483.5985 \times 10^{12} \text{ Hz/volt} \pm 3.4 \text{ ppm from radiation emission}$$

$$= 483.5978 \times 10^{12} \text{ Hz/volt} \pm 0.4 \text{ ppm from radiation absorption}$$

Work to improve the accuracy of the measurement of $2e/h$ is being continued in various laboratories and, minimally, should lead to greater consistency in the values of the fundamental constants of physics.

Problems

11-1. Prove that the moment of inertia of a rigid diatomic molecule about its center of mass is given by

$$I = \mu r_0^2$$

where r_0 is the interatomic separation and μ is the reduced mass.

11-2. The interatomic separation in the CO molecule is 1.13 Å. (a) Find the reduced mass. (b) Find the moment of inertia. (c) What is the energy difference between the ground state and the first excited rotational level?

11-3. (a) Estimate the moment of inertia for the CO molecule for rotation about the axis which passes through both atoms. (b) Estimate the excitation energy of the first rotational excited state. Give your answer in electron volts and compare it with 11.1 eV, the energy required to break up the CO molecule.

11-4. Sketch the normal modes of motion for the CO_2 molecule.

11-5. One of the empirical parameters in the van der Waals equation of state for a gas is essentially the molecular volume. Obtain from the literature the appropriate van der Waals constant for CO, estimate its molecular size, and compare it with the interatomic spacing from rotational spectra.

11-6. Prove by differentiation that the Morse function has its minimum at $r = r_0$.

11-7. Near the point $r = r_0$ the Morse potential has nearly the same shape as a harmonic oscillator potential. Use this fact to calculate the spring constant k_s and the natural frequency ω_0 for a diatomic molecule as functions of D and a, the parameters in the Morse potential.

11-8. For H_2 gas at room temperature and at atmospheric pressure (a) estimate the De Broglie wavelength of a molecule and (b) estimate the intermolecular separation. Compare these values.

11-9. (a) Estimate the De Broglie wavelength of a conduction electron in copper at room temperature. (b) Estimate the distance between two nearby electrons. Compare these values.

11-10. Perform the long divisions needed to establish the sums of the infinite series used in Appendix V-8.

11-11. Prove that the point of inflection for the Fermi-Dirac distribution function $f(\mathcal{E})$ occurs at the Fermi energy.

11-12. Using $n(\mathcal{E}) \, d\mathcal{E}$ for a Maxwell-Boltzmann ideal gas, obtain the velocity distribution $n(v) \, dv$.

11-13. Calculate $<v>$ for a Maxwell-Boltzmann gas.

11-14. Calculate the root-mean-square velocity for a molecule in a Maxwell-Boltzmann gas.

11-15. The sun's corona is thought to consist of hydrogen at a temperature of about 10^6 K. (a) The dissociation energy of the H_2 molecule is about 4.5 eV. Use this fact in conjunction with Maxwell-Boltzmann statistics to estimate the fraction of molecules that will be dissociated into atomic hydrogen at this temperature. (b) What is the average kinetic energy of a hydrogen atom at this temperature? (c) Does this energy imply a relativistic speed?

11-16. Derive Wien's law by differentiation of the wavelength spectrum for blackbody radiation.

11-17. Integrate the energy distribution to derive the Stefan-Boltzmann law.

11-18. Estimate the temperature of a piece of red-hot iron.

11-19. Prove the identity

$$\int_0^\infty \frac{x^3}{e^x - 1}\, dx = \frac{\pi^4}{15}.$$

(*Hints:* Expand $1/(e^x - 1)$ by long division into powers of e^{-x}; integrate term-by-term; use the definition of the Bernoulli number B_2.)

11-20. (a) Use the Einstein expression for the average energy per atom of a solid to calculate the specific heat C_v. (b) Show that $\lim_{T \to 0} C_v = 0$. (c) Show that $\lim_{T \to \infty} C_v = 3R$.

11-21. Define a Debye wavelength by $\lambda_D = v_s/\nu_D$. Compare this wavelength with the interatomic spacing for a typical crystal.

11-22. The Debye temperature for gold is 164 K; for silver it is 229 K. Which metal has the higher specific heat at room temperature?

11-23. Given that the Debye temperature of Cu is 315 K, (a) how much heat is required to raise the temperature of 1 kg of Cu from 2 to 8 K? (b) How much heat is required to raise the temperature of 1 kg of Cu from 400 K to 450 K?

11-24. Obtain the Debye specific heat for a solid by differentiating Equation (11-73).

11-25. If in Equation (11-76) we assume that x^3 grows slowly enough that we can ignore the 1 in the denominator, how much error do we make?

11-26. Derive Equation (11-80) from (11-78) and (11-79).

11-27. Show that $\langle \mathcal{E} \rangle = \frac{3}{5}\mathcal{E}_F$ for valence electrons in a crystal at absolute zero.

11-28. At what temperatures does the most probable energy for an electron in a crystal coincide with the Fermi energy?

11-29. Show that for a potential spike of infinite height, zero width, but finite "area," the first integral of the Schroedinger equation yields the following information: the discontinuity in ψ' is proportional to the value of the wave function at the point of discontinuity.

11-30. Prove Equation (11-89) for Bloch wave functions.

11-31. Graph curves like the one in Figure 11-11 for other values of K. How does the value of K affect (a) the number of bands (b) their width?

11-32. (a) Referring to Equation (11-96) and to Figure 11-12, obtain an expression for the group velocity of a wave which propagates in the periodic structure discussed in the text. (b) Calculate its value for $\alpha L = \pi$ and for $\alpha L = 2\pi$, assuming that $2KL = 5$. It is a property of periodic structures generally that the phase velocity and the group velocity need not have the same sign.

References

Molecular Physics

Born, M., *Atomic Physics*. New York: Hafner Publishing Company, 1959, Chapter IX.
Herzberg, G., *Molecular Spectra and Molecular Structure*. Englewood Cliffs: Prentice-Hall, 1939.
Pauling, L. and E. B. Wilson, *Introduction to Quantum Mechanics*. New York: McGraw-Hill Book Company, 1935, Chapter X.

Statistical Mechanics

Born, *op. cit.*, Chapter VIII.
Desloge, E. A., *Statistical Physics*. New York: Holt, Rinehart and Winston, 1966.

Solid-State Physics

Clarke, J., "Josephson Effect and *e/h*," *Am. J. Phys.*, **38,** 1071 (1970).
Goldsmid, H. J., *The Thermal Properties of Solids*. New York: Dover Publications, 1965.
Kittel, C., *Introduction to Solid State Physics*. New York: John Wiley & Sons, 1966.
Livesey, D. L., *Atomic and Nuclear Physics*. Waltham: Blaisdell Publishing Company, 1966, Chapter 8.
Nussbaum, A., *Electric and Magnetic Behavior of Materials*. Englewood Cliffs: Prentice-Hall, 1967.
Nussbaum, A., *Semiconductor Device Physics*. Englewood Cliffs: Prentice-Hall, 1962.
Slater, J. C., *Insulators, Semiconductors and Metals*. New York: McGraw-Hill Book Company, 1967.
Smith, R. A., *Wave Mechanics of Crystalline Solids*. London: Chapman and Hall, 1963.
Sproull, R. L., *Modern Physics*. New York: John Wiley & Sons, 1963, Chapters 8–12.
Stringer, J., *An Introduction to the Electron Theory of Solids*. Oxford: Pergamon Press, 1967.
Taylor, B. N., W. H. Parker, and D. N. Langenberg, *Rev. Mod. Phys.*, **41,** 375 (1969).

PART THREE

Nuclear Physics

12   Particle Accelerators

12-1   Introduction

The remaining chapters of this book are devoted to the subject of nuclear physics, including discussions of nuclear structure and the fundamental particles involved in nuclear processes. Before considering the phenomena that involve nuclei, it would be well to gain an understanding of some of the equipment used to measure quantities on a nuclear scale. The emphasis here will be on the principles underlying the equipment. The design, construction, and operational details are engineering problems, and data are available on these aspects in the literature; references to some of this literature are given at the end of this chapter and the one that follows.

The typical experiment in nuclear physics is a scattering experiment. The types of equipment needed generally fall into three categories:

1. Accelerators
2. Beam transport
3. Detectors

This chapter is concerned with accelerators; the following chapter considers beam transport devices and detectors. The purpose of the accelerator is to provide a beam of projectiles for the experiment in such a way that the energy and intensity of the beam are capable of being controlled. Before the invention of accelerators, the only sources of projectiles were samples of radioactive materials and cosmic radiation. Control of these sources was inadequate for most purposes, and the development of many varieties of accelerator has been essential to progress in understanding of nuclei. The purpose of beam transport equipment is simply to convey the particles from the accelerator to the target in a way that will maintain reasonably high beam intensity, combined with a separation of desired projectiles from those that are not desired. After the beam impinges on the target, various detecting devices are needed to discern the number of particles (and their properties) which have been scattered.

Particle accelerators may be grouped into four categories. The first type provides simply a large difference in potential, but when combined with appropriate ion tubes they are capable of accelerating ions to energies of some millions of electron volts. The particle effectively falls through an electrostatic potential difference, gaining kinetic energy in the process.

The second group of accelerators, exemplified by the cyclotron (for positive ions) and the betatron (for electrons), consists of devices in which the charged particle makes many trips around an essentially circular path, receiving additional energy during each trip. Although the energy acquired during each revolution may be comparatively small, after a sufficient number of revolutions the particle will acquire a very large amount of energy. When the particle has acquired the desired energy, it is extracted from the device and used in nuclear experiments.

The third group contains the *linear accelerators* in which the particles travel in straight paths, receiving additions to their energies at given positions along the path. Linear accelerators have been designed for operation with electrons and with protons.

The fourth group is a modification of the second group which utilizes the phenomenon of *phase stability* in its operation. Among these devices are the synchrocyclotron, the electron synchrotron, and the proton synchrotron. The third and fourth groups are capable of producing much higher energies than those in the first two groups.

12-2 Electrostatic Accelerators

The basic principle of electrostatic acceleration was used for generation of x-rays, with the high voltage being produced by rectifiers connected in a voltage-multiplying circuit. The first application of this principle to nuclear physics was achieved by Cockcroft and Walton (1930); their names are frequently used in connection with this kind of accelerator. Electrostatic breakdown imposes a limit of about 1.5 million volts to this type of voltage generator when coupled to a tube suitable for evacuation so that the particles can travel across the potential difference without being lost to interactions with gas molecules. The principal modern application of the Cockcroft-Walton device is as a preaccelerator, giving enough energy to particles so that they can be injected into a larger accelerator.

Van de Graaff (1931) designed a direct-current electrostatic generator which can develop a potential difference of several million volts and when used with positive ion tubes can impart energies of several MeV to the ions. This type of generator consists essentially of a continuous belt B, made of some insulating material such as rubber, silk, linen, or paper, which passes over two pulleys P_1 and P_2, as shown in Figure 12-1. Pulley P_1 is at ground potential and is driven by a motor; pulley P_2 is mounted inside a hollow metal sphere S of

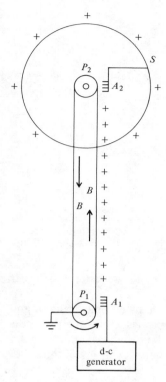

Figure 12-1 Schematic diagram of a Van de Graaff electrostatic generator.

large radius or a metal cylinder with curved bases. The hollow metal body S is insulated from the rest of the apparatus. In the operation of this generator an electric charge is sprayed onto the side of the belt which is moving upward. The source of this charge is usually a small transformer-rectifier set capable of developing a potential difference of about 10 to 100 kV. The charge is sprayed on the belt from a set of sharp points A_1 and then removed from the upper end of the belt by another series of sharp points A_2 and conducted to the surface S. If Q is the charge on this surface at any instant, its potential with respect to ground is simply

$$V = \frac{Q}{C} \tag{12-1}$$

where C is the capacitance of the system.

The conditions that limit the charge on the surface S, hence its potential V, are the nearness of other objects, such as the walls and ceilings of the laboratory and the breakdown of the air near the surface due to the intense electric field around it. To improve the operation of the Van de Graaff generator, and to reduce its size for a given maximum potential difference, the entire generator is placed inside a steel container in which the air is maintained at a high

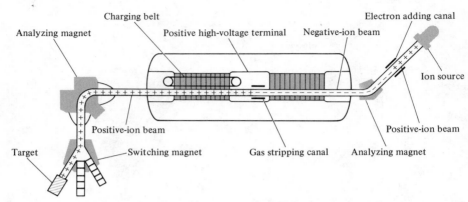

Figure 12-2 Schematic drawing of a 12-MeV tandem Van de Graaff accelerator. Particles begin at the ion source, are accelerated as negative ions until they reach the stripping canal, which is at a high voltage, and then are accelerated back to ground potential. (Courtesy of Florida State University, Tallahassee, Florida.)

pressure, about 150 psi. At the higher pressures air can withstand stronger electric fields before breakdown occurs.

Also, since the air is in a closed steel tube, it may be dried and cleaned, and other gases (such as CCl_2F_2 or SF_6) may be mixed with it to improve the operating conditions. Van de Graaff generators can now be obtained commercially to provide voltages up to 10 million volts and proton beam currents of the order of 6 to 8 μA. The limitation in energy comes from electrostatic breakdown. Within the energy region they can achieve, Van de Graaff accelerators are capable of a continuous range of energy settings and are unexcelled for energy resolution; for instance, at energies below 3 MeV it is possible to obtain a resolution better than 400 eV.

A modification that preserves most of the good features of the Van de Graaff is the tandem accelerator, illustrated in Figure 12-2. Positive ions are accelerated to an energy that is higher than the charge of the ion multiplied by the voltage of the terminal. The method consists of adding electrons to neutral atoms to form negative ions, which are accelerated from ground potential to the positive terminal. At that point the electrons are removed by causing the ions to pass through gas in a "stripping canal." The resulting positive ions are repelled from the positive terminal and receive more energy as they go to a grounded region. For the case of a proton beam the ion source provides H^- ions (two electrons bound to a proton); both electrons are removed in the stripping canal, so that the kinetic energy of the protons is twice as great as it would have been if a single stage of acceleration had been used. The increased energy of tandem accelerators is achieved primarily at the cost of beam intensity, since negative ions are not formed in great abundance. Tandem accelerators are now the standard tool for investigation of proton scattering in the region of 6 to 20 MeV. A wide variety of positive ions can be accelerated to energies that depend on the electric charges involved.

12-3   The Cyclotron

Another device for producing high-energy particles which has come into fairly common use is the *cyclotron,* developed by E. O. Lawrence and M. S. Livingston (1931). It consists essentially of a short hollow cylinder divided into two sections, D_1 and D_2, as shown in Figure 12-3. Each section is usually referred to as a "dee" because of its resemblance to the letter D. These dees are placed between the poles of a very large electromagnet. Some of the cyclotrons presently in use have magnets whose pole pieces are 30 to 60 in. in diameter; the diameters of the dees are approximately the same as those of the pole pieces. The dees are placed in another metal cylinder, as shown in Figure 12-4, and the whole assembly is placed between the poles of the electromagnet so that the magnetic field is perpendicular to the base of the cylinder and parallel to its axis.

There are two methods in general use for producing ions inside the dees. One method introduces a gas such as hydrogen into the system at a low pressure. A filament situated in the center of the chamber just outside the dees is heated, and a small difference of potential is applied between the filament and the cylindrical metal box to give the electrons from the filament sufficient energy to ionize some of the hydrogen atoms. This produces a vertical column of positive ions, some of which travel into the space between the dees. Another method produces the ions in a separate small source with a narrow opening into the space between the dees; this is usually referred to as a capillary ion source. The pressure of the gas in the ion source can be adjusted for optimum conditions while the rest of the system is maintained at a pressure as low as possible. This avoids electric discharges inside the dees and also makes it possible to use narrower dees and a smaller air gap between the poles of the magnet.

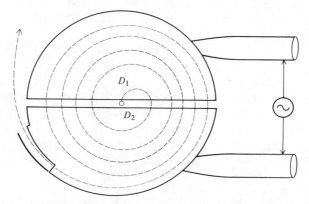

Figure 12-3   Schematic diagram showing the path of a charged particle in a cyclotron. Magnetic field is perpendicular to the paper.

Figure 12-4 Shop assembly photograph of the M.I.T. cyclotron chamber showing the construction of the chamber and dees. (From *J. Appl. Phys.*, January 1944. Courtesy of the Radioactivity Center at the Massachusetts Institute of Technology.)

The two dees, D_1 and D_2, are connected to the terminals of a high-frequency ac circuit so that the charge on each half changes a few million times per second. When D_1 is positive, the protons are accelerated toward D_2. The magnetic field causes each proton inside D_2 to travel in a circle of radius r given by

$$Bqv = \frac{Mv^2}{r} \quad \text{(mks units)} \tag{12-2}$$

where B is the magnetic induction, q is the electric charge of the proton (or other positive ion), and M is its mass. After it has traversed half a circle the proton comes to the edge of D_2. If, in the meantime, the potential difference between D_1 and D_2 has changed direction so that D_2 is now positive and D_1 negative, the proton will receive an additional acceleration, go across the gap from D_2 to D_1, and then travel in a circular path of larger radius inside D_1. After traversing a half-circle in D_1, it will reach the edge of D_1 and receive an additional acceleration from D_1 to D_2 because, in the meantime, the potential difference between D_1 and D_2 will have changed sign. The proton will continue traveling in semicircles of increasing radii each time it goes from D_1 to D_2 and from D_2 to D_1. This is due to the fact that the time required by the proton to travel half a circumference is independent of the radius of the circle. This can be shown very simply, since the time t required to travel the distance πr when the particle is moving with velocity v is

$$t = \frac{\pi r}{v}$$

but

$$v = \frac{Bqr}{M}$$

so that

$$t = \frac{\pi M}{Bq} \qquad (12\text{-}3)$$

For any given value of M/q the time required to traverse half a circumference is determined by the magnetic induction B. By adjusting B this time can be made the same as that required to change the potentials of D_1 and D_2.

After the protons have traversed many semicircular paths and are approaching the circumference of the cylinder, an auxiliary electric field is used to deflect them from the circular path and make them come out through a thin window. The target to be bombarded by the protons is placed near the window, and the investigation of the results of this bombardment can then be performed in a suitable manner.

Instead of ordinary hydrogen, heavy hydrogen or deuterium can be introduced into the chamber of the cyclotron so that deuterons become available as bombarding particles; or, if helium is used in this chamber, we have an artificial source of alpha particles.

The voltage between the sections D_1 and D_2 may have any value from about 10,000 to 200,000 volts. The particle, as it emerges from the cyclotron, may have an energy of several million electron volts due to the successive accelerations it experiences in going from one section to the other. The cyclotron is thus a comparatively low-voltage source of high-energy particles.

The upper limit to the energy that a particle can obtain from a cyclotron comes from the fact that Equation (12-2) is nonrelativistic. As a particle gains more and more energy, its speed does not increase at the proper rate to stay in phase with the oscillations. For protons the highest practicable energy is about 22 MeV.

12-4   The Betatron

Electrons may be accelerated to high energies by having them move in a circular path of constant radius and at the same time increasing the magnetic flux through the circular orbit in such a way that the electrons acquire addi-

Figure 12-5 Photograph showing the original 2.3-MeV betatron in front of a 20-MeV betatron. Each machine has its doughnut-shaped vacuum tube in place between the poles of the magnet. (From a photograph supplied by Professor D. W. Kerst.)

tional energy during each revolution. Such an accelerator is known as a *betatron*. The first successful betatron was designed and built by D. W. Kerst (1940) from the theory worked out by Kerst and R. Serber. The original betatron, as shown in Figure 12-5, which accelerated electrons up to 2.3 MeV, was operated as an x-ray tube; the x-rays were produced in the conventional manner by allowing the high-energy electrons to strike a target. Most betatrons built since then have also been used as x-ray sources, but some have been used as sources of high-energy electrons for nuclear experiments. Betatrons that accelerate electrons up to about 300 MeV have been built.

In the operation of the betatron electrons from the heated filament F are injected into the circular or doughnut-shaped tube by applying a difference of potential between the filament and the plate P, as shown in Figure 12-6. The electrons are focused with the aid of the grid G. When an alternating magnetic field is applied parallel to the axis of the tube, two effects are produced: an electromotive force is produced in the electron orbit by the changing magnetic flux that gives the electrons additional energy; a radial force is produced by the action of the magnetic field whose direction is perpendicular to the electron velocity which keeps the electron moving in a circular path. The magnetic flux through the orbit has to be chosen in such a way that the electrons will move in a stable orbit of fixed radius R. The electrons make several hundred thousand revolutions in this circular path while the alternating magnetic field is increasing in intensity from zero to a maximum—that is, during a quarter of a cycle. With each revolution they gain additional energy. When the electrons have acquired the desired amount of energy, a capacitor is discharged through two coils of wire, one directly above and the other directly below the stable orbit, thus producing a sudden addition to the magnetic flux. This destroys the condition for the stability of this orbit and the electron beam moves out to larger radii until it strikes the back of the injector P which acts as the x-ray target.

We can think of the circular electron path of fixed radius R as a circuit; the emf V induced in this circuit by the changing magnetic flux is, according to Faraday's law,

$$V = \frac{d\Phi}{dt}$$

where Φ is the instantaneous value of the magnetic flux which is perpendicular to the plane of the circuit. The work done on an electron of charge e in one revolution is therefore

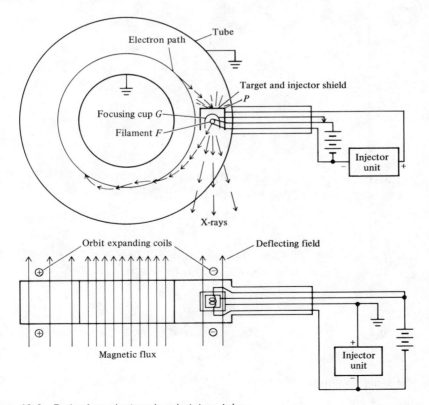

Figure 12-6   Path of an electron in a betatron tube.

$$Ve = e\frac{d\Phi}{dt}$$

This work can also be expressed in terms of the tangential force F which, acting on the electron over a distance ds, does an amount of work dW given by

$$dW = \mathbf{F} \cdot \mathbf{ds}$$

from which

$$F = \frac{dW}{ds}$$

Thus the tangential force acting on the electron is equal to the work done per unit length of path. Evaluating this force for one revolution for which the path length is $2\pi R$, we get

$$F = \frac{Ve}{2\pi R} = \frac{e}{2\pi R}\frac{d\Phi}{dt}$$

Now, from Newton's second law,

$$F = \frac{dp}{dt}$$

hence

$$\frac{dp}{dt} = \frac{e}{2\pi R}\frac{d\Phi}{dt}$$

or

$$dp = \frac{e}{2\pi R}\, d\Phi \qquad (12\text{-}4)$$

Because of the presence of the magnetic flux perpendicular to the plane of the electron orbit, the electron will experience a radial force inward given by

$$Bev = \frac{m\gamma v^2}{R}$$

where $\gamma = (1 - v^2/c^2)^{-1/2}$ and B is the value of the magnetic induction at the electron orbit of constant radius R. From the above equation

$$m\gamma v = BeR$$
$$p = BeR$$

If R is kept constant, then differentiation of this equation yields

$$dp = eR\, dB \qquad (12\text{-}5)$$

Comparing Equations (12-4) and (12-5), we see that

$$\frac{e}{2\pi R}\, d\Phi = eR\, dB$$

from which

$$d\Phi = 2\pi R^2\, dB$$

Integrating this equation between the limits of zero and B, respectively, we get

$$\Phi = 2\pi R^2 B \tag{12-6}$$

for the instantaneous relationship between the total magnetic flux Φ and the magnetic induction B at a distance R from the center. This equation shows that the magnetic flux within the orbit of radius R is always proportional to the intensity of the magnetic field at the orbit and, furthermore, that the magnetic flux through the orbit is twice what it would have been if the magnetic induction were uniform throughout the orbit at the value B. This distribution of magnetic flux is obtained in an air gap between specially shaped pole faces of an electromagnet.

Most modern betatrons are operated from a 60-Hz ac source. Since the magnet and its coils constitute a large inductance, a very large capacitance is introduced into the circuit to bring the power factor closer to unity for efficient operation. In a 100-MeV betatron the electrons are accelerated during a quarter of a cycle—that is, during 1/240 sec. The energy acquired by an electron per revolution is 400 eV; hence it has to make 250,000 revolutions to acquire the maximum energy. In practice the energy of the electrons, hence that of the x-ray photons, can be varied from about 10 to 100 MeV by applying the orbit-shifting magnetic field at different times during the quarter-cycle that the field is increasing.

12-5 Linear Accelerators

The development of the linear accelerator for charged particles was started at about the same time as that of the cyclotron. The success of the cyclotron and the unavailability of high-powered high-frequency sources shifted the interest from linear accelerators. Interest in linear accelerators was renewed about 1945 as the result of the development of magnetrons, klystrons, and other tubes capable of delivering power of several megawatts for periods of several microseconds with repetition rates up to a few hundred pulses per second. Among the advantages of a linear accelerator are that fewer magnets are needed to guide the particles and that the particles readily emerge from the apparatus, since they travel in straight lines down the length of the vacuum tube. Linear accelerators have been developed for both electrons and protons, though the actual number in operation at present is not large.

One of the earliest of the modern linear accelerators is the proton accelerator designed by L. Alvarez (1947) at Berkeley. It consists essentially of a long steel vacuum chamber containing a 12-sided copper tube which is about 40 ft long and 1 meter in diameter and which acts as a resonant cavity for a radiofrequency wave of 202.5×10^6 Hz. Protons with an energy of 4 MeV

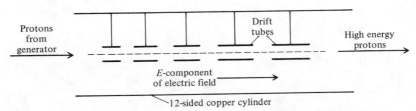

Figure 12-7 Simplified diagram showing arrangement of drift tubes within copper cavity resonator of the linear proton accelerator.

obtained from a Van de Graaff generator are injected along the axis of the copper cylinder. The protons travel through a set of 46 drift tubes aligned with their axes coincident with the axis of the copper tube, as shown in Figure 12-7. The drift tube shields the protons from the electric field of the wave traveling down the tube during the time that the phase of the wave is such that it produces a negative acceleration of the protons. The protons are accelerated in the forward direction only while they pass from one drift tube to the other. As the velocity of the proton increases, the length of the drift tube must also increase so that the proton will remain in the same phase with respect to the electric field. The drift tubes vary in length from 4.4 to 10.9 in. and the distance between their centers along the axis also increases. Power is supplied to the resonator from 26 triode oscillators delivering a peak power of 2.15×10^6 watts. The power is delivered in pulses of 3×10^{-4} sec duration with a repetition rate of 15/sec. This linear accelerator delivers protons with an energy of 32 MeV.

Linear accelerators for electrons are, in a sense, simpler than those for protons because the electrons acquire a speed almost equal to the speed of light at a comparatively low energy, in the neighborhood of a few MeV. Hence, if electrons are supplied to a linear accelerator from an outside source with an energy of 3 or 4 MeV, and in the correct phase with the electric field traveling down the tube, they will remain in phase and acquire energy from it. The problem in the design of a linear accelerator for electrons is to produce an electromagnetic wave that progresses along the tube with a controlled phase or wave velocity slightly less than or equal to the velocity of light. In order to supply energy to the electrons, the electric field of the wave must have a component in the direction of propagation of the wave. Methods of accomplishing this are known from microwave techniques, and several linear electron accelerators are now in operation.

Electrons with energies up to 1.0 GeV have been obtained with the 220-ft Stanford linear accelerator. This accelerator is built of sections 3 meters long, each supplied with power by klystrons operated as amplifiers and producing a traveling wave in the tube having a frequency of 3×10^9 Hz. The accelerator is operated at the rate of 60 pulses/sec, each pulse lasting 1 μsec. The maximum number of electrons per pulse is about 5×10^{10}; these electrons have a narrow energy spread of 0.5 percent about the mean energy.

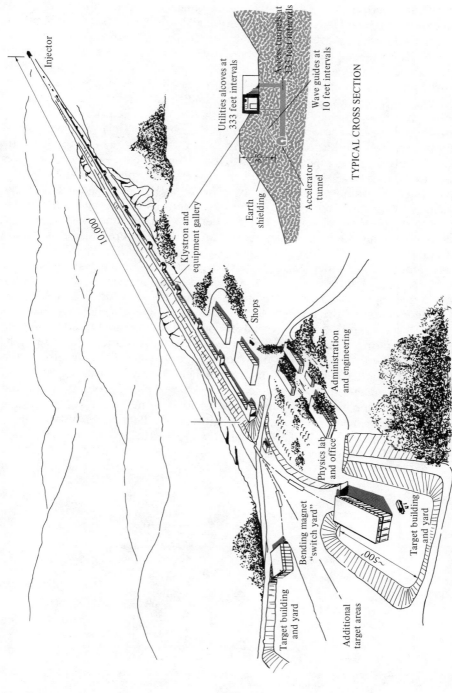

Injector

10,000'

Klystron and
equipment gallery

Shops

Utilities alcoves at
333 feet intervals

Access tunnels at
333 feet intervals

Wave guides at
10 feet intervals

Earth
shielding

Accelerator
tunnel

35'

TYPICAL CROSS SECTION

Administration
and engineering

Physics lab
and office

Bending magnet
"switch yard"

Target building
and yard

Additional
target areas

~500'

Target building
and yard

Figure 12-8 The Stanford two-mile linear accelerator. (Courtesy Project M, Stanford University, Stanford, California.)

A linear electron accelerator, two miles long, has been built at Stanford University (see Fig. 12-8). The laboratory which includes this device is called the Stanford Linear Accelerator Center, or SLAC. The machine itself consists of 3-meter long sections powered by klystrons. There are 960 such sections, and in Stage I of the operation 240 klystrons are being used, one klystron supplying power to four sections, as shown in Figure 12-9. The klystrons are operated at an 18-MW level, and the accelerator can deliver an electron beam of 20 GeV. Power is supplied to the accelerator by the klystrons at 360 pulses/sec with the duration of each pulse 1.6 μsec. The traveling wave in the tube has a frequency of 2.856×10^9 Hz. The average beam current is about 12 μamps at about 20 GeV. Higher beam energy is possible if klystrons that can operate at higher power should become available. By increasing the number of klystrons by a factor n the energy is increased by a factor of $\sqrt{n}$; therefore the energy of the electrons can be increased in the future to 40 GeV by increasing the number of klystron tubes to 960 so that each one supplies power to a 3-meter section of the accelerator tube as shown in Figure 12-9.

The Stanford two-mile linear accelerator is built inside two parallel tunnels; the lower one, 3 meters wide and 4 meters high, contains the accelerator proper; the upper one, 7.2×6.6 meters, contains the klystrons and associated equipment. Parts of the high-energy electron beam are deflected by magnets at the "switchyard" and delivered to several experimental stations. An additional feature of the accelerator is its ability to produce a beam of high-energy positrons. At a position one-third of the way from the injector to the switchyard a target can be stationed on which the electron beam impinges, producing positrons that are accelerated by the remaining two-thirds of the machine. By using either a rotating target or a vibrating "wand," it is possible to provide experiments with beams of positrons and electrons simultaneously. Typical energies for simultaneous operation are 4–10 GeV for positrons and 10–16 GeV for electrons.

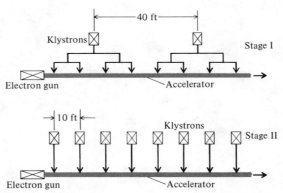

Figure 12-9   Klystron connections to the Stanford linear accelerator in Stage I and Stage II. (Courtesy of Project M, Stanford University, Stanford, California.)

Linear accelerators (often referred to as "linacs") have an advantage over other types in that they produce a high-intensity beam. This feature makes them very useful as preaccelerators or injectors for the large cyclic accelerators to be described in the following sections. For accelerating electrons the linac has an additional advantage over circular machines in that the energy lost in the form of electromagnetic radiation from the accelerated charge is less for straightline motion than for circular motion. For protons the radiation from centripetal acceleration is negligible at all energies obtained so far in the laboratory, and the cost per MeV is lower for cyclic accelerators than for linacs, with the result that proton linear accelerators are not used at very high energies. For acceleration of heavy ions intensity is often of more importance than energy, and linacs are widely used for work of this kind.

12-6 Frequency-Modulated Cyclotron

The energy limitation on the conventional cyclotron, imposed by the onset of relativistic effects with increasing energy, can be circumvented by means of a variation of the frequency applied to the dees. It can readily be shown that the total energy $\mathcal{E}$ of a charged particle moving in a circular orbit with constant angular velocity ω in a magnetic induction B is given by

$$\mathcal{E} = \frac{Bec^2}{\omega} \qquad \text{(mks units)} \tag{12-7}$$

(see Problem 12-3). At nonrelativistic speeds, the rest energy dominates the kinetic energy, making energy and frequency independent. As the speed increases to a value large enough to cause a significant change in $\mathcal{E}$, it becomes necessary to *decrease* ω to maintain orbital stability as expressed by Equation (12-7). A device that uses this principle is called a *frequency-modulated cyclotron*, an *FM cyclotron*, or a *synchrocyclotron*.

The procedure, then, for using a cyclotron to achieve relativistic speeds is to inject a group of particles at the center and accelerate them, but to decrease the frequency as the radius and energy increase. It is clear that once the frequency has been lowered it is no longer fruitful to inject low-energy particles, since they will not satisfy the stability condition. When the energy has reached its final value, the particles are ejected from the machine into the experiment, the frequency is raised back to its starting value, and a new group of particles can be injected. Since many revolutions occur between injection and ejection, the FM cyclotron produces particles at time intervals that are widely spaced compared with the period of oscillation of the dee voltage; by contrast the unmodulated cyclotron can produce particles during each cycle of the dee voltage.

The FM cyclotron is the first type of accelerator considered here to use the principle of *phase stability* possessed by charged particles moving in unidirectional magnetic fields and alternating electric fields of varying fre-

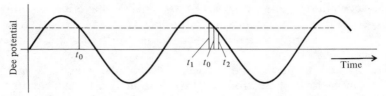

Figure 12-10

quency. This property was discovered independently in 1945 by V. Veksler in the Soviet Union and E. M. McMillan at Berkeley. Suppose, for example, that a positive ion is moving within the dees of a cyclotron at the proper frequency and that during each cycle of the alternating potential it arrives at the edge of the dee at a time t_0, as shown in Figure 12-10. This time is shortly after the time that the dee potential has reached its maximum value. In the ordinary cyclotron the period of motion of an ion is independent of its velocity and is the same as that of the alternating voltage applied to the dees, so that the ion arrives in the same phase during each cycle—that is, at the same time t_0. Suppose, however, that the increment in velocity is smaller during the transit time across the gap between the dees because of relativity (the velocity is approaching c); then the ion will move in a circle of smaller radius and arrive at the edge of the dee in a shorter time, represented by t_1, on the diagram; it will thus receive a larger acceleration from the ac field and may get into phase again during the next cycle or cycles. If, on the other hand, an ion receives more than the normal increment of velocity, it will move in a circle of larger radius and arrive at the edge of the dee at a later time t_2, at which time it will receive less energy from the ac field and fall into phase during the next cycle or cycles. Once a particle gets into phase it can continue in the same orbit as long as the magnetic field and the frequency of the voltage applied to the dees remain constant. In order to increase the energy of the ion it must be made to move in an orbit of larger radius. Both Veksler and McMillan showed that this could be accomplished either by increasing the magnetic field, by

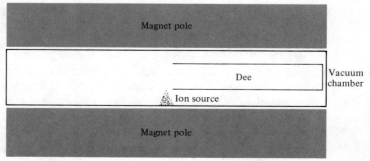

Figure 12-11 Schematic diagram showing a single dee inside the vacuum chamber between the poles of a synchrocyclotron.

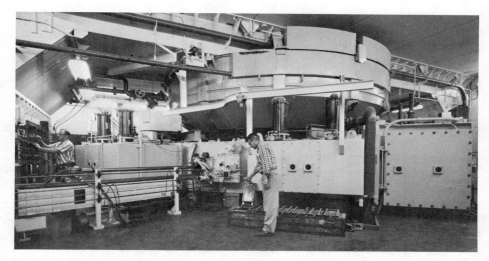

Figure 12-12 A photograph of the 184-in. LRL synchrocyclotron with the shielding removed. (Courtesy of University of California Lawrence Radiation Laboratory, Berkeley.)

decreasing the frequency of the applied voltage, or both. In the FM cyclotron the magnetic field is kept constant in time while the frequency of the applied voltage is decreased slowly.

An FM cyclotron consists of a single dee placed inside a vacuum chamber between the poles of a very large electromagnet, as shown in Figure 12-11. The pole piece of the magnet of the Berkeley synchrocyclotron (shown in Fig. 12-12) is 184 in. in diameter, that at the University of Chicago is 170 in., and the one at Columbia University (Nevis) is 164 in. The pole faces are specially shaped or contoured to provide a field that decreases almost linearly from the center out to the position of maximum orbital radius; for example, the magnetic field of the Nevis synchrocyclotron decreases from a maximum of about 17,400 gauss at the center to about 16,600 gauss at a radius of 74 in., whereas the magnetic field at the Chicago synchrocyclotron decreases from a maximum of about 18,600 gauss at the center to 17,600 gauss at a radius of 76 in. The reason for contouring the magnet is to provide a magnetic field that will focus the ions in the median plane. The contoured pole faces of the magnet go through circular holes in the top and bottom of the vacuum chamber.

The potential difference between the dee and the grounded plane is comparatively low, about 10 to 30 kV. The radio frequency is supplied by a tube oscillator circuit that is modulated by being coupled to a variable capacitor. The capacitance is varied by rotating one set of plates by means of a motor. This variable capacitor is sometimes placed inside the vacuum system between the pole faces, as in the Nevis synchrocyclotron, or it may be outside the magnet to avoid effects due to eddy currents. In the Chicago synchrocyclotron the frequency of the voltage applied to the dee varies from 28.6 to 18.0 MHz to accelerate protons to 450 MeV with a repetition rate of 60 pulses/sec; the oscillator of the Nevis synchrocyclotron goes through a frequency range of 28 to 16.9 MHz with a repetition rate that can be varied from 60 to 120 pulses/sec when accelerating protons to about 435 MeV.

Targets for nuclear experiments can be introduced through a vacuum lock in the vacuum chamber into the region in which the ion beam is circulating. A variety of particle energies is thus available by simply placing the target at appropriate positions in the chamber. Furthermore, the magnetic field of the synchrocyclotron will also act on any charged particles produced by the bombardment of the target. There is also room enough for suitable detectors to be placed inside the vacuum chamber. This method has been widely used for the production and detection of charged π mesons (see Chapter 18). Neutrons produced by bombarding a target with protons will come out of the walls of the chamber; channels are usually provided in the shielding around the synchrocyclotron so that a collimated beam of neutrons is available at a large distance from the magnet for experimental purposes.

Frequency-modulated cyclotrons are commonly operated with protons that are obtained from a suitable ion source placed at the center of the vacuum chamber. These cyclotrons can be modified to operate with more massive ions, such as deuterons, helium ions, and beryllium ions. Chou, Fry, and Lord (1952) introduced CO_2 in the ion source of the Chicago synchrocyclotron and observed carbon nuclei, C^{6+}—that is, carbon atoms stripped of their electrons. They also observed stripped beryllium atoms—that is, beryllium nuclei, Be^{4+} —probably produced by the evaporation of an internal beryllium target bombarded by protons circulating in the vacuum chamber. The carbon nuclei were observed in orbits up to 65 in. in radius and the beryllium nuclei were observed in orbits of radii up to 50 in. These authors estimated that the carbon ions had energies up to 1.1 GeV. Nuclear emulsion photographic plates were placed in the cyclotron and nuclear disintegrations produced in the emulsions by these heavy ions were studied.

The upper limit to the energy obtainable from FM cyclotrons is set by the fact that the magnetic field must extend throughout the accelerator, with the result that the weight of steel involved can be formidable; for example, an FM cyclotron at Dubna in the Soviet Union has a magnet with a pole diameter of 236 in. By using a magnetic induction of 17 kG, a proton energy of 680 MeV can be obtained. The weight of the magnet is 7200 tons. To achieve higher proton energies it is necessary to employ principles that do not require such a large magnet. The last two sections of this chapter describe synchrotrons —accelerators that eliminate the need for massive magnets.

12-7 Electron Synchrotron

The total energy $\mathcal{E}$ of a particle in a phase-stable orbit moving with angular velocity ω at a place where the magnetic induction is B is given by

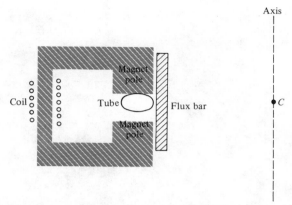

Figure 12-13 Cross section of one side of a betatron-synchrotron showing the position of the vacuum tube between the poles of a C-shaped magnet with a flux bar. This synchrotron may be visualized by rotating this section through 360 degrees about the vertical axis through the center C.

$$\mathscr{E} = \frac{Bec^2}{\omega} \tag{12-7}$$

An electron with an energy of a few million electron volts has a linear velocity almost equal to c, so that an increase in its energy produces no significant change in ω. Thus to increase the total energy of the electron it is necessary to increase the magnetic induction proportionally.

The condition for a stable electron orbit, given by Equation (12-6) for a betatron, must still be fulfilled, but it is no longer necessary to have a heavy solid magnet over the entire area of the tube. The electromagnet need be built with its poles above and below the tube only and with some steel bars, called *flux bars,* inside the hollow center near the tube. This device is illustrated schematically in Figure 12-13, in which a section of one side of the magnet, the tube, and a flux bar are shown. The steel flux bars become saturated magnetically early in the magnetic cycle, and further time variation in the magnetic induction B then takes place in the region of the electron orbit. Figure 12-14 is a photograph of a 70-MeV betatron-synchrotron in operation.

The doughnut-shaped vacuum tube shown in Figure 12-15 differs from the regular betatron tube in that a coating of silver is placed on a small section of the inside of the tube to form a metallic cavity, but a small break in the coating separates it into two parts. A high-frequency electric field from a radio-frequency oscillator is applied across this gap at the proper time in the magnetic cycle. This frequency is in synchronism with the angular velocity of the electrons so that the electrons will be accelerated when they go through the cavity and cross the gap. The property of phase stability is also utilized in its operation so that electrons which have deviated from the normal orbit get the necessary decrease or increase in energy to bring them back to the normal orbit.

When the electrons are injected at the beginning of the magnetic cycle, the

Figure 12-14   Photograph of a 70-MeV synchrotron. The brilliant spot of light just left of the center of the picture is caused by visible radiation emitted by the accelerated electrons. (Courtesy of the General Electric Company.)

device operates as a betatron for a short time, of the order of microseconds, until the electrons have acquired an energy of about 2 MeV. The high-frequency field is then automatically applied across the gap. It remains on while the magnetic field is increasing and is automatically cut off when the electrons have acquired their maximum energy or, if desired, at an earlier time in the cycle. Once the electrons have acquired the desired energy, the magnetic flux condition for a stable orbit is destroyed by sending a large current through additional coils to make the electrons spiral toward the target. In the case of the 70-MeV synchrotron the electrons spiral inward to the target, which consists of a tungsten wire. X-rays come from the target in a narrow cone in the forward direction.

Electron synchrotrons (sometimes called betatron synchrotrons) that can accelerate electrons up to 6 GeV and produce photons of this energy are now in operation. However, there is a practical limit to the size of a synchrotron because of the radiation of electromagnetic waves, a large amount of it in the visible region, coming from the accelerated electrons. The light that is observed coming from the synchrotron in Figure 12-14 is evidence of this radiation. Electrons can be accelerated to higher energies without appreciable radiation losses by means of a *linear accelerator* (see Section 12-5).

The power radiated from accelerated charges moving with speeds v comparable to the speed of light c was calculated by Schwinger (1946), who used the theory of relativity. For the case of a circular trajectory, such as exists in a betatron, cyclotron, and similar devices, the power S radiated was shown to be given by

$$S = k\tfrac{2}{3}\omega \frac{e^2}{R}\left(\frac{v}{c}\right)^3\left(\frac{\mathcal{E}}{mc^2}\right)^4 \tag{12-8}$$

in which R is the radius of the trajectory, ω is the angular velocity of the particle, m is its rest mass, $\mathcal{E}$ is its total energy, and k is $1/(4\pi\epsilon_0)$ in mks units or 1 in electrostatic units. For the case of particles in a betatron or synchrotron whose speed $v = c$ this equation becomes

$$S = k\tfrac{2}{3}\omega \frac{e^2}{R}\left(\frac{\mathcal{E}}{mc^2}\right)^4 \tag{12-9}$$

The energy $\mathcal{E}_1$ radiated during one revolution is

$$\mathcal{E}_1 = S\frac{2\pi}{\omega} \tag{12-10}$$

or

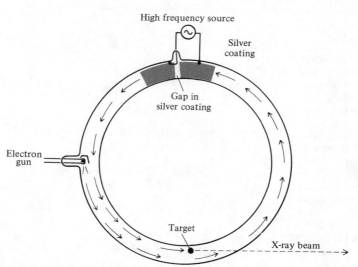

Figure 12-15 Schematic diagram of the vacuum tube of a betatron-synchrotron showing the method of applying a high-frequency potential difference across a gap in the silver-coated section of the tube.

$$\mathcal{E}_1 = k\,\frac{4\pi}{3}\,\frac{e^2}{R}\left(\frac{\mathcal{E}}{mc^2}\right)^4 \tag{12-11}$$

Hence, for relativistic charged particles the energy radiated per revolution varies directly with the fourth power of the total energy and inversely as the fourth power of the rest mass.

12-8 Proton Synchrotron

At present the device which accelerates particles to highest energies is the *proton synchrotron*. Its principle of operation is the same as that of the electron synchrotron, although the details are different. The protons are held by a magnetic field in an orbit of fixed radius and accelerated by means of a radio-frequency electric field, just as in the case of the electron synchrotron. Because the proton is quite massive, Equation (12-11) indicates that energy loss due to synchrotron radiation will be much less for protons than for electrons. For this reason it is possible to obtain much higher energies with proton synchrotrons. Table 12-1 contains a list of some of the proton synchrotrons which have been built, along with their energies and approximate sizes. The cosmotron (see Figs. 12-16 and 12-17) was the first accelerator of its kind, completed in 1952. In 1967 it was retired from service in order to concentrate the available economic resources on more modern machines. In 1968 the

TABLE 12-1 PROTON SYNCHROTRONS

Location	Name	Energy (GeV)	Radius (meters)
Brookhaven National Laboratory, Long Island	Cosmotron	3	9
Lawrence Radiation Laboratory, Berkeley, California	Bevatron	6.2	15
Harwell, U. K.	Nimrod	7	24
Dubna, U.S.S.R.	Synchrophasatron	10	28
Argonne National Laboratory, Illinois	ZGS (zero gradient synchrotron)	12.5	38.5
CERN (Centre Européenne pour la Recherche Nucléaire), Geneva, Switzerland	PS (proton synchrotron)	28	70
Brookhaven	AGS (alternating gradient synchrotron)	33	85
Serpukhov, U.S.S.R.		76	235

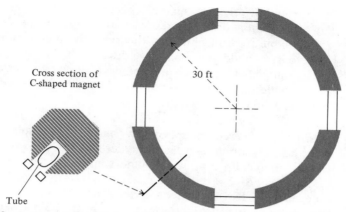

Cross section of
C-shaped magnet

30 ft

Tube

Figure 12-16 Arrangement of C-shaped magnets and straight sections of the vacuum chamber of the cosmotron shown schematically.

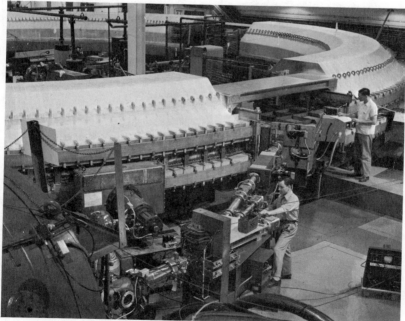

Figure 12-17 View of the cosmotron at Brookhaven National Laboratory, Upton, New York, from the angle at which protons enter the machine from the Van de Graaff generator. The proton beam coming from the generator tank passes through the magnetic field of a small magnet which analyzes the beam. Protons of the proper energy then enter the vacuum chamber through the straight section. Note the C magnets in which the vacuum chamber rests. Twelve large cylindrical vacuum pumps, one of which can be seen to the left of the small analyzing magnet, are attached to the vacuum tube. (Courtesy of Brookhaven National Laboratory.)

synchrotron at Serpukhov, then completed, became the most energetic accelerator of any kind in the world. At present no theoretical upper limit exists to the energy obtainable from machines of this type, the practical limit being determined by the cost of building such a large device. The National Accelerator Laboratory in Batavia, Illinois, is proceeding with the construction of a 500-GeV proton accelerator, and synchrotrons in the 1000 GeV (or TeV) range have been suggested.

All proton synchrotrons with energies above 20 GeV have used the principle of *strong focusing,* discovered independently by N. Christophilos and by M. S. Livingston, E. D. Courant, and H. S. Snyder (1952). As a proton travels around the ring of the synchrotron, it encounters magnetic fields which have radial gradients, that is, $\partial B/\partial r \neq 0$. In one magnet $\partial B/\partial r$ will be positive, in the next it will be negative, and so on around the ring. The combined effect of these alternating gradients is to focus the protons into a much thinner beam than is possible for the "weak-focusing" accelerators in which the gradients are small. Confinement of the beam to smaller volume permits the use of smaller and less expensive magnets, leading to lower cost per GeV for the alternating-gradient machines. The principle of strong focusing has also been applied to electron synchrotrons and linear accelerators.

By way of example let us consider the operation of the AGS at Brookhaven. Protons are produced by an ion source, accelerated to 750 keV by a Cockcroft-Walton generator (see Fig. 12-18) and injected into a linac (Fig. 12-19) in which they are accelerated to 50 MeV. Then they are deflected to move into

Figure 12-18 The Cockcroft-Walton generator which serves as the first stage in the acceleration of protons at the alternating gradient synchrotron. The protons receive an energy of 0.75 MeV from this device. (Courtesy of Brookhaven National Laboratory.)

Figure 12-19   The 50-MeV linac at the AGS. Protons from the Cockcroft-Walton generator are accelerated in this linac and are then deflected into the ring of the synchrotron. (Courtesy of Brookhaven National Laboratory.)

an orbit in the synchrotron itself (Fig. 12-20). There are 12 rf stations, each of which causes a proton to gain 8 keV as it goes past, so that about 0.1 MeV per turn is gained. As the protons gain energy, the magnetic field must be increased at the correct rate to keep the radius constant. Since protons at 50 MeV are not relativistic, their velocity also increases as they gain energy, thus requiring an increase in the frequency of the accelerating field. It is to alleviate this problem somewhat that the preaccelerators are used. When the protons reach their peak energy, they are ejected from the ring as a result of being deflected by a small pulsed magnet. An alternative procedure is to steer the protons onto a target inside the ring where they can interact with nuclei to form a variety of particles for use in experiments. After the protons have been removed from the ring, the magnetic field is reduced to its original strength (120 gauss) for the injection of the next pulse of protons. The entire cycle requires 1.5 sec.

It is interesting and instructive to consider the question of energy flow to and from the magnets of a synchrotron. At the instant that the protons possess

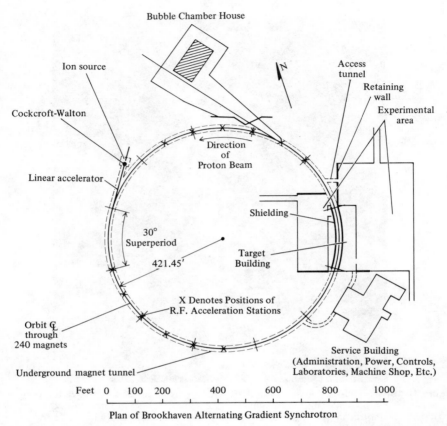

Bubble Chamber House

Ion source

Cockcroft-Walton

Linear accelerator

Direction
of
Proton Beam

N

Access
tunnel

Retaining
wall

Experimental
area

Shielding

30°
Superperiod

421.45'

Target
Building

X Denotes Positions of
R.F. Acceleration Stations

Orbit ℄
through
240 magnets

Service Building
(Administration, Power, Controls,
Laboratories, Machine Shop, Etc.)

Underground magnet tunnel

Feet 0   100 200   400   600   800   1000

Plan of Brookhaven Alternating Gradient Synchrotron

Figure 12-20   Plan of Brookhaven AGS. (Courtesy of Brookhaven National Laboratory.)

their maximum energy, the magnetic field is at its maximum value, as is the current in the windings of the magnets. At the AGS this maximum field is about 11 kG, corresponding to a current of several thousand amperes. The amount of energy stored in the magnetic field is too large to be wasted when the field is lowered to the injection value. As the field decreases the magnetic flux lines cross the windings and cause a back emf which is used to drive a motor attached to a large flywheel. To increase the magnetic field during acceleration some of the rotational energy of the flywheel is used to drive a generator of current to energize the magnets, thus completing the cycle.

The protons receive their energy from the 12 rf stations, each of which is made of a nonmetallic ferromagnetic material, called ferrite, which has poor electrical conductivity so that eddy-current losses are small, but which has a relative permeability of about 500 at high frequencies. In essence this ferrite tube forms the core of a high-frequency transformer. The primary consists of one turn of wire supplied with rf current from an oscillator; the secondary is the proton beam itself. At the AGS the frequency of revolution of the protons is 125,000 rps at injection and 375,000 rps at peak energy. Earlier accelerators used radiofrequencies equal to the revolution frequency, but the AGS uses the twelfth harmonic of this fundamental frequency, operating between 1.5 MHz at injection and 4.5 MHz at peak energy. Approximately 10^{12} protons are accelerated at each pulse. They travel in a vacuum chamber which is elliptical in cross section with major and minor axes of 18 and 7.5 cm, respectively.

A development which holds promise for the future of existing proton synchrotrons is the construction of storage rings to contain protons that have been accelerated to their peak energy. Subsequent pulses of protons can be made to collide head-on with those that are circulating in storage, enabling experimenters to study processes at much higher energies than those obtainable from the accelerator alone.

Another technical advance which holds much promise for accelerators is

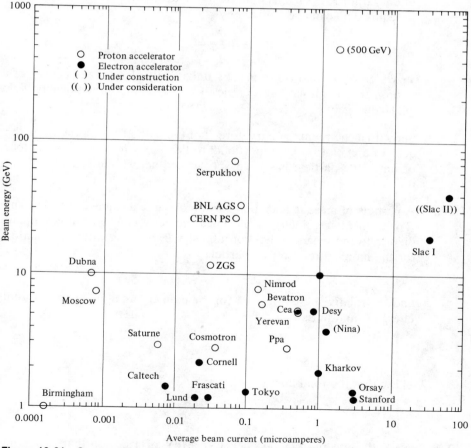

Figure 12-21 Comparative graph of various accelerators. (Courtesy Stanford Linear Accelerator Center.)

the development of superconducting magnets. By cooling electrical conductors to temperatures of a few kelvins it becomes possible to lower the resistance to zero, permitting very large currents to flow. Much higher currents give rise to much higher magnetic fields, making it possible to construct an accelerator in much less space than would be required with conventional magnets because the charged particles can be bent into circular arcs with smaller radii.

Figure 12-21 presents a summary of the energies and intensities of high-energy accelerators throughout the world.

Problems

12-1. (a) Show that the energy of an ion of mass M and charge q which is circulating within the dees of a cyclotron along a path with a radius of curvature R is such as to correspond to a total acceleration through an equivalent voltage V given by

$$V = \tfrac{1}{2}B^2R^2\frac{q}{M}$$

in which V, B, and q are in mks units. (b) Calculate the energy of a proton that is moving in a circle of 60-cm radius in a magnetic field of 10,000 gauss.

12-2. Show that the radius of curvature R of the path of a particle inside the dees of a cyclotron is proportional to the $\sqrt{n}$, where n is the number of times the particle has been accelerated across the space between the dees.

12-3. A particle of mass m and charge e is moving in a stable orbit of radius R with angular velocity ω and linear velocity v. The orbit is perpendicular to the magnetic field of intensity B. Starting with the equation for the momentum p of the particle

$$p = BeR$$

and the relativistic expression for the momentum of a particle of total energy $\mathcal{E}$ and mass m,

$$p^2 = \frac{\mathcal{E}^2}{c^2} - m^2c^2$$

show that the total energy is given by

$$\mathcal{E} = \frac{Bec^2}{\omega}$$

12-4. (a) Calculate the kinetic energy in MeV of a proton that is moving in a circular orbit in a magnetic field of 17,000 gauss in resonance with the applied dee frequency of 18 MHz. (b) Determine the momentum of this proton in units of MeV/c. (c) Determine the linear velocity of the proton. (d) Determine the radius of this orbit.

12-5.   (a) Calculate the kinetic energy, in MeV, of a deuteron that is moving in a circular orbit in a magnetic field of 17,000 gauss in resonance with the applied dee frequency of 12 MHz. (b) Calculate the momentum of the deuteron in units of MeV/c. (c) Determine the linear velocity of the deuteron. (d) Determine the radius of this orbit.

12-6.   Assume that in the 70 MeV betatron synchrotron the radius of the stable electron orbit is 28 cm. Calculate (a) the angular velocity of the electrons, (b) the frequency of the applied electric field, and (c) the value of the magnetic field intensity at the orbit for this energy.

12-7.   The radius of the stable electron orbit of a small betatron-synchrotron is 12.0 cm and the maximum magnetic field available at this position is 9000 gauss. Calculate (a) the angular velocity of the electron in the stable orbit, (b) the frequency of the applied electric field in synchronism with these electrons, and (c) the maximum electron energy produced.

12-8.   In Problem 12-7, the energy of the electron is increased from 2 to 32 MeV in a quarter of the magnetic cycle which is operated from a 50-Hz ac line. Determine the average energy given to the electron by the high-frequency electric field during each revolution.

12-9.   Show that the radius of a positive ion of mass m moving in a circular orbit within a synchrocyclotron is given by

$$R = \left[\frac{c^2}{\omega^2} - \left(\frac{mc}{Be} \right)^2 \right]^{1/2}$$

where ω is its angular velocity, B is the magnetic induction in mks units, and e is the charge of the ion in coulombs.

12-10. (a) Calculate the ratio of ω/B for a stripped carbon atom, $A = 12$, to move in a stable orbit in a synchrocyclotron with a kinetic energy of 1 GeV. (b) If the value of B at the orbit is 10,000 gauss, calculate the frequency of the electric field applied to the dee. (c) Determine the radius of this orbit.

12-11. (a) Calculate the power radiated from an electron synchrotron operated at 70 MeV when the electron orbit radius is 30 cm. (b) Calculate the energy lost by radiation during a single revolution of an electron. (c) Determine the ratio of the radiated energy to the total energy of the electron.

12-12. (a) Calculate the power radiated from an electron synchrotron operated at 300 MeV when the electron orbit radius is 100 cm. (b) Calculate the energy lost by radiation during a single revolution of an electron. (c) Determine the ratio of the radiated energy to the total energy of the electron.

12-13. (a) Calculate the power radiated from a proton synchrotron operated at 350 MeV when the orbit radius is 200 cm. (b) Calculate the energy lost by radiation during a single revolution of the proton. (c) Determine the linear and angular velocities of the protons in this orbit.

12-14. Using the data in Section 12-5 on the Stanford 220-ft linear accelerator, determine (a) the maximum electron-beam current, (b) the average beam current, and (c) the maximum beam power.

12-15. A proton linear accelerator with an energy of 800 MeV and a current of 1 mA, is being built at Los Alamos. How many protons per second are produced?

12-16. Pulsed accelerators often use a flywheel to store energy for magnet excitation. By what fraction should the angular velocity be decreased in order to decrease its energy by 10 percent?

12-17. What is the length of the Stanford two-mile accelerator as viewed by a 20-GeV electron? The answer to this question has a bearing on the problem of alignment of the structure.

12-18. An electrostatic accelerator contains two concentric metal cylinders. The outer cylinder has a fixed radius R_2. For a given voltage V between the two conductors what value of R_1 (the radius of the inner cylinder) will minimize the electric field?

12-19. Before World War II a cyclotron was built at Berkeley for which the magnetic field was 15 kG. (a) Calculate the frequency in megahertz needed to accelerate protons. (b) The radius of the cyclotron was 15 in. Calculate the energy of the protons produced. Give your answer in MeV.

12-20. The two-mile linac at Stanford is aligned with the aid of a laser. How many wavelengths of the red laser light (He–Ne) are there in two miles?

References

Blewett, J. P., "Resource Letter on Particle Accelerators," *Am. J. Phys.*, **34**, 742 (1966).

Chu, E. L., and L. I. Schiff, "Recent Progress in Accelerators," *Ann. Rev. Nucl. Sci.*, **2**, 79–92 (1953). Contains extensive references to earlier literature.

Encyclopedia of Physics (Handbuch der Physik), S. Fluegge, Ed. Berlin: Springer-

Verlag, 1959, Vol. XLIV. General reference for all types of accelerator; seven out of eight articles are in English.

Fremlin, J. H., and J. S. Gooden, "Cyclic Accelerators," *Repts. Progr. Phys.* London: The Physical Society of London, 1950, XIII, 295.

Judd, D. L., "Conceptual Advances in Accelerators." *Ann. Rev. Nucl. Sci.,* 9, 181–216 (1959).

Livingston, M. S., "High Energy Accelerators," *Ann. Rev. Nucl. Sci.,* 1, 157–174 (1952).

Livingston, M. S., and J. P. Blewett, *Particle Accelerators.* New York: McGraw-Hill Book Company, 1962.

Review of Scientific Instruments, September 1953. Entire issue devoted to articles on the cosmotron.

Ridenour, L. N., and W. A. Nierenberg, *Modern Physics for the Engineer,* second series. New York: McGraw-Hill Book Company, 1961, Chapter 14.

13 | Beam Transport and Detecting Devices

13-1 Introduction

In the preceding chapter we considered various types of devices for imparting the required energy to numbers of particles to be used in nuclear experiments. For various reasons (including safety) experiments are usually performed at some distance from the accelerator, and it is necessary to convey the beam of particles across the required distance without undue loss of either energy or intensity. The equipment used for this purpose of beam transport often must fulfill other functions at the same time; for instance, an experiment at a proton synchrotron may require a beam consisting of particles that are unstable (such as mesons) or that do not lend themselves to easy production in the ion source of the accelerator (e.g., positrons). For these beams the usual procedure is to direct the proton beam onto a metal target. The resulting nuclear reactions cause a variety of particles to be produced at different energies and moving in different directions. It is then necessary to use the beam transport equipment to separate particles with acceptable energies from those that have energies either too high or too low. It may also be necessary to separate particles of different masses.

As particles leave an accelerator they usually tend to spread out, making the beam more diffuse. The number of particles per unit area will decrease according to the inverse square of the distance from their source, if no particles are lost and if they move in straight lines. An improvement on the inverse-square law can be obtained by applying forces (often with magnets) to focus the particles in a manner analogous to the focusing of beams of light in optics. An additional effect to be overcome is the absorption of beam particles as a result of their interaction with matter, leading to an exponential decrease, as in the case of x-rays (see Chapter 5). The usual remedy for this problem is the construction of an evacuated pipe through which the beam can travel.

Once the particles have left the general region of the accelerator, experi-

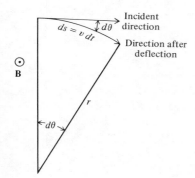

Figure 13-1   Deflection of a positively charged particle in a magnetic field; **B** is directed out of the paper.

mental equipment is needed to detect the presence of particles and to count how many of a given type are involved. Equipment is also needed to measure the momentum, velocity, and energy of the particles. A wide variety of types of equipment is available for these functions and will be described in subsequent sections of this chapter. It will be noticed that the various detectors are sensitive to electrically charged particles. Neutral particles can be detected, but not directly; rather, one observes the charged particles that result from the decay of the neutral or from its interaction with matter.

13-2   Bending Magnets

Bending magnets (often called deflecting magnets) are constructed to provide a region of nearly uniform magnetic induction B through which charged particles move. The particles experience a deflecting force,

$$\mathbf{F} = q\mathbf{v} \times \mathbf{B} \tag{13-1}$$

where q is the charge and $\mathbf{v}$ is the velocity. In the usual arrangement $\mathbf{v}$ and $\mathbf{B}$ are perpendicular so that the equation can be written in scalar form:

$$F = qvB \tag{13-2}$$

By looking at Figure 13-1 we see that the small angle of deflection $d\theta$ is given by the expression

$$d\theta = \frac{ds}{r} \tag{13-3}$$

where r is the instantaneous radius of curvature, and ds is the distance traveled. Since the momentum is given by

$$p = Bqr \tag{13-4}$$

we may write Equation (13-3) in the form

$$d\theta = \frac{qB\ ds}{p}$$

or

$$d\theta = \frac{B\ ds}{p/q} \tag{13-5}$$

Since the deflecting force is always perpendicular to the motion, the energy of the particle is not changed by the magnet; likewise the magnitude of the momentum remains constant. Since the charge on the particle is also constant, we notice that the denominator of Equation (13-5) is constant. We next integrate both sides of this equation to obtain

$$\theta = \frac{q}{p} \int B\ ds \tag{13-6}$$

where θ is the total angle by which the particle is deflected as a result of its passage through the magnet. Notice that we have not assumed that B is constant, and Equation (13-6) is valid even for a magnetic induction that varies from point to point. In practice B is held as nearly constant as possible, but a variable magnitude in the fringe field near the edge of the magnet is always present, and if its functional form is known the deflecting angle can be calculated.

The most important function of bending magnets is momentum separation. The principle involved is easily seen by inspecting Equation (13-6). A beam of particles (of charge q) which have different momenta will undergo angular dispersion on going through a bending magnet, and a slit can be constructed that will absorb all particles except those that leave the magnet with the correct value of θ within a small tolerance. This use of a bending magnet is quite analogous to the use of a prism to disperse white light into its component colors. A slit can be arranged to pass certain wavelengths and block the others. The analogy is close enough that in a schematic diagram of a charged-particle beam a bending magnet is often drawn as a prism. It should be noted that the analogy fails quantitatively—a magnet imparts more deflection to particles of low momentum, whereas a glass prism at optical frequencies deflects more strongly the high-frequency light for which the photons have higher momentum.

Deflecting magnets have uses other than momentum separation. They find an obvious practical use in causing the beam to avoid hitting various obstacles.

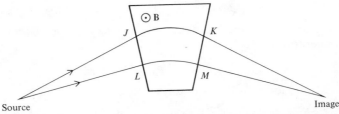

Figure 13-2　Wedge focusing of a bending magnet; **B** is directed out of the paper. Arc _JK_ has the same radius of curvature as arc _LM._

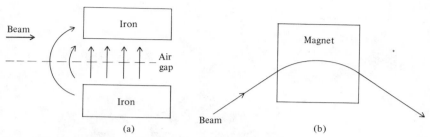

Figure 13-3 Edge focusing of a bending magnet with parallel sides: (a) vertical section, (b) horizontal section.

They have another less obvious use by virtue of the fact that every bending magnet acts to focus the beam to some extent. There are two separate ways in which this effect occurs: wedge focusing and edge focusing. The former occurs for all magnets whose edges are not parallel. The principle, as illustrated in Figure 13-2, is that two particles diverging from the same source will have different path lengths but equal radii of curvature. Then from Equation (13-6) the longer arc will be deflected more, leading to focusing for the configuration shown. Wedge focusing affects the particles only in the plane shown in the diagram. Components of the motion along the magnetic field are not changed.

Edge focusing is a more complicated phenomenon; it cannot be understood without reference to all three dimensions. Figure 13-3 shows two sections of a magnet with parallel edges. In this configuration the beam particles will experience a downward component of force if they enter (as shown) above the plane of symmetry. This downward force acts only in the region of the fringe field, and it applies at both entrance to and exit from the central region where the field is nearly uniform. This edge effect is present for all bending magnets in which the entry (and exit) of the beam are not perpendicular to the sides of the magnet. It is apparent from this condition and from the condition for wedge focusing that every bending magnet must possess focusing properties. An arbitrary magnet will possess both kinds of focusing, but in practical use simplicity is often obtained by choosing one kind of focusing to the exclusion of the other. Sometimes both types of focusing are purposely designed into the same magnet in order to increase the solid angle in which beam particles are accepted. Such design is often more compact and more economical than would be possible if the focusing in both directions were performed by separate magnets. Magnets for mass spectrometers are often designed to incorporate double focusing.

13-3 Quadrupole Magnets

Although bending magnets possess useful focusing properties for beams of charged particles, their focusing strength is usually insufficient in practice.

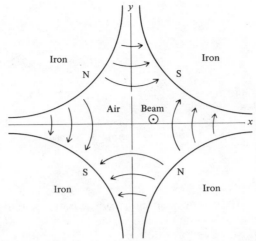

Figure 13-4 Cross section of a quadrupole magnet. The z axis and the beam are both coming out of the paper.

The required focusing can be obtained by using quadrupole magnets, such as the one shown schematically in Figure 13-4. Four pole-pieces of alternating polarity are arranged so that the magnetic induction is given by

$$B_x = ay$$

$$B_y = ax \tag{13-7}$$

$$B_z = 0$$

where a is a constant that depends on the size of the magnet, the number of turns in the windings, and the current flowing through the windings.

A particle with charge q traveling exactly along the z axis will experience no force at all. A particle traveling in the z direction but displaced in its x coordinate (as shown in Figure 13-4) will undergo a deflecting force that will tend to reduce the size of the x coordinate. If it had been displaced in the negative x direction, it would likewise have undergone a restorative deflection. We can estimate the strength of this focusing effect by using Equation (13-6), the expression for the angle of deflection which results from the passage of a charged particle through a magnetic field. Let us assume that the quadrupole magnet

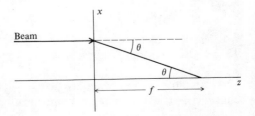

Figure 13-5

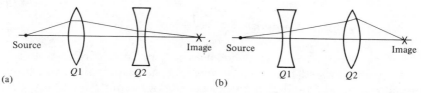

Figure 13-6   Strong focusing of a quadrupole doublet: (a) horizontal plane, (b) vertical plane.

has length L and that this length is short enough so that the particle does not change its x coordinate appreciably within the magnet—only the x component of its momentum changes. Then $\int B \, ds$ becomes simply BL and

$$\theta \approx \frac{qB_yL}{p} = \frac{qaxL}{p} \tag{13-8}$$

By looking at Figure 13-5 we see that the focal length of the quadrupole is related to x and θ by

$$\frac{x}{f} = \tan \theta \approx \theta \tag{13-9}$$

The small-angle approximation $\tan \theta \approx \theta$ is often good enough; in practice it is usually more nearly true than our earlier assumption of short L (essentially we assumed that $L \ll f$). Then in these approximations

$$f \approx \frac{x}{\theta} = \frac{p}{qaL} \tag{13-10}$$

The student is referred to the literature cited at the end of the chapter for a more accurate treatment. The result obtained here shows that for a given beam (p and q fixed) an increase in the length (L) or strength (a) of the quadrupole will lead to stronger focusing.

Referring back to Figure 13-4, let us consider a beam particle that again moves in the z direction, but with $x = 0$ and a displacement in the positive y direction. Application of the right-hand rule shows that such a particle will be deflected upward, away from the center of the system. Similarly, a particle displaced along the negative y axis will be forced downward. It is apparent that the quadrupole which we have been considering will focus in the horizontal plane but defocus in the vertical plane. In order to achieve focusing in both planes it is necessary to have two (or more) quadrupoles of opposite polarity. With ordinary light and lenses, two lenses of equal strength, one converging and the other diverging, will provide over-all focusing if they are separated in space. A pair of quadrupoles, $Q1$ and $Q2$, represented as lenses in Figure 13-6, produce simultaneous focusing in *both* horizontal and vertical planes. The principle illustrated here is that of *strong focusing*, mentioned in Chapter 12 in connection with the design of synchrotrons.

13-4   Velocity Spectrometers

A system for beam transport which uses only magnets is referred to in practice as an unseparated beam, since magnetic fields provide a measure only
of the momentum of the beam particles. Such a beam designed for a certain
momentum will transmit particles which have that momentum (and which have
the correct charge) without regard to their mass. The devices known variously
as velocity spectrometers or separators are used to distinguish among the
various masses that may be present. As the name implies, they use the fact
that the velocities of equal-momentum particles will depend on the mass. Two
general types of velocity spectrometer are used: static (or dc) separators and
radiofrequency (or rf) separators.

The static separator in its simplest form consists of crossed electric and magnetic fields, with the magnitudes of these fields adjusted to result in no deflection for the wanted particles. The electric force is given by

$$F_E = qE \tag{13-11}$$

and the magnetic force by

$$F_B = qvB \tag{13-12}$$

where E and B are electrostatic and magnetostatic fields, q is the charge of a
beam particle, and v is its velocity. When these forces cancel, no deflection
will occur, and we will have

$$F_E = -F_B$$

leading to

$$E = v|B| \tag{13-13}$$

with **E**, **B**, and **v** mutually perpendicular. This equation points directly to the
velocity as the important physical quantity. After E and B have been chosen
to satisfy Equation (13-13) for the desired particles the forces will cancel,
but for a heavier particle of the same momentum the velocity will be less and
the condition (13-13) will not be satisfied. The unwanted particles will be de-

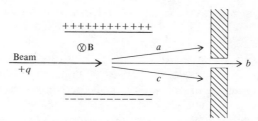

Figure 13-7   Static separator. The electric force is down, the magnetic force is up. For the
wanted particles (b), no deflection occurs. Unwanted light particles are deflected upward (a); unwanted heavy particles are deflected downward (c).

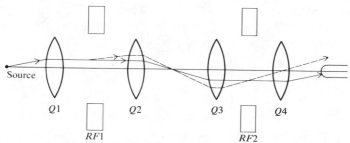

Figure 13-8   Velocity spectrometer which uses radio-frequency electromagnetic fields to
deflect particles; Q1, Q2, Q3, Q4 are quadrupole doublets; *RF*1 and *RF*2 are
microwave cavities where the deflections occur. The solid trajectory is typical
for the case in which the magnets are turned on and the rf is turned off. The
dotted trajectory between *RF*1 and *RF*2 is that of a deflected particle. At *RF*2
it receives another deflection. Unwanted particles are deflected back into the
solid trajectory; wanted particles are deflected into the dotted path around
the metal stopper.

flected to collide with a metal shield that will remove them from the beam. The
desired particles will pass through a slit in the shield and will continue down-
stream, as shown in Figure 13-7. This arrangement is sometimes called a
Wien filter.

In practice E and B are made as nearly uniform as practicable. The value
of E is set as large as possible, limited by the fact that sparks will jump from
one plate to the other if E is too large. By maximizing E and B the amount of
deflection undergone by unwanted particles will be maximized and will lead to
greater spatial separation of the components at the slit. When this method is
used at high energies, the deflections become smaller, requiring more stringent
tolerances for all of the equipment. Small deflections could in principle be
overcome by increasing the path length within the parallel plates of the spectrometer; unfortunately, however, some of the more interesting experiments
at high energy involve beams consisting of unstable particles, and a very long
path between accelerator and detectors would result in serious loss in beam
intensity because of radioactive decay of the beam. In spite of this restriction
on the total length of the beam and in spite of the problems involved in separation with small deflections, the method outlined here has been used successfully for momenta in excess of 6 GeV/c.

A different principle, that of rf separation, is applicable for use in velocity
filters at momenta too high for dc separation. A schematic diagram of such a
system is shown in Figure 13-8. Two microwave cavities $RF1$ and $RF2$ contain an oscillating electromagnetic field with a definite phase relation between
the two cavities. The beam of unwanted particles is modulated at $RF1$, de-
modulated at $RF2$, and brought to focus on a stopper which consists of a
wedge-shaped piece of brass or copper. The time of flight for the unwanted
particles is given by

$$\Delta t_u = \frac{L}{\beta_u c} \qquad\qquad (13\text{-}14)$$

where L is the distance between $RF1$ and $RF2$, and $\beta_u c$ is the velocity of the unwanted particle. The beam of wanted particles is modulated at $RF1$ in the same manner as were the unwanted particles, but their time of flight will be different,

$$\Delta t_w = \frac{L}{\beta_w c} \tag{13-15}$$

and they will arrive at $RF2$ during a different part of its oscillatory cycle; they will receive an additional deflection and will not hit the stopper. It is important to realize why the unwanted particles are effectively not deflected, in contrast to the usage with static separators. Since the deflection at $RF1$ arises from an oscillatory field, there will be two times in each cycle when the field is exactly zero. A particle of any mass that comes through $RF1$ at one of those instants when the field is zero will not be deflected and will follow the solid trajectory in Figure 13-8. The result is that an rf separator will eliminate the unwanted particles and will transmit some but not all of the wanted particles.

The first rf separators to be built used two deflecting cavities as described here. Other arrangements are possible. If the source of the beam is a linear accelerator, the rf cavities of the accelerator itself modulate the beam and play the role of $RF1$ in our example, so that mass separation requires only one cavity, of which the phase must be related to the oscillations in the accelerator.

In practice it is always rather difficult to move a deflecting cavity. Therefore, if only two rf cavities are used, velocity separation can be obtained only for a discrete spectrum of momenta, corresponding to a fundamental mode of separation and its harmonics. Systems with three or more deflectors are more complicated but can provide more flexibility in the choice of momentum.

13-5   Photographic Emulsions

The remaining sections of this chapter are devoted to discussion of the various types of apparatus used to detect the presence of charged particles. It is fitting that we begin with photographic emulsions because the first experiment in nuclear physics, the discovery of radioactivity by Becquerel in 1896 (see Section 3-1), employed photographic emulsions. Becquerel was expecting to observe ordinary light rather than charged particles, so his emulsion was not of a special design. Modern emulsions for nuclear work differ from standard photographic film in that the emulsion is much thicker so as to contain a larger percentage of the paths of particles that travel at an angle with respect to the plane of the emulsion. They also have a higher percentage of silver, divided into smaller grains; the grain size is more nearly uniform for nuclear work than for photographic work. Some quantitative information is listed in Table 13-1.

The mechanism of detection depends on the fact that charged particles disrupt the silver halides in the emulsion, reducing silver ions to metallic silver.

TABLE 13-1   COMPARISON OF THICK EMULSION WITH ORDINARY PHOTOGRAPHIC
EMULSIONS

	Ordinary Emulsions for Optical Photography	Emulsions for Detection of Nuclear Particles
Composition by weight	50% AgBr 50% gelatine	83% AgBr 17% gelatine
Average diameter of AgBr crystals in microns	0.5–3	0.07–0.3
Typical thickness in microns	10	400–600

Composition of a Typical Emulsion

Element	Density (gm/cm³)
Ag	1.82
Br	1.34
I	0.01
C	0.28
H	0.53
O	0.25
N	0.07
S	0.01

Grains of silver are left behind after development, marking the tracks of the particle.

The advantages of emulsions are numerous enough so that their use continues to the present. They are conveniently small in size and inexpensive, compared with other detectors. They provide the possibility of measuring the momentum and the velocity independently for a high-energy track. Some of the more accurate mass measurements of mesons have been made with emulsions. Their light weight is used to advantage in the film badges which are worn by personnel who must work in areas where radiation is present.

Disadvantages of emulsions are serious enough that they are not used so much as they once were in experimental work. They tend to be less useful at low energies because slow particles cannot penetrate far into the emulsion. Manufacture and development of thick films involve many tedious difficulties. In a nuclear scattering experiment it is desirable to know the identity of the target struck by the projectile; emulsions contain such a variety of chemicals (see Table 13-1) that identification of the target is a problem. Another disadvantage is that all tracks must be viewed through a microscope, both for finding the tracks and for making measurements on them.

It should be noted that these devices are often called "nuclear emulsions," a name that has led to much misunderstanding by postal employees and others who erroneously believe the emulsions to be radioactive. They do not emit particles; they merely detect them.

13-6   Gas-Filled Detectors

Whenever a charged particle passes through a gas, it collides with some of the molecules of the gas and disrupts these molecules. Often an electron may be driven away from the molecule, which is then said to be ionized. As a result of the presence of these ions and electrons the gas gains electrical conductivity. The gas-filled detectors which we will consider in this section all depend on this increase in conductivity to detect the passage of ionizing radiation. In each instance the equipment consists of a container to hold the gas and two electrodes across which there is a potential difference. The details of construction and the voltage used differ according to the specific use of the detector.

A gas-filled chamber with parallel-plate electrodes and voltage of the order of 100 volts can be used as an *ionization chamber* (see Section 5-3). An ammeter connected externally can measure the flow of electric current through the chamber. This current will be proportional to the number of charged particles that enter the chamber per unit time. The current will not depend on the energy of these particles.

If the potential difference is raised to several hundred volts, the phenomenon of gas multiplication will occur; electrons that were liberated in collisions of an incident particle with gas molecules will be accelerated by the electric field in the chamber, gaining enough energy so that they too will cause ionizing collisions. The result is that instead of a direct-current flow (as in the ionization chamber) each incident charged particle will give rise to a current pulse. The strength of each pulse will be approximately proportional to the kinetic energy of the original charged particle. Under these conditions the gas-filled detector is called a *proportional counter*. The geometry for this type of counter often consists of a single wire as the positive electrode, surrounded by a cylindrical negative electrode.

The *Geiger-Mueller counter* (often called simply a Geiger counter) is one of the most renowned devices of nuclear physics. It has a wire anode usually made of wolfram (tungsten), surrounded by a cylindrical cathode made of some material such as conducting glass, copper, brass, or stainless steel. The potential difference depends on the specific tube, of course, but it is often in the vicinity of 1000 volts. With a voltage that high, the phenomenon of gas multiplication is so predominant that it causes an avalanche of ions and electrons and leads to complete electrical breakdown inside the tube. The result is that the Geiger-Mueller counter is very sensitive: even weakly ionizing particles will yield an authoritative pulse. However, the strength of the pulse will be independent of the energy of the particle that caused the pulse. Typical gases used in Geiger-Mueller counters are neon and argon, chosen because the voltage required is lower than that for most other gases; another advantage is that they are chemically inert and cannot corrode the interior of the tube. With these noble gases there is a problem associated with the fact that the avalanche of

Figure 13-9 A large spark chamber which was used for a neutrino experiment. The plates are made of aluminum with a gap of 1 cm between plates. (Courtesy of Brookhaven National Laboratory.)

electrons and ions tends to be self-perpetuating. Therefore some means must be provided to "quench" the discharge within the tube. This end can be accomplished by appropriate external circuitry or by the addition of a quenching impurity to the gas within the counter. Bromine, chlorine, and ethanol vapor have been used with success.

The gas counters described so far have the advantage of being relatively inexpensive. However, they all tend to be too bulky for use in cramped conditions. Further, a Geiger-Mueller counter which has just discharged is no longer sensitive until the interelectrode voltage has climbed back to its necessary high value. A second particle that comes through the counter during this "dead time" (typically about 10^{-4} sec) will not be counted, a fact that leads to inaccuracy in counting rates. Some types of Geiger-Mueller counter have the property that if they are subjected to a very large flux of particles they count none of them because the voltage never returns to its high value.

A more recent device which uses the principle of conduction in a gas is the spark chamber, a detector that is useful at high energies. In its simplest form this detector consists of a series of parallel plates with alternate high and low voltages. A photograph of a spark chamber appears in Figure 13-9; the edges

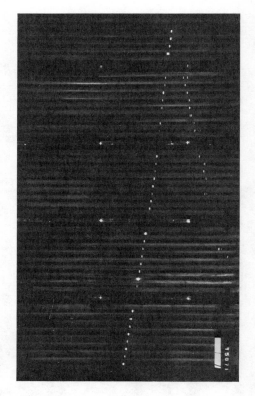

Figure 13-10 Tracks from the spark chamber shown in Figure 13-9. The long track was caused by a muon. (Courtesy of Brookhaven National Laboratory.)

of the plates are visible. The region between the plates is filled with gas, usually neon; here, as in the Geiger-Mueller tube, the noble gases are valued because they undergo electrical breakdown at comparatively low voltage. When a charged particle passes through a spark chamber, the conductivity of the gas is increased along the path of the particle and a trail of sparks appears. It is a simple matter to photograph the sparks. Figure 13-10 shows the typical appearance of tracks in a spark chamber.

Spark chambers have many advantages. They can be "triggered" by other counters; in other words, the high voltage is applied to the plates only when a signal from a counter indicates that a particle has passed through the spark chamber. This technique of triggering is quite standard practice. It prevents needless photographing of uninteresting track arrangements. The fact that the spark chamber is a visual detector is an advantage because it gives the physicist a certain satisfaction and additional confidence in his results. Momenta of charged particles can be measured by measuring the curvature of the tracks in a known magnetic field.

There are disadvantages in using spark chambers for certain types of experiment. Because of the rather large size of the sparks, the space resolution is not always good enough. A kind of experiment in which this problem arises might be a study of a particle that decays after traveling less than a centimeter. Since the interplate separation is usually of this same approximate size, there will be no possibility of detecting or measuring the unstable particle. A second (less serious) disadvantage is that the interior of a spark chamber always contains nuclei of at least two types: those in the plates and those in the gas. For this reason scattering experiments can be performed only by arranging to have the target outside the spark chamber. For many experiments the external target is no disadvantage, but for a few types of study this restriction is in effect prohibitive.

The spark chamber is a device that lends itself well to extensions of the original concept. A simple way to modify a spark chamber to suit the purposes of a specific experiment is to use plates that are not flat. Concentric cylinders are sometimes used. Other shapes would be hard to fabricate; they would make it hard to see the sparks, and they would lead to spurious sparks in regions where the electric field is more intense.

A recent extension of the spark chamber is the streamer chamber, which is a wide-gap chamber with only two plates. When a particle passes through the chamber, a pulse of very high voltage is applied to the plates for a brief instant, not long enough for a spark to go the distance between the plates. A streamer

Figure 13-11   A streamer chamber showing a wire grid across the top serving as an electrode. (Courtesy of Stanford Linear Accelerator Center.)

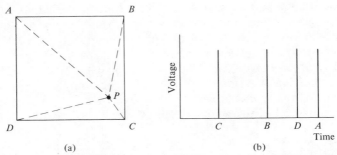

Figure 13-12 (a) A plate of an acoustic spark chamber. Microphones are located at the corners; a charged particle causes a spark at *P*. (b) Output caused by this spark.

discharge occurs in the chamber. Viewed along a perpendicular to the electric field, this discharge looks rather like a curtain. Viewed along the electric field lines, the streamers look like small dots, forming a measurable track. To look along the electric field lines requires that one of the plates be both transparent and electrically conductive. This need is met by using a wire mesh for one of the electrodes (see Figure 13-11).

Two additional modifications that have been of value for narrow-gap chambers are directed toward elimination of the photographic film as an intermediate stage between the sparks and the final data of the experiment. One of these ideas is the wire chamber. The plates are replaced by grids of thin wires. A spark which jumps between two wires will cause a pulse in those two wires but not in the others. These pulses are sent directly to a digital computer and stored. The computer can then reconstruct the position of the spark because it knows which wires provided the pulses. The result is that the data are in a form that is convenient for calculation without the intermediate steps of exposing, developing, and measuring photographic film. The other idea for rapid acquisition of spark-chamber data is the acoustic spark chamber. Conventional flat plates are used. In each corner of a gap there is a microphone which detects the sound of the spark and gives rise to an electrical pulse that is sent to a computer. The relative times of arrival of the four pulses determine the location of the spark. The situation is shown schematically in Figure 13-12.

13-7 Scintillation Counters and Solid-State Detectors

One of the early methods used for the detection and counting of alpha particles was to allow them to strike a zinc sulfide screen and then observe the scintillations on the screen through a magnifier. In the modern adaptation of this method the zinc sulfide screen is placed in close contact with a photomultiplier tube (described in Section 4-12). An alpha particle incident on the phosphor causes a scintillation from which the light strikes the photocathode and produces one or more photoelectrons. The action of the photomultiplier then increases the total charge by a factor of 10^6 or greater; this charge, when fed to a capacitor, will build up a potential difference whose value can be measured with a suitable circuit. The circuits used are such that a pulse of a given amplitude or "height" is produced when the capacitor is charged by the electrons from the photomultiplier. An oscilloscope can be incorporated in the circuit for viewing the pulses produced as a result of the scintillations of the phosphor. These pulses are also fed into a counting circuit and scaler so that the number of scintillations can be counted. At low energies the height of the pulse produced by a scintillation counter is proportional to the kinetic energy of the incident particles. The reason for this proportionality is that the particle stops in the phosphor. A particle with high enough energy to traverse the phosphor will, of course, carry kinetic energy with it as it leaves. For such particles the proportionality no longer holds.

Although a thin zinc sulfide screen is suitable for alpha-particle counting, phosphors having a large volume must be used for the more penetrating particles and radiation, such as beta rays and gamma rays. Further, these phosphors must be transparent so that the scintillations can reach the photomultiplier. Since the light produced in the phosphor can travel in all directions, the phosphor is frequently covered with a thin layer of aluminum, except for the face near the photocathode. The incident radiation will penetrate the aluminum very readily, while the visible radiation produced in the phosphor will be reflected from it one or more times until it reaches the photocathode, as illustrated in Figure 13-13.

A wide variety of substances has been found to be suitable as phosphors. In addition to zinc sulfide, alkali-metal iodides activated with thallium have found wide application. These are designated by LiI(Tl), NaI(Tl), KI(Tl), and CsI(Tl). Lithium iodide can be used with activators other than Tl; it is especially useful for detecting slow neutrons. For charged particles at high energies

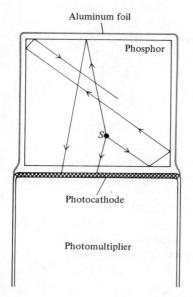

Figure 13-13 Light coming from a scintillation S may be reflected many times to reach the photocathode of the photomultiplier tube.

cesium fluoride (CsF) has been used with success. In addition to the inorganic materials already mentioned, many organic materials—some crystalline, some liquid, and some plastic—have been found to be suitable. The crystals include anthracene ($C_{14}H_{10}$) and *trans*-stilbene ($C_6H_5CH{=}CHC_6H_5$); liquids include xylene ($C_6H_4(CH_3)_2$), toluene ($C_6H_5CH_3$), and phenylcyclohexane ($C_6H_5C_6H_{11}$). Plastics such as polystyrene or polyvinyltoluene are excellent for use as scintillators; sometimes one of the previously mentioned organic compounds is mixed with the plastic during its manufacture, combining the properties of the two media.

Applications of scintillation counters have been similar to those of Geiger-Mueller counters, and the comparison of the properties of the two types has led to a decline in the use of the Geiger counter in recent years. Scintillators can be manufactured in a great variety of sizes and shapes, a fact that is not true of Geiger tubes. The scintillator-photomultiplier combination produces a sharper pulse than that from a Geiger counter, leading to several orders of magnitude better time resolution for scintillators. The problem of dead time, so troublesome in the use of Geiger counters, is similarly reduced by several orders of magnitude.

There are only a few disadvantages to the use of scintillation counters for nuclear experiments. For certain kinds of work the smallest scintillators available do not provide enough spatial resolution. For work in cramped conditions it is a disadvantage that the scintillator requires the presence of a sizeable photomultiplier tube to be effective. These disadvantages are more serious at low energies than at high energies, where scintillation counters are still a standard tool of research.

Recent years have seen the development of a new type of detector, referred to as the solid-state detector or the semiconductor detector. As the name implies, the detector consists of a semiconducting crystal in which electrons are not free to move, even though a voltage difference is applied to different parts of the crystal. The passage of ionizing radiation through the crystal provides mobile electrons which are collected by the wires that apply the external voltage. The result, of course, is an electrical pulse which can be amplified by external circuits. Many types of crystal have been used with varying degrees of success; a choice in wide use is a silicon crystal with lithium atoms diffused into one face of crystal.

The advantages of solid-state detectors are marked in low-energy physics research. They are small in size, leading to ease in placement inside other pieces of equipment. They require amplifiers which are more convenient than photomultipliers. Their small size enables the experimentalist to achieve good spatial resolution.

For research in low-energy nuclear physics the strength of the output pulse from a "lithium-drifted" silicon detector is proportional to the energy of the ionizing particle, just as in the case of scintillation counters. For a variety of reasons the uncertainty in an energy measurement is much greater from a scintillator than from a semiconductor detector. It is this advantage that has led in recent years to the replacement of scintillators by semiconductors in low-energy work. For the proportionality to remain valid it is necessary that the ionizing particle stop in the sensitive region of the crystal. It is difficult to manufacture semiconductors in which this sensitive region is thicker than a few millimeters; for this reason the good energy resolution is attainable only at low energy.

13-8 Cerenkov Counters

A new method of producing visible radiation was discovered by P. A. Cerenkov in 1934. He observed that a beam of fast electrons, such as beta particles from radioactive substances, when moving in a transparent medium caused the emission of visible radiation, provided that the velocity of the electrons was greater than the velocity of light in the same medium. The theory developed by I. M. Frank and I. E. Tamm predicts that the light should be propagated at an angle θ to the direction of motion of the electron given by

$$\cos \theta = \frac{c}{nv} \qquad (13\text{-}16)$$

where n is the index of refraction of light and v the velocity of the electron in this medium. An electron moving through a substance loses most of its energy in ionization and excitation of the atoms. In these processes the electron itself experiences small accelerations, hence radiates energy in the form of electro-

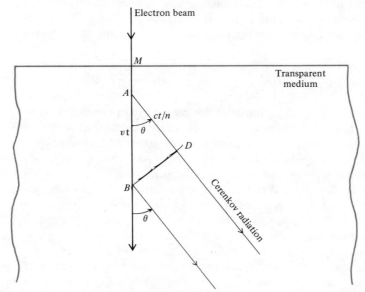

Figure 13-14   Direction of propagation of Cerenkov radiation relative to the direction of motion of the electrons in a transparent medium.

magnetic waves. Since these waves originate at different points along the path of the electron, radiant energy will be observed only if the waves from the different points reinforce each other. The condition for reinforcement of the waves by interference can readily be derived with the aid of Figure 13-14. The electron beam enters the transparent medium at M and continues along the path MAB. The electron radiates energy in all directions from the points along this path. Using the Huygens construction, we obtain the wave front BD, where the distance AD is ct/n and $AB = vt$; t is the time taken by the electron to reach point B. From the figure it follows that the radiation will travel in a direction such that

$$\cos \theta = \frac{c}{nv}$$

The Cerenkov radiation was investigated by G. B. Collins and V. G. Reiling (1938), who used a beam of electrons of 1.9-MeV energy incident on a series of thin films of various transparent substances such as glass, water, mica, and cellophane. They photographed the pattern of the emitted light and found the intensity to be a maximum in the direction given by the above equation. They also examined the emitted light spectroscopically and found the spectrum of the Cerenkov radiation to be continuous and to extend from the long wavelength limit of the apparatus (usually > 5000 Å) down to the ultraviolet absorption limit of the medium in which it was produced. H. O. Wyckoff and J. E. Henderson (1943) extended the work of Collins and Reiling to slower electrons, with energies ranging from 240 to 815 keV. Using mica as the transparent medium, they obtained results in agreement with the predictions of Frank and Tamm concerning the direction of emission of the Cerenkov radiation. The intensity of Cerenkov radiation emitted is given by the expression

$$I(\nu) \, d\nu = \frac{4\pi^2 k e^2 Z^2}{c^2} \left(1 - \frac{c^2}{v^2 n^2} \right) \nu \, d\nu \tag{13-17}$$

where Ze is the electric charge of the particle in coulombs; $I \, d\nu$ is the number of joules emitted per meter of path for frequencies between ν and $\nu + d\nu$.

Cerenkov radiation is useful in the detection of high-energy charged particles. The only condition imposed on these particles is that their speeds must be greater than the speed of light in the same medium. In any given medium the angle of emission θ is determined only by the velocity of the particle; hence the detection of the Cerenkov radiation is a convenient method for measuring the velocities of high-energy charged particles. As Equation (13-17) shows, the intensity of the radiation is proportional to the square of the electric charge, so that an absolute charge measurement is possible.

In practice many substances have been used as Cerenkov detectors. In conjunction with a number of photomultiplier tubes, the entire arrangement is called a Cerenkov counter. Lead glass and polymers such as lucite have been used as the medium, as have liquids such as water, CS_2, and CCl_4. In each case convenience and the index of refraction are the important considerations. At very high energies appreciable radiation can be obtained even when n is close to unity because v/c is becoming even closer to unity. Under such circumstances various gases are suitable for Cerenkov counters; N_2, CO_2, and SF_6 are often used; for instance, n is about 1.1 for CO_2 at 200 atmospheres of pressure. By varying the pressure, n can be adjusted within a continuous range of values, leading to a highly desirable flexibility of the instrument.

13-9 Cloud Chambers

Historically, one of the most important instruments for the detection of charged particles was the cloud chamber, invented by C. T. R. Wilson (1912). The device consists of an enclosure containing a gas and a saturated vapor under pressure. When the pressure is suddenly released, the vapor will cool, becoming supersaturated when the temperature drops below the dew point. The passage of a charged particle through this vapor will give rise to ion pairs, which in turn act as local centers around which droplets of liquid will form. So a string of droplets will mark the track of the charged particle in a way that is convenient to photograph.

Cloud chambers offered several advantages over their early competitors. They are a visual device. They work well in a strong magnetic field, enabling one to measure the momentum of the charged particles. Another advantage is that the cloud chamber can be "triggered"; in other words, a set of counters external to the cloud chamber can detect the presence of the desired particles, and the pulses from these other detectors can be caused to initiate the expansion of the chamber. The reason that triggering is possible is the fact that the gas and vapor are not very dense; therefore the ions and electrons formed in the passage of a charged particle will not recombine immediately. Rather, the ion pairs are persistent, and the expansion of the chamber need not take place until some time after the particle has passed through the chamber. Geiger-Mueller counters were used for many years for the purpose of triggering. The combination was quite effective and led to sets of photographs in which each picture contained tracks of particles.

However, cloud chambers have several disadvantages that have caused their use to be greatly diminished in recent years. In nuclear scattering experiments the cloud chamber represents a target in which the scattering centers are too diffuse. Too many beam particles must be sent into the chamber in order to observe each interaction because the collision probabilities are small; many beam particles will pass through the chamber without interacting at all. Further, the nuclei of the vapor are usually not the same kind as the nuclei of the gas in the chamber, so that when a collision occurs there is a problem in identifying the target nucleus, a problem similar to that which plagues emulsion experiments.

Another serious difficulty, especially in the case of large cloud chambers, is a result of the persistence of ion pairs. The very attribute that enables triggering to occur becomes a detriment; after a track has been photographed, it is necessary to remove all the electrons and ions from the vapor before re-expanding the chamber, so as not to rephotograph the same track. The usual method for removing these charges is to apply an electric field inside the chamber, using electrodes to which electrons and ions are attracted. Unfortunately the mobility of the charges is low enough that it takes many seconds to prepare the chamber for the next expansion. The larger the chamber, the more serious the problem of clearing away charges that formed previous tracks.

The cloud chambers referred to so far depend on the expansion principle. A different type, the *diffusion cloud chamber,* was developed by A. Langsdorf (1936). It consists of a container with a cold liquid at the bottom, a reservoir of warmer liquid near the top, and a gas-vapor mixture in the intermediate region. There is a layer in which the vapor is near enough to the dew point so that tracks are formed whenever a charged particle traverses this layer. No expansion is necessary, and the chamber is continuously sensitive. This technique has been successful with methanol (CH_3OH) as the liquid and vapor, with dry ice (solid CO_2), as a cooling agent at the bottom of the chamber. Diffusion chambers have also been built with hydrogen as the liquid and vapor; hydrogen has a single proton in its nucleus, leading to certainty in knowledge of the target in a nuclear scattering process.

The diffusion cloud chamber has three important disadvantages that have prevented its widespread use as a detector.

1. Like the expansion cloud chamber, it requires an electric field to remove old tracks—a slow process.

2. The sensitive layer is only 5 or 6 cm in thickness.

3. Since the chamber contains thermal gradients, there is a tendency toward turbulent flow of vapor, leading to distortions in the tracks and consequent inaccuracies in measurement.

13-10   Bubble Chambers

The disadvantages of the cloud chamber have been overcome to a great extent as a result of the invention of the bubble chamber by D. A. Glaser (1953). This device is a closed container containing a superheated liquid under pressure. On the release of the pressure boiling begins in the liquid. If a charged particle travels through the liquid while the pressure is decreasing, boiling will begin, preferentially along the path of the charged particle. If a flash photograph is made within a few milliseconds of the passage of the particle, the film will display the track of the particle as a string of bubbles, as seen in Figure 13-15.

Various liquids have been used successfully in bubble chambers. Table 13-2 lists and compares the properties of some of the more frequently used liquids. The properties of hydrogen, deuterium, and neon are sufficiently similar that a bubble chamber built for one will also operate successfully with the other

Figure 13-15 Example of a photograph obtained from a bubble chamber. The parallel tracks are a collimated beam of antiprotons. The spirals are electron tracks. An example of electron-positron pair production is clearly visible. (Courtesy of Brookhaven National Laboratory.)

two. Liquid hydrogen and liquid neon are miscible, and chambers have been operated with various proportions of the two substances. In similar fashion propane bubble chambers are able to operate with freon or with a mixture of propane and freon of any percentage.

The most significant liquid for use in bubble chambers is liquid hydrogen.

TABLE 13-2 PROPERTIES OF LIQUIDS USED IN BUBBLE CHAMBERS

Liquid	Composition	Operating Temperature (kelvins)	Operating Absolute Pressure (atmospheres)	Density (gm cm^{-3})
Hydrogen	H_2	27	5	0.06
Deuterium	D_2	32	7	0.13
Helium	He	3.4	1	0.124
Neon	Ne	32	7	1.2
Propane	C_3H_8	331	21	0.44
Freon	CF_3Br	301	18	1.5
Xenon	Xe	252	25	2.2

The reason is that the nucleus of hydrogen contains a single proton, and when a nuclear scattering occurs in a hydrogen bubble chamber, one can be certain of the identity of the target. Deuterium is used in the chamber for scattering experiments in which a neutron is to serve as the target. The deuteron—the nucleus of the deuterium atom—consists of a proton and a neutron; approximately half the interactions observed in a given experiment will occur with a neutron target, the other half with a proton target. When neon is added to liquid hydrogen in a bubble-chamber experiment, the purpose is usually to raise the effective atomic number (Z) of the liquid; the probability that a high-energy photon will produce an electron-positron pair is proportional to Z^2, and therefore the admixture of neon facilitates the detection of photons.

Bubble chambers using propane or freon are much easier to construct and operate than the low-temperature chambers. For many experiments this advantage is offset because the use of organic molecules leads to considerable uncertainty in the knowledge of the identity of the target in a scattering experiment. Nevertheless, propane chambers are often useful because they provide a high density of protons as targets. It is a fact that propane possesses more hydrogen atoms per unit volume than are present in pure hydrogen. Hence for a given beam intensity and a given chamber size more interactions of a given type will occur in propane than in hydrogen, even after subtracting the interactions for which the target is a carbon nucleus. Experimentally there is often a great deal of difficulty in distinguishing interactions with hydrogen nuclei from interactions with carbon nuclei. Admixture of freon to a propane chamber will increase the probability of pair production in similar fashion to the addition of neon to hydrogen. In the freon that is most often used (CF_3Br) the bromine nucleus contributes the most to the increase in pair production. Occasionally, when still greater probability is needed, NaI can be dissolved in a mixture of propane and freon.

The bubble chamber has advantages that have led to its wide use in nuclear-physics experiments at high energy. It shares with cloud chambers and most spark chambers the advantage of being a visual detector. It provides a direct way of measuring the momentum of a charged particle; nearly all bubble chambers are built inside a magnet, so that charged tracks are curved to permit momentum determination (see Fig. 13-15 for an example of curved tracks). In contrast to the cloud chamber, the bubble chamber can possess a short cycle time; as soon as a set of tracks is photographed the chamber undergoes re-compression and the bubbles disappear. The close-packed nature of the liquid ensures that any stray ions and electrons will soon recombine, and no externally applied electric field is needed to remove them. No additional waiting is required before re-expansion of the chamber. In actual practice bubble chambers are pulsed with a period determined by the accelerator which produces the beam particles. Rates of several expansions per second are possible, even for rather large chambers.

Another advantage of the bubble chamber over the cloud chamber is that the former possesses a high density of targets for scattering experiments, enabling more rapid collection of data.

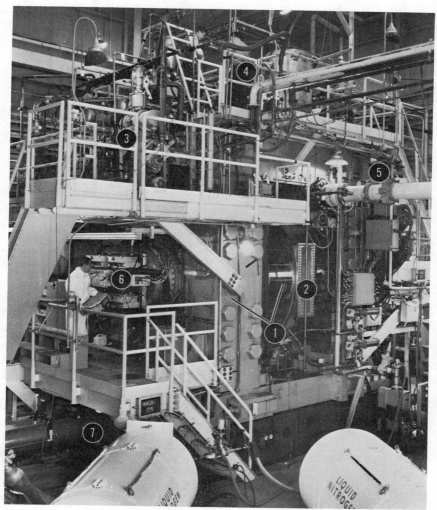

Figure 13-16 The 80-in. bubble chamber showing (1) copper coils and steel yoke of magnet, (2) beam window of the vacuum chamber surrounding the bubble chamber, (3) vacuum pump system, (4) chamber expansion system, (5) liquid hydrogen refrigerator (located out of sight on far side of the chamber), (6) illumination and camera system, and (7) undercarriage for moving entire 450-ton chamber assembly. (Courtesy of Brookhaven National Laboratory.)

There are some disadvantages associated with bubble chambers. Because the liquid is rather dense (even in the case of hydrogen), a low-energy particle cannot go far in the liquid. Therefore bubble chambers are of no use in low-energy experiments. The early recombination of electrons and ions after the passage of a charged particle makes it impossible to trigger a bubble chamber in the way that serves so well for cloud chambers and spark chambers. The expansion of the chamber must occur within 10^{-4} sec of the passage of the particles, and the machinery associated with the expansion system is too ponderous to set in motion within such a short time. For this reason bubble

TABLE 13-3   BROOKHAVEN 80-IN. HYDROGEN BUBBLE CHAMBER*

Dimensions	Visible by cameras	$80 \times 27 \times 26$ in.
Volume	Total hydrogen	1500 liters
	Visible by cameras	900 liters
Expansion	Piston diameter	36 in.
	Piston weight	250 lb
	Actuated by hydrogen gas	
Cycling rate	Maximum	One per second
Illumination	Retrodirective	Scotch lite
Window	Glass, partially tempered	$6\frac{1}{2}$ in. thick
Cameras	Number	3
Film	1000-ft rolls	35-mm width
Refrigerator	Hydrogen, capacity at present	2500 watts
	Possible future capacity	4000 watts
Insulation	Vacuum	
	Aluminized Mylar	200 layers
Magnet	Field	20,400 gauss
	Power	4 MW
	Total weight	450 tons
	Copper weight	30 tons
Magnet carriage	Translation	Sliding
	Rotation	360°
	Elevation	24 in.

*Information courtesy of Brookhaven National Laboratory.

chambers have not found wide use in detecting cosmic rays (which arrive at unpredictable times); rather, they are typically used in conjunction with accelerators that produce pulses of particles at regular intervals.

It may well be counted as a disadvantage that recent years have seen a trend toward larger and more expensive bubble chambers. At energies greater than several GeV the extra size is necessary to enable accurate measurement of tracks that are nearly straight, even though the charged particle travels in a magnetic induction of the order of 20 kG. Figure 13-16 shows the 80-in. hydrogen bubble chamber at Brookhaven National Laboratory, as an illustration of the size and complexity of a single detecting instrument. Table 13-3 lists some of the relevant parameters of this device. When the photograph in Figure 13-16 was made (1965), this was the largest bubble chamber in the world. Since then efforts have proceeded at nearly all the major synchrotrons for the construction of even larger bubble chambers. The cost of such large devices clearly limits their use to the largest laboratories, to which physicists from smaller laboratories must travel if they wish to perform bubble-chamber experiments. The rate of data production by a bubble chamber at a synchrotron is large enough that time is made available for university users, who can gather a large number of photographs in a few weeks; they perform measurement and analysis later at their own laboratories.

13-11 Principles of Particle Identification

In any nuclear experiment with equipment such as we have described it is important not only to count the number of particles but also to be able to identify the particles that are being counted. Many schemes have been used successfully, and we shall not attempt to list them all. Rather, we shall mention a few principles on which some of the methods depend; we shall concentrate on identification of charged particles.

In order to identify a particle it is often adequate to measure its mass. In macroscopic physics mass measurements usually depend on gravity in some way, but for nuclear particles it is necessary to use other methods because electromagnetic effects are so much stronger than gravitation. It is therefore necessary to use less direct methods. If the momentum p and the total energy $\mathcal{E}$ of a particle can be measured independently, the mass can be obtained from the equation

$$m^2c^4 = \mathcal{E}^2 - p^2c^2 \tag{13-18}$$

Alternatively, it is sufficient to measure either the momentum or the energy and combine this information with a measure of the velocity.

In order to understand some of the methods for measuring energy or velocity we need to consider the mechanism for energy loss as a charged particle travels through matter. As we have already seen, charged particles collide with atomic electrons and remove them from atoms. The theory of energy loss from ionization was derived from classical principles by N. Bohr (1915). Improvements that include quantum mechanics were made by H. Bethe (1930) and F. Bloch (1933). Experimental measurements have been made and the agreement is

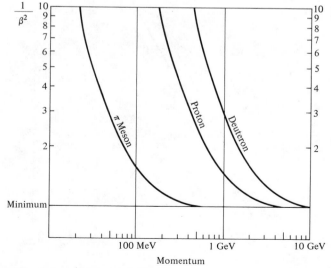

Figure 13-17   Energy loss in liquid hydrogen.

rather good. We consider a particle traveling in the x direction with a kinetic energy $\mathcal{E}_k$. Then the energy loss is usually written $-d\mathcal{E}_k/dx$, where the negative sign means that the particle is losing energy to electrons. To a first approximation

$$-\frac{d\mathcal{E}_k}{dx} \sim \frac{1}{\beta^2} \qquad (13\text{-}19)$$

where $\beta = v/c$. Figure 13-17 shows a graph of energy loss as a function of momentum for several kinds of particle in liquid hydrogen. In this figure it is clear that the energy loss decreases with increasing momentum until it attains a minimum value. For liquid hydrogen this minimum value is about 0.29 MeV/cm. The shape of the curve does not depend much on the material through which the particles travel, but, of course, the minimum value does depend on the material. The relativistic form of the theory predicts that $(-d\mathcal{E}_k/dx)$ should increase slightly for energies much higher than those shown in the diagram. This relativistic increase depends on the details of the matter; for liquid hydrogen it does not occur at all.

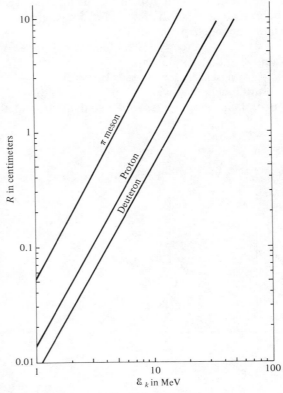

Figure 13-18   Range versus energy in liquid hydrogen.

The expression for energy loss can be manipulated to provide the distance that a particle can travel before it loses all its energy. This distance is called the range R:

$$R = \int_{\mathcal{E}_k}^{0} \left(-\frac{d\mathcal{E}_k}{dx}\right)^{-1} d\mathcal{E}_k \qquad (13\text{-}20)$$

The ranges of certain particles as a function of their kinetic energies are shown in Figure 13-18.

We have now developed enough ideas to provide several suggestions for mass identification. If the range and energy are independently measured, it is clear that the mass is determined. If the momentum is measured with a magnetic field, then knowledge of the velocity will determine the mass. The velocity can be measured in a variety of ways: (a) the energy loss $(-d\mathcal{E}_k/dx)$ can provide the velocity; (b) the Cerenkov effect can be used; (c) a Wien filter can be used; (d) the time of flight between two scintillators can be measured. The tools we have described can be combined in many ways to assist capable experimenters in making the required measurements.

Problems

13-1.   (a) Show that the rest energy of a particle is given by

$$mc^2 = \frac{p^2 c^2 - \mathcal{E}_k{}^2}{2\mathcal{E}_k}$$

where p is its momentum and $\mathcal{E}_k$ is its kinetic energy. Measurements on the track of a meson in a photographic emulsion yield the values of 215 MeV for its kinetic energy and 330 MeV/c for its momentum. Determine (b) its mass and (c) its velocity. Identify the particle.

13-2.   The kinetic energy of a positively charged particle is found to be 62 MeV and its momentum, 335 MeV/c. Determine (a) its mass, and (b) its velocity. Identify the particle.

13-3.   A proton has a radius of curvature of 0.8 meter in a magnetic induction of 0.6 tesla (1 tesla = 1 weber/meter^2, abbreviation T). What percentage error would result from using nonrelativistic formulas to compute (a) the kinetic energy, (b) the velocity of the proton?

13-4.   In a single event four gamma rays are produced. For each photon separately the detecting system has an efficiency of 90 percent. (a) What is the probability that all four will be detected? (b) What is the probability that three out of the four will be detected? (c) What is the probability that two or fewer will be detected?

13-5.   Use the thin-lens approximation to show that two glass lenses of equal strength but opposite sign will give net focusing if they are separated by a short distance.

13-6.   Choose appropriate parameters and find the focal length of a wedge-shaped bending magnet.

13-7.   Show graphically that the beam passing through the magnet of Figure 13-2 will be defocused if the current in the windings is reversed.

13-8.   Calculate the deflection angle for a beam of protons which passes through a 2-meter bending magnet with a field of 1.5 T, assuming a beam momentum of (a) 1 GeV/c (b) 10 GeV/c.

13-9.   A static velocity spectrometer has a gap of 10 cm between plates. The voltage is 360 kV. (a) What value of the magnetic induction in the separator is required so that particles with a mass equal to that of a proton and a momentum of 2 GeV/c will be undeflected? (b) If the spectrometer is 10 meters long, through how large an angle will π mesons ($mc^2 = 140$ MeV) be deflected?

13-10.   Show graphically that both the entrance and exit of a parallel-face magnet will cause focusing.

13-11.   How many free protons per cubic centimeter are there in (a) a liquid hydrogen bubble chamber, (b) a propane bubble chamber, (c) a hydrogen diffusion cloud chamber, (d) emulsion?

13-12.   In the 30-in. bubble chamber at Brookhaven National Laboratory it is possible to maintain a magnetic induction of 1.5 T. It is desired to study protons whose momentum is 1.0 GeV/c. The protons move in a direction perpendicular to the magnetic field. What will be the radius of curvature of the tracks which the protons leave in the chamber?

13-13.   In the original experiment in which the Ω^- hyperon was discovered, a liquid-hydrogen bubble chamber of length equal to 2 meters was bombarded with K^- mesons. About 0.167×10^6 beam particles were observed in the experiment and one Ω^- was produced. Assume that liquid hydrogen has a density of 0.06 gm cm^{-3}. Estimate the cross section for production of Ω^- hyperons from this experiment.

13-14.   A proton passes through a Cerenkov counter (C) and then through a spark chamber (S). A measurement of the curvature of the track in S yields 1 GeV/c for the momentum of the proton. What range of values for the index of refraction of C will give a signal from C?

13-15.   (a) Calculate the minimum velocity that electrons must possess in order to produce visible radiation when passing through glass whose index of refraction is 1.50. (b) Determine the kinetic energy of such electrons. (c) If a beam of electrons of this energy passes through a thin sheet of such glass, at what angle will the light be emitted relative to the direction of the beam?

13-16.   A beam of electrons of kinetic energy 0.8 MeV travels through a transparent substance whose index of refraction is 1.40. Calculate the angle at which Cerenkov radiation will be observed relative to the direction of the beam.

13-17.   A beam of electrons traveling through a glass of index of refraction 1.50 is observed to emit Cerenkov radiation at an angle of 40 degrees to the electron beam. Determine the kinetic energy of the electrons.

13-18.   A stream of protons of 200 MeV kinetic energy is traveling through a transparent crystal whose index of refraction is 1.80. (a) Calculate the velocities of the protons. (b) Determine the angle that the beam of Cerenkov radiation will make with the proton beam.

13-19.   A stream of protons is sent through a transparent crystal whose index of refraction for light of 5000 Å is 1.85. Cerenkov radiation of this wavelength is observed at an angle of 14 degrees to the proton beam. Determine (a) the velocity of the protons and (b) their kinetic energy.

13-20.   A stream of protons traverses a piece of glass whose index of refraction for light of 5000 Å is 1.70. Cerenkov radiation of this wavelength is observed at an angle of 25 degrees to the proton beam. Determine the kinetic energy of the protons.

13-21.   The two rf cavities of a high-energy beam are located 40 meters apart. What frequency of oscillation would be best for separating protons from deuterons at a momentum of 10 GeV?

13-22.   A proton begins to move in liquid hydrogen with an energy of 10 MeV. How much energy has it lost after traveling 2 mm?

References

General

Encyclopedia of Physics (Handbuch der Physik), S. Fluegge, Ed. Berlin: Springer-Verlag, 1958, Vol. XLV.

Finkelnburg, Wolfgang, *Structure of Matter.* New York: Academic Press, 1964, Chapter V.

Livesey, Derek L., *Atomic and Nuclear Physics.* Waltham, Mass.: Blaisdell Publishing Company, 1966, Chapter IX.

Specific

Sections 13-2, 3, 4

Chamberlain, O., "Optics of High-Energy Beams." *Ann. Rev. Nucl. Sci.,* **10,** 49–104 (1960).

Penner, S., "Calculations of Properties of Magnetic Deflection Systems," *Rev. Sci. Instr.,* **32,** 150 (February 1961).

Section 13-5

Barkas, W. H., *Nuclear Research Emulsions*. New York: Academic Press, 1963.

Friedlander, M. W., "Resource Letter on Nuclear Photographic Emulsions," *Am. J. Phys.,* **12,** 1105 (December 1967).

Section 13-6

O'Neill, G. K., "The Spark Chamber," *Scientific American,* **207,** 36 (August 1962).

Sachs, A. M., "Spark Chambers," *Am. J. Phys.,* **35,** 582 (July 1967).

Shutt, R. P., Ed., *Bubble and Spark Chambers*. New York: Academic Press, 1967.

Section 13-7

Akimov, Iu. K., *Scintillation Counters in High-Energy Physics*. New York: Academic Press, 1964.

Taylor, J. M., *Semiconductor Particle Detectors*. London: Butterworths, 1963.

Section 13-8

Jelley, J. V., *Cerenkov Radiation and Its Applications*. New York: Pergamon Press, 1958.

Section 13-10

Bradner, Hugh, "Bubble Chambers," *Ann. Rev. Nucl. Sci.,* **10,** 109 (1960).

Shutt, R. P., *loc. cit.,* Section 13-6.

14 | Radioactivity

14-1 Résumé of Some Known Properties of Nuclei

Many of the important properties of atomic nuclei, as well as the experimental evidence for these properties, were discussed in previous chapters. It was shown that the total number of nucleons in a nucleus is equal to the mass number A of the particular isotope of the element. A nucleon thus has a mass number 1 and may be a proton or a neutron. It was further shown that the number of protons in the nucleus is equal to the atomic number (Z) of the element and that the number (N) of neutrons in the nucleus is A-Z.

From the results of the experiments on the scattering of alpha particles, it was concluded that the nucleus occupies only a very small fraction of the volume of an atom and that nuclear radii do not exceed 10^{-12} cm. It was further shown that the nucleus possesses angular momentum I due to spin and also possesses a magnetic moment.

In this and the following chapters we shall discuss many nuclear processes and transformations which not only are interesting in themselves but also provide additional information concerning the nucleus. Among these processes and transformations are (1) the natural radioactivity of some of the heavier elements, (2) the disintegration of nuclei by bombardment with particles and radiation, (3) artificial radioactivity induced by the bombardment of nuclei with particles and radiation, and (4) nuclear fission, (5) nuclear fusion, (6) the production of new elements, (7) the production of many new elementary particles, and (8) the production of particle-antiparticle pairs.

14-2 Natural Radioactive Transformations

An isotope that is naturally radioactive usually is found to emit either alpha particles or beta particles. Sometimes gamma rays accompany the emission of these particles. When the nucleus of an atom emits an alpha particle, the atom is transformed into a new atom, since its atomic mass is decreased by four units and its atomic number is decreased by two units; for example, radium, with $A = 226$ and $Z = 88$, is known to emit alpha particles. Hence the product of this transformation, known as radon or radium emanation, will have $A = 222$ and $Z = 86$. That we are dealing with nuclear transformations is confirmed by the fact that radium, which is a solid, is in the same chemical group as barium, whereas radon is one of the inert gases. In this particular case it is easy to separate the product from its parent substance. When a beta particle is emitted by a nucleus of atomic number Z, the atomic number of the new atom formed becomes $Z + 1$, but the mass remains almost unaltered, since the mass of a beta particle is negligible in comparison with that of a nucleus. Thus in beta disintegration the mass number remains the same, and parent and product atoms form a pair of *isobars*.

The rate at which a particular radioactive material disintegrates or *decays* is a constant that is almost completely independent of all physical and chemical conditions. Given a large number of atoms of any one radioactive isotope, the average number dN that will decay in a small time interval dt is found to be proportional to the number of atoms N present at the time t; that is,

$$-dN = \lambda N \, dt \tag{14-1}$$

where λ is a constant for the particular radioactive isotope. Integrating this equation, we get

$$N = N_0 \exp(-\lambda t) \tag{14-2}$$

where N_0 represents the number of atoms present at the time $t = 0$. Equation (14-2) shows that the number of atoms of a given radioactive substance decreases exponentially with time, provided that no new atoms are introduced. Half the material will have decayed at the end of a certain time interval T, which can be determined by setting $N = N_0/2$ and $t = T$ in Equation (14-2), yielding

$$\lambda T = \ln 2$$

so that

$$T = \frac{0.693}{\lambda} \tag{14-3}$$

T is called the *half-life* of the isotope. It can be seen that at the end of a time interval equal to $2T$ one-quarter of the original material will still be in existence. The number of atoms still in existence at any time t is shown in Figure 14-1. It is impossible to tell just when one particular atom will decay because radioactive disintegrations follow the laws of chance or probability. At the end of a certain time t, N of the original atoms will still be in existence. In the next interval of time dt, dN of these atoms will have decayed. The *average lifetime* T_a of a single atom may be computed by multiplying dN, the number

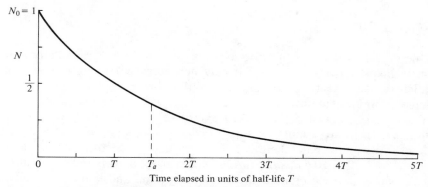

Figure 14-1 Exponential decay of a radioactive element with time.

of atoms decaying, by the time t during which they existed, summing these products over all the atoms and then dividing by the total number of atoms at the start, N_0. Thus

$$T_a = \frac{\int_0^{N_0} t \, dN}{N_0}$$

Now, from Equation (14-2)

$$dN = -N_0 \lambda \exp(-\lambda t) \, dt$$

Substitution of this value for dN in the above equation yields

$$T_a = -\int_\infty^0 t\lambda \exp(-\lambda t) \, dt = +\int_0^\infty t\lambda \exp(-\lambda t) \, dt$$

which yields

$$T_a = \frac{1}{\lambda} \tag{14-4}$$

The reciprocal of the decay constant is thus the average lifetime (often called *mean life*) of a radioactive atom. It is easy to convert from mean life to half-life or vice versa by using Equation (14-3). In using tables of lifetimes it is important to note· which of these two measures is used, since carelessness in this matter will often lead to erroneous results. The usual practice is to use half-life for tables involving nuclei and mean life for the various elementary particles.

The half-lives of radioactive elements occurring in nature vary considerably. Radium, for example, has a half-life of 1620 yr, whereas that of radon is 3.82 days. Thorium C′ (the original name of ^{212}Po) has the shortest naturally occurring half-life, 3×10^{-7} sec. Very long-lived elements require careful work to measure their lifetimes or even to detect that radioactivity exists. One of the longest lifetimes measured is that of ^{204}Pb: 1.4×10^{17} yr.

14-3 The Curie

One method commonly used to express the activity of a radioactive sample is in terms of its rate of decay. It may be expressed simply in terms of the number of disintegrations per unit time of the sample or of a given mass of the sample. A unit commonly used is the *curie*. It was originally based on the rate of decay of a gram of radium. Very extensive experiments have yielded the result that there are about 3.7×10^{10} disintegrations per second per gram of radium. This number is taken as a standard and is called the curie; *the curie is equal to* 3.70×10^{10} *disintegrations per second*. This is independent of the nature of the particle emitted in the disintegration and is now applied to all types of nuclear disintegration.

A curie represents a very strong source of radiation. Submultiples such as the millicurie (= 10^{-3} curie, abbrev., mCi), and the microcurie (= 10^{-6} curie, abbrev., μCi) are more convenient for the quantities usually found in physics laboratories.

14-4 Radioactive Series

Nearly all the naturally occurring radioactive elements lie in the range of atomic numbers, $Z = 81$ to $Z = 92$. These elements have been grouped into three genetically related series: the uranium-radium series, the thorium series, and the actinium series. Many of these elements have two or more isotopes in these series. In the early work in radioactivity many of the isotopes were given names indicative of the manner of their discovery or their formation rather than those of the appropriate elements; for example, radium A is the daughter formed in the alpha decay of radon, but it is not an isotope of radium. It is an isotope of the element polonium with $Z = 84$ and $A = 218$. Similarly uranium X_1 (UX$_1$) is the daughter of uranium I, but it is actually an isotope of thorium with $Z = 90$ and $A = 234$. Since these names appear in the extensive literature of radioactivity, they are shown in Figures 14-2, 3, and 4; the appropriate elements are shown at the top of each figure. Wherever it is more convenient to retain the older name, the appropriate isotopic identification will be made.

A long-lived isotope is at the head of each series and some stable isotope of lead ends each one. The uranium series originates with the uranium isotope $A = 238$ with a half-life of 4.51 billion years, as shown in Figure 14-2, and goes through a series of transformations that involves the emission of alpha and beta particles, giving rise successively to radioactive isotopes of thorium, protoactinium, uranium, thorium again (ionium Io in the figure), radium, . . . down to lead (radium G) ($A = 206$, $Z = 82$). In the figure the mass number A is plotted against the atomic number Z as abscissae. An emission of an alpha particle, or *alpha-particle decay,* is indicated by a displacement down by four

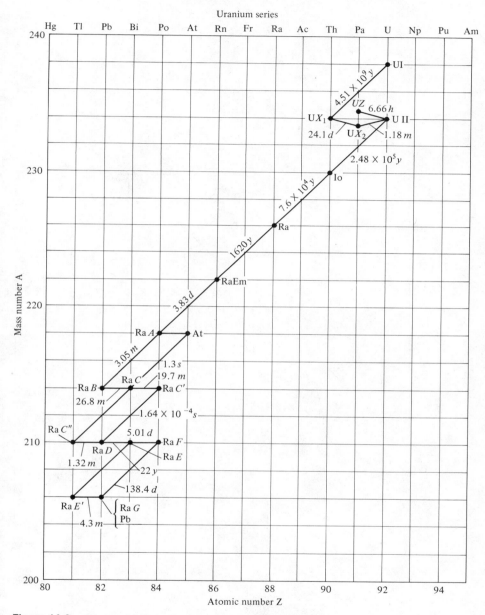

Figure 14-2   The naturally radioactive uranium series ($4n + 2$)

units and to the left by two units; an emission of a beta particle, or *beta decay*, is indicated by a displacement to the right by one unit.

The thorium series starts with a long-lived isotope of thorium ($A = 232$) with a half-life of 14.1 billion years and goes through a series of alpha- and

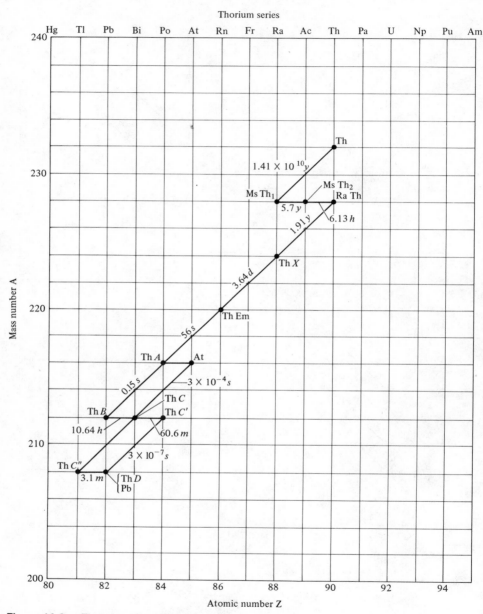

Figure 14-3   The naturally radioactive thorium series (4n).

beta-particle decays similar in many respects to those of the uranium series and terminates with the isotope of lead of mass number $A = 208$. The actinium series, (Figure 14-4) was at one time believed to be an independent series,

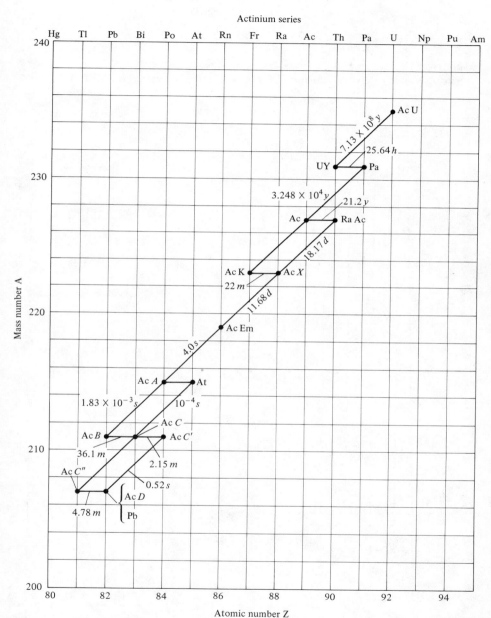

Figure 14-4 The naturally radioactive actinium series ($4n + 3$).

but its origin has been traced to the rarer isotope of uranium of mass number $A = 235$ and is sometimes called *actino-uranium* (Ac U). The end product of the actinium series is an odd-numbered isotope of lead, $A = 207$.

14-5 The Neptunium Series

Each of the three naturally occurring radioactive series discussed above
starts with a long-lived isotope. Physicists have often speculated about the

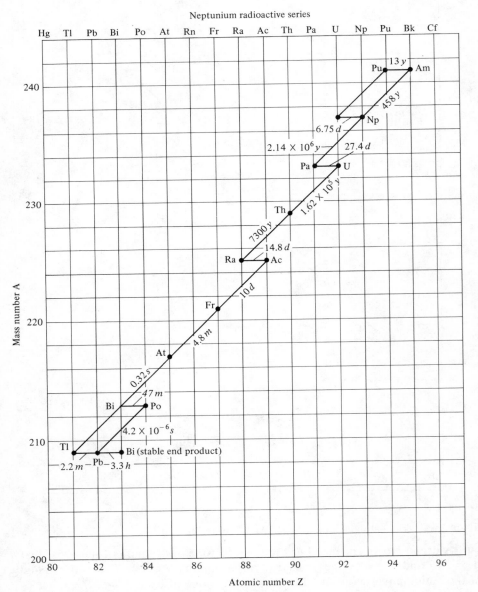

Figure 14-5 The neptunium radioactive series ($4n + 1$).

possibility of the occurrence of other radioactive series, the isotopes of which may have disappeared or may be in such extremely small concentrations that they are undetectable by common methods. One type of speculation revolved around the fact that the mass numbers of the heads of the three known series could be represented by the following set of numbers: $4n$ (for the thorium series), $4n + 2$ (for the uranium series), and $4n + 3$ (for the actinium series), where n is an integer. It was felt that there might have been a $4n + 1$ series and that perhaps traces of it still exist.

With the discovery of *transuranic elements*—that is, elements of atomic number greater than 92 and with the ability to produce many different isotopes of both old and new elements—it was possible to trace a fourth radioactive series, a $4n + 1$ series. This series is called the *neptunium series,* after the longest-lived isotope, neptunium, $Z = 93$, $A = 237$, of this series. This series is represented in Figure 14-5. It will be noted that the origin of this series can be traced back to americium and plutonium. The series ends, not in a stable isotope of lead but in the stable isotope of bismuth, $Z = 83$, $A = 209$. Recently very small amounts of neptunium and plutonium have been separated from pitchblende, a uranium-bearing ore. We shall discuss some of the methods for producing the isotopes of this series in Chapter 16.

14-6   Branching

A given species of nuclei will, in general, decay by one particular mode, say by the emission of a beta particle. But many cases have been found in which a smaller percentage of the nuclei will decay by a different mode such as alpha-particle emission. This phenomenon is known as *branching;* for example, 99.96 percent of the nuclei of the isotope of bismuth, $Z = 83$, $A = 214$ (also known as RaC), undergo beta decay to form the isotope of polonium $A = 214$, whereas 0.04 percent of the nuclei undergo alpha decay to form an isotope of thallium ($Z = 81$, $A = 210$) during the same time interval. The daughter isotopes then decay by the alternate modes to form the same end product, radium D, a radioactive isotope of lead, $Z = 82$, $A = 210$. This sequence of events shows up as a parallelogram in Figure 14-2. Many such examples of branching appear in the figures of the four radioactive series and many others that occur in the artificially produced radioactive isotopes will be described in later chapters.

14-7   Nuclear Isomers

In the early days of radioactive investigations, when chemical identification and separation of the minute quantities of newly formed isotopes was an extremely difficult task, it was frequently common to assign a separate isotope to a particular half-life; hence if two half-lives were observed in a given sample,

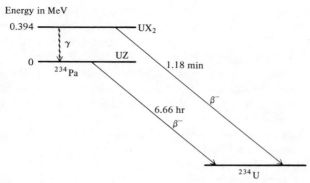

Figure 14-6   Nuclear isomers.

it was assumed that two different isotopes were present, each decaying with a particular half-life. One particularly interesting example is the isotope of protactinium, $Z = 91$, $A = 234$, which is formed in the beta decay of its parent, thorium $Z = 90$, $A = 234$ (UX$_1$). The isotope of protactinium was found to decay by the emission of beta particles of two distinct half-lives, one of 1.18 min, the other of 6.66 hr. It was assumed that these two half-lives were due to two different isotopes and these were given separate names, UX$_2$ and UZ, respectively. In 1921 Hahn showed that these two substances form a pair of *nuclear isomers;* that is, they are *different energy states* of the same nucleus. Feather and Bretscher later showed that these nuclear isomers are *genetically related;* that is, one type of nucleus is formed from the other. This is shown diagrammatically in Figure 14-6, in which the concept of energy levels is carried over into nuclear phenomena. The nucleus called UX$_2$ is an isomeric state of ^{234}Pa at an energy of 0.394 MeV above the ground state, called UZ. The nucleus may decay by beta-particle emission directly from the isomeric state of higher energy with a half-life of 1.18 min or it may first emit a gamma-ray photon of 0.394 MeV, going to the ground state of Pa, and then undergo beta decay to ^{234}U with a half-life of 6.66 hr.

Many examples of nuclear isomers have since been found mostly among the artificially produced nuclei. These will be discussed in greater detail in later sections. An isomeric state differs from an ordinary excited state of a nucleus in that it lasts for a measurable time. Hence the number of known nuclear isomers has increased as the experimental techniques for measuring short time intervals has improved.

14-8   Radioactive Isotopes of the Lighter Elements

The lowest atomic number among the isotopes of the radioactive series found in nature is 81. Radioactive isotopes of lower atomic numbers have also been found in nature; the number of such isotopes has been increasing

TABLE 14-1

Atomic Number	Element	Mass Number	Type of Radioactivity	Half-Life (years)
1	Hydrogen	3	Beta	12.26
6	Carbon	14	Beta	5730
19	Potassium	40	Beta; electron capture	1.3×10^9
23	Vanadium	50	Beta; electron capture	6×10^{15}
37	Rubidium	87	Beta	4.7×10^{10}
38	Strontium	90	Beta	27.7
49	Indium	115	Beta	5×10^{14}
57	Lanthanum	138	Beta; electron capture	1.1×10^{11}
58	Cerium	142	Alpha	5×10^{15}
60	Neodymium	144	Alpha	2.4×10^{15}
62	Samarium	147	Alpha	1.06×10^{11}
71	Lutetium	176	Beta; electron capture	2.2×10^{10}
75	Rhenium	187	Beta	4×10^{10}
78	Platinum	190	Alpha	6×10^{11}
82	Lead	204	Alpha	1.4×10^{17}

with improvements in detecting and measuring devices. Table 14-1 is a list of some of the more prominent ones. Most of these isotopes have very long half-lives and very low abundances so that the amount of radioactivity from any one sample of a given element is extremely small. A few have comparatively short half-lives; this means that their supply must be continually replenished by processes that are now occurring in nature. For example, it has been definitely established that ^{14}C is being produced by the bombardment of ^{14}N in the atmosphere by cosmic-ray neutrons (see Section 15-13).

Many new radioactive isotopes have been added to the natural scene as a result of the extensive testing of nuclear weapons. Most of them have very short half-lives but others, like strontium, ^{90}Sr, have long half-lives. We shall consider these isotopes in our discussion of *nuclear fusion* and *nuclear fission*.

14-9 Alpha-Particle Disintegration Energy

We have already shown that alpha particles emitted by radioactive substances have velocities of the order of 10^9 cm/sec and kinetic energies of about 5 to 10 MeV. In discussing a nuclear transformation, several conservation laws must be taken into consideration, such as the conservation of energy, the conservation of momentum, and the conservation of charge. An alpha-particle disintegration can be represented by means of a *nuclear reaction equation* of the type

$$_Z^A\text{El} \rightarrow {}_{Z-2}^{A-4}\text{El} + {}_2^4\text{He} + Q_\alpha \tag{14-5}$$

where El stands for the chemical symbol representing the particular element under discussion and $^4_2$He represents the alpha particle emitted. This equation shows that charge is conserved; the element of nuclear charge Ze is transformed into a new element of nuclear charge $(Z - 2)e$ with the emission of an alpha particle of charge $2e$. The quantity Q_α is called the *disintegration energy* and represents the total energy released in this process. This energy consists of the kinetic energy of the alpha particle and the kinetic energy of the product nucleus and comes from the difference in mass between the parent nucleus and the product nuclei. The disintegration energy Q_α can be readily evaluated in terms of the kinetic energy $\mathcal{E}_\alpha$ of the alpha particle with the aid of the principles of conservation of energy and momentum.

Suppose that the mass of the parent atom is M_1; let us assume it to be at rest. When it emits an alpha particle of mass M and velocity v, the residual atom of mass M_2 will recoil with a velocity v_2 such that

$$M_2 v_2 = Mv$$

Now

$$Q_\alpha = \tfrac{1}{2} M_2 v_2{}^2 + \tfrac{1}{2} Mv^2$$

Eliminating v_2 from these two equations, we obtain

$$Q_\alpha = \frac{1}{2}\frac{M}{M_2} Mv^2 + \tfrac{1}{2}Mv^2$$

and, calling the kinetic energy of the alpha particle $\mathcal{E}_\alpha = \tfrac{1}{2}Mv^2$, we can write

$$Q_\alpha = \mathcal{E}_\alpha \left(\frac{M}{M_2} + 1\right) \tag{14-6}$$

To a very close approximation we can replace the ratio of the masses with the ratio of the mass numbers; thus

$$\frac{M}{M_2} = \frac{4}{A - 4}$$

where A is the mass number of the parent atom. Hence the disintegration energy can be written as

$$Q_\alpha = \frac{A}{A - 4} \mathcal{E}_\alpha \tag{14-7}$$

14-10 Range of Alpha Particles

Several methods have been useful for studying the alpha particles that are emitted by radioactive nuclei. Their velocities may be measured by the magnetic spectrograph method described in Section 3-11. Another method for investigating the alpha particles is the determination of the range of the particle

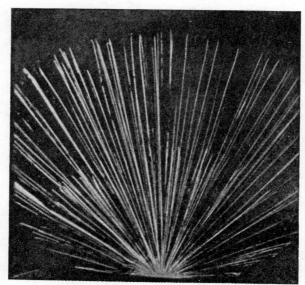

Figure 14-7   Tracks of alpha particles from thorium ($C + C'$) in a Wilson cloud chamber showing two distinct ranges. (From Rutherford, Chadwick, and Ellis, *Radiations from Radioactive Substances*. By permission of Cambridge University Press, publishers.)

in a gas such as hydrogen, nitrogen, or air by using a Wilson cloud chamber (see Sections 13-9 and 13-11).

Typical alpha-ray tracks are shown in Figure 14-7. These tracks are, in general, straight lines almost up to the end of the range. Occasionally a track is bent sharply or it branches off into two tracks. These are usually ascribed to collisions with nuclei; they will be discussed in detail later (see Chapter 15).

Another method for determining the range of alpha particles in a gas is to measure the ionization produced in a gas at different distances from the source of the alpha particles. A typical arrangement for this type of measurement is shown in Figure 14-8. The source of alpha particles A is placed in a recess in a

Figure 14-8   Schematic diagram of apparatus for measuring the range of alpha particles.

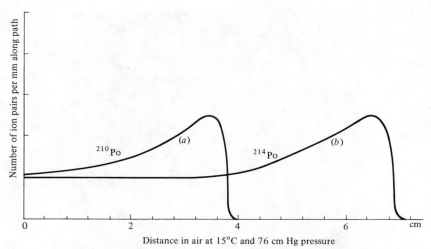

Figure 14-9 Graphs showing the specific ionization along the path of alpha particles from
(a) ^{210}Po, (b) ^{214}Po.

block of lead, thus providing a fairly well-collimated beam of alpha rays. The
ionization chamber consists of a wire grid G and a plate P connected to an
electrometer for measuring the ionization produced in the narrow region between P and G. The distance between the alpha-particle source and the ionization chamber is usually varied by moving the source. Typical curves showing
the ionization produced at different distances from the source are shown in
Figure 14-9. It will be noticed that for the greater part of the range the ionization current in the first part of the range is almost constant, then increases, and

TABLE 14-2 RANGE AND ENERGY OF ALPHA PARTICLES

| | *Isotope* | | | *Mean Range in Centimeters in* | *Energy* |
Z	*Symbol*	A	*Older Name*	*Air at 15° C*	*(MeV)*
84	Po	210	Polonium	3.842	5.298
86	Rn	222	Radium emanation	4.051	5.486
84	Po	218	Radium A	4.657	5.998
86	Rn	220	Thoron	5.004	6.2818
84	Po	216	Thorium A	5.638	6.774
84	Po	214	Radium C′	6.907	7.680
84	Po	214	Radium C′	7.792	8.277
84	Po	214	Radium C′	9.04	9.066
84	Po	214	Radium C′	11.51	10.505
84	Po	212	Thorium C′	8.570	8.776
84	Po	212	Thorium C′	9.724	9.488
84	Po	212	Thorium C′	11.580	10.538

reaches its maximum value just before the end of the range. The peak near the end of the range is due to an increase in the efficiency of ionization by slow alpha particles.

It has been found that in most cases the alpha particles from a given isotope have a very definite range. This range R is usually expressed in centimeters of air at $15°C$ and at a pressure of 76 cm of mercury. The ranges of the alpha particles from some of the isotopes are given in Table 14-2 together with their energies as determined by Holloway and Livingston. The alpha particles from some of the isotopes, such as ^{212}Po, fall into several groups of ranges corresponding to their velocity spectrum.

The velocity spectrum of the alpha particles from any isotope is always a sharp line spectrum; hence we may think of the emission of an alpha particle as taking place between two definite nuclear energy states, the initial state being that of the parent nucleus and the final state, that of the daughter nucleus. If the alpha-particle spectrum is complex, as in the case of polonium, $Z = 84$, $A = 214$ (Ra C′), there are undoubtedly several alpha-particle energy levels in the parent nucleus, daughter nucleus, or both nuclei. One of the problems of nuclear physics is to determine the energy levels of nuclei and the types of

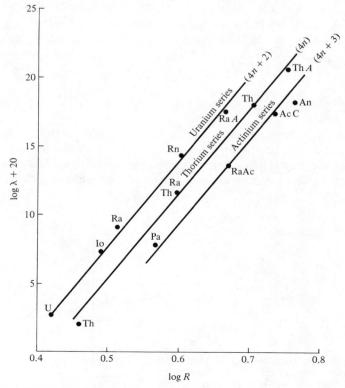

Figure 14-10   The Geiger-Nuttall law.

transition that occur between them and to compare these determinations with theoretical predictions. The name *nuclear spectroscopy* is used to designate this type of study.

The relationship between the range of an alpha particle and its velocity cannot be expressed by any one simple formula, but those decays that happen to be of medium range are found to follow Geiger's empirical formula

$$R = av^3 \tag{14-8}$$

where a is a constant numerically equal to 9.6×10^{-28} when R is expressed in centimeters and v in centimeters per second.

An important empirical relationship exists between the range of an alpha particle and the average lifetime of the emitter, known as the *Geiger-Nuttall law,* which is usually written in the form

$$\log R = A \log \lambda + B \tag{14-9}$$

where A is a constant with almost the same value for each of the three radioactive series and B is a constant with a different value for each series. This relationship is plotted for each series in Figure 14-10; the range R is expressed in centimeters and the disintegration constant λ is expressed in sec^{-1}. This equation has been used to estimate the half-lives of some of the products of disintegration which could not be easily determined by direct measurements. Although the Geiger-Nuttall law is only an approximation, it can be used as a check on the validity of any theory of alpha-particle decay (see Section 14-11).

14-11 Alpha Decay

The radioactive disintegration of a nucleus by alpha-particle emission was first successfully explained on the basis of wave mechanics by Gamow, Condon, and Gurney (1928). Some assumptions have to be made concerning the nature of the forces acting on the alpha particle. Although we have postu-

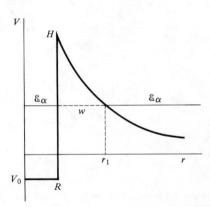

Figure 14-11 Potential energy $V(r)$ of a nucleus and an alpha particle as a function of the distance r from the center of the nucleus. The width of the potential barrier is w.

lated that only nucleons exist in the nucleus, we may imagine that two protons and two neutrons come together to form an alpha particle. There is constant interchange of energy between the alpha particle and the remaining nucleons. Although we do not know the exact nature of the forces acting on the alpha particle, we do know that there are repulsive forces due to the charges and some strong attractive forces of a specifically nuclear character. The effect of these forces may be considered as giving rise to a negative potential energy V_0 out to a distance R from the center of the nucleus. This distance R may be considered the radius of the nucleus. The specifically nuclear forces are of such short range that outside the nucleus only the Coulomb force acts on the alpha particle; the potential energy V of an alpha particle outside the nucleus varies inversely as its distance r from the center of the nucleus, where $r > R$, as shown in Figure 14-11. The Coulomb potential and the constant potential energy V_0 are joined, for the sake of simplicity, by a straight line at $r = R$.

The value of the potential energy at $r = R$ is designated by the letter H and is called the *height of the potential barrier* for an alpha particle. In the case of ^{238}U, for example, the height of the potential barrier for an alpha particle is about 30 MeV and is of this order of magnitude for most of the heavy elements. This represents the minimum energy that an alpha particle must have, according to classical ideas, in order to escape from the nucleus. A glance at Table 14-2 shows that the alpha particles emitted by radioactive nuclei have energies of the order of 5 or 6 MeV and, in rare cases, about 11 MeV. These values are all much lower than the heights of the potential barriers of the respective nuclei. Furthermore, the alpha-particle energy spectrum usually consists of only one or a few very sharp lines. Hence we may imagine that the alpha particle receives enough energy in its interaction with other nuclei to raise it to an energy level $\mathcal{E}_\alpha$ equal to the kinetic energy that it has when at a large distance from the nucleus. Since it is less than the height of the potential barrier, the problem still remains of getting through a potential wall of thickness w. This problem was considered in detail in Section 7-9 in which we showed that the wave properties of a particle incident on a rectangular potential barrier made it possible to penetrate, even though the energy might be too low for transmission according to classical mechanics. For a single encounter between a particle and a barrier of arbitrary shape the transparency, or transmission probability, was shown to be approximately

$$\exp\left(-\frac{2}{\hbar} \int_R^{r_1} \sqrt{2M\left(V(r) - \mathcal{E}_\alpha\right)}\, dr\right) \qquad (14\text{-}10)$$

where $V(r)$ is the potential energy as a function of r and M is the mass of the alpha particle.

The problem of alpha-particle decay is to determine the lifetime τ of the alpha particle in the nucleus; this is the reciprocal of the probability per unit time, λ, for the escape of the alpha particle from the nucleus. Inside the nucleus the particle oscillates and strikes the potential wall periodically. If we assume that it moves with an average speed $\bar{v}$, then the frequency with which it strikes the wall is $\bar{v}/2R$. The product of this frequency and the transparency will be practically equal to the probability per unit time of escape of the alpha particle, thus

$$\lambda = \frac{\bar{v}}{2R} \exp\left(-\frac{2}{\hbar} \int_R^{r_1} [2M\,(V(r) - \mathcal{E}_\alpha)]^{1/2}\,dr\right) \qquad (14\text{-}11)$$

The exponent, which is usually designated by G, can be evaluated for the given potential barrier. The calculations of λ give qualitative agreement with experiment. The equation has been used to compute nuclear radii by substituting the measured values of λ and $\mathcal{E}_\alpha$. This is one of the methods for obtaining R and the results so obtained are in good agreement with those values of R obtained by other methods (see Section 3-9). We shall adopt the value

$$R = 1.2 \times 10^{-13}\,A^{1/3}\,\text{cm}$$

for the radius of a nucleus of mass number A.

It is also worth noting that Equation (14-11) written as

$$\lambda = \text{const exp}\,(-G) \qquad (14\text{-}12)$$

is one form of the Geiger-Nuttall law, since G is a function of the energy and hence of the range of the alpha particle. A fair approximation for G yields

$$G = \frac{4\pi k e^2}{\hbar}\frac{Z}{v} \qquad (14\text{-}13)$$

where v is the velocity of the alpha particle outside the nucleus. The constant factor of Equation (14-12) has a different value for each of the radioactive series.

14-12 Beta-Ray Spectra

The most commonly used method for determining the energies of the beta particles emitted by radioactive isotopes is the measurement of the radii of curvature of their paths in a magnetic field of known flux density B. Various experimental arrangements have been used in making these measurements. One arrangement is almost identical to that used by Robinson, as shown in Figure 10-6, for the determination of the velocities of the electrons ejected by x-rays from an element placed at C. In the magnetic spectrum analysis of beta rays C is replaced either by a fine wire on which the radioactive substance has been placed or by a small thin-walled glass tube containing the radioactive substance. The beta rays are recorded on a photographic plate and their velocity distribution determined.

Another method for detecting the beta rays is shown in Figure 14-12, in which the photographic plate is replaced by a Geiger counter G and the beta

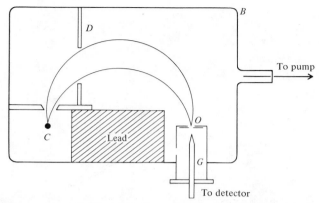

Figure 14-12   Robinson's magnetic spectrograph using a Geiger counter for detecting beta particles.

rays from the source at C are bent around by the magnetic field and focused on the aperture O. In this type of experiment the number of beta particles entering the aperture O is counted at a given value of the magnetic field of intensity B.

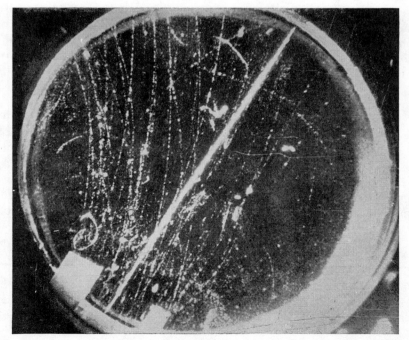

Figure 14-13   Cloud-chamber photograph of beta-ray tracks in a magnetic field of 1000 gauss. The beta rays come from the disintegration of ^{12}B; their energies range from about 6 to 12 MeV. The heavy track across the diameter of the chamber is that of a proton of about 9 MeV energy. (Photograph taken by H. R. Crane.)

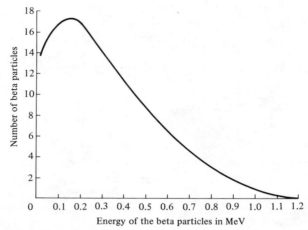

Figure 14-14 Distribution of energy among the beta particles of radium E.

The intensity of the magnetic field is then changed to a new value and the number of beta particles entering the aperture *O* is again counted. In this manner the velocity distribution of the beta particles is determined. The radius of curvature of the path of a beta particle in a magnetic field can also be measured by photographing the track of the beta particle in a Wilson cloud chamber, as shown in Figure 14-13.

The results of these measurements show that there are apparently two distinct types of beta-ray spectrum, one a *sharp line spectrum* and the other a *continuous spectrum*. It has definitely been shown, however, that the sharp line spectra are due to electrons that have been ejected from the *K, L, M,* and *N* shells of the atom by the process of internal conversion (see Section 14-15). The continuous beta-ray spectrum is that produced by the electrons that have been ejected from the nuclei of radioactive atoms. The curve in Figure 14-14 shows the continuous beta-ray spectrum of bismuth, $Z = 83$, $A = 210$ (radium E). The number of particles with a given energy is plotted as ordinate and the energy of these particles, expressed in MeV, as abscissa. It will be noted that the curve has a definite upper limit for the energy of the disintegration electrons

TABLE 14-3 END-POINT ENERGIES OF SOME BETA-RAY SPECTRA

Z	Element	A	Radioactive Isotope	End-Point Energy (MeV)	Half-Life
82	Pb	214	RaB	0.72	26.8 min
83	Bi	210	RaE	1.17	5.0 days
87	Fr	223	AcK	1.20	22 min
89	Ac	228	Ms Th$_2$	1.55	6.13 hr
90	Th	231	UY	0.21	25.65 hr

and also passes through a maximum toward the low-energy part of the spectrum. It will be shown that the end-point energy represents the energy released in this radioactive disintegration. Table 14-3 gives some of the end-point energies of the continuous spectra of some of the beta-ray emitters.

The beta-ray spectrum of an element differs remarkably from most of the other spectra characteristic of the same element in that these characteristic spectra, optical, x-ray, alpha ray, and gamma ray, are line spectra, whereas the beta-ray spectrum is a continuous one. It was found possible to interpret the optical and x-ray line spectra in terms of the changes in atomic energy states due to changes in the extranuclear electronic configurations. It seems reasonable to try to extend this interpretation to the line spectra of the particles emitted by radioactive nuclei.

14-13 Beta Decay

The energy available for a beta (β^-) disintegration is provided by the difference in mass between the initial nucleus of mass number A and atomic number Z and the sum of the masses of the final particles—that is, the nucleus of atomic number $Z + 1$ and mass number A and the beta particle. The disintegration energy is thus

$$Q_\beta = {}^A_Z M_n - [{}_{(Z+1)}^A M_n + m_e]$$

Both the parent and daughter nuclei are assumed to be in their lowest or ground states. The subscript n refers to the mass of the nucleus only.

Since the tables generally give the masses of the neutral atoms, we can convert the nuclear masses to atomic masses by adding the masses of Z electrons to each term; thus

$$Q_\beta = [{}^A_Z M_n + Z m_e] - [{}_{Z+1}^A M_n + Z m_e + m_e]$$

hence

$$Q_\beta = {}^A_Z M - {}_{Z+1}^A M \qquad (14\text{-}14)$$

where ${}^A_Z M$ is the atomic mass of the isotope of mass number A and atomic number Z and ${}_{Z+1}^A M$ is the atomic mass of the isotope of mass number A and atomic number $Z + 1$. Hence a definite amount of energy Q_β is released in this beta (β^-) disintegration. It will be noted that the beta decay occurs between a pair of *isobars*—that is, two atoms with the same mass number A but different atomic numbers. The atomic numbers of the isobars involved in beta decay always differ by unity.

If there are only two particles involved in the radioactive disintegration of a nucleus—the emitted particle and the recoil nucleus—these two particles have equal and opposite momenta. Their kinetic energies are in the inverse ratio of their masses. If the energy available for a given nuclear disintegration is a constant, all the particles emitted in such a disintegration should have the same energy—that is, they should yield a sharp line spectrum. A continuous spectrum under such conditions must necessarily involve more than two particles. W. Pauli (1931) suggested that the end-point energy of the beta-ray spectrum is equal to the disintegration energy. To account for the continuous distribution of energy Pauli introduced the hypothesis that a third particle, a neutral particle, is also emitted in beta decay. This particle has the generic name of *neutrino*. The total energy is shared by the three particles—the recoil nucleus, the electron, and the neutrino. Because of its comparatively great mass, the recoil energy of the nucleus is very small and nearly all the kinetic energy is shared between the beta particle and the neutrino. Thus, when the beta particle has a very large amount of kinetic energy, that of the neutrino is very small, whereas the neutrino will carry away most of the kinetic energy when the kinetic energy of the electron is small.

In addition to the laws of conservation of charge and energy, we must also apply the laws of conservation of linear and angular momentum to every nuclear process. Taking our reference system as the parent nucleus at rest, the vector sum of the linear momenta of the recoil nucleus, the beta particle, and the neutrino must be zero. Since it will be very difficult to observe the track of a neutrino, measurements of the momenta of the recoil nucleus and the associated beta particle can be used to infer its existence. Such experiments have been performed and will be described later (see Section 17-10). To conserve angular momentum in beta decay, we note that the parent and daughter nuclei are isobars; that is, they have equal numbers of nucleons. Hence the total change in nuclear angular momentum will be either zero or an integral multiple of $\hbar$. The beta particle has an intrinsic spin angular momentum of $\frac{1}{2}\hbar$; hence the neutrino must also have a spin angular momentum of $\frac{1}{2}\hbar$. The vector sum of the angular momenta of the neutrino and the beta particle will be either zero or one in units of $\hbar$.

The present accepted theory, which is supported by experimental evidence, (see Chapter 18) shows that there are two types of neutrino or two components of the neutrino. It has been found that the axis of spin of the neutrino is parallel to its direction of motion; one type spins according to the left-hand rule with respect to its direction of motion as its axis, the other component spins according to the right-hand rule. The first type is usually called the neutrino, represented by the symbol ν, the second type is called the *antineutrino* and is represented by the symbol $\bar{\nu}$ (nu bar). These are shown in Figure 14-15, in which the lighter line represents the direction of motion of the two types of neutrino and the heavier line represents the direction of spin. The spin vector of the neutrino points opposite to the direction of its motion, and the spin vector

Figure 14-15   The two components of the neutrino.

of the antineutrino is directed parallel to its direction of motion. Another way of saying this is that the *helicity* of the neutrino is negative and that of the antineutrino is positive, or one has right-handed helicity, the other, left-handed helicity.

The remaining problem is to determine the origin of the electron and antineutrino. The present view, stated in an elementary way, is that the neutron breaks up into a proton which remains in the nucleus and an electron and an antineutrino which are ejected from the nucleus. This is analogous to the emission of a light quantum when an atom goes from one energy state to another. There are no photons in the atom, but when an atom goes from a state characterized by one set of quantum numbers to another state, a photon is emitted. We may imagine that the neutron represents one quantum state of a nucleon and the proton represents another quantum state of the same nuclear particle. When this particle goes from the neutron quantum state to the proton quantum state, it does so with the emission of a pair of particles, an electron and an antineutrino. This process can be represented by the equation

$$n \rightarrow p + \beta^- + \bar{\nu} \qquad (14\text{-}15)$$

where n represents the neutron, p, the proton, and $\bar{\nu}$, the antineutrino.

The beta-ray disintegration energy Q_β, given by Equation (14-14), is equal to the end-point energy of the beta-ray spectrum only when the transition is from the *ground state*, or state of lowest energy, of the parent nucleus to the ground state of the product nucleus. The rest mass of the antineutrino is zero.

14-14 Gamma Decay

Early in the history of research on radioactivity it was discovered that neutral radiation was emitted, known as gamma radiation (see Section 3-2). In many cases gamma rays accompany the emission of alpha or beta particles. It has been shown that gamma rays are of the same nature as x-rays; wavelengths of certain low-energy gamma rays have been measured by means of a single-crystal spectrometer; wavelengths of gamma rays at higher energy have been measured with the curved-crystal focusing spectrometer (Section 5-9). Like x-rays, gamma rays are photons with particlelike properties in addition to the wave property of diffraction. Since they are photons, gamma rays can be studied by using electrons that are liberated from atoms in the photoelectric effect or in the Compton effect. At energies above 1 MeV an additional process, the production of electron-positron pairs, becomes important; the phenomenon of pair production will be discussed in Chapter 18.

When gamma radiation accompanies alpha decay, the gamma rays typically possess definite frequencies, immediately suggesting a structure of the nucleus with discrete energy states, analogous to the case of atomic spectra. We illustrate this interpretation of the energy levels by considering the decay of ^{226}Ra into ^{222}Rn by means of alpha emission. Experimentally, the decay spectrum shows alpha particles with energies of 4.78 and 4.59 MeV. The gamma-ray spectrum accompanying this process consists of a single line of wavelength equal to 0.0662 Å, corresponding to an energy of 0.187 MeV. The fact that the energy of the photon is equal to the difference in energies of the alpha particles suggests that the decay sometimes proceeds by a two-step process as shown in Figure 14-16. In effect, there are two competing decay paths for the ^{226}Ra nucleus. It can decay into the ground state or into the first excited state of ^{222}Rn. In the latter case the first excited state decays into the ground state by emitting a photon.

Gamma radiation can also accompany beta decay. The spectrum from this combination can be somewhat more complicated than in the case of alpha decay. There are two types of gamma-ray spectrum: the sharp-line spectrum and the continuous spectrum. This situation is analogous in many respects to x-ray spectra. The sharp-line spectrum is caused by the same sort of energy-level structure that occurs with alpha decay. The continuous spectrum (or bremsstrahlung) may have two different origins. One type may be due to the radiation emitted when a beta particle, in its passage through a substance, is accelerated whenever it passes close to any of the nuclei of the substance. This radiation, called *outer bremsstrahlung*, is analogous to the continuous x-rays in an x-ray tube. The acceleration may occur in the passage of the electrons through the beta-emitting substance itself or it may be produced in some nearby material that is different from the emitter. Theory predicts that the intensity of the outer bremsstrahlung should be proportional to the square of the atomic number of the material through which the electron passes. This type of radiation was first discovered by J. A. Gray (1911), who bombarded various substances with beta rays from RaE (^{210}Bi) and measured the intensity of the electromagnetic radiation thus produced.

The second type of bremsstrahlung has its origin in the readjustment of the charges in the nucleus as a result of the emission of a beta particle. This type is

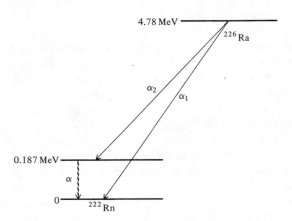

Figure 14-16   Nuclear energy-level diagram for the decay of radium.

called *inner* or *internal bremsstrahlung*. C. S. Wu (1941) measured the intensity of the inner bremsstrahlung emitted by a radioactive isotope of phosphorus ^{32}P. This isotope of phosphorus emits beta rays with an end-point energy of 1.71 MeV but does not emit any characteristic gamma rays. The intensity of the radiation was measured by means of an ionization chamber and the results found to be in agreement with theoretical predictions of Knipp and Uhlenbeck and F. Bloch. Wu also compared the relative intensities of the inner brems-strahlung with the outer bremsstrahlung by completely stopping the beta rays in an aluminum absorber. The intensity of the inner bremsstrahlung was found to be about a quarter of that of the outer bremsstrahlung.

14-15 Selection Rules for Gamma Decay

We have seen that there is evidence of discrete energy levels in nuclei, just as in atomic electrons. We expect that the ideas of quantum mechanics ought to apply to nuclei as well as to atomic or molecular systems. If so, then the process of gamma decay should be governed by the same selection rules used for atoms in Chapters 8 and 9. These rules were obtained by calculating the matrix element of a dipole operator, leading to the results

$$\Delta J = 0, \pm 1 \tag{14-16}$$

for allowed transitions and

$$J = 0 \nrightarrow J = 0 \tag{14-17}$$

where J is the total angular momentum of the atom.

For gamma decay we shall be using I to denote the total angular momentum of the nucleus. It produces dipole selection rules like those involving J but with a significant difference. In atoms a transition that is forbidden by the dipole selection rules is unimportant under usual experimental conditions because an excited atom can be de-excited by ordinary thermal collisions with other atoms. An excited nucleus which has the wrong angular momentum to decay by a dipole transition cannot usually undergo collisional de-excitation because it is shielded from nuclear encounter by its electrons. Therefore the excited nucleus can remain in its excited state long enough for gamma decay to occur by some means other than by dipole radiation. Table 14-4 summarizes the properties of multipole radiation of an order higher than dipole. The parity of the wave function is either +1 or −1, depending on the behavior of the wave function under the transformation $(x, y, z) \rightarrow (-x, -y, -z)$. This concept will be discussed further in Chapters 17 and 18. It will be noticed that the crude estimates of relative intensity always favor the lower order transitions. This fact is true of atoms as well as nuclei, but the factor of 10^3 between electric dipole and the M1 and E2 transitions is sufficient to prevent these higher transitions from competing successfully against collisional de-excitation.

Higher order gamma transitions are not the only way for an excited nucleus

TABLE 14-4 MULTIPOLE RADIATION

Type	Symbol	max $\|\Delta I\|$	Change in Parity	Estimate of Relative Intensity
Electric dipole	E1	1	yes	1
Magnetic dipole	M1	1	no	10^{-3}
Electric quadrupole	E2	2	no	10^{-3}
Magnetic quadrupole	M2	2	yes	10^{-6}
Electric octupole	E3	3	yes	10^{-6}
Magnetic octupole	M3	3	no	10^{-9}
Electric hexadecapole	E4	4	no	10^{-9}
Magnetic hexadecapole	M4	4	yes	10^{-12}

to get to the ground state with a large change in I. The process of *internal conversion* is an important mechanism for states in which the gamma decay is forbidden by dipole rules; in this phenomenon an atomic electron is ejected from its orbit as the nucleus changes from an excited state into a lower state. It was formerly believed that internal conversion is a two-step process in which a gamma photon is emitted by the nucleus, followed by the absorption of the photon by the orbital electron, which is then emitted as in the photoelectric effect. That this two-step model is incorrect can be inferred from the experimental evidence that internal conversion occurs for transitions between states in which $I = 0$ for both. A zero-zero transition is forbidden for all multipole orders, and so the first step of the older theory cannot occur. The correct explanation for internal conversion is that an orbital electron has a wave function that penetrates the nucleus to some extent, permitting a direct interaction of the orbital electron with the nucleus. Internal conversion electrons can be ejected from K, L, M, or N shells. The energies of these electrons appear as a sharp-line beta spectrum.

14-16 Isomeric States

In Section 14-7 we introduced the concept of nuclear isomers. We are now in a better position to understand why certain excited states have long lifetimes; their gamma decay into the ground state can proceed only by a high-multipole transition. As an example we shall discuss the ^{80}Br nucleus. Bromine is of historical interest, since it was the first case in which nuclear isomers were produced artificially and their existence made certain by chemical separation of the nuclear isomers.

The experimental procedure consisted of bombardment of a bromine target with slow neutrons. Bromine, as found in nature, is an isotopic mixture of ^{79}Br and ^{81}Br; the reactions are

$$n + {}^{79}_{35}Br \rightarrow ({}^{80}_{35}Br^*) \rightarrow {}^{80}_{35}Br + \gamma \qquad {}^{80}_{35}Br \rightarrow {}^{80}_{36}Kr + \beta^- + \bar{\nu}$$

and

$$n + {}^{81}_{35}Br \rightarrow ({}^{82}_{35}Br^*) \rightarrow {}^{82}_{35}Br + \gamma \qquad {}^{82}_{35}Br \rightarrow {}^{82}_{36}Kr + \beta^- + \bar{\nu} \qquad (14\text{-}18)$$

The asterisk refers to an excited nuclear state. These reactions led to beta activities with three different half-lives, 18 min, 4.5 hr, and 36 hr. Chemical tests showed that all the radioactivity came from isotopes of bromine.

Bromine can also be bombarded with gamma-ray photons to produce the following reactions:

$$\gamma + {}^{79}Br \rightarrow ({}^{79}Br^*) \rightarrow {}^{78}Br + n \qquad {}^{78}Br \rightarrow {}^{78}Se + \beta^+ + \nu$$

and

$$\gamma + {}^{81}Br \rightarrow ({}^{81}Br^*) \rightarrow {}^{80}Br + n \qquad {}^{80}Br \rightarrow {}^{80}Kr + \beta^- + \bar{\nu} \qquad (14\text{-}19)$$

Again three activities were observed, 6.3 m, 18 m, and 4.5 h. Two of these half-lives are common to both sets of reactions and must therefore be assigned to the isotope that is common to both—namely ${}^{80}Br$.

The existence of two different lifetimes for the same nucleus can be explained by assuming the existence of an isomeric or *metastable* state, denoted by the letter m after the atomic mass: ${}^{80m}Br$. More recent work on ${}^{80}Br$ has shown that two gamma-ray photons are emitted in cascade with energies of 0.048 and

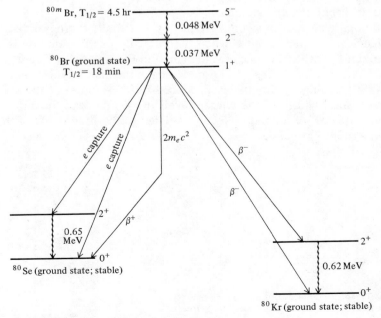

Figure 14-17 Energy-level diagram for $A = 80$.

0.037 MeV. The angular momentum and the parity have been determined for the states involved, and the energy-level diagram is shown in Figure 14-17. The topmost level in the diagram is the metastable state ^{80m}Br, with angular momentum equal to $5\hbar$ and with odd parity. If the Br nucleus were to make a transition to its ground state, emitting a photon, the transition would be M4 (see Table 14-4). The transition to the first excited state is M3 and therefore strongly favored over M4 but still sufficiently forbidden so that the lifetime is long. The second part of the cascade from ^{80m}Br to ^{80}Br is an E1 transition with a very short lifetime. From the ground state of ^{80}Br there are several possible decays, as shown in Figure 14-17.

14-17   Resonance Absorption of Gamma Rays

The phenomenon of resonance is so well established in all branches of physics, and particularly in atomic and nuclear physics, that resonance experiments with gamma rays should be part of the normal routine procedure. The usual type of experiment is one in which gamma rays of frequency ν emitted by a given type of nucleus in going from an excited state of energy $\mathscr{E}_E$ to the ground state of energy $\mathscr{E}_G$ are incident on identical nuclei most of which are in the ground state. The incident photons should be readily absorbed and thus raise the absorbing nuclei to the higher energy state; the latter, in returning to the ground state, will reradiate gamma-ray photons of the same frequency in all directions. However, such occurrences are rare—that is, there is little resonance absorption of the gamma rays.

Although we speak of a single frequency ν, actually there is a distribution of frequencies centered about ν when the intensity of the radiation is measured as a function of the frequency, as indicated in Figure 14-18. There are two important reasons for this spectral broadening of the line: (a) the *Doppler* effect— that is, the change in frequency produced by the motion of the source—and (b) the natural widths of the energy levels involved in the transition.

The width $\Delta\mathscr{E}$ of an energy level is related to the lifetime Δt of the state by the uncertainty principle

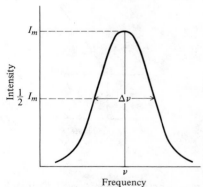

Figure 14-18   The width of a line is measured at the position of one-half its maximum intensity.

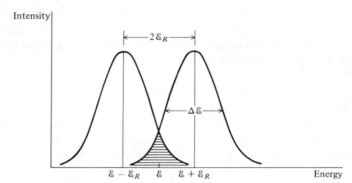

Figure 14-19   Centers of emission and absorption lines shifted by $\pm\mathcal{E}_R$; small region of overlap.

$$\Delta\mathcal{E} \cdot \Delta t \gtrsim h \qquad (14\text{-}20)$$

The excited state contributes a frequency width

$$\Delta\nu = \frac{\Delta\mathcal{E}}{h} \qquad (14\text{-}21)$$

to the gamma-ray resonance line; the contribution from the ground state is almost zero, since its lifetime is very long and its width is almost zero. Measured lifetimes of excited states involved in resonance radiation range from about 10^{-11} sec upward; the contribution to the width of the line is thus about 10^{-4} eV or smaller.

The gamma-ray photon is emitted with a momentum p and the nucleus will recoil with an equal and opposite momentum. If M is the mass of the nucleus, the energy of recoil $\mathcal{E}_R$ will be given by

$$\mathcal{E}_R = \frac{p^2}{2M} \qquad (14\text{-}22)$$

The gamma-ray photon will thus be emitted with an energy $\mathcal{E} - \mathcal{E}_R$ so that the center of the spectral line will be shifted from its expected position at $\mathcal{E}$ to the new position at $\mathcal{E} - \mathcal{E}_R$. Similarly, when the gamma-ray photon is to be absorbed, it will impart an equal momentum to the absorbing nucleus so that the energy of this system will have to be $\mathcal{E} + \mathcal{E}_R$. The value of $\mathcal{E}_R$ may be of the order of about 0.05 eV; in order that gamma-ray photons may be readily absorbed, the separation, $2\mathcal{E}_R$, between the centers of the emission and absorption lines should be less than $\Delta\mathcal{E}$, the width of each line at half its maximum intensity (see Fig. 14-19). The shaded region in the figure is a measure of the probability of resonance absorption of the gamma-ray photons of energy $\mathcal{E}$. To increase the amount of resonance absorption it is necessary to increase the amount of overlap between the two curves. This can be done and has been done by several ingenious methods.

One method for increasing the amount of resonance absorption is to impress a Doppler broadening on the source, on the absorber, or both. It will be re-called that if a source of radiation of frequency ν is moving with a nonrelativistic velocity v in the direction of emission of the radiation, the frequency of the radiation will be increased to a value ν_1 given by

$$\nu_1 = \nu \frac{c}{c - v} \tag{14-23}$$

where c is the velocity of the radiation. A similar equation holds for the frequency received by an absorber moving with velocity v toward a stationary source emitting radiation of frequency ν. The above equation can be rewritten as

$$\nu_1 = \nu \frac{1}{1 - v/c}$$

and to a first approximation can be expanded to

$$\nu_1 = \nu \left(1 + \frac{v}{c}\right) \tag{14-24}$$

so that the Doppler shift $\Delta\nu_D$ becomes

$$\Delta\nu_D = \nu_1 - \nu = \nu\frac{v}{c} \tag{14-25}$$

The energy change produced by the Doppler effect is thus

$$\Delta\mathcal{E}_D = h\Delta\nu_D = \frac{v}{c}\mathcal{E} \tag{14-26}$$

The Doppler effect can be produced in several ways. One is to heat the source, the absorber, or both to higher temperatures so that the emitting and absorbing nuclei will have higher velocities. Temperatures should be of the order of $T = \mathcal{E}_R/k$ where k is the Boltzmann constant. Thus if $\mathcal{E}_R = 0.1$ eV, $T \approx 1200$K. Another method is to give the source, the absorber, or both sufficiently high velocities by mechanical means—for example, by putting the substance in a centrifuge. The order of magnitude of the linear speeds needed is about 10^4 to 10^5 cm/sec. A third method is to select as a source nuclei that emit gamma-ray photons while recoiling from a previous decay. An entirely different and novel method was discovered and developed by R. L. Moess-bauer in 1958. The success of this method has opened up a whole new field of investigation with applications to many other fields extending from nuclear physics to general relativity. The Moessbauer effect will be discussed below.

14-18 The Moessbauer Effect

It will be recalled that if a nucleus emits a gamma-ray photon of momentum p the recoil energy imparted to the nucleus is given by

$$\mathcal{E}_R = \frac{p^2}{2M} \qquad\qquad (14\text{-}22)$$

When the atoms emitting the radiation are in the vapor state, M is the mass of the atom. Moessbauer discovered that if the source of gamma rays is part of a crystalline solid the recoil energy is practically zero in a significant fraction of the decays. The reason for this is that the mass M is now comparable to the mass of a microcrystal of the solid. The mass to which the recoil momentum is transferred can be considered infinite in comparison with that of an atom, so that the velocity of recoil is zero. This phenomenon is sometimes called *recoil-less emission* of radiation.

In his original experiment Moessbauer measured the transmission of the 129-keV gamma-ray photons from radioactive iridium 191 as they passed through a metallic iridium absorber. At ordinary temperatures metallic iridium is a crystalline solid containing about 37 percent of the isotope of mass number 191; most of the nuclei are in the ground state. The source of the gamma rays was ^{191}Ir formed in an isomeric state in the beta decay of osmium 191 (see Fig. 14-20). This isomeric state has a half-life of about 4.9 sec and decays to the first excited state, $I = \frac{5}{2}+$; it then goes to the ground state $I = \frac{3}{2}+$, with the emission of a gamma-ray photon of 129 keV. Photons of this energy should be readily absorbed by nuclei of metallic ^{191}Ir.

In one experiment Moessbauer determined the absorption cross section of ^{191}Ir for radiation of 129 keV as a function of temperature. The metallic iridium absorber was kept at 88K; the temperature of the sample containing the radioactive source of ^{191}Ir was cooled from about 400 to 88K. He found that the absorption cross section increased sharply as the temperature of the source was decreased.

In another experiment Moessbauer kept both the source and the absorber at 88K and imposed a very low velocity of the order of cm/sec on the source. This

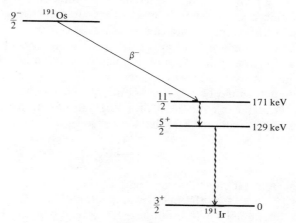

Figure 14-20   Decay scheme of osmium→indium. The positive and negative signs after the quantum numbers refer to the parity of the states.

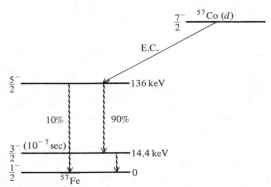

Figure 14-21   Cobalt 57 decays by electron capture to iron 57 in an excited state. The latter decays in 90 percent of the cases to 14.4 keV level. A gamma ray is emitted when nucleus goes to ground state. Half-life of excited state is 10^{-7} sec.

was done by mounting the source on a turntable and rotating it very slowly. The percentage absorption was a maximum at zero relative velocity and dropped rapidly with a width at half maximum of 2 cm/sec corresponding to an energy width of about 10^{-5} eV. This would indicate that the gamma-ray photons producing resonance absorption were emitted by nuclei that received negligible recoil energy.

Once the Moessbauer effect was announced, physicists not only repeated these experiments but investigated other nuclei that might be suitable for this type of investigation. One of the best found so far is iron 57, formed in the decay of cobalt 57 by electron capture with a half-life of 267 d (see Fig. 14-21). The first excited state of ^{57}Fe is at 14.4 keV above the ground state; with this low value of $\mathcal{E}$ the maximum recoil energy is only $\mathcal{E}_R = 0.002$ eV. The lifetime of the first excited state is relatively long and yields a very narrow line. This makes it possible to measure very small energy changes induced in the source or the absorber by means of an external agency. Such energy changes may be induced by temperature changes, by changes in the magnetic field at the nucleus, and by changes in the gravitational field.

One of the interesting applications of the Moessbauer effect is the determination of the strength and direction of the magnetic induction B at the nucleus of iron 57 produced by the external electrons. Since iron is ferromagnetic, it was assumed that B would be large and could produce a splitting of the energy levels; this is the nuclear equivalent of a Zeeman effect. We can introduce a nuclear g factor (see Section 17-3) defined by

$$g = \frac{\mu}{IM_n} \tag{14-27}$$

where μ is the nuclear magnetic moment of the state for which the angular momentum is $I\hbar$ and M_n is the nuclear magneton. The effect of the magnetic induction B at the nucleus is to produce a splitting of each of the energy levels into several levels, depending on the value of I. The vector $\mathbf{I}$ behaves in a manner similar to that of the quantum number $\mathbf{J}$ in the electronic case; that is, each energy level will be split into several levels, depending on the value of m_I, the projection of $\mathbf{I}$ in the direction of the magnetic field. The energy separation of these levels will be given by

$$\Delta\mathscr{E} = Bgm_I M_n \qquad (14\text{-}28)$$

In the case of ^{57}Fe the value of I for the ground state is $\frac{1}{2}$ and for the first excited state, $\frac{3}{2}$; thus the ground state should be split into two levels by the magnetic field, the first excited state into four levels, as shown in Figure 14-22. The selection rule

$$\Delta m_I = 0, \pm 1 \qquad (14\text{-}29)$$

predicts the splitting of the gamma-ray line into six components.

The experimental arrangement usually consists of a source containing radioactive ^{57}Fe and an absorber containing natural ^{57}Fe; the latter occurs with a natural abundance of 2.19 percent in ordinary iron, but this may be enriched if desirable. If the ^{57}Fe is the constituent of the ferromagnetic substance, the same splitting of the energy levels will occur in both the emitter and absorber, and if there is no relative motion between them the absorption curve will be typical of the unresolved line. If, however, either the source or the absorber is moving and their relative velocity is v, there will be a change in the energy of the radiation due to the Doppler effect given by

$$\Delta\mathscr{E}_D = \frac{v}{c}\mathscr{E} \qquad (14\text{-}26)$$

Since the energy spacings are very small, the relative velocity v must be very small. If both the absorber and source are ferromagnetic, however, there will be some velocities for which the energy change will be equal to the separations

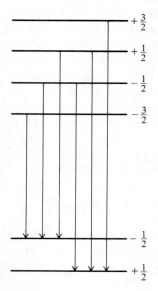

Figure 14-22   The splitting of the excited and ground states of ^{57}Fe and the transitions allowed by the selection rule: $\Delta m_I = \pm1, 0$.

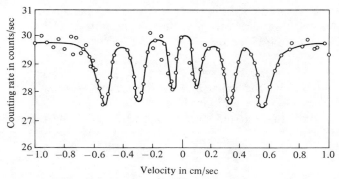

Figure 14-23   The hyperfine spectrum of ^{57}Fe in iron metal, obtained with a stainless steel source and a natural iron absorber 0.001 in. thick. [After G. K. Wertheim, *Journal of Applied Physics,* Suppl. to Vol. **32,** 110S–117S (1961).]

of some of the energy levels, so that several components may overlap. One method of overcoming this difficulty is to use a nonferromagnetic material containing the isotope ^{57}Fe as either the source or the absorber. Stainless steel is a substance that is not ferromagnetic and has become widely used in these experiments. G. K. Wertheim (1961) performed a series of experiments by using the Moessbauer effect to study the magnetic properties of materials. In one experiment he prepared a source by evaporating ^{57}CoCl$_2$ solution on a piece of stainless steel and allowed the ^{57}Co to diffuse into the steel by heating it to 950°C in a vacuum. The ^{57}Co is, of course, the source of the radioactive ^{57}Fe. Figure 14-23 shows the hyperfine structure or nuclear Zeeman pattern obtained with a stainless steel source containing radioactive ^{57}Fe and an absorber consisting of metallic iron 0.001 in. thick containing natural ^{57}Fe. The six components are easily resolved.

In an earlier experiment by Hanna et al. (1960), at the Argonne Laboratory, an external magnetic field was applied to both the source and the absorber in order to determine the direction of polarization of the Zeeman components of ^{57}Fe. These measurements yielded values of the g's for the ground state and the first excited state. Using the known value of $\mu = 0.0903$ nuclear magneton for the magnetic moment of the ground state, they found that B at the nucleus was equal to 3.33×10^5 gauss. They were also able to determine the direction of the internal magnetic field at the nucleus by applying an external field as well and measuring the change in separation of the components; they found that the energy separation was decreased, indicating that the internal field was antiparallel to the applied external field, a result contrary to all predictions.

14-19   Application of Moessbauer Effect to General Relativity

One of the predictions of the general theory of relativity is the so-called *red shift;* that is, a clock or an atomic system that radiates energy of a definite frequency will have a longer period or emit a longer wavelength when in a strong gravitational field than it has in a weak field. The wavelength is thus shifted toward the red or longer wavelength end of the spectrum. The only verification of this prediction was a determination of the wavelength of light emitted from a very dense star. It occurred to several groups of physicists that the Moessbauer effect provided a means for checking this prediction by a terrestrial experiment.

It will be recalled that the intensity I_G of the gravitational field at a distance r from a large mass M is given by

$$I_G = \frac{GM}{r^2} \qquad (14\text{-}30)$$

where G is the universal constant of gravitation. The intensity I_G, by definition, is simply the force per unit mass at the given point. In the case of the earth's gravitational field

$$I_G = g \qquad (14\text{-}31)$$

the well-known acceleration due to gravity. The gravitational potential Φ at the point r due to the mass M is simply

$$\Phi = -\frac{GM}{r} \qquad (14\text{-}32)$$

The concept involved in the terrestrial experiment is the equivalence between gravitational mass and inertial mass, a fact that has been verified to a high degree of accuracy by the experiment of Eötvös (Section 2-13). Another aspect of this concept is the equivalence of an accelerated system and a gravitational field; that is, there is no way of distinguishing locally between the effects produced by a gravitational field and an acceleration of a system in which an experiment is being performed. A system with an acceleration a produces an effect equivalent to that of a gravitational acceleration $-a$. The difference of potential between two points near the surface of the earth, one at a height H above the other is, to a first approximation,

$$\Delta\Phi = gH \qquad (14\text{-}33)$$

where g is the average value of the gravitational acceleration over the height H. If the two points are close together near the surface of the earth, then $g = 980$ cm/sec^2. To determine the difference in frequencies between two identical atomic (or nuclear) systems a distance H apart in a gravitational field let us replace the latter with a system moving with an acceleration $a = -g$ but in a horizontal direction.

Figure 14-24 shows a rigid rod of length H in an inertial frame of reference; the rod is moving with a constant acceleration a to the right. Identical atoms (nuclei) are situated at the ends A and B of the rod. Suppose that at some particular instant when the velocity of the rod is v_1 the nucleus at A emits a

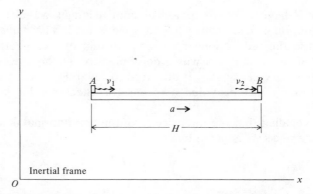

Figure 14-24   Rigid rod *AB* has constant acceleration *a* in inertial frame.

gamma-ray photon of frequency ν_1 in the direction of its motion. Because of its velocity v_1 at the instant of emission, the frequency as measured by a stationary observer will have been shifted by the Doppler effect to a value

$$\nu_{\text{obs}} = \nu_1 \left(1 + \frac{v_1}{c}\right) \tag{14-34}$$

This photon moves toward *B* with constant velocity *c*, and at a time

$$t = \frac{H}{c} \tag{14-35}$$

it will reach an identical nucleus that will act as the absorber. But the velocity v_2 of the rod and nucleus at this time is

$$v_2 = v_1 + at \tag{14-36}$$

The radiation that can be absorbed by *B* will have an apparent frequency ν_2; a stationary observer will report it as shifted by the Doppler effect to a value

$$\nu_{\text{obs}} = \nu_2 \left(1 + \frac{v_2}{c}\right) \tag{14-37}$$

For the identical gamma-ray photon to be the one emitted and absorbed we must have

$$\nu_1 \left(1 + \frac{v_1}{c}\right) = \nu_2 \left(1 + \frac{v_2}{c}\right) \tag{14-38}$$

The difference in frequency $\Delta\nu$ is therefore

$$\Delta\nu = \nu_1 - \nu_2 = \frac{(\nu_2 v_2 - \nu_1 v_1)}{c} \tag{14-39}$$

We can set $\nu \doteq \nu_1 = \nu_2$ to a very good approximation and set

$$v_2 - v_1 = at = \frac{aH}{c} \tag{14-40}$$

so that

$$\Delta\nu = \frac{\nu a H}{c^2} \tag{14-41}$$

Thus in an accelerated system or, from the principle of equivalence, in a gravitation field in which the acceleration $g = -a$, identical atomic (nuclear) systems will emit or absorb radiation that will differ in frequency by an amount

$$\Delta\nu = \frac{\nu g H}{c^2} \tag{14-42}$$

where H is the height of B above A. Stated another way, if $\Delta\Phi$ is the potential difference between two points in a gravitational field, the difference in frequency of identical clocks at these points will be

$$\Delta\nu = \frac{\nu\Delta\Phi}{c^2} \tag{14-43}$$

A series of experiments to verify this prediction was performed by R. V. Pound and G. A. Rebka, Jr., at Harvard (1959–1960) with ^{57}Fe as the emitter and absorber. A simplified schematic diagram of their apparatus is sketched in Figure 14-25. The difference in height between A and B was 74 ft. The temperatures of the iron foils at A and B were measured acccurately and corrections were made for the Doppler shift produced by the thermal motion

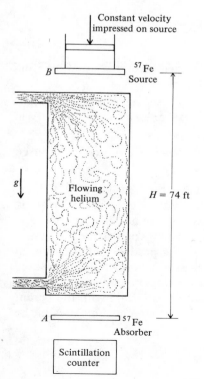

Figure 14-25   Simplified diagram of experimental arrangement for determining red shift.

of the crystal lattice. A constant velocity was impressed on the foil at B to provide the necessary Doppler shift to compensate for the gravitational red shift $\Delta\nu$ and to permit resonance absorption of the 14.4-keV gamma ray from the source. A bag containing flowing helium was placed between the source and absorber to reduce the absorption of these photons that would occur in air. To improve the accuracy of the experiment the positions of the absorber and source were interchanged, thus providing a base line of $2H = 148$ ft. The expected fractional frequency shift was

$$\frac{\Delta\nu}{\nu} = \frac{2gH}{c^2}$$

$$= 4.92 \times 10^{-15} \tag{14-44}$$

The observed fractional frequency shift was

$$5.13 \times 10^{-15}$$

or a ratio of 1.05 between the observed and theoretically expected frequency shift. The frequency increased in falling, as expected. There is thus good agreement between theory and experiment.

Another way of viewing this result is that the equality between gravitational and inertial mass is now known to be correct to an accuracy of 1 part in 10^{14}, an improvement by a factor of 10^5 on the Eötvös experiment.

Problems

14-1.   Calculate the number of alpha particles per second emitted by 3 mg of radium. Express this activity in millicuries.

14-2.   Calculate the number of alpha particles per second emitted by 2 cm³ of radon at normal temperature and pressure. Express this activity in millicuries.

14-3.   Determine the amount of polonium 210 necessary to provide a source of alpha particles of 5 mCi.

14-4.   Radium disintegrates with a comparatively long half-life, $T = 1620$ yr, into radon, which in turn disintegrates with a comparatively short half-life, $T = 3.823$ days. In such cases the rate at which radium disintegrates can be considered almost constant, and if the radium is kept in a closed container the amount of the product radon builds up to a steady value; that is, just as much radon is formed during a short interval as the amount that disintegrates during the same time interval. The product is then said to be in *secular equilibrium* with the parent substance. (a) Show that the rate at which radon accumulates in the presence of radium is given by

$$\frac{dN}{dt} = \lambda_1 N_1 - \lambda_2 N$$

where N is the amount of radon present at any instant, N_1 is the original amount of radium, and λ_2 and λ_1 are their respective disintegration constants. (b) If the amount of radium is assumed to remain constant, show that the amount of radon present after a time t is given by

$$N = \frac{\lambda_1}{\lambda_2} N_1 [1 - \exp(-\lambda_2 t)]$$

(c) Show that after a sufficient lapse of time for secular equilibrium to be established, the amount of radon present is

$$N_2 = \frac{\lambda_1}{\lambda_2} N_1$$

14-5.   (a) From Problem 14-4 plot a curve with N/N_2 as ordinates and the time (in units of T) as abscissae. (b) Suppose that after a long time t, the amount of radon N_2 is pumped off; if we start measuring the amount of radon present from this instant, then $N = N_2 \exp(-\lambda_2 t)$. Plot this equation with N/N_2 as ordinates on the same axes as part (a). (c) Sum the ordinates of the two curves and discuss the results.

14-6.   (a) From the data of Table 14-1 compute the value of the disintegration constant λ in $\sec^{-1}$ for ^{40}K. (b) Using the above value of λ, determine the number of beta particles per second emitted by 1 gram of ^{40}K. (c) Using the value of the normal abundance of ^{40}K in a sample of potassium, compute the number of beta particles per second emitted by 1 gram of potassium.

14-7.   Using the value of the half-life of ^{238}U from Figure 14-2, (a) compute the disintegration constant in $\sec^{-1}$, and (b) calculate the number of alpha particles per second emitted by 1 gram of uranium.

14-8.   Using the value of the half-life of radon from Figure 14-2, (a) compute its disintegration constant in $\sec^{-1}$; (b) calculate the number of alpha particles per second emitted by 1 cm^3 of radon at 1 atmosphere pressure and 0°C.

14-9.   Using the value of the abundance of rubidium 87, calculate the number of beta particles per second emitted by 100 grams of rubidium.

14-10.   Assuming that 90 percent of the disintegrations of potassium 40 result in beta emission and using the known abundance of this isotope, calculate the number of beta particles per minute emitted by a crystal of potassium chloride whose mass is 200 grams.

14-11.   (a) Determine the constants A and B of the Geiger-Nuttall law. (b) How do these constants depend on the particular radioactive series? (c) Using the data from Figure 14-4, calculate the range of the alpha particles from AcU.

14-12. From the data in Table 14-2 plot a curve of energy against range of the alpha particles.

14-13. From the data in Table 14-2 plot a curve of the range of the alpha particles against the cube of their velocities and compare this curve with Geiger's law.

14-14. Using the data of Figure 14-5, plot a graph of $\log \lambda + 20$ against $\log R$ for the neptunium series. Determine the constants A and B of the Geiger-Nuttall equation for this series.

14-15. (a) Using the data of Table 14-2, compute the disintegration energy of the alpha-particle decay of polonium 210. (b) Determine the recoil energy of the nucleus.

14-16. Polonium 212 emits an alpha particle whose kinetic energy is 10.54 MeV. (a) Determine the disintegration energy. Determine (b) the momentum and (c) the energy of the recoil nucleus.

14-17. (a) Write an expression for the potential energy of an alpha particle after it has been emitted by a nucleus of atomic number Z. (b) Determine the height of the potential barrier of uranium for alpha particles. (c) Uranium 238 emits an alpha particle with a kinetic energy of 4.2 MeV; calculate the width w of the potential barrier through which the alpha particle must pass.

14-18. (a) Compute the disintegration energy in the alpha-particle decay of radium, $Z = 88$, $A = 226$, from the data on the atomic masses. (b) From this calculate the kinetic energy of the alpha particle and compare it with the measured value of 4.795 MeV.

14-19. The K conversion electron from ^{137}Cs produces a sharp line in its beta-ray spectrum whose momentum, measured with a magnetic spectrometer, yields a value of $Br = 3381$ gauss cm. The binding energy of this K electron is 37.44 keV. (a) Determine the kinetic energy of the conversion electron. (b) Determine the energy of the converted gamma-ray photon.

14-20. The gamma-ray photon from ^{137}Cs, when incident on a piece of uranium, ejects photoelectrons from its K shell. The momentum, measured with a magnetic beta-ray spectrometer, yields a value of $Br = 3083$ gauss cm. The binding energy of a K electron in uranium is 115.59 keV. Determine (a) the kinetic energy of the photoelectrons and (b) the energy of the gamma-ray photons.

14-21. Bismuth 210 emits a gamma-ray photon whose energy is 47 keV. Determine (a) the recoil momentum of the nucleus and (b) its recoil energy.

14-22. Lead 208 emits a gamma-ray photon whose energy is 2.62 MeV.

Determine (a) the recoil momentum of the nucleus and (b) its recoil energy.

14-23. Consider $N(t)$ from Equation (14-2) as an unnormalized probability density function and calculate $<t>$.

14-24. In 1921 the American people gave Marie Curie 1 gram of radium. How much of it is left now?

14-25. (a) In a sample of 1 kg of ^{238}U, how many disintegrations occur per day? (b) If all of the alpha particles could escape from the uranium and the sample were electrically insulated, how much negative charge would accumulate on the sample?

14-26. The K meson has a mean life of 1.24×10^{-8} sec. A beam of K mesons traverses a bubble chamber which is 30 in. in length. How many K mesons decay in the chamber if their momentum is 700 MeV/c?

14-27. Particle A is stable and particle B has a mean life $1/\lambda$. The interaction cross section for a beam of particles of type A is three times as great as for particles of type B, independent of momentum. At what momentum will the attenuation be the same for both beams?

14-28. A gamma-ray photon of energy $h\nu$ is emitted from the nucleus of an atom of mass M. (a) Derive an expression for the kinetic energy of the recoiling nucleus. (b) ^{137}Ba emits a gamma-ray photon of 0.66 MeV during an isomeric transition. Calculate the recoil kinetic energy of the atom in electron volts.

14-29. (a) Derive an expression for the recoil kinetic energy of an atom that emits a conversion electron of momentum p. (b) The momentum of the conversion electron in its path of radius R in a uniform magnetic induction of strength B is BeR. Show that the recoil kinetic energy is given by

$$\mathscr{E}_k = \frac{e^2}{2} \frac{(BR)^2}{M}$$

where all quantities are in mks units. (c) The measured value of the momentum of the conversion electrons from ^{137}Cs is given as 3380 gauss cm. Determine the recoil energy in electron volts.

14-30. The measured value of the width of the resonance gamma-ray line of iron 57 is 3.4×10^{-8} eV. Determine the lifetime of the excited state.

14-31. The lifetime of one of the excited states of mercury 203 is 8×10^{-13} sec. Determine the width of the gamma-ray line emitted when the nucleus goes to the ground state.

14-32. A moving nucleus with velocity v emits a photon for which the direction of motion is the same as that of the original nucleus. (a) Obtain the relativistic expression for the Doppler shift for this case. (b) Show that your expression reduces to Equation (14-23) in the nonrelativistic limit.

14-33. If one starts with 1 kg of ^{3}H, how long will it be until only 1 gram is left?

14-34. Two isotopes of a certain element have half-lives T_1 and T_2. They are observed in nature on the earth with abundances A_1 and A_2, respectively. Assuming that these two isotopes were equally abundant at the formation of the earth, calculate the age of the earth.

References

Most of these references contain information that is relevant to all the remaining chapters of this book.

Arya, A., *Fundamentals of Nuclear Physics*. Boston: Allyn and Bacon, 1966.

Bethe, H. A., and P. Morrison, *Elementary Nuclear Theory*. New York: John Wiley and Sons, 1956.

Elton, L. R. B., *Introductory Nuclear Theory*. New York: Interscience Publishers, 1959.

Evans, R. D., *The Atomic Nucleus*. New York: McGraw-Hill Book Company, 1955.

Fermi, E., *Nuclear Physics*. Notes compiled by J. Orear, A. H. Rosenfeld, and R. A. Schluter from a course given by E. Fermi. Chicago: The University of Chicago Press. 1950.

Kaplan, I., *Nuclear Physics*. Cambridge, Mass.: Addison-Wesley Publishing Company, 1955.

Meyerhof, W. E., *Elements of Nuclear Physics*. New York: McGraw-Hill Book Company, 1967.

Rutherford, E., J. Chadwick, and C. B. Ellis, *Radiations from Radioactive Substances*. London: Cambridge University Press, 1930.

Segrè, E. *Nuclei and Particles*. New York: W. A. Benjamin, 1965.

15 | Nuclear Reactions

15-1 Introduction

The earliest research in nuclear physics was centered around materials that are naturally radioactive. The processes of radioactivity per se were studied as indicated in the preceding chapter. Radioactive samples were also used as sources of alpha particles which in turn were used to bombard other nuclei. Such experiments, performed by Rutherford and his co-workers, gave an accurate idea of the small size of the nucleus and led to the nuclear theory of the atom. These experiments were immediately followed by attempts to disrupt the nucleus. The first successful experiment on the disintegration of the nucleus was performed by Rutherford in 1919 with alpha particles as projectiles. Progress was very slow during the next decade, partly because very few laboratories had adequate amounts of radioactive substances but mostly because the energies of the alpha particles, although extending from about 4 to about 10 MeV, were too small to overcome the potential barrier or Coulomb repulsion of the nucleus. Beginning about 1930, physicists began designing and building particle accelerators for imparting high energies to particles such as electrons, protons, deuterons (nuclei of deuterium), and helium ions (alpha particles), as well as for the production of high-energy x-rays. The physical principles of these accelerators were discussed in Chapter 12. Improvements in accelerators (higher energy and greater intensity and extension to different types of particle as projectiles) led to the opportunity to perform a wide variety of experiments in which nuclei were bombarded, often with the result that the target nucleus was changed into a different element.

In this chapter we shall describe some of the historical experiments on nuclear disintegration with comparatively low-energy particles—that is, particles with energies less than 40 MeV, without specific reference to the manner in which these particles were produced. However, the process of nuclear fission produced by low-energy neutrons and other particles will be reserved for a

467

separate chapter. In addition we shall describe the discovery of some new particles such as the neutron and the positron and the phenomenon of artificially produced radioactivity.

The use of intermediate-energy (> 40 MeV) or high-energy (> 1 GeV) particles leads to entirely different types of nuclear processes, particularly the production of entirely new particles such as mesons and hyperons; these will be treated in a separate chapter.

15-2   Discovery of Artificial Disintegration

The artificial transmutation of one element into another, the dream of alchemists for centuries, was first definitely accomplished by Rutherford in 1919 in a very simple type of experiment. A diagram of the apparatus used by Rutherford is shown in Figure 15-1. The chamber C was filled with a gas such as nitrogen, and alpha particles from a radioactive source at A were absorbed in the gas. A sheet of silver foil F, itself thick enough to absorb the alpha particles, was placed over an opening in the side of the chamber. A zinc sulfide screen S was placed outside this opening and a microscope M was used for observing any scintillations occurring on the screen S. Scintillations were observed when the chamber was filled with nitrogen, but when the nitrogen was replaced by oxygen or carbon dioxide no scintillations were observed on the screen S. Rutherford concluded that the scintillations were produced by high-energy particles which were ejected from nitrogen nuclei as a result of the bombardment of these nuclei by the alpha particles. Magnetic deflection experiments indicated that these particles were hydrogen nuclei, or protons. Later experiments by Rutherford and Chadwick showed that these ejected protons had ranges up to 40 cm in air. Other light elements in the range from boron to potassium were also disintegrated by bombardment with alpha particles. Since then alpha particles, used as projectiles, have been successful in causing the disintegration of many elements.

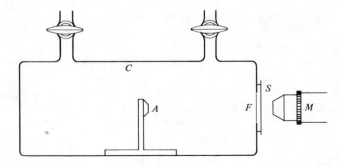

Figure 15-1   Diagram of the apparatus used by Rutherford in the first successful experiment on artificial disintegration of nuclei.

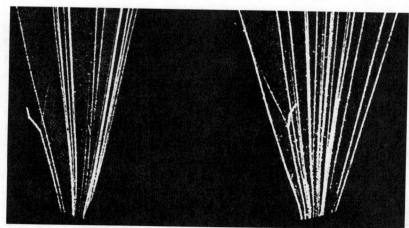

Figure 15-2   A pair of stereoscopic photographs of alpha-particle tracks in a cloud chamber containing nitrogen. One of the alpha particles is captured by a nitrogen nucleus resulting in the disintegration of the compound nucleus into a proton (thin track) and an oxygen ion (thick track). (From Rutherford, Chadwick, and Ellis, *Radiations from Radioactive Substances*. By permission of the Cambridge University Press, publishers.)

The disintegration of nuclei has also been studied with the Wilson cloud chamber. One of the first of these investigations was that of Blackett, who photographed the tracks of alpha particles in a Wilson cloud chamber containing about 90 percent nitrogen and 10 percent oxygen. The majority of the tracks photographed were straight tracks typical of alpha-particle tracks. Many of the tracks were observed to be forked tracks, indicating that an elastic collision had taken place between an alpha particle and a nitrogen nucleus. It is an easy matter to distinguish between the track made by an alpha particle and that made by a nitrogen nucleus. The heavier particle produces more ion pairs per centimeter of path and thus forms a thicker track. Of about 500,000 tracks photographed, eight were of an unusual type. Each of these was a forked track containing two branches, as shown in Figure 15-2, one a very thick track, the other a very thin track. The thick track is ascribed to a slow-moving oxygen ion, the thin track to a fast-moving proton.

In order to be able to measure accurately the lengths of the tracks and the angles that the forked components make with the original direction of the alpha particle it is necessary to photograph them from two different positions to be able to determine the plane in which the tracks are formed. A common method is to use two cameras at right angles to each other, thus obtaining a pair of stereoscopic photographs from which the correct space relationships of the several tracks can be determined. Measurements on the type of tracks illustrated in Figure 15-2 showed that the momentum of the system was conserved but that the sum of the kinetic energies of the particles after impact was less than the kinetic energy of the alpha particle before impact. On the basis of a theory of the nucleus advanced by Bohr (1936), the disintegration of nitrogen by bombardment with alpha particles may be thought of as consisting of two separate parts. The first is the capture of the alpha particle by the nitrogen nucleus which resulted in the formation of a new *compound nucleus;* the second is the breaking up of the compound nucleus into two particles, one of which is a proton. These two processes can be represented by means of a *nuclear reaction* equation analogous to one representing a chemical reaction. The nuclear reaction equation for this process is

$$\ce{^4_2He} + \ce{^{14}_7N} \rightarrow (\ce{^{18}_9F}^*) \rightarrow \ce{^{17}_8O} + \ce{^1_1H} \tag{15-1}$$

The alpha particle, since it is a helium nucleus, is represented by the symbol $\ce{^4_2He}$. In order to satisfy the principle of the conservation of charge, the atomic number of the compound nucleus must be the sum of the atomic numbers of the helium and nitrogen nuclei. The compound nucleus formed in this case is fluorine, $Z = 9$. The symbol representing the compound nucleus will always be enclosed in parentheses. The star indicates that the ^{18}F nucleus is in an excited state rather than the ground state. Since this unstable fluorine disintegrates with the emission of a proton, the remaining part, or product nucleus, must be oxygen, $Z = 8$.

The guiding principle in determining which isotope of an element is formed during a nuclear reaction is that the mass number of the compound nucleus must equal the sum of the mass numbers of the initial particles and also the sum of the mass numbers of the final particles. This is not the same as the principle of the conservation of mass, since the mass numbers differ slightly from the actual values of the atomic masses. The principle of conservation of mass is no longer a separate and independent principle but part of the more general principle of the conservation of energy, since, as has already been noted, a mass m is equivalent to an amount of energy mc^2, where c is the speed of light. Equation (15-1) can be rewritten to satisfy the general principle of the conservation of energy as

$$\ce{^4_2He} + \ce{^{14}_7N} \rightarrow (\ce{^{18}_9F}^*) \rightarrow \ce{^{17}_8O} + \ce{^1_1H} + Q \tag{15-2}$$

where Q represents the energy evolved or absorbed during the nuclear reaction. If Q is positive, energy has been evolved; if Q is negative, energy has been absorbed. Q is called the *nuclear reaction energy* or the *disintegration energy* and is equal to the difference in the masses of the initial and final particles in their ground states. If the sum of the masses of the final particles exceeds that of the initial particles, Q must be negative; the energy absorbed in such a nuclear reaction must have been obtained from the kinetic energies of the particles. If $\mathcal{E}_1$ is the kinetic energy of the alpha particle just before capture, $\mathcal{E}_2$ the kinetic energy of the proton, and $\mathcal{E}_3$ the kinetic energy of the product nucleus, then

$$Q = \mathcal{E}_2 + \mathcal{E}_3 - \mathcal{E}_1 \tag{15-3}$$

In those cases in which Q is positive the sum of the kinetic energies of the final particles will be greater than the kinetic energy of the incident alpha particle. In nearly all cases the kinetic energy of the nucleus which captures the alpha particle is comparatively small and may be neglected in this type of calculation.

In the above reaction, Equation (15-2), the best value of Q obtained from measurements of the kinetic energies of the particles is

$$Q = -1.26 \text{ MeV}$$

The value of Q could have been predicted by using the table of mass excesses from Appendix IV.

Initial Particles		Final Particles	
^{4}He:	+2.425	^{1}H:	+7.289
^{14}N:	+2.864	^{17}O:	−0.808
	5.289 MeV		6.481 MeV

The Q value is just the difference of these two numbers, or

$$Q = -1.192 \text{ MeV}$$

These two results agree very well within the limits of error of the experiment.

15-3   The (α,p) Reaction

The disintegration of the nitrogen nucleus by alpha particles is historically the first of a series of nuclear reactions in which an alpha particle is captured by a nucleus forming a compound nucleus which immediately disintegrates into a new nucleus by the ejection of a proton. Such reactions are known as the (α,p) type of reaction in which the first letter, α, designates the bombarding particle and the second letter, p, designates the ejected particle. The (α,p) reaction has been observed with most of the lighter elements up to selenium. This type of artificial disintegration may be represented by the nuclear reaction equation

$$^A_Z X + {}^4_2\text{He} \rightarrow ({}^{A+4}_{Z+2}\text{Cp*}) \rightarrow {}^{A+3}_{Z+1}Y + {}^1_1\text{H} + Q \qquad (15\text{-}4)$$

where Cp represents the compound nucleus formed as a result of the capture of an alpha particle by the atom X of mass number A and atomic number Z. The compound nucleus is always formed in an excited state. The ejection of a proton from this compound nucleus results in the formation of a new atom Y of mass number $A + 3$ and atomic number $Z + 1$. In the majority of these cases the reaction energy, or the disintegration energy Q, has been found to be negative. A few of these (α,p) reactions are listed below.

$$^{10}_{5}\text{B} + ^4_2\text{He} \rightarrow (^{14}_{7}\text{N}^*) \rightarrow ^{13}_{6}\text{C} + ^1_1\text{H} \qquad Q = +4.04 \text{ MeV}$$

$$^{19}_{9}\text{F} + ^4_2\text{He} \rightarrow (^{23}_{11}\text{Na}^*) \rightarrow ^{22}_{10}\text{Ne} + ^1_1\text{H} \qquad Q = +1.58 \text{ MeV}$$

$$^{27}_{13}\text{Al} + ^4_2\text{He} \rightarrow (^{31}_{15}\text{P}^*) \rightarrow ^{30}_{14}\text{Si} + ^1_1\text{H} \qquad Q = +2.26 \text{ MeV}$$

$$^{28}_{14}\text{Si} + ^4_2\text{He} \rightarrow (^{32}_{16}\text{S}^*) \rightarrow ^{31}_{15}\text{P} + ^1_1\text{H} \qquad Q = -1.92 \text{ MeV} \qquad (15\text{-}5)$$

$$^{32}_{16}\text{S} + ^4_2\text{He} \rightarrow (^{36}_{18}\text{A}^*) \rightarrow ^{35}_{17}\text{Cl} + ^1_1\text{H} \qquad Q = -2.10 \text{ MeV}$$

$$^{39}_{19}\text{K} + ^4_2\text{He} \rightarrow (^{43}_{21}\text{Sc}^*) \rightarrow ^{42}_{20}\text{Ca} + ^1_1\text{H} \qquad Q = -0.89 \text{ MeV}$$

$$^{45}_{21}\text{Sc} + ^4_2\text{He} \rightarrow (^{49}_{23}\text{V}^*) \rightarrow ^{48}_{22}\text{Ti} + ^1_1\text{H} \qquad Q = -0.3 \text{ MeV}$$

A simpler notation is frequently used to represent nuclear reactions. In this notation the symbols for the bombarding particle and the particle released in the reaction are placed in parentheses in the order stated; the parentheses are written between the symbol for the target nucleus and that for the product nucleus. Thus the above reactions (15-5) would be written as $^{10}\text{B}(\alpha,p)^{13}\text{C}$, $^{19}\text{F}(\alpha,p)^{22}\text{Ne}$, $^{27}\text{Al}(\alpha,p)^{30}\text{Si}$, and so forth. It is generally not necessary to write the atomic number with the chemical symbol, since one is uniquely determined by the other.

The compound nuclei formed in the first five nuclear reactions listed in (15-5) have mass numbers identical with those of stable isotopes of the respective elements. These compound nuclei, however, are not in their *ground state;* each compound nucleus is in an *excited state.* The mass of the compound nucleus is greater than the mass of the stable form. This can be verified readily by a calculation similar to that made for reaction (15-2).

In most of these artificial disintegration experiments it is desirable to have the bombarded element in the form of a solid, so that when a stream of alpha particles is directed against this solid target a larger fraction of them will be absorbed in a very small volume. The effectiveness of the alpha particles in producing this type of disintegration is dependent on the energies of the alpha particles and on the nuclear charge. A measure of this effectiveness, sometimes called the *yield,* is the ratio of the number of protons produced to the number of alpha particles completely stopped in the target. The yield for the (α,p) reaction ranges in values from 10^{-7} to 10^{-5} for alpha particles of 3 to 8 MeV incident on elements of small atomic number.

One of the (α,p) reactions which has been studied very carefully is that in which aluminum formed the target for the alpha particles. The energies of the protons emitted in this reaction have been studied for different values of the energy of the incident alpha particles. One method of presenting these results is shown in the curve in Figure 15-3. This curve is known as a distribution-in-range curve. It is obtained by plotting the number of protons penetrating a certain thickness of air, or its equivalent, as ordinate against the corresponding value of the absorber thickness. It will be noticed that the protons produced

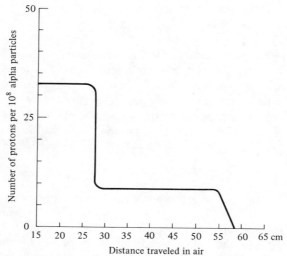

Figure 15-3   Distribution-in-range curve of protons.

in this particular experiment form two homogeneous groups, one of about 28 cm range and the other of about 58 cm range. Other homogeneous groups of protons have been observed, using alpha particles of different energies. Of the protons observed in this reaction those with the longest range had a range of about 66 cm.

The fact that protons are ejected with definite ranges, or definite energies, indicates that the product nucleus, in this case silicon, possesses several energy levels. The product nucleus is left in the ground state by the ejection of the proton of longest range, and it is left in one of its excited states by the ejection of a proton of smaller range. One might then expect gamma rays to be emitted during this reaction by those nuclei that go from the excited states to the normal or ground state, and the energies of these gamma rays should be equal to the differences in the energies of the various proton groups. Gamma rays have actually been observed in the above reaction and in several of the other

The value of the reaction energy Q, listed with each (α,p) reaction, is the largest reaction energy and corresponds to the emission of protons of maximum range. In each case, therefore, the product nucleus is left in its normal state. Reactions in which Q has been measured accurately have been used for the determination of the masses of the product nuclei. The values of the atomic masses obtained in this way can be used as independent checks on the measurements made with the mass spectrograph. In some cases in which such data are not available the nuclear reaction equations form the only reliable methods for determining the masses of the isotopes formed in the reaction.

15-4 Discovery of the Neutron

The capture of an alpha particle by a nucleus does not always result in the emission of a proton by the compound nucleus formed as a result of this capture. In one particular reaction studied, that resulting from the bombardment of beryllium by alpha particles, a very penetrating type of radiation was found to be emitted by the newly formed compound nucleus. It was at first assumed that this radiation was of the nature of gamma rays resulting from the nuclear reaction

$$\ce{^{9}_{4}Be} + \ce{^{4}_{2}He} \rightarrow (\ce{^{13}_{6}C}^*) \rightarrow \ce{^{13}_{6}C} + h\nu$$

where $h\nu$ is the energy of the gamma-ray photon. The measurements of Bothe and Becker (1930) of the absorption of these rays in lead showed that each photon should possess an energy of about 7 MeV. The Curie-Joliots (1932) showed that these rays had the very interesting property of being able to knock out protons from paraffin and other substances containing hydrogen. The protons knocked out of paraffin by these rays had a range in air of about 40 cm, or an energy of about 6 MeV. Assuming that these protons were produced as the result of elastic collisions with the gamma-ray photons, calculations showed that each photon must have possessed an amount of energy of about 55 MeV. These results were entirely inconsistent with the results from the experiments on the absorption of these rays in lead. Furthermore, the amount of energy available for gamma radiation, when computed for the above reaction from the known mass excesses (see Appendix IV) of the particles, is much less than 55 MeV. If an alpha particle of 5-MeV energy is captured by a beryllium nucleus, the energy available for the emission of a gamma ray from the carbon nucleus can be obtained as follows:

$$\ce{^{9}_{4}Be} + \ce{^{4}_{2}He} + \mathcal{E}_1 \rightarrow (\ce{^{13}_{6}C}^*) \rightarrow \ce{^{13}_{6}C} + h\nu$$

$$11.350 + 2.425 + 5.0 = 3.125 + h\nu$$

or

$$h\nu = 15.65 \text{ MeV}$$

that is, the maximum amount of energy available for the gamma-ray photon is 15.65 MeV. Chadwick (1932) performed a series of experiments on the recoil of nuclei that were struck by the rays coming from beryllium bombarded by alpha particles and showed that if these rays were assumed to be gamma rays the results of the experiments led to values for the energies of these rays that depended on the nature of the recoil nucleus. For example, the protons ejected from paraffin had energies of 5.7 MeV, which led to a value of 55 MeV for the energy of the gamma ray; nitrogen recoil nuclei had energies of about 1.2 MeV, indicating that the photon that struck this nucleus must have had an

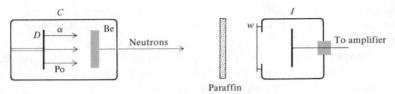

Figure 15-4 Arrangement of apparatus for the detection of neutrons.

energy of about 90 MeV. In general, if the recoil atoms are to be attributed to collisions with photons, the amount of energy that has to be assigned to the photon will increase with the increase in mass of the recoil atom. This is contrary to the principles of conservation of energy and momentum during collisions. Chadwick, however, showed that all these difficulties disappear completely if we adopt the hypothesis that the radiation coming from the beryllium bombarded with alpha particles does not consist of photons but of particles of mass very nearly equal to that of the proton but having no charge. These particles are called *neutrons* and are formed as a result of the reaction

$$\ce{^9_4Be + ^4_2He} \rightarrow (\ce{^{13}_6C^*}) \rightarrow \ce{^{12}_6C} + \ce{^1_0}n \qquad (15\text{-}6)$$

where ^1_0n is the symbol representing the neutron, showing that it has zero charge and mass number unity.

One arrangement used by Chadwick for demonstrating the existence and properties of neutrons is shown in Figure 15-4. The source of alpha particles is a disk D on which polonium has been deposited. This disk and the beryllium target are placed in an evacuated chamber C. The neutrons coming from the beryllium pass through the thin wall of this chamber and enter the ionization chamber I through the thin window w. This ionization chamber is connected to an amplifier and then to a recording device, such as an oscillograph, a loudspeaker, or an electrical counter.

Since the neutrons possess no electric charge, they produce no ionization directly in their passage through the chamber. But some neutrons that strike the walls of the ionization chamber cause the ejection of nuclei which then produce ions in the chamber and are thus recorded on the film of the oscillograph or on the electrical counter; if a loudspeaker is used, a "click" is heard for every nucleus that produces intense ionization. The results of these experiments show that when the neutrons from beryllium go directly into the ionization chamber, a few counts per minute are recorded. If thin sheets of lead are placed in front of the ionization chamber, the number of counts produced is not reduced appreciably. If, however, a thin slab of paraffin is placed in front of the window w, the number of counts per minute increases markedly. This increase is due to the fact that the neutrons, in collisions with the nuclei of the hydrogen atoms contained in paraffin, give up a considerable fraction of their energy to those nuclei or protons, and those protons that enter the ionization chamber are then recorded. If the paraffin is removed and the neutrons are allowed to enter the chamber directly, the number of counts falls immediately to its former low value. This is just the opposite of what would happen if radiation of the nature of gamma rays were used; that is, the introduction of any absorbing material in the path of the gamma radiation produces a decrease in the intensity of the transmitted radiation. The radiation from beryllium therefore cannot be of the nature of gamma rays.

When neutrons pass through matter, they lose energy as a result of collisions with other nuclei and so give rise to the recoil atoms. If the mass of the neutron is approximately unity, then, in collision with hydrogen nuclei, the ejected protons will have velocites of all values up to a maximum that is the same as the velocity of the neutrons. The mass M of the neutron can be calculated, to a first approximation, from the measured values of the maximum velocities of the hydrogen and nitrogen recoil atoms. It can be shown, on the basis of mechanics (see Problem 15-1), that the maximum velocity of the recoil nucleus is given by

$$v = \frac{2M}{M + M_r} V \tag{15-7}$$

where V is the velocity of the incident neutron, v the velocity of the recoil atom, and M_r the mass of the recoil atom. If two experiments are performed, one in which the maximum velocity of the recoil protons is measured and the other in which the maximum velocity of the recoil nitrogen atoms is determined, then

$$\frac{v_H}{v_N} = \frac{M + M_N}{M + M_H} \tag{15-8}$$

In Chadwick's experiment the measured value of the maximum velocity of the recoil proton, v_H, was 3.3×10^9 cm/sec and of the recoil nitrogen atoms, $v_N = 4.7 \times 10^8$ cm/sec. The assumption that the mass of nitrogen is 14 times that of hydrogen yields $M = 1.15$ as the approximate mass of the neutron. More accurate measurements yield, for the mass of the neutron, $M = 1.008665$ (Section 15-12).

Because of their lack of charge, neutrons should be able to penetrate atomic nuclei very easily, and a study of these nuclear reactions should yield valuable information concerning nuclear properties and nuclear structure (Section 15-13).

15-5 The (α,n) Reaction

The bombardment of beryllium by alpha particles with the subsequent emission of neutrons is one of many nuclear reactions of the type designated as an (α,n) type and is given by the formula

$$\ce{^A_Z X + ^4_2 He} \rightarrow \ce{(^{A+4}_{Z+2} Y^*)} \rightarrow \ce{^{A+3}_{Z+2} Y + ^1_0 n} + Q \tag{15-9}$$

A few of these reactions are listed below.

$$
\begin{aligned}
\ce{^7_3 Li + ^4_2 He} &\rightarrow (\ce{^{11}_5 B^*}) &&\rightarrow \ce{^{10}_5 B + ^1_0 n} \\
\ce{^9_4 Be + ^4_2 He} &\rightarrow (\ce{^{13}_6 C^*}) &&\rightarrow \ce{^{12}_6 C + ^1_0 n} \\
\ce{^{11}_5 B + ^4_2 He} &\rightarrow (\ce{^{15}_7 N^*}) &&\rightarrow \ce{^{14}_7 N + ^1_0 n} \\
\ce{^{14}_7 N + ^4_2 He} &\rightarrow (\ce{^{18}_9 F^*}) &&\rightarrow \ce{^{17}_9 F + ^1_0 n} \\
\ce{^{19}_9 F + ^4_2 He} &\rightarrow (\ce{^{23}_{11} Na^*}) &&\rightarrow \ce{^{22}_{11} Na + ^1_0 n} \\
\ce{^{27}_{13} Al + ^4_2 He} &\rightarrow (\ce{^{31}_{15} P^*}) &&\rightarrow \ce{^{30}_{15} P + ^1_0 n}
\end{aligned}
\tag{15-10}
$$

In many (α,n) reactions the product nuclei are left in excited states, as evidenced by the fact that gamma rays have been observed in some of these reactions. For example, in the beryllium reaction, gamma rays consisting of three definite lines with energies of 2.7, 4.47, and 6.7 MeV have been observed.

The energies of the neutrons emitted in the (α,n) reaction can be investigated in several different ways. One method is to measure the ranges of the protons that are ejected from paraffin by the action of the neutrons. Another method is to irradiate the gas in a cloud chamber with the neutrons and to measure the ranges of the nuclei that are set in motion as a result of collisions with the neutrons. A third method is to allow the neutrons to fall on a photographic emulsion, parallel or at a small angle to the plane of the photographic plate. Although the neutrons do not leave tracks in the emulsion, the protons ejected by them from the light elements in the emulsion do leave developable grains. After the plates have been developed, the ranges of the protons in the emulsion can be measured with a microscope; these ranges can then be converted to energies by proper calibration of such plates with protons of known energies.

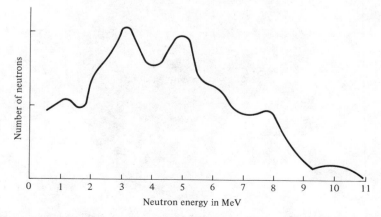

Figure 15-5 Energy distribution of neutrons from the $\ce{^9Be}(\alpha,n)$ reaction, using alpha particles from polonium. [After Whitmore and Baker, *Phys. Rev.*, **78**, 799 (1950).]

From such experiments it has been found that the neutrons from an (α,n) reaction possess high energies; in many cases the energies of the neutrons have been found to consist of several sharp energy groups.

Whitmore and Baker (1950) measured the energy distribution of the neutrons from the (α,n) reaction of beryllium, using polonium as the source of alpha particles. They used the photographic emulsion technique, making a total of nearly 7000 observations on three separate plates. Figure 15-5 shows the energy distribution curve they obtained. It will be observed that there are maxima at 3.2, 4.8, and 7.7 MeV and an upper limit to the energy of the neutrons at 11 MeV. There are indications of smaller maxima at 1.2, 5.8, and 9.7 MeV.

15-6 Discovery of the Positron

Shortly after the discovery of the neutron, another new particle, the *positron,* was discovered by C. D. Anderson (1932) in his experiments on the particles produced by the action of the very penetrating rays known as *cosmic rays,* which come to the earth from all directions in space. Anderson was taking Wilson cloud-chamber photographs of the tracks of the particles in a strong magnetic field. A few tracks were found to be curved, showing that they were formed by charged particles passing through the gas in the cloud chamber. From the appearance of these tracks (see Chapter 13) they were judged to be due to particles of electronic mass and electronic charge, but from the direction

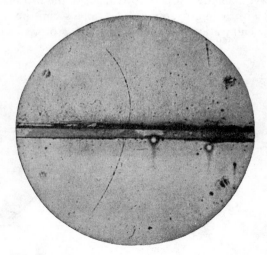

Figure 15-6 Cloud-chamber photograph of the path of a positron in a magnetic field. The positron originated at the bottom of the chamber and passed through a sheet of lead 6 mm thick. The magnetic field is directed into the paper. (Photograph by Carl D. Anderson.)

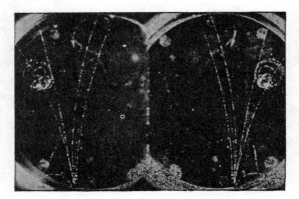

Figure 15-7 A pair of stereoscopic photographs of the tracks of a group of charged particles produced in a cloud chamber by the action of cosmic rays. The picture on the left is the direct image; the one on the right is a reflected image. The magnetic field of 7900 gauss is directed into the paper. In the picture on the left the three tracks on the left are electron tracks, and the three on the right are positron tracks. The energies of these particles, from left to right, are 3.5, 55, 190, 78, 70, and 90 MeV. (Photograph by Carl D. Anderson.)

of the curvature of these tracks it was evident that the particles producing them must have been positively charged. Anderson called these particles *positrons*. One of the first photographs to establish definitely the existence of a positron is shown in Figure 15-6. The particle originated at the bottom of the chamber, passed through a lead plate 6 mm thick, and then continued with a smaller amount of energy. From the curvatures of these two parts of the track in the magnetic field of known strength, in this case directed into the plane of the figure, and from the amount of ionization along the paths, it was concluded that the positron had an energy of 63 MeV before entering the lead and emerged from it with 23-MeV energy. Figure 15-7 shows a pair of stereoscopic pictures of the tracks of a group of charged particles produced at the top of the cloud chamber by the action of cosmic rays. Three of the tracks are produced by electrons and three by positrons.

About a year after the discovery of the positron by Anderson, sources of positrons became plentiful and easily obtainable as a result of the discovery by the Curie-Joliots of the phenomenon of artificial or induced radioactivity.

15-7 Discovery of Artificial or Induced Radioactivity

One of the most important discoveries in nuclear physics came from experiments on the bombardment of light nuclei by alpha particles. In the course of such experiments with boron and aluminum as targets M. and Mme Curie-Joliot (1934) observed that the bombarded substances continued to emit radiations even after the source of alpha particles had been removed. Ionization measurements and magnetic deflection experiments showed that the radiations consisted of positrons. Further, the intensity of the radiation was found to decrease exponentially with time, just as in the case of the naturally radioactive elements. The half-life T of the positron disintegration was measured in each case. The explanation of this phenomenon given by the Curie-Joliots was that the product nucleus formed in the (α, n) reaction in each case was an unstable isotope, which then disintegrated with the emission of a positron. The nuclear reactions for these elements are

$$^{10}_{5}B + {}^{4}_{2}He \rightarrow ({}^{14}_{7}N^{*}) \rightarrow {}^{13}_{7}N + {}^{1}_{0}n$$

then

$$^{13}_{7}N \rightarrow {}^{13}_{6}C + \beta^{+} + \nu \qquad T = 9.96 \text{ min} \qquad (15\text{-}11)$$

$$^{27}_{13}Al + {}^{4}_{2}He \rightarrow ({}^{31}_{15}P^{*}) \rightarrow {}^{30}_{15}P + {}^{1}_{0}n$$

then

$$^{30}_{15}P \rightarrow {}^{30}_{14}Si + \beta^{+} + \nu \qquad T = 2.5 \text{ min} \qquad (15\text{-}12)$$

The symbol β^{+} is used to represent the positron when it is emitted by a nucleus. Occasionally the symbol $_{+1}^{0}e$ is used to represent the positron, since its charge is equal to that of a proton and its mass number is zero. A glance at the table of isotopes (Appendix IV) reveals that the product nuclei formed in the (α,n) reactions listed above are not among the known stable isotopes but that the nuclei formed after the emission of the positron are known stable isotopes.

One of the best methods for the identification of an element is chemical analysis. Because of the minute amount of material that is made radioactive by alpha-particle bombardment, it is necessary to use a somewhat indirect chemical test to identify the radioactive isotope. The general method used is to dissolve the irradiated substance and then to add to this solution small quantities of neighboring elements in the ordinary inactive form. The various elements are then separated by chemical methods, generally the precipitation of an insoluble salt and sometimes the formation of a gaseous compound. These materials are put in different tubes and each one is tested for radioactivity. The chemical identification of the radioelement is then easily made. The Curie-Joliots made chemical tests on each of the materials investigated and in each case they were able to identify the radioactive isotope; for example, in the boron reaction they made a target of boron nitride, BN, irradiated it with alpha particles for several minutes, and then heated it with caustic soda. One of the products of this chemical reaction was gaseous ammonia, NH_3. The fact that this ammonia was the only one of the chemical substances which was radioactive indicated that the nitrogen, $^{13}_{7}N$, was the radioelement produced in this experiment. Its half-life was found to be the same as that produced in other boron targets, whereas no radioactivity was observed when ordinary nitrogen was used as a target.

Many nuclei formed in (α,n) reactions are unstable and disintegrate with the emission of positrons. A few of these are given below, together with the measured half-lives.

$$
\begin{aligned}
{}^{17}_{9}\text{F} &\rightarrow {}^{17}_{8}\text{O} + \beta^{+} + \nu & T &= 66 \text{ sec} \\
{}^{22}_{11}\text{Na} &\rightarrow {}^{22}_{10}\text{Ne} + \beta^{+} + \nu & T &= 2.58 \text{ yr} \\
{}^{26}_{13}\text{Al} &\rightarrow {}^{26}_{12}\text{Mg} + \beta^{+} + \nu & T &= 6.5 \text{ sec} \\
{}^{27}_{14}\text{Si} &\rightarrow {}^{27}_{13}\text{Al} + \beta^{+} + \nu & T &= 4.2 \text{ sec} \\
{}^{30}_{15}\text{P} &\rightarrow {}^{30}_{14}\text{Si} + \beta^{+} + \nu & T &= 2.5 \text{ min} \\
{}^{34}_{17}\text{Cl} &\rightarrow {}^{34}_{16}\text{S} + \beta^{+} + \nu & T &= 32.0 \text{ min}
\end{aligned}
\tag{15-13}
$$

15-8   Induced Beta Decay

After the discovery of induced radioactivity by the Curie-Joliots, physicists throughout the world began a program of producing radioactive isotopes by bombarding nuclei of the different elements with a variety of projectiles obtained both from natural radioactive sources and from particle accelerators. As we shall see, there are now radioactive isotopes for all the elements. Most of them are beta-ray emitters, alpha-particle emission being confined almost exclusively to the heavier elements.

Although the beta-particle emitters produced in the (α,n) reactions described above are positron emitters, induced beta decay can also take place by negative beta-particle emission. A few (α,p) reactions lead to the formation of radioactive isotopes that decay by negative beta-particle emission. One of these, for example, is the following:

$$
{}^{11}_{5}\text{B} + {}^{4}_{2}\text{He} \rightarrow ({}^{15}_{7}\text{N}^{*}) \rightarrow {}^{14}_{6}\text{C} + {}^{1}_{1}\text{H} \qquad Q = 0.75 \text{ MeV}
$$

followed by

$$
{}^{14}_{6}\text{C} \rightarrow {}^{14}_{7}\text{N} + \beta^{-} + \bar{\nu}, \qquad T = 5730 \text{ yr}
\tag{15-14}
$$

The beta-ray spectra, both positive and negative, of the artificially produced beta emitters are similar to those of natural beta emitters; that is, each spectrum shows a continuous distribution of energy up to the maximum end-point energy. We have already shown that the negative beta-ray emission can be explained by assuming a neutron to disintegrate into a proton, an electron, and an antineutrino, thus:

$$
n \rightarrow p + \beta^{-} + \bar{\nu}
\tag{15-15}
$$

To account for the emission of positrons it is assumed that the other type of nucleon, the proton, disintegrates into a neutron, a positron, and a neutrino, thus:

$$
p \text{ (in nucleus)} \rightarrow n \text{ (in nucleus)} + \beta^{+} + \nu
\tag{15-16}
$$

However, since the mass of the proton is less than that of the neutron, the process given by Equation (15-16) can occur only if sufficient energy is supplied to the proton by the other particles of the nuclear system. The disintegration of the proton can thus occur only for a proton that is in the nucleus.

We have shown (Section 14-13) that the disintegration energy in negative beta decay is equal to the difference in masses between the initial atom and the product atom. However, this is not the case for positron emission. The energy available for a β^+-disintegration is the difference in mass between the initial nucleus $_Z^A M_n$ and the sum of the masses of the final particles—that is, the nucleus $_{Z-1}^A M_n$ and the positron, m_e, thus:

$$\mathcal{E} = {}_Z^A M_n - [{}_{Z-1}^A M_n + m_e] \tag{15-17}$$

To convert to atomic masses let us add Z electronic masses to each term to obtain

$$\mathcal{E} = [{}_Z^A M_n + Z m_e] - [{}_{Z-1}^A M_n + (Z-1)m_e + 2m_e]$$

from which

$$\mathcal{E} = {}_Z^A M - {}_{Z-1}^A M - 2m_e \tag{15-18}$$

where $_Z^A M$ is the atomic mass of the parent atom and $_{Z-1}^A M$ is the atomic mass of the product atom. The maximum kinetic energy of the positron disintegration is thus less than the difference in atomic masses of the parent and product atoms by the energy equivalent of two electronic masses—that is, by $2m_e c^2$, where c is the speed of light.

In both positive and negative beta decay the parent and daughter (product) nuclei form a pair of isobars. In the case of β^--decay the disintegration energy is equal to the maximum kinetic energy of the beta-ray spectrum, but in the case of β^+-decay the disintegration energy is the sum of the end-point energy of the beta-ray spectrum plus twice the rest-mass energy of the electron. In other words, in β^+-decay the difference in energy between the parent *atom* and the daughter *atom* must be at least $2m_e c^2$—that is, approximately 1 MeV. Another way of describing this is to say that when the nucleus emits a positron, an electron must also be released from the external part of the atom in order that the product atom may be electrically neutral.

15-9   Simple Alpha-Particle Capture; Radiative Capture

Both the (α,p) and (α,n) reactions described previously involve the emission of a nuclear particle by the compound nucleus. It is possible, however, for the compound nucleus formed by the capture of an alpha particle to go to a more stable configuration without emitting a particle but emitting a gamma-ray photon instead. This phenomenon of the simple capture, or radiative capture, of alpha particles was first observed by Bennet, Roys, and Toppel (1950). They used comparatively low-energy alpha particles produced by accelerating helium ions by means of an electrostatic generator. The energies of the alpha particles could be accurately controlled and measured. Various light elements, such as lithium, beryllium, and boron, were used as targets for these projectiles. The radiative capture process can be called an (α, γ) process.

For the case of lithium, $Z = 3$, $A = 7$, the reaction is

$$^7\text{Li} + {}^4\text{He} \rightarrow ({}^{11}\text{B*}) \rightarrow {}^{11}\text{B} + \gamma \qquad (15\text{-}19)$$

Figure 15-8 shows the yield of gamma-ray photons as a function of the bombarding energy of the alpha particles. It will be noted that there is a sharp increase in yield at about 0.4 MeV, after which the intensity levels off until the energy of the alpha particle reaches about 0.82 MeV, at which point there is another sharp increase in the yield of gamma-ray photons. Another sudden increase in yield was observed at about 0.96 MeV, but this was not quite so sharp as the earlier two increases in yield.

The sharp increases in the yield of gamma rays at alpha-particle energies of 0.4, 0.82, and 0.96 MeV are interpreted as *resonance capture* of the alpha

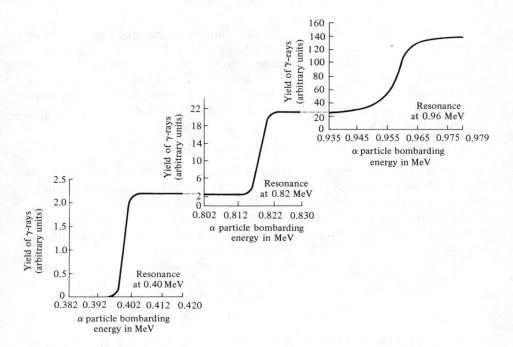

Figure 15-8   Yield of gamma rays in radiative capture of alpha particles by ^{7}Li. [After Bennett, Roys, and Toppel, *Phys. Rev.,* **82**, 20 (1951).]

particle—that is, the probability of capture is comparatively large at these energies. That the yield curve presents a steep slope at these energies rather than sharp maxima is due to the fact that a thick target was used rather than a thin one. These resonances are ascribed to the compound nucleus and indicate the existence of energy levels in this nucleus.

The energies plotted as abscissae in Figure 15-8 are those of the alpha particles incident on the surface of the target. If the target is thick, those alpha particles with the resonant energy of, say, 0.4 MeV may be captured by nuclei near the surface; alpha particles of greater energy will lose energy in penetrating the target and may be resonantly captured by nuclei inside the target when their energy is reduced to the resonant energy, 0.4 MeV. Hence in a thick target resonance capture at 0.4 MeV will occur at different distances from the surface for alpha particles whose initial energy was greater than the resonance energy, giving rise to a steep slope at this value rather than a sharp line.

The gamma rays emitted in a radiative capture process are sometimes called *capture γ-rays*. The radiative capture process is one type of evidence confirming the existence of a compound nucleus.

15-10 Disintegrations by Proton Bombardment

The first successful disintegration experiments utilizing protons as bombarding particles were performed by Cockcroft and Walton (1932). The protons were produced in a hydrogen discharge tube operated at voltages up to 500,000 volts. When the protons were used to bombard a lithium target, alpha particles were observed on a fluorescent screen. This experiment was later repeated by Dee and Walton, using a Wilson cloud chamber for detecting the alpha particles. This nuclear reaction is given by the equation

$$\ _{3}^{7}\mathrm{Li} + \ _{1}^{1}\mathrm{H} \rightarrow (\ _{4}^{8}\mathrm{Be^{*}}) \rightarrow \ _{2}^{4}\mathrm{He} + \ _{2}^{4}\mathrm{He} + Q \qquad (15\text{-}20)$$

that is, the compound nucleus formed as a result of the capture of the proton by the lithium nucleus breaks up into two alpha particles that travel in almost opposite directions. Each alpha particle has a range of 8.31 cm in air corresponding to an energy of about 8.63 MeV. The reaction energy Q is found to be 17.28 MeV and checks very well with the value obtained from the difference in the masses of the initial and final particles.

There are many other interesting disintegrations produced by bombardment with protons. The nuclear reaction which occurs with the lithium target containing only the isotope of mass number 6 is given by the equation

$$\ _{3}^{6}\mathrm{Li} + \ _{1}^{1}\mathrm{H} \rightarrow (\ _{4}^{7}\mathrm{Be^{*}}) \rightarrow \ _{2}^{4}\mathrm{He} + \ _{2}^{3}\mathrm{He} + Q \qquad (15\text{-}21)$$

in which two helium atoms, one of mass number 4 and the other of mass number 3, are produced with ranges of 0.8 and 1.2 cm, respectively. The measured value of Q is 3.94 MeV. A few of the other reactions observed are given below:

$$^{11}_{5}\text{B} \; + \; ^1_1\text{H} \rightarrow (^{12}_{6}\text{C*}) \;\; \rightarrow \;\; ^8_4\text{Be} + ^4_2\text{He}$$

$$^{23}_{11}\text{Na} + ^1_1\text{H} \rightarrow (^{24}_{12}\text{Mg*}) \rightarrow ^{20}_{10}\text{Ne} + ^4_2\text{He} \tag{15-22}$$

$$^{35}_{17}\text{Cl} + ^1_1\text{H} \rightarrow (^{36}_{18}\text{A*}) \;\; \rightarrow \;\; ^{32}_{16}\text{S} \;\; + ^4_2\text{He}$$

The disintegrations listed above can be classified as (p,α) reactions, since one of the final products is an alpha particle.

The $^{11}\text{B}(p,\alpha)^8\text{Be}$ reaction has been studied very extensively, and it has been found that in the majority of cases the beryllium nucleus is left in an excited state and then disintegrates with the emission of two alpha particles.

Not all reactions in which the proton is the bombarding particle are of the (p,α) type. In the case of beryllium two different reactions have been observed, one in which an alpha particle is emitted and another in which a deuteron is emitted. The latter is called a (p,d) reaction, where d represents the deuteron. These reactions are given by the equations

$$^9_4\text{Be} + ^1_1\text{H} \rightarrow (^{10}_{5}\text{B*}) \rightarrow ^6_3\text{Li} + ^4_2\text{He}$$

$$^9_4\text{Be} + ^1_1\text{H} \rightarrow (^{10}_{5}\text{B*}) \rightarrow ^8_4\text{Be} + ^2_1\text{H} \tag{15-23}$$

Gamma rays have also been observed as the result of the bombardment of an element by protons. In some of these cases the gamma-ray photons possess such high energies that the only possible explanation seems to be that the proton is simply captured by the nucleus and that the compound nucleus thus formed is in an excited state. This nucleus then returns to its normal state with the emission of a gamma-ray photon of definite energy. This process is designated as a (p,γ) reaction and is a simple proton-capture process; the emitted gamma rays are capture gammas. One such example is the reaction

$$^7_3\text{Li} + ^1_1\text{H} \rightarrow (^8_4\text{Be*}) \rightarrow ^8_4\text{Be} + \gamma \tag{15-24}$$

The capture gamma-ray spectrum has been investigated by Walker and McDaniel, who found a very intense sharp line of 17.6 MeV energy and a less intense broad line of 14.4 MeV energy. The interpretation of this spectrum is that the ^8Be nucleus is left in an excited state in the (p,γ) reaction with ^7Li; it then can decay to the ground state by emitting a gamma-ray photon of 17.6 MeV, or it may emit a photon of 14.4 MeV and then disintegrate into two alpha particles. The latter process has been studied by Burcham and Freeman. They measured the alpha-particle energies and found them centered about an energy of 1.38 MeV when the bombarding protons had an energy of 440 keV. These results lead to a value of the excited state of ^8Be as 2.9 MeV above the ground state.

The three processes in which ^8Be is formed, namely,

$$^{11}\text{B}(p,\alpha)^8\text{Be}, \quad ^9\text{Be}(p,d)^8\text{Be}, \quad \text{and} \quad ^7\text{Li}(p,\gamma)^8\text{Be}$$

can be represented in a simplified energy-level diagram shown in Figure 15-9. It will be noted that ^{8}Be, even in the ground state, is unstable and disintegrates into two alpha particles. Its half-life is very short, probably less than 2×10^{-14} sec.

15-11 Disintegration by Deuteron Bombardment

A great many nuclear reactions have been observed with deuterons as the bombarding particles. Their energies have usually been of the order of several MeV. The reactions involving deuterons may be classified according to the type of particle or particles emitted by the compound nucleus that is formed as a result of the capture of the deuteron. Alpha particles, tritons, protons, and neutrons have been produced in these processes. In some cases the product nucleus, formed as a result of the emission of one of these particles, is radioactive and disintegrates with the emission of a positron or an electron. Gamma rays have been observed in some of the reactions that involve the emission of a particle. In a few cases the compound nucleus has been observed to break up into three particles. Only a few typical examples of each of these reactions resulting from the capture of a deuteron will be considered.

One of the simplest and most important of these reactions is the one in which the deuterons are used to bombard a target containing deuterons.

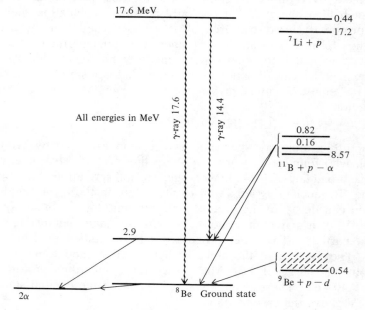

Figure 15-9 Simplified energy-level diagram of ^{8}Be showing its formation by three methods using protons as projectiles together with the modes of decay of three of the levels of ^{8}Be. [After Hornyak, Lauritsen, Morrison, and Fowler, *Revs. Modern Phys.*, **22**, 310 (1950).]

Deuteron targets have been made by freezing heavy water (deuterium oxide) onto a surface kept cold by means of liquid air. Other deuterium targets have been made out of compounds such as ammonium sulfate, in which the ordinary hydrogen was replaced by deuterium. Two different nuclear reactions have been observed as a result of the bombardment of deuterium by deuterons:

$$^2_1\text{H} + {}^2_1\text{H} \rightarrow ({}^4_2\text{He}^*) \rightarrow {}^3_1\text{H} + {}^1_1\text{H} + Q \qquad (15\text{-}25)$$

and

$$^2_1\text{H} + {}^2_1\text{H} \rightarrow ({}^4_2\text{He}^*) \rightarrow {}^3_2\text{He} + n + Q \qquad (15\text{-}26)$$

The first of these reactions has been studied with the aid of a Wilson cloud chamber, which enabled the particles to be identified as isotopes of hydrogen of mass numbers 1 and 3. The ranges of these particles in air have been found to be 14.7 and 1.6 cm, respectively, yielding a value for $Q = 4.03$ MeV. With the value of Q and the known masses of ^{1}H and ^{2}H, the mass of ^{3}H, sometimes called *tritium,* can be determined very accurately.

The dependence of the cross section for the production of tritium ^{3}H, in the ^{2}H (d,p) ^{3}H reaction, on the energy of the incident deuteron has been determined by Cook and Smith (1953) by bombarding a thin target of deuterium absorbed in zirconium films with deuterons of comparatively low energy, 50 to 100 keV. The results showed a linear dependence of total cross section on deuteron energy.

Hydrogen of mass number $A = 3$ is unstable. O'Neal and Goldhaber showed that it disintegrates with the emission of a beta particle as follows:

$$^3_1\text{H} \rightarrow {}^3_2\text{He} + \beta^- + \bar{\nu} \qquad T = 12.26 \text{ yr} \qquad (15\text{-}27)$$

The beta rays emitted by this radioactive isotope of hydrogen have a maximum kinetic energy of only 18 keV.

The energy of the neutrons formed in the bombardment of deuterium by deuterons has been investigated by observing the recoil tracks of the atoms of the gas in a cloud chamber. It was observed that the neutrons were almost homogeneous in energy. Recent measurements show that the energy of the neutrons emitted at right angles to the direction of motion of the deuterons is 2.45 MeV plus one-quarter of the deuteron energy. (See problem 15-2.) This reaction forms a very convenient source of neutrons of known energy. Furthermore, neutrons have been observed for comparatively low values of incident deuteron energies—that is, of the order of 6 keV. The neutron yield increases rapidly with the deuteron energy. The value of the reaction energy has been found to be $Q = 3.269$ MeV, and with this value of Q the mass of ^{3}He can be determined.

Some of the nuclear reactions produced by the capture of a deuteron by lithium that have been observed are as follows:-

$$\text{}^{6}_{3}\text{Li} + \text{}^{2}_{1}\text{H} \rightarrow (\text{}^{8}_{4}\text{Be*}) \rightarrow \text{}^{4}_{2}\text{He} + \text{}^{4}_{2}\text{He} \qquad Q = 22.23 \text{ MeV}$$

$$\text{}^{6}_{3}\text{Li} + \text{}^{2}_{1}\text{H} \rightarrow (\text{}^{8}_{4}\text{Be*}) \rightarrow \text{}^{7}_{3}\text{Li} + \text{}^{1}_{1}\text{H}$$

$$\text{}^{7}_{3}\text{Li} + \text{}^{2}_{1}\text{H} \rightarrow (\text{}^{9}_{4}\text{Be*}) \rightarrow \text{}^{4}_{2}\text{He} + \text{}^{4}_{2}\text{He} + n \qquad (15\text{-}28)$$

$$\text{}^{7}_{3}\text{Li} + \text{}^{2}_{1}\text{H} \rightarrow (\text{}^{9}_{4}\text{Be*}) \rightarrow \text{}^{8}_{3}\text{Li} + \text{}^{1}_{1}\text{H}$$

followed by

$$\text{}^{8}_{3}\text{Li} \rightarrow \text{}^{8}_{4}\text{Be} + \beta^{-} + \bar{\nu} \qquad T = 0.85 \text{ sec}$$

The lithium isotope of mass number 8 is radioactive and disintegrates with the emission of a beta particle. The beta rays liberated in the disintegration of ^{8}Li have a continuous energy distribution up to a maximum; the end point is about 12 MeV. The ^{8}Be nucleus formed in this reaction is unstable and breaks up into two alpha particles.

In the case of carbon bombarded by deuterons two reactions have been observed:

$$\text{}^{12}_{6}\text{C} + \text{}^{2}_{1}\text{H} \rightarrow (\text{}^{14}_{7}\text{N*}) \rightarrow \text{}^{13}_{6}\text{C} + \text{}^{1}_{1}\text{H}$$
$$\text{}^{12}_{6}\text{C} + \text{}^{2}_{1}\text{H} \rightarrow (\text{}^{14}_{7}\text{N*}) \rightarrow \text{}^{13}_{7}\text{N} + n \qquad (15\text{-}29)$$

followed by

$$\text{}^{13}_{7}\text{N} \rightarrow \text{}^{13}_{6}\text{C} + \beta^{+} + \nu$$

The half-life of the radionitrogen formed in this reaction is identical with that observed in the reaction produced by the capture of an alpha particle by boron. The positron spectrum shows a continuous range of energies. The maximum value of the positron energy in this case is 1.24 MeV.

Reactions with deuterons as projectiles have led to (d,p), (d,n), and (d,α) processes with elements throughout the periodic table. Of some interest is the disintegration of sodium by deuteron bombardment. The following two reactions have been observed:

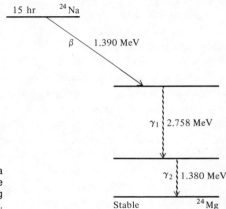

Figure 15-10 Energy-level diagram of the beta decay of ^{24}Na to an excited state of ^{24}Mg. The latter decays to the stable ground state of ^{24}Mg by emitting two gamma-ray photons in cascade.

$$_{11}^{23}\text{Na} + _{1}^{2}\text{H} \rightarrow (_{12}^{25}\text{Mg*}) \rightarrow _{12}^{24}\text{Mg} + n$$
$$_{11}^{23}\text{Na} + _{1}^{2}\text{H} \rightarrow (_{12}^{25}\text{Mg*}) \rightarrow _{11}^{24}\text{Na} + _{1}^{1}\text{H}$$

(15-30)

followed by

$$_{11}^{24}\text{Na} \rightarrow _{12}^{24}\text{Mg} + \beta^- + \bar{\nu} \qquad T = 15.0 \text{ hr}$$

The magnesium formed in the (d,n) reaction is left in a stable state. The magnesium formed in the decay of the radioactive sodium is left in an excited state and decays to the ground state by emitting two gamma-ray photons in cascade as shown in Figure 15-10.

Radioactive sodium has been used as a "tracer" in many physiological experiments.

The gamma-ray spectrum emitted by the disintegration of ^{24}Na has been investigated by R. Hofstadter and J. A. McIntyre (1950) with a scintillation counter utilizing a thallium-activated sodium iodide crystal, NaI(Tl). A narrow beam of gamma rays from a sample of ^{24}Na (about 1 mCi) traveled through the middle of the crystal. In Figure 15-11 the number of counts per minute of the scintillations produced by the gamma rays is plotted against the pulse height in volts. The origin of each line on the graph is indicated thereon. When gamma rays traverse matter, they eject electrons from it either by a Compton process or by a photoelectric process. The electrons thus ejected give up some of their energy to produce visible light, or scintillations, within the crystal. The lines of the graph marked "Compton" are produced by scintillations that have their origin in the Compton electrons; the lines marked "photoelectric" have their origin in the scintillations produced by the action of the photoelectrons. In addition, if the energy of the gamma-ray photons exceeds $2m_e c^2$, where m_e is the mass of the electron, the gamma-ray photon may, in the presence of a strong electric field, form an electron and a positron (see Section 18-6). The sum of the kinetic energies of this pair is $h\nu - 2m_e c^2$ where $h\nu$ is the energy of the gamma-ray photon. The line marked "pairs due to 2.76 MeV" is due to scintillations produced by the electron-positron pairs formed by the 2.76-MeV photons.

The energy to be assigned to each of the lines obtained with the scintillation spectrometer is determined by calibrating it with gamma-ray photons of known energy. In this case Hofstadter and McIntyre used the 1.17- and 1.33-MeV gamma-ray photons of ^{60}Co for calibration purposes; these are shown in the insert on the graph. The lines obtained with the scintillation spectrometer are ascribed to two gamma-ray photons from ^{24}Na, one of 1.38 MeV and the other of 2.76 MeV.

15-12 Disintegration of Nuclei by Photons

Atomic nuclei have been disintegrated by high-energy photons. In most cases this process of *photodisintegration* results in the emission of neutrons

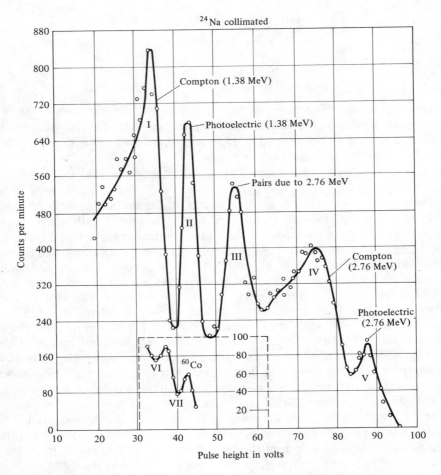

Figure 15-11 Graph of the gamma-ray spectrum of ^{24}Na taken with a NaI scintillation spectrometer. The peaks are due to the two gamma-ray photons of 1.38 MeV and 2.76 MeV. The ^{60}Co curve was used for calibrating the spectrometer. (Courtesy of R. Hofstadter and J. A. McIntyre.)

by the nuclei that have been raised to excited states by the absorption of these photons. In the early experiments high-energy gamma-ray photons were used, but with the development of the betatron, x-ray photons of sufficiently high energy have become available for these experiments. Of very great interest is the disintegration of the deuteron, the lightest of the complex nuclei, by the action of the gamma rays. This nuclear reaction is

$$^2_1H + \gamma \rightarrow (^2_1H^*) \rightarrow \,^1_1H + n \qquad (15\text{-}31)$$

According to our present view of the structure of the nucleus, the deuteron consists of a proton and a neutron held together by some force of attraction.

A measurement of the minimum amount of energy necessary to disrupt the deuteron would also give the binding energy of the proton and neutron in the nucleus. Chadwick and Goldhaber, who discovered the above reaction, used gamma rays from ^{208}Pb and measured the energy of the protons liberated by the amount of ionization they produced. The energy of the gamma-ray photon is known to be 2.62 MeV. The binding energy of the neutron and proton will be the difference between the energy of the incident photon and the sum of the kinetic energies of the proton and neutron. Since the mass of the neutron is almost the same as that of the proton, the total kinetic energy of the two particles will be twice that of the proton: the value of the latter is known from the ionization measurements. The total kinetic energy was found to be 0.45 MeV, yielding a binding energy of $2.62 - 0.45 = 2.17$ MeV. Most recent experiments yield a value of 2.226 MeV for the binding energy of the deuteron. Since the binding energy per nucleon is of the order of 7 to 8 MeV for most nuclei, we can infer that the deuteron is a loosely bound state of the neutron and proton.

Since the masses of the proton and deuteron are known very accurately from measurements with the mass spectrograph, the photodisintegration of the deuteron affords the most accurate means of determining the mass of the neutron. The value so determined is 1.008665u.

To produce the photodisintegration of the heavier nuclei the energy of the incident photon must exceed the binding energy of the particle to be ejected. In many cases in which photodisintegration results in the emission of a neutron the product nucleus is radioactive; the reaction can then be studied by observing this radioactivity and determining its half-life. The particular isotope involved in this photodisintegration can then more readily be identified. Baldwin and Koch (1945) performed such experiments using the high-energy x-ray photons emitted from the betatron. By varying the energy of the x-ray photons used in irradiating the samples of material, they were able to determine the minimum amount of energy—that is, the *threshold* for the photodisintegration of several different nuclei in the range of atomic numbers $Z = 6$ to $Z = 47$. In each case the effect produced by the photodisintegration was measured by the intensity of the beta rays emitted by the irradiated sample. Figure 15-12 shows a typical curve obtained by irradiating flat plates of iron, 10 cm × 8 cm × 1 cm, with x-rays of varying energies for 10 min and then placing the iron plate near a counter sensitive to beta rays. From the curve the photodisintegration threshold of iron, $A = 53$, is found to be 14.2 MeV.

The (γ,n) reaction has since been studied very extensively and threshold values have been determined for a very large number of nuclei. Sher, Halpern, and Mann determined the thresholds of many (γ,n) reactions by detecting the emitted neutrons with a boron trifluoride-filled proportional counter. Just as the (γ,n) reaction with deuterium gave the binding energy of the neutron to the proton, so the threshold value of the (γ,n) reaction with any isotope of mass number A will give the binding energy of the neutron in the nucleus of the isotope of mass number $A - 1$. We shall consider the threshold values again in our discussion of the structure of the nucleus (Section 17-17).

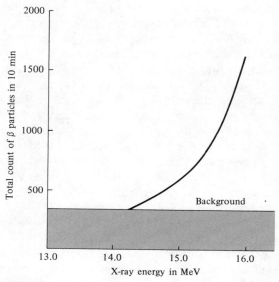

Figure 15-12 Excitation curve for the photodisintegration of ^{53}Fe showing the threshold at 14.2 MeV.

In addition to the (γ,n) reactions, other reactions have been observed in the photodisintegration of nuclei. Among these are the (γ,p) reaction in which a proton is emitted, the (γ,d) reaction in which a deuteron is emitted, and the (γ,t) reaction, in which *triton* is emitted, the nucleus of $^{3}_{1}$H, and the (γ,α) reaction in which an alpha particle is emitted. The following are typical of the above nuclear photodisintegrations:

$$
\left.
\begin{aligned}
&^{9}\text{Be}(\gamma,p)^{8}\text{Li} \\
&^{10}\text{B } (\gamma,d)^{8}\text{Be} \\
&^{11}\text{B } (\gamma,t)^{8}\text{Be} \\
&^{12}\text{C } (\gamma,\alpha)^{8}\text{Be}
\end{aligned}
\right\}
\tag{15-32}
$$

The above reactions may be considered as the reverse of the capture gamma processes mentioned previously. The last three photodisintegrations show that different nuclear reactions can be used to produce a given nucleus, in this case ^{8}Be.

15-13 Disintegration by Neutron Bombardment

Neutrons, because they possess no electric charge, have proved to be very effective in penetrating the positively charged nuclei, thereby producing nuclear transformations. Not only are high-energy neutrons capable of penetrating the nucleus but comparatively *slow* neutrons have also been found to be extremely effective. A great deal of work has been done with slow neutrons, and the information so obtained is the basis of the nuclear model proposed by Bohr. The simplest method of obtaining slow neutrons is to allow the fast neutrons from some source, such as the alpha-particle-beryllium reaction, to pass through some hydrogen-containing substance such as paraffin or water. A neutron gives up a large fraction of its energy in a collision with a hydrogen nucleus, and after many collisions it will come to thermal equilibrium with the material; that is, its average energy will be equal to the energy of thermal agitation, which is equivalent to $\frac{1}{40}$ eV at room temperature. In one type of reaction (n,α) the capture of a slow neutron results in the emission of an alpha particle. Two such cases that have been studied extensively are

$$_{3}^{6}\text{Li} + n \rightarrow (_{3}^{7}\text{Li}^{*}) \rightarrow {_{1}^{3}}\text{H} + {_{2}^{4}}\text{He}$$
$$_{5}^{10}\text{B} + n \rightarrow (_{5}^{11}\text{B}^{*}) \rightarrow {_{3}^{7}}\text{Li} + {_{2}^{4}}\text{He} \tag{15-33}$$

The (n,α) reaction with boron is widely used as a sensitive detector of neutrons, particularly in ionization chamber work. In some cases the ionization chamber is lined with a boron compound; more frequently the ionization chamber is filled with a boron gas such as BF_3, boron trifluoride. The ionization in the chamber is produced by the alpha particle released in the (n,α) reaction with boron.

Neutron-induced nuclear disintegrations can also be studied with a cloud chamber. A. B. Lillie (1952) investigated the disintegration of oxygen and

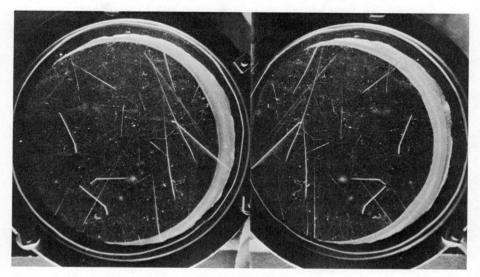

Figure 15-13   Stereoscopic photographs of disintegrations produced by bombarding oxygen in a cloud chamber with high energy neutrons. Neutrons enter from the top of the picture. The forked tracks are produced by alpha particles (longer track) and ^{13}C ions (shorter track). [Photograph by Lillie, *Phys. Rev.*, **87**, 716 (1952).]

Figure 15-14 Stereoscopic photograph of disintegrations produced by bombarding nitrogen in cloud chamber with high energy neutrons. The neutrons enter from the top of the picture. In addition to a forked track, there are two cases of a three-particle disintegration. There is also a disintegration in which a hydrogen nucleus, either *p, d,* or *t,* is emitted; this is the long track that crosses chamber and leaves. [Photograph by Lillie, *Phys. Rev.,* **87,** 716 (1952).]

nitrogen by 14-MeV neutrons obtained from the ^{3}H (d,n) ^{4}He reaction. The deuterons had an energy of 200 keV, and a monoenergetic beam of neutrons was obtained at an angle of 90 degrees to the deuteron beam. The neutrons entered a cloud chamber filled with oxygen or nitrogen at a pressure of 1 atmosphere and containing water vapor. Figure 15-13 is a stereoscopic photograph showing the disintegrations produced with oxygen in the chamber. The forked tracks are produced by the products of the reaction ^{16}O (n,α) ^{13}C. The neutron paths are not visible, since neutrons are nonionizing particles. The alpha-particle track is the longer one, and the track of the product nucleus ^{13}C is the shorter one. Figure 15-14 is an interesting photograph showing several different processes taking place when nitrogen is used in the cloud chamber. There is a typical forked track produced by the ^{14}N (n,α) ^{11}B reaction. In addition, there are two examples of the disintegration of the compound nucleus $(^{15}$N*$)$ into three particles, two of which are alpha particles. This reaction equation is

$$^{14}_{7}\text{N} + n \rightarrow (^{15}_{7}\text{N*}) \rightarrow {}^{7}_{3}\text{Li} + {}^{4}_{2}\text{He} + {}^{4}_{2}\text{He} \qquad (15\text{-}34)$$

When fast neutrons are captured by heavier nuclei, resulting in the emission of an alpha particle, the product nucleus is usually radioactive. Some typical reactions are

$$^{23}_{11}\text{Na} + n \rightarrow (^{24}_{11}\text{Na*}) \rightarrow {}^{20}_{9}\text{F} + {}^{4}_{2}\text{He}$$

followed by

$$^{20}_{9}F \rightarrow {}^{20}_{10}Ne + \beta^- + \bar{\nu} \qquad T = 10.7 \text{ sec}$$
$$^{27}_{13}Al + n \rightarrow ({}^{28}_{13}Al^*) \rightarrow {}^{24}_{11}Na + {}^{4}_{2}He \qquad (15\text{-}35)$$

followed by

$$^{24}_{11}Na \rightarrow {}^{24}_{12}Mg + \beta^- + \bar{\nu} \qquad T = 15.0 \text{ hr}$$

The capture of a neutron may sometimes result in the emission of a proton by the compound nucleus. This (n,p) process has been observed with slow neutrons in the case of nitrogen in the reaction

$$^{14}_{7}N + n \rightarrow ({}^{15}_{7}N^*) \rightarrow {}^{14}_{6}C + {}^{1}_{1}H$$

followed by $\qquad\qquad\qquad\qquad\qquad\qquad\qquad\qquad\qquad\qquad$ (15-36)

$$^{14}_{6}C \rightarrow {}^{14}_{7}N + \beta^- + \bar{\nu} \qquad T = 5730 \text{ yr}$$

The above reaction is assumed to be the source of the ^{14}C found in nature. This reaction undoubtedly takes place in the atmosphere where nitrogen nuclei are bombarded by neutrons produced by the interaction of cosmic rays with other particles in the atmosphere. The earth is being continually bombarded by high-energy cosmic-ray particles that reach the atmosphere from regions outside the earth.

Methods have been developed by W. F. Libby and others for dating archaeological and geological samples by determining the ^{14}C content. The basic assumption of the carbon dating method is that the relative abundance of the carbon isotopes has remained unchanged for the last few thousand years. When a living organism dies, its intake of carbon ceases and the amount of radioactive ^{14}C in it decreases continually. A determination of its present ^{14}C content per gram of substance, together with a knowledge of its half-life, yields the age of the sample. Wherever other reliable methods for determining the ages of such samples were available, the two methods gave consistent results. Ages of such samples extend up to 5000 yr. One conclusion that can be drawn from this is that the neutron flux in cosmic radiation has been fairly constant for the extent of about 10,000 to 15,000 yr—that is, about two or three times the half-life of ^{14}C.

Tritium is also found in the atmosphere; it is most likely being produced by (n,t) reactions involving high-energy cosmic-ray neutrons and ^{14}N according to the following reaction:

$$^{14}_{7}N + n \rightarrow ({}^{15}_{7}N^*) \rightarrow {}^{12}_{6}C + {}^{3}_{1}H \qquad (15\text{-}37)$$

Tritium, as an isotope of hydrogen, should be widely distributed over the surface of the earth, but partly because of its short lifetime it occurs in minute concentrations. The relative abundance of tritium in Norwegian surface water has been measured by Grosse, Johnston, Wolfgang, and Libby (1951) and found to be about one atom of tritium to 10^{18} atoms of ordinary hydrogen.

The samples of water tested were all highly enriched with the heavier isotope D_2O, by electrolytic methods. The enrichment factor was of the order of 10^6. The ratio of tritium to deuterium in these samples was found to be of the order of 10^{-12}. The presence of tritium was determined by counting the rate at which beta rays were emitted by the sample. The natural abundance was then calculated from the measured beta-ray activity, the probable enrichment factor of the sample, its age, and the known half-life of tritium. It was noted that with a concentration of 3×10^{-18} gram of tritium per gram of ordinary hydrogen, it is the rarest atomic species discovered in nature. It was further concluded that most, if not all, of the ^3He found in the atmosphere has its origin in the disintegration of cosmic-ray-produced tritium.

Other (n,p) reactions with fast neutrons have usually resulted in the production of nuclei that are radioactive, emitting β^--particles. Some typical reactions are

$$^{32}_{16}S + n \rightarrow (^{33}_{16}S^*) \rightarrow ^{32}_{15}P + ^1_1H$$

followed by

$$^{32}_{15}P \rightarrow ^{32}_{16}S + \beta^- + \bar{\nu} \qquad T = 14.3 \text{ d}$$

(15-38)

$$^{65}_{29}Cu + n \rightarrow (^{66}_{29}Cu^*) \rightarrow ^{65}_{28}Ni + ^1_1H$$

followed by

$$^{65}_{28}Ni \rightarrow ^{65}_{29}Cu + \beta^- + \bar{\nu} \qquad T = 2.56 \text{ hr}$$

The above (n,p) reaction has several interesting consequences. It will be noticed that the bombarded nucleus and the final stable nucleus are identical, whereas the incident neutron has been transformed, apparently, into a proton, an electron, and an antineutrino. If the mass of the intermediate radioactive nucleus—for example, ^{32}P—is equal to the mass of the initial nucleus ^{32}S, the energy available for the radioactive disintegration must come from the mass difference between a neutron and a hydrogen atom plus the initial energy of the neutrons. If slow neutrons are used as bombarding particles, the energy available is equivalent to 0.000842 u or 0.784 MeV. Now the end-point energy of the beta-ray spectrum is equal to the disintegration energy, and if this end-point energy is less than 0.784 MeV, then slow neutrons will be effective in producing the above reaction; but if the end-point energy exceeds 0.784 MeV the additional energy must come from the kinetic energy of the incident neutrons; hence only fast neutrons can then be effective in producing this reaction. For example, the end-point energy of the beta-ray spectrum from ^{32}P is 1.70 MeV; hence only fast neutrons bombarding ^{32}S will be effective in producing this reaction. This reaction can then also be used to differentiate between slow and fast neutrons.

Another interesting conclusion is that if the mass of the nucleus formed in an (n,p) reaction exceeds the mass of the bombarded nucleus by more than 0.000842 u, then only fast neutrons will be effective in producing this reaction.

Sometimes the capture of a neutron is not accompanied by the emission of a particle but by the emission of a gamma-ray photon. In this case the compound nucleus is evidently raised to one of its excited states as the result of this capture and then returns to its normal state with the emission of a gamma-ray photon. Such gamma-rays have been observed coming from the paraffin used to slow down neutrons. The ultimate capture of some of the slow neutrons by the hydrogen nuclei in the paraffin yields the reaction

$$p + n \rightarrow (^2H^*) \rightarrow {}^2H + \gamma \tag{15-39}$$

This process is just the reverse of the photodisintegration of the deuteron discussed in the preceding section.

A measurement of the energy of the gamma-ray photon emitted in an (n,γ) reaction with a nucleus of mass number A will yield the binding energy of the neutron in the nucleus of mass number $A + 1$. The results so obtained can be used as a check on the binding energies obtained in the (γ,n) reactions. Of great importance for nuclear theory is the binding energy of the neutron and proton. Bell and Elliott (1950) performed an experiment for the determination of this binding energy, using the thermal neutrons from the Chalk River nuclear reactor. The neutrons were allowed to strike a block of paraffin, and the energy of the capture gamma rays was measured by means of the photoelectric effect they produced in a thin sheet of uranium. The energies of the photoelectrons thus produced were measured in a magnetic-lens beta-ray spectrometer, using a Geiger counter to detect the electrons. The spectrometer was calibrated by using the ^{208}Pb γ-ray line of 2.615 MeV and measuring the photoelectrons produced by it. The K photoelectrons gave a very sharp peak, making precise measurements possible. The value of the K ionization energy of uranium was taken as 116 keV., Using the above data, Bell and Elliott obtained a value of 2.230 MeV for the binding energy of a neutron and a proton in a deuteron. The current value is 2.2256 MeV.

The (n,γ) reaction has been observed with a great many elements, particularly the heavier ones. In most cases the isotopes formed by the capture of a neutron have been found to be radioactive. A few typical cases are

$$^{65}_{29}\text{Cu} + n \rightarrow (^{66}_{29}\text{Cu}^*) \rightarrow {}^{66}_{29}\text{Cu} + \gamma$$

followed by

$$^{66}_{29}\text{Cu} \rightarrow {}^{66}_{30}\text{Zn} + \beta^- + \bar{\nu} \qquad T = 5.10 \text{ min}$$

$$^{197}_{79}\text{Au} + n \rightarrow (^{198}_{79}\text{Au}^*) \rightarrow {}^{198}_{79}\text{Au} + \gamma \tag{15-40}$$

followed by

$$^{198}_{79}\text{Au} \rightarrow {}^{198}_{80}\text{Hg} + \beta^- + \bar{\nu} \qquad T = 2.7 \text{ days}$$

There are many cases in which the capture of a fast neutron has resulted in the emission of two neutrons by the compound nucleus. In most cases the product nucleus is unstable. A few such cases follow:

$$_{19}^{39}\text{K} + n \rightarrow (_{19}^{40}\text{K}^*) \rightarrow _{19}^{38}\text{K} + n + n$$

followed by

$$_{19}^{38}\text{K} \rightarrow _{18}^{38}\text{A} + \beta^+ + \nu \qquad T = 7.7 \text{ min}$$

$$_{51}^{121}\text{Sb} + n \rightarrow (_{51}^{122}\text{Sb}^*) \rightarrow _{51}^{120}\text{Sb} + n + n \qquad (15\text{-}41)$$

followed by

$$_{51}^{120}\text{Sb} \rightarrow _{50}^{120}\text{Sn} + \beta^+ + \nu \qquad T = 15.9 \text{ min}$$

The ability to produce a radioactive isotope of almost any element has made a new tool available to the chemist, biologist, and physiologist for the detailed study of various processes. By introducing the radioactive isotope along with the nonradioactive isotopes of an element, the progress of this element can be traced in the process under investigation by means of the beta rays emitted by the radioactive isotope. In many cases a choice of several different half-lives is available to suit the needs of the particular experiment. A great deal of work has already been done with the radioactive isotopes of carbon, sodium, iron, phosphorus, and iodine used as *tracers* in various processes.

15-14   Radioactive Decay of the Neutron

Shortly after the discovery of the neutron by Chadwick, the first accurate determination of its mass by Chadwick and Goldhaber showed that its mass was greater than that of a proton. This led them to suggest that a neutron should be unstable and should decay radioactively into a proton, an electron, and a neutrino. The Fermi theory of beta decay indicated that its half-life should be of the order of 30 min. Early experiments to find this radioactivity were unsuccessful primarily because of the unavailability of a sufficiently intense source of neutrons. With the development of nuclear reactors, intense sources of neutrons became available.

A. H. Snell and his co-workers (1948), using the neutrons from the Oak Ridge pile, were able to detect protons coming out laterally from a beam of thermal neutrons. They were later able to show the presence of electrons that were emitted simultaneously with the protons and estimated the half-life as between 10 and 30 min. J. M. Robson (1951), using the much more intense source of neutrons available from the Chalk River pile, made a more accurate determination of the half-life of the radioactive disintegration of the neutron and determined its beta-ray spectrum.

The apparatus used by Robson is sketched in Figure 15-15. A collimated beam of thermal neutrons from the reactor enters a vacuum chamber. The neutron beam was about 3 cm in diameter and had an intensity of about 1.5 $\times$ 10^{10} thermal neutrons per second. Some of the neutrons decayed in their flight through the vacuum chamber. The protons released in this decay were

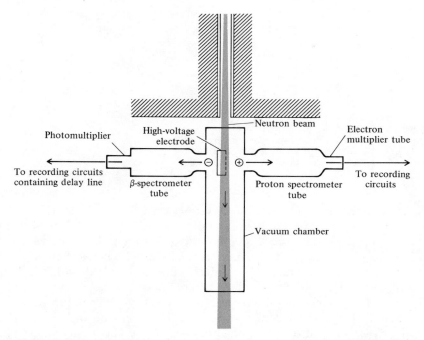

Figure 15-15 Simplified diagram of the apparatus used by J. M. Robson to show the radio-active decay of the neutron.

deflected out of the neutron beam by means of an electrostatic field into a magnetic-lens spectrometer and detected with an electron-multiplier tube. The beta rays emitted during this disintegration were deflected to the left into a beta-ray magnetic-lens spectrometer and were detected with a scintillation counter using anthracene crystals as the phosphor. The two detectors, one for protons and one for beta rays, were connected to a coincidence circuit which would register only when pulses from the two detectors arrived there simultaneously. Because of its greater mass, however, the proton would move more slowly through its spectrometer system and would be delayed with

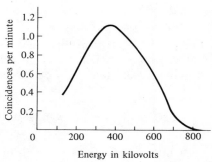

Figure 15-16 Graph showing the coincidence rate between beta particles and protons resulting from the neutron decay, plotted against beta-particle energy. [After Robson, *Phys. Rev.*, **83**, 352 (1951).]

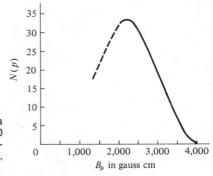

Figure 15-17 Momentum spectrum of the beta particles emitted in neutron decay. Below $B\rho = 2000$ gauss cm, the curve is unreliable because of instrumental effects. [After Robson, *Phys. Rev.*, **83**, 352 (1951).]

respect to the electrons; calculations gave this delay as about 0.9 μsec. A delay line was built into the beta-ray detecting system which would delay the beta-ray pulses by 0.9 μsec. Figure 15-16 shows the graph of the coincidence rate between the beta particles and protons plotted against the energy of the beta particles; the values of the latter were determined with the beta-ray spectrometer. The distribution in momentum of the beta rays is shown in Figure 15-17. The end-point energy of this spectrum is found to be 785 keV; the results of the coincidence curve are in agreement with this end-point energy. This value of the end-point energy is in excellent agreement with the mass difference of 784 keV between a neutron and a hydrogen atom.

The value of the half-life of the neutron was calculated from a determination of the density of the neutron beam and the number of neutrons decaying per unit time per unit volume. The neutron density was determined by inserting manganese foils in the beam and measuring their activities against the activities of standard manganese foils of identical thickness which had previously been calibrated. The value so obtained was 1.16×10^4 neutrons/cm^3 at the center of the beam. The number of neutrons decaying per unit time per unit volume was estimated from the number of protons per unit time striking the first electrode of the electron multiplier and the volume of the beam where the protons originate. The value so determined was 630 neutrons min^{-1}cm^{-3} of the neutron beam. The half-life T is obtained from these values is

$$T = \frac{1.16 \times 10^4}{630} \times 0.693 = 12.8 \text{ min}$$

The probable error in this value given by Robson is 18 percent.

The present accepted value for the mean life of the neutron is 932 sec. yielding a half-life of $T = 10.8$ min.

15-15 Electron Capture by Nuclei

In nuclear reactions the extranuclear electrons may be ignored, except in a few special cases; for example, we saw that the sharp line beta-ray spectrum was due to internal conversion. In the case of β^+ decay we found (Section 15-7) that in addition to the positron emitted by the nucleus, an electron must be emitted from the external part of the atom in order that the product atom may be neutral, thus involving the release of a minimum amount of energy equal to $2m_e c^2$ or about 1 MeV. There is, however, an alternative method to β^+ decay in which a given parent atom forms the same product atom and that is by the *capture* of one of the external electrons by the nucleus. This process of decay is known as *electron capture*. Usually a K electron will be the one that is captured by the nucleus, although one of the others may be captured instead; hence this process is sometimes designated simply as K *capture*.

The energy available for the electron-capture process is simply the difference in mass between the parent atom and the daughter atom, thus

$$\mathcal{E} = {}_Z^A M - {}_{Z-1}^A M$$

This energy exceeds that for positron decay by 2 electron masses. Electron capture and positron decay are competing processes; in those cases in which the mass difference between parent and daughter nuclei is less than 2 electron masses only electron capture can occur.

The capture of the K electron by the nucleus will leave the product atom in an excited state with one electron missing from the K shell; it will then return to the normal state by the emission of x-rays characteristic of the product atom. This type of nuclear reaction is often detected by means of the x-rays emitted during this process.

An early clear-cut example of K-electron capture is the radioactive disintegration of vanadium, ${}_{23}^{49}V$, into titanium, ${}_{22}^{49}Ti$, with the capture of a K electron by the vanadium nucleus to form a titanium atom in the K state. This reaction was investigated carefully by Walke, Williams, and Evans (1939), who found that vanadium decays only by K-electron capture with a half-life $T = 600$ days. The radioactive vanadium was formed by bombarding titanium with deuterons, the following reactions taking place:

$$\text{${}_{22}^{48}$Ti} + {}_1^2\text{H} \rightarrow ({}_{23}^{50}\text{V}^*) \rightarrow {}_{23}^{49}\text{V} + n$$

then   (15-42)

$$\text{${}_{23}^{49}$V} + {}_{-1}^0 e \rightarrow {}_{22}^{49}\text{Ti} + \nu \qquad T = 600 \text{ days}$$

The active product was separated chemically from the titanium and found to be vanadium, but no radiations of any kind other than Ti K_α radiation were found to be emitted by the vanadium precipitate. The only conclusion, therefore, is that the excited vanadium nucleus captured one of its K electrons to form titanium with one electron missing from its K shell with the subsequent emission of x-rays characteristic of titanium. Furthermore, the intensity of these x-rays diminishes exponentially with the time, so that at the end of 600 days the intensity has dropped to half its original value. A more recent determination of the half-life by Hayward and Hoppes (1956) yields a value of $T = 330$ days.

The fact that no radiation other than x-rays is emitted in the radioactive

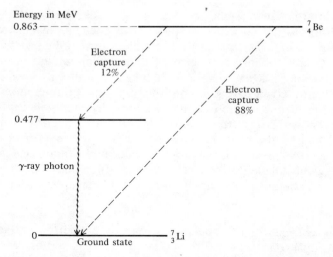

Figure 15-18 Nuclear energy-level diagram showing decay of ^{7}Be by K-electron capture.

disintegration of vanadium by K-electron capture to titanium indicates that the product titanium nucleus is in the ground state or state of lowest energy. Since the capture of an electron by the nucleus results in the change of a proton into a neutron, this must be accompanied by the emission of a neutrino in order to conserve the angular momentum of the nucleus.

In many cases of radioactive disintegration by K-electron capture, gamma rays are emitted; in other cases the disintegration can take place either by positron emission or by K-electron capture, so that both processes are observed in the same radioactive sample; for example, the isotope of beryllium of mass number $A = 7$, which disintegrates by K-electron capture, also emits gamma rays of 0.477 MeV. The reaction is

$$^7_4\text{Be} + {}_{-1}^{0}e \rightarrow ({}_3^7\text{Li*}) + \nu$$

$$({}_3^7\text{Li*}) \rightarrow {}_3^7\text{Li} + \gamma \qquad T = 53.6 \text{ days} \tag{15-43}$$

This radioactive beryllium isotope can be produced by any of the following reactions:

$$\left.\begin{array}{l} {}_3^7\text{Li} + {}_1^1\text{H} \rightarrow ({}_4^8\text{Be*}) \rightarrow {}_4^7\text{Be} + n \\[4pt] {}_3^6\text{Li} + {}_1^2\text{H} \rightarrow ({}_4^8\text{Be*}) \rightarrow {}_4^7\text{Be} + n \\[4pt] {}_5^{10}\text{B} + {}_1^1\text{H} \rightarrow ({}_6^{11}\text{C*}) \rightarrow {}_4^7\text{Be} + {}_2^4\text{He} \end{array}\right\} \tag{15-44}$$

The nuclear energy-level diagram illustrating the decay of ^{7}Be by K-electron capture is shown in Figure 15-18. In 88 percent of the cases the decay is to the ground state of ^{7}Li, but in 12 percent of the cases the decay is to an excited state from which the ^{7}Li nucleus goes to the ground state by emitting a gamma-ray photon of 0.477 MeV energy. The energy released in the K-capture process in which the ^{7}Li nucleus is formed in the ground state is 0.863 MeV. This value has been determined from the disintegration energy or Q value of the ^{7}Li$(p,n)^7$Be reaction which is 1.646 MeV and the energy equivalent of the neutron-proton mass difference of 0.784 MeV. The energy equivalent of the ^{7}Be-^{7}Li mass difference is thus 0.862 MeV; since this is less than the minimum energy required for positron decay, the ^{7}Be nucleus can decay only by K-electron capture. Its half-life is 53.0 days.

One of the many examples in which both positron emission and K capture has been observed is found in the disintegration of ^{107}Cd. This isotope of cadmium can be formed by any one of the following reactions:

$$^{106}\text{Cd}(n,\gamma)^{107}\text{Cd}, \quad ^{107}\text{Ag}(d,2n)^{107}\text{Cd}, \quad \text{and} \quad ^{107}\text{Ag}(p,n)^{107}\text{Cd}$$

The disintegration is given by either of the following equations:

$$^{107}_{48}\text{Cd} \rightarrow {}^{107}_{47}\text{Ag} + \beta^+ + \nu$$

or (15-45)

$$^{107}_{48}\text{Cd} + {}^{0}_{-1}e \rightarrow {}^{107}_{47}\text{Ag} + \nu \qquad T = 6.5 \text{ hr}$$

As shown in Figure 15-19, the greatest probability is for a transition by K capture to an excited state of ^{107}Ag, the latter then decaying to the ground state

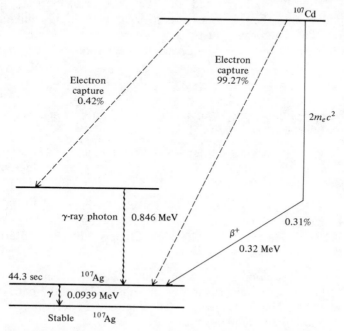

Figure 15-19 Nuclear energy-level diagram of the modes of decay of ^{107}Cd to an isomeric state of ^{107}Ag. (After *Nuclear Data*, National Bureau of Standards, Circular 499.)

by the emission of a gamma-ray photon of 0.0939 MeV. A process with much smaller probability is K capture leading to a higher excited state of silver followed by the emission of a gamma-ray photon of 0.846 MeV leading to the lower excited state of silver mentioned above. The smallest probability is the emission of a positron by ^{107}Cd leading to the lower excited state of ^{107}Ag. The maximum kinetic energy of the positron is 0.32 MeV; hence the disintegration energy is $2m_ec^2 + 0.32$ MeV, or about 1.33 MeV.

Problems

15-1.   A particle of mass M moving initially with velocity V makes an elastic collision with a particle of mass m initially at rest. Using the principles of the conservation of energy and momentum, show that if the particle of mass m is given a velocity v in the same direction as V, then

$$v = \frac{2M}{M + m} V$$

15-2.   A particle of mass M_1, kinetic energy E_1, velocity V_1, and momentum P_1 is captured by a nucleus (M_0, E_0, V_0, P_0) at rest. A light particle (M_2, E_2, V_2, P_2) is ejected and the heavy particle (M_3, E_3, V_3, P_3) recoils. Using the principles of conservation of energy and momentum, show that when the light particle is emitted at an angle of 90 degrees with the path of the incident particle, the energy of the light particle is

$$E_2 = \frac{M_3}{M_2 + M_3}\left(Q - \frac{M_1 - M_3}{M_3}E_1\right)$$

where Q is the energy equivalent of the difference in mass between the initial particles and the final particles.

15-3.   A beam of 0.40 MeV deuterons is directed against a deuterium target. Calculate the energy of the neutrons emitted at 90 degrees with respect to the incident beam; use the atomic mass data of Appendix IV.

15-4.   Using the equation derived in Problem 15-2, calculate the energy of the neutrons emitted at an angle of 90 degrees to the deuteron beam in a ^{3}H $(d,n)^4$He reaction. The kinetic energy of the deuterons is 200 keV.

15-5.   When $^{11}_5$B is bombarded with 200-keV protons, alpha particles of 4.41-cm range in air are emitted at right angles to the path of the incident beam. Using the result of Problem 15-2, determine the mass of the residual $^8_4$Be nucleus.

15-6.   Using the values of mass excesses from Appendix IV, calculate the Q value for the reaction ^{14}N $+ n \rightarrow (^{15}$N*$) \rightarrow {}^{14}$C $+ p + Q$.

15-7.   Calculate the Q value for the reaction $^6Li + p \rightarrow\ ^4He + \ ^3He + Q$.

15-8.   (a) Using the principles of conservation of energy and momentum, show that the reduction in energy $\Delta\mathscr{E}$ of a neutron in a central elastic collision with a nucleus of mass number A is given by

$$\Delta\mathscr{E} = \mathscr{E}\left[1 - \left(\frac{A-1}{A+1}\right)^2\right]$$

where $\mathscr{E}$ is the original energy of the neutron. Calculate the fractional energy loss of a neutron colliding with (b) a proton, (c) a deuteron, and (d) a carbon nucleus, $A = 12$.

15-9.   Calculate the Q value for the formation of ^{31}P in the ground state in the reaction $^{30}Si\ (d,n)^{31}P$ from the following cycle of nuclear reactions:

   (1) $^{30}Si + d \rightarrow\ ^{31}Si + p + 4.370$ MeV
   (2) $^{31}Si \rightarrow\ ^{31}P + \beta^- + \bar{\nu} + 1.476$ MeV
   (3) $n \rightarrow p + \beta^- + \bar{\nu} + 0.783$ MeV

15-10. Calculate the Q value for the formation of ^{30}P in the ground state in the reaction $^{29}Si\ (d,n)^{30}P$ from the following cycle of nuclear reactions:

   (1) $^{31}P + \gamma \rightarrow\ ^{30}P + n - 12.313$ MeV
   (2) $^{31}P + p \rightarrow\ ^{28}Si + \ ^4He + 1.916$ MeV
   (3) $^{28}Si + d \rightarrow\ ^{29}Si + p + 6.251$ MeV
   (4) $2d \rightarrow\ ^4He + 23.847$ MeV

15-11. The following data were taken with a sample of nitrogen in a $^{14}_{7}N\ (\gamma, n)$ reaction:

Time after irradiation in minutes	1	2	3	4	5	8	10	12	14	16	18	20	24
Counts per minute	520	485	450	410	390	310	290	240	210	190	160	140	100

(a) Plot the logarithm of the number of counts per minute against the time in minutes. (b) Determine the half-life of this activity.

15-12. When chlorine, $A = 37$, is bombarded with neutrons, gamma rays are emitted and the resultant nucleus is radioactive, emitting beta particles. A target containing chlorine was bombarded with neutrons for a short time, after which counts were taken of the number of beta particles emitted in successive 10-min intervals. These counts, after correction for background readings, were as follows: 344, 267, 235, 189, 160, 145, 112, 97, 81, 65, 55, 50, 39, 37, 27, 25, 20, 14, 14, 11, 9. (a) Write the equation for each reaction. (b) Plot a curve of the logarithm of the activity (number of counts per 10 minutes), against the time in minutes. (c) From the slope of the above curve determine the half-life of the beta-ray emitter.

15-13. Calculate the Q value for the ^{22}Ne(d,p) reaction.

15-14. (a) Calculate the Q value for the ^{19}F(n,p) reaction. (b) Determine the threshold value of the neutron energy for this reaction.

15-15. (a) Determine the Q value for the ^{23}Na(n,α) reaction. (b) Determine the threshold value of the neutron energy.

15-16. Calculate the Q value for each of the following reactions:

- (a) ^{18}O(p,α)
- (b) ^{15}N(p,α)
- (c) ^{6}Li(d,n)
- (d) ^{2}H(d,p)
- (e) ^{16}O(γ,n)
- (f) ^{16}O(γ,p)

15-17. Complete the following:

- (a) ^{235}U $\rightarrow \alpha +$ _____
- (b) ^{23}Na(d, p) _____
- (c)   _____ $\rightarrow e^- + \bar{\nu} + {}^3$He
- (d) ^{20}Na $\rightarrow e^+ +$ _____ $+$ _____
- (e) ^{16}O(d, p) _____
- (f) ^{14}N(_____, n) ^{17}F

15-18. A nonrelativistic particle of mass M collides elastically with a particle also of mass M which was at rest before the collision. Show that the angle between the paths of the two particles after the collision is 90°.

15-19. A target is bombarded with a beam of which the intensity is constant in time. A radioactive isotope is formed at a constant rate: R nuclei per second. The isotope decays with decay constant λ (see Section 14-2). (a) Write the differential equation for this irradiation process. (b) Solve it to get $N(t)$, the number of nuclei of the isotope present as a function of time; use $N = 0$ at $t = 0$ as the initial condition. (c) Graph the solution.

15-20. Calculate the binding energy per nucleon for ^{2}H, ^{16}O, ^{56}Fe, ^{109}Ag, ^{127}I, ^{197}Au, and ^{238}U. Display the results in a graph.

15-21. Devise at least four processes for making ^{60}Co by bombarding stable nuclei.

15-22. Devise at least five processes for making ^{36}Cl by bombarding stable nuclei.

15-23. In bombarding copper with deuterons, J. M. Cork (1941) observed characteristic x-rays coming from the target at an angle of 90 degrees to the beam. To determine whether the radiation came from the copper or from zinc formed in the process of K-electron capture, he put thin strips of copper, nickel, and iron adjacent to one another over a photographic plate and exposed it to the radiation. He found that the iron foil almost completely absorbed the radiation, whereas the copper and nickel foils reduced the intensity only slightly. Using the following data, (a) determine the source of the radiation. (b) What would have been the source of the radiation if nickel and iron foils had shown strong absorption, while the copper foil showed weak absorption?

Data:

$$CuK_\alpha \text{ line} = 1.541 \text{ Å}$$
$$ZnK_\alpha \text{ line} = 1.438 \text{ Å}$$

K-absorption edge of $\left\{ \begin{array}{l} \text{Fe} = 1.739 \text{ Å} \\ \text{Ni} = 1.484 \text{ Å} \\ \text{Cu} = 1.377 \text{ Å} \end{array} \right.$

15-24.   Complete the following:

$$^{10}\text{B } (\alpha,p)\text{_____}$$
$$^{10}\text{B } (\alpha,d)\text{_____}$$
$$^{10}\text{B } (\alpha,\alpha)\text{_____}$$
$$^{10}\text{B } (\alpha,n)\text{_____}$$
$$^{10}\text{B } (\alpha,\gamma)\text{_____}$$
$$^{10}\text{B } (\alpha,^{3}\text{H})\text{_____}$$

15-25.   Using data from Appendix IV on the abundance of isotopes, estimate the atomic weight of copper as found in nature.

References

Beyer, R. T., *Foundations of Nuclear Physics*. New York: Dover Publications, 1949. Facsimiles of 13 fundamental papers and an extensive bibliography to 1949.

Bleuler, E., and G. J. Goldsmith, *Experimental Nucleonics*. New York: Rinehart & Company, 1952.

Kaplan, I., *Nuclear Physics*. Cambridge: Addison-Wesley Publishing Company, 1955.

Pollard, E., and W. L. Davidson, Jr., *Applied Nuclear Physics*. New York: John Wiley & Sons, 1951.

Rutherford, E., J. Chadwick, and C. B. Ellis, *Radiations from Radioactive Substances*. London: Cambridge University Press, 1930.

Strominger, D., J. M. Hollander, and G. T. Seaborg, "Table of Isotopes," *Revs. Mod. Phys.*, **30**, 585, (1958).

See also the references cited at the end of Chapter 14.

16

Fission and Fusion of Nuclei

16-1 Discovery of Nuclear Fission

By 1934 enough data had been accumulated on the disintegration of nuclei bombarded by neutrons to lead Fermi and his co-workers to try to produce elements of atomic number greater than 92 by bombarding uranium with neutrons. In their early experiments they found four beta-ray activities with different half-lives as a result of such bombardment. These beta-ray activities indicated that new elements were being formed, since normal uranium decays by alpha-particle emission with a long half-life. These new elements were assumed to be *transuranic* elements—that is, elements of atomic number greater than 92. Some chemical tests were tried to verify this hypothesis. Element 93, for example, was at that time thought to be a chemical homologue of manganese and should have similar chemical properties. In one chemical experiment a manganese salt was added to a solution of a uranium salt that had been bombarded with neutrons; the manganese was then precipitated out as managanese dioxide. Two of the beta-ray activities came down with the precipitate. Other chemical tests were tried which ruled out the possibility that these beta-ray activities could be due to any of the elements in the range of atomic numbers 86 to 92, inclusive. It was therefore concluded that element 93 has been produced. A chemist, Noddack, criticized this conclusion on the basis that many elements are precipitated with MnO_2 and suggested the possibility that the bombarded nuclei were split into nuclei of elements of lower atomic number. This suggestion was apparently ignored, and other workers, particularly I. Curie and her co-workers and Hahn, Strassmann, and Meitner, entered this field in search of the transuranic elements. Uranium, thorium, and protactinium were bombarded with neutrons and many different beta-ray activities were discovered. These were carefully checked by both physical and chemical methods to determine the nature of the emitters, and until early in 1939 they were generally ascribed to radioactive substances of atomic number greater than 92. Several new radioactive series were suggested to account for these activities; each one started with uranium and by a series of beta-particle disintegrations led to elements of atomic numbers as high as 96 and 97. It must be pointed out that chemical analysis in this region of the periodic table is extremely difficult and becomes even more difficult because of the very small samples of newly formed radioactive material available.

In 1939 Hahn and Strassmann found, after a series of very careful chemical experiments, that one of the radioactive elements formed by the bombardment of uranium with neutrons was an isotope of the element barium, $Z = 56$. Another of the radioactive elements thus formed was the rare-earth element lanthanum, $Z = 57$. It is obvious that the lanthanum is formed by the beta decay of the barium. Hahn and Strassmann therefore suggested that the beta-ray activities previously ascribed to transuranic elements are probably produced by radioactive isotopes of elements of lower atomic number. The process that is started by bombarding uranium with neutrons is one in which the new uranium nucleus becomes unstable and splits up into two nuclei of medium atomic masses; if one nucleus formed is barium, $Z = 56$, the other nucleus must be krypton, $Z = 36$. This type of disintegration process in which a heavy nucleus splits up into two nuclei of nearly comparable masses is called *nuclear fission.*

As soon as the discovery of nuclear fission was announced early in 1939, physicists in laboratories throughout the world where neutron sources were available immediately repeated and confirmed these experiments. Within the next two years the results were extended to include the nuclear fission of thorium and protactinium, the measurement of the energies of the bombarding neutrons necessary to produce fission in the particular isotope of the heavy element, the amount of energy released in nuclear fission, and analyses of the products of nuclear fission together with their genetic relationships.

16-2 Fission of Uranium

In the process of nuclear fission a very much greater amount of energy is released than that ever previously encountered in any nuclear or atomic process. In addition to the release of energy, the fission of uranium was accompanied by the emission of several neutrons. It was immediately apparent that these neutrons could be utilized, under proper conditions, to produce fission of other uranium nuclei and thus a *chain reaction* could be initiated that could result in the release of tremendous amounts of energy. The first successful *nuclear reactor* that utilized the fission of uranium nuclei in a self-sustaining chain reaction was constructed at Chicago and operated successfully on December 2, 1942. This reactor, sometimes known as a *uranium graphite pile* because of the manner in which it was constructed, was built under the direction of E. Fermi by groups headed by W. H. Zinn and H. L. Anderson. Since then many other nuclear reactors of various designs have been built throughout the world.

An idea of the amount of energy released in nuclear fission can be obtained by considering one example of the fission of uranium, $Z = 92$, $A = 235$, assuming that the fission products are barium, $Z = 56$, $A = 141$, and krypton, $Z = 36$, $A = 92$, with the prompt release of three neutrons. We can write the equation for this reaction as

$$ _{0}^{1}n + {}_{92}^{235}\text{U} \rightarrow ({}_{92}^{236}\text{U}^{*}) \rightarrow {}_{56}^{141}\text{Ba} + {}_{36}^{92}\text{Kr} + 3\,{}_{0}^{1}n + Q \qquad (16\text{-}1)$$

where Q represents the energy released in this reaction. The energy is provided by the difference in mass between the initial ingredients and the final products. In the initial state the mass excesses in MeV (from Appendix IV) are

$$
\begin{aligned}
&8.072 \ (\text{neutron}) \\
+&40.906 \ (^{235}\text{U}) \\
\hline
&48.978
\end{aligned}
$$

The isotopes ^{141}Ba and ^{92}Kr are rather far from the region of stable nuclei and their mass excesses have not been measured well. Good estimates have been provided by W. D. Myers and W. J. Swiatecki (1966):

$$
\begin{aligned}
-&80.060 \ (^{141}\text{Ba}) \\
-&68.784 \ (^{92}\text{Kr}) \\
+&24.216 \ (\text{three neutrons}) \\
\hline
-&124.628
\end{aligned}
$$

The difference between these signed numbers is almost 175 MeV—a value 20 times as great as that released in the average alpha-particle disintegration (5 to 10 MeV), and millions of times greater than that released in the process of combustion (~ 4 eV).

Figure 16-1   Cloud-chamber photograph of the tracks produced by two fission particles; these particles are the products of the fission of uranium that has captured a neutron. The uranium is on the foil, going from top to bottom of page in the center of the chamber. The two fission tracks are on opposite sides of the foil and are almost horizontal. (From a photograph by J. K. Bøggild, K. J. Brøstrom, and T. Lauritsen.)

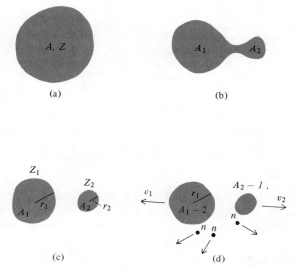

Figure 16-2   Schematic representation of nuclear fission: (a) compound nucleus considered as a drop of liquid; (b) intermediate stage before fission; (c) nuclei formed after fission in highly excited states; (d) motion of fission products and release of prompt neutrons.

The fission of a nucleus results in the production of two massive particles having equal and opposite momenta. Figure 16-1 is a cloud-chamber photograph of the fission of uranium showing the tracks of two massive fragments traveling in almost opposite directions. The kinetic energies of these fission particles are in the inverse ratio of their masses. These fission products are unstable, each having an excess number of neutrons. Each particle may emit one or two neutrons shortly after fission but in a time too short to be measured, probably in less than 10^{-15} sec. Such neutrons are called *prompt neutrons;* they have been included in Equation (16-1). Other neutrons may be emitted later; such neutrons are called *delayed neutrons* (see Section 16-7).

Uranium fission can be produced by both slow and fast neutrons. Bohr and Wheeler (1939) employed a *liquid drop* model of the nucleus in their theory of nuclear fission. According to this theory, nuclear fission takes place in two steps: (a) the formation of the compound nucleus in which the energy is temporarily stored among the different degrees of freedom of the nuclear particles in a manner similar to that of thermal agitation of a liquid and (b) the transformation of a sufficient portion of this energy into potential energy of deformation of the compound nucleus which will lead to its fission.

The process of fission may be pictured as shown in Figure 16-2 in which (a) shows the compound nucleus of mass number A and atomic number Z as a drop of liquid in a highly excited state. As a result of the forces between the nucleons, the liquid drop will change its shape and at some stage will assume the shape shown in (b). This is an unstable configuration and if the

TABLE 16-1

Bombarded Nucleus	Compound Nucleus	Critical Energy $\mathcal{E}_c$ (MeV)	Binding Energy W_n (MeV)	$\mathcal{E}_c - W_n$ (MeV)
92 U 234	92 U 235	5.0	5.4	−0.4
92 U 235	92 U 236	5.3	6.4	−1.1
92 U 238	92 U 239	5.9	5.2	0.7
91 Pa 231	91 Pa 232	5.5	5.4	0.1
90 Th 232	90 Th 233	6.9	5.2	1.7
90 Th 230	90 Th 231	6.5	5.3	1.2

excitation energy is sufficiently great, this liquid drop will separate into two drops by contracting at some point along the thin neck, thus forming two nuclei, A_1Z_1 and A_2Z_2 of radii r_1 and r_2 as shown in (c). The distance r between their centers will probably be greater than $r_1 + r_2$. At this stage these nuclei will promptly release one or more neutrons as shown in (d).

The distance r between the nuclei Z_1 and Z_2 at the instant of fission is probably greater than the range of nuclear forces so that these nuclei are under the influence only of the electrostatic repulsion because of their electric charges. This can be inferred from the fact that the sum of the kinetic energies of these nuclei is less than the energy released in the fission process (see Section 16-3). The nuclei A_1Z_1 and A_2Z_2 must therefore be in highly excited states at the instant of fission. Some of this energy is released almost immediately and carried away by the prompt neutrons. The remaining excess energy will be released as these nuclei go to more stable configurations by the ejection of neutrons and even heavier particles such as alpha particles by beta decay and by gamma-ray emission.

On the assumption that a compound nucleus behaves as a drop of liquid, Bohr and Wheeler calculated the critical energy $\mathcal{E}_c$ necessary to produce the unstable deformation such as that sketched in Figure 16-2. Values of the critical energy for some compound nuclei are listed in Table 16-1. The addition of a slow neutron to a target nucleus produces a compound nucleus, which, as we have seen, is always in an excited state. The excess energy is simply the binding energy W_n of the neutron and is listed in the adjacent column of Table 16-1. If W_n is greater than $\mathcal{E}_c$, the addition of a slow neutron to a target nucleus should be sufficient to produce the unstable configuration of the compound nucleus and hence result in fission. If W_n is less than $\mathcal{E}_c$, fission cannot be produced by the capture of slow neutrons; however, the additional energy necessary for fission can be supplied by increasing the kinetic energy of the incident neutron. A glance at the last column of Table 16-1 shows that it should be possible to produce the fission of ^{235}U with slow neutrons but that it is necessary to use fast neutrons with kinetic energies greater than 0.7 MeV to produce the fission of ^{238}U.

This was confirmed experimentally by Nier, Booth, Dunning, and Grosse.

They first separated small quantities of the uranium isotopes, $A = 235$ and $A = 238$, by means of the mass spectrometer, and then bombarded each of these isotopes with slow neutrons. They observed almost no fission with uranium, $A = 238$, but did get a fairly large number of fissions with $A = 235$. Furthermore, the rate of fission per microgram of uranium 235 observed in this experiment was in good agreement with the number obtained under the same experimental conditions from unseparated samples of uranium containing the normal percentage of uranium 235.

16-3   Energies of the Fission Fragments

The measurement of the kinetic energy of the fission fragments and the distribution in energy among these fragments has been carried on over a period of years, some by calorimetric methods, but mostly by ionization methods. Fowler and Rosen (1947), using a specially designed ionization chamber, determined the distribution in energy of the fragments from the fission of ^{235}U by both slow and fast neutrons. Figure 16-3 shows the distribution in energy among the fission fragments of the slow neutron fission of ^{235}U. There are two definite peaks to the curve, one at 61.4 MeV and the other at 93.1 MeV. They obtained a similar curve for the distribution in energy of the fragments from the fission of ^{235}U by fast neutrons with a range of energy from 1 keV to about 1 MeV, with the peaks shifted slightly to higher energies. The kinetic energy of a fission fragment was calculated from the ionization current on the assumption that the energy required to produce an ion pair in the gas of the chamber is the same for a fission product as for an alpha particle.

The assumption that the energy required to produce an ion pair in a gas is the same for a highly charged fission fragment as for an alpha particle is open to question. This problem was investigated by Leachman (1952), who mea-

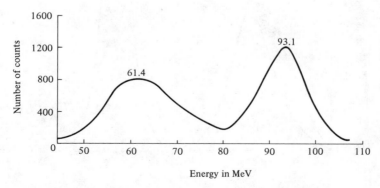

Figure 16-3   Energy spectrum of the fission fragments produced in the slow neutron fission of ^{235}U. (After Fowler and Rosen, *Phys. Rev.*, **72**, 928, 1947.)

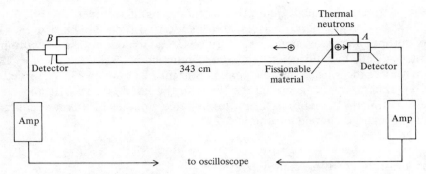

Figure 16-4 Schematic diagram of the apparatus used for measuring the velocities of fission fragments.

sured the velocities of the fission fragments from the slow neutron fission of ^{233}U, ^{235}U, and ^{239}Pu, using a time-of-flight method and comparing these results with the velocity distribution calculated from ionization methods. A simplified diagram of the experiment is shown in Figure 16-4. The fissionable material, such as UO_3, is put on a thin foil of nickel, which is placed 1 cm from one end A of a long tube, the other end B being 343 cm away. A scintillation detector is placed at each end, the pulses being fed through separate amplifiers to an oscilloscope. Fission is induced by slow neutrons from a nuclear reactor. When fission occurs, one fragment travels 1 cm and produces a pulse, the other fragment travels 343 cm and produces a pulse; the time interval between these

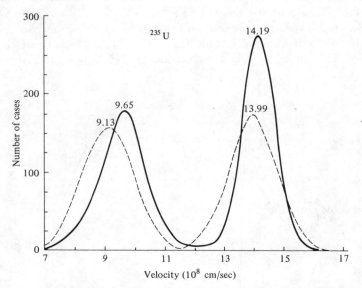

Figure 16-5 Distribution in velocity of the fission fragments from the thermal neutron fission of ^{235}U. (After Leachman, *Phys. Rev.,* **87**, 444, 1952.)

pulses is then determined from the oscillograph record. The results for the distribution in velocity of the fission fragments for ^{235}U are shown in the full curve of Figure 16-5; the dashed curve is that computed from ionization measurements. The curves have the same general features; the peaks of the curve obtained by the time-of-flight method are both shifted to higher velocities relative to the curve obtained from ionization methods. From these data the shift in the peak for the light fragment is 5.7 MeV and that for the heavy fragment is 6.5 MeV.

Using the previously measured kinetic energies of the fission fragments, 61.4 and 93.1 MeV, and adding the above values of 5.7 and 6.5 MeV, we obtain for the total kinetic energies of the fission fragments 166.7 MeV.

Similar results were obtained for the velocity distributions of the fragments from the thermal neutron fission of ^{233}U and ^{239}Pu.

16-4 Some Products of Nuclear Fission

Many different atomic nuclei have been produced by the fission of uranium and thorium as a result of bombarding them with neutrons. Most of these fission products have been identified by chemical tests; others, by means of the x-rays emitted by the excited atoms produced during fission. In many cases the particular isotopes produced have been identified by comparing their half-life periods with those produced by other types of nuclear reactions. The fact that so many different fission products have been produced indicates that the excited uranium or thorium nucleus can split up in many different ways. All of the presently known fission products are elements in the middle of the periodic table with atomic numbers ranging from $Z = 30$ to $Z = 66$.

The relative distribution of the different nuclides among the fission products depends on the energy of excitation available for the fission process. Figure 16-6 is a graph showing the fission yield (in percentages) plotted against the mass number of the fission fragment for the fission of uranium 235. From this graph it is seen that the most probable values for the mass numbers of the two fission fragments are about 95 and 139 when two prompt neutrons are emitted simultaneously. It will be observed that the yield passes through a minimum at mass number 117, corresponding to fragments of equal mass. The yield curve drops rapidly at mass numbers 72 and 162.

The ratio of the number of neutrons to the number of protons—that is N/Z—for uranium is about 1.6. For stable elements in the range of atomic numbers 30 to 63 the range of maximum values of N/Z is 1.3 to 1.5; hence the fission fragments will have an excess of neutrons. The fission products will therefore be unstable; they will thus go to a more stable form either by beta disintegration or by the emission of one or more excess neutrons. Neutrons emitted a measurable time after the fission process are called *delayed neutrons;* they play an important role in the control of nuclear reactors. The beta-decay chain leading to a stable nuclide has been followed for a great many fission products;

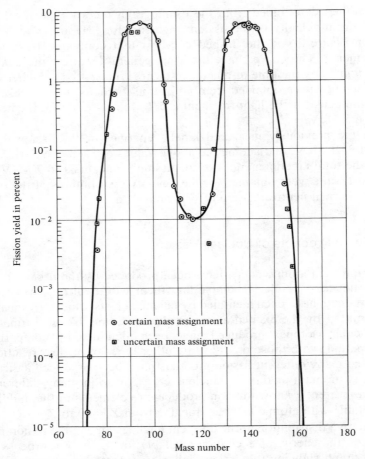

Figure 16-6 Graph showing yields of fission product chains of ^{235}U as a function of mass number. [After "Plutonium Project Report on Nuclei Formed in Fission," *Revs. Modern Phys.*, **18**, 539,(1946).]

for example, the heaviest stable isotope of barium has a mass number 138; hence barium 141, one of the fission products of uranium, is unstable because it has an excess of three neutrons. The beta-decay chain of barium 141 has been found to be

$$^{141}_{56}Ba \xrightarrow[18\,min]{} {}^{141}_{57}La \xrightarrow[3.9\,hr]{} {}^{141}_{58}Ce \xrightarrow[32.5\,da]{} {}^{141}_{59}Pr \qquad (16\text{-}2)$$

The end product of this chain is a stable isotope of praseodymium. Similarly, krypton 92 is unstable and starts a beta-decay chain that ends in a stable isotope of zirconium:

$$^{92}_{36}Kr \xrightarrow[3\,sec]{} {}^{92}_{37}Rb \xrightarrow[5\,sec]{} {}^{92}_{38}Sr \xrightarrow[2.7\,hr]{} {}^{92}_{39}Y \xrightarrow[3.53\,hr]{} {}^{92}_{40}Zr \qquad (16\text{-}3)$$

The assignment of the mass number to the particular isotope produced in the fission process is aided by the production of the same beta activity by other methods; for example, the 3.53-hr activity of yttrium 92 can be produced in an (n,p) reaction with zirconium. Similarly, the 32.5-day activity of cerium can be produced by an (α,n) reaction with barium as well as by several other reactions.

16-5   Neutrons from Thermal Fission of ^{235}U

Shortly after the discovery of the fission of uranium it was realized that a new source of energy would become readily available if more than one neutron was emitted for each neutron that produced fission. In 1939 von Halban, Joliot, and Kowarski made a determination of the average number of neutrons released per fission of uranium and obtained the value of 3.5 ± 0.7. That same year Szilard and Zinn performed an experiment in which they bombarded uranium oxide with slow neutrons and estimated that two fast neutrons were emitted per fission. In a refinement of this experiment, they bombarded natural metallic uranium with slow neutrons and detected the neutrons produced in fission by means of a high-pressure ionization chamber containing hydrogen at 10 atmospheres pressure and argon at 8 atmospheres pressure. The ionization chamber was connected to an oscillograph, and the pulses due to the recoil protons produced by collisions with high-energy fission neutrons were recorded photographically. The height of a pulse was a measure of the energy of the recoil proton. They were thus able to obtain the distribution of energy among the fission neutrons as well as average number of neutrons released per fission. For the latter they obtained the value of 2.3 neutrons/fission.

The year 1939 saw the beginning of World War II; Physicists in the United States decided to withhold further information on the results of experiments on nuclear fission because of its obvious importance in producing a nuclear chain reaction. After 1945 some information on this subject began to be released. In 1950 information was released giving the average number of neutrons emitted per fission of uranium 235 by slow or thermal neutrons as 2.5 ± 0.1. The present accepted value of this quantity, designated by ν, is

$$\nu = 2.47 \pm 0.03 \text{ neutrons/fission} \qquad (16\text{-}4)$$

for the fission produced in ^{235}U by a beam of slow or thermal neutrons whose velocity is 2200 meters/sec. Such neutrons have a kinetic energy of 0.0253 eV. Equating this value of the kinetic energy to kT, where k is the Boltzmann constant and T is the temperature of a Maxwellian distribution, we get

$$T = 293.6\text{K} = 20.5°\text{C}$$

for the corresponding temperature of these slow neutrons.

16-6 Energy of Neutrons from Thermal Fission of ^{235}U

The energy spectrum of the neutrons emitted in the thermal fission of ^{235}U has been determined by measuring the energy of the recoil protons ejected by the neutrons from hydrogen or a substance containing hydrogen. The energy spectrum was found to extend from about 0.05 to about 17 MeV. In the low-energy range Bonner, Ferrell, and Rinehart (1952) used a cloud chamber containing hydrogen at a pressure of $\frac{1}{3}$ atmosphere. Neutrons from the thermal fission of uranium entered the cloud chamber and produced recoil protons by collisions with hydrogen. The energy spectrum of the fission neutrons was determined from the number of tracks produced by these recoil protons and the lengths of these tracks. This spectrum is shown in Figure 16-7. The graph shows the number of neutrons $N(\mathcal{E})$ per 100 keV energy interval as a function of the neutron energy from about 75 to 600 keV.

The spectrum of the fission neutrons in the energy range of 0.4 to 7 MeV was measured by D. L. Hill (1952). The fission neutrons emitted in the thermal fission of ^{235}U ejected protons from a layer of paraffin; these protons then passed through an absorbing material and were detected in a series of counters. The amount of absorbing material between the paraffin and the counters yielded the range of the protons. The energy of the protons, and hence the energy of

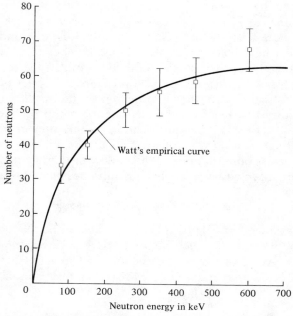

Figure 16-7 Energy distribution of neutrons from the thermal neutron fission of ^{235}U in the low energy range. [After Bonner, Ferrell, and Rinehart, *Phys. Rev.*, **87**, 1033 (1952).]

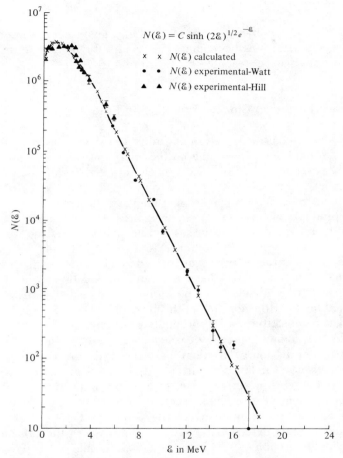

Figure 16-8   Energy distribution of fission neutrons from the thermal neutron fission of ^{235}U. [After Watt, *Phys. Rev.*, **87**, 1040 (1952).]

the neutrons, could be determined from the range-energy relationship of the protons. A different technique was used by B. E. Watt (1952) to determine the energy spectrum of the neutrons in the range of 3.3 to 17 MeV. The fission neutrons ejected protons from a polyethylene foil and were detected by a set of counters. Different thicknesses of aluminum absorbers were placed in the path of the protons; the range of the protons reaching the counters could then be determined from the amount of absorbing material in their path. The energy of the protons was determined from the known range-energy relationship and the energy of the incident neutrons calculated from it. The results of the measurements of both Hill and Watt are shown on the single graph of Figure 16-8. This curve shows a maximum in the region of 0.75 MeV. This distribution in energy is well represented by the equation

$$N(\mathcal{E}) = C \sinh (2\mathcal{E})^{1/2} e^{-\mathcal{E}} \qquad (16\text{-}5)$$

where $N(\mathcal{E})$ is the number of neutrons within a given energy interval and C is an empirically determined constant. The results obtained by Bonner, Ferrell, and Rinehart are also represented by Equation (16-5).

16-7 Delayed Neutron Emission by Fission Fragments

A nucleus which has an excess of neutrons may decay either by β^- emission or by neutron emission, depending on the energy of the nucleus with respect to each of these processes. Bohr and Wheeler (1939), in their theory of nuclear fission, predicted the possibility of neutron emission from some of the fission products. Neutron emission would most likely occur if, in the process of beta decay, the product nucleus were left in an excited state with an energy in excess of the binding energy of a neutron in that nucleus. The neutron would then be emitted promptly, that is, with an immeasurably short lifetime. However, since the nucleus emitting the neutron was formed in a beta-decay process, the neutron activity of a sample will have the same period or half-life as the beta activity of the parent nuclide. Such neutron emission is called *delayed neutron emission* and the neutrons emitted in this process are termed *delayed neutrons*.

Delayed neutrons from fission products of uranium were first detected by Roberts, Meyer, and Wang (1939), who observed the continued emission of neutrons after the fission of uranium had ceased. Booth, Dunning, and Slack (1939) also observed delayed neutron emission and measured two different half-lives of the neutron activity. Because of the importance of the delayed neutrons in the control of a chain reaction, they were intensively studied for the design of the first uranium pile. Later Hughes, Dabbs, Cahn, and Hall (1948) made accurate determinations of the half-lives of the delayed neutrons accompanying the fission of ^{235}U in the heavy-water pile of the Argonne Laboratory. Their results are given in Table 16-2.

The origins of some of the delayed neutrons have been established by

TABLE 16-2 DELAYED NEUTRONS FROM FISSION OF ^{235}U

Half-Life (sec)	Energy of Neutrons (keV)	Yield (Relative to Total Neutron Emission) in Percentages
55.	250	0.025
24.	560	0.166
4.51	430	0.213
1.52	620	0.241
0.43	420	0.085
0.05	—	0.025
Total yield		0.755

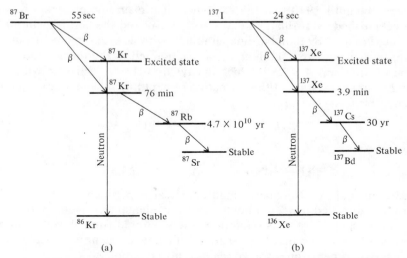

Figure 16-9 Delayed neutron emitters. [After Snell, Levinger, Meiners, Sampson, and Wilkinson, *Phys. Rev.*, **72**, 545 (1947).]

chemical tests. A. H. Snell and his co-workers (1947) showed that the 55-sec delayed-neutron activity came from an excited state of krypton 87 formed by the beta decay of bromine 87 with a half-life of 55 sec. Figure 16-9a shows the nuclear disintegration scheme suggested by them for the delayed neutron emission. Bromine 87 may disintegrate with the emission of a beta particle of small energy, forming ^{87}Kr in an excited state; the latter promptly emits a neutron forming ^{86}Kr in the ground state. A branching process also occurs in which ^{87}Br emits a beta particle with a larger energy, forming ^{87}Kr in a lower energy state. This is followed by a beta decay to ^{87}Rb with a half-life of 76 min. This nuclide of rubidium is found in nature and has a very long half-life of 4.7×10^{10} yr, undergoing beta decay to a stable form of ^{87}Sr. An exactly analogous decay scheme was found by Snell and his co-workers for the 24-sec delayed-neutron activity. This follows the beta decay of ^{137}I to ^{137}Xe as shown in Figure 16-9b.

16-8 Transuranic Elements—Neptunium and Plutonium

The capture of a neutron by uranium does not necessarily lead to the fission of the compound nucleus. Uranium 238, for example, has a small capture cross section for neutrons of low energy. For certain values of the neutron energy, the capture cross section becomes very large. These values are known as the resonance energies. Sharp resonances have been observed at energies of 6.7, 21, 38, and 50 eV. The importance of the capture of neutrons by ^{238}U is that it leads, through a series of disintegrations given below, to the formation of two transuranic elements, *neptunium* (Np), $Z = 93$, and *plutonium* (Pu), $Z = 94$.

It was not until 1940 that the existence of transuranic elements was definitely established. McMillan and Abelson discovered the first transuranic element, neptunium, by bombarding uranium with low-energy neutrons. The capture of neutrons by ^{238}U in an (n,γ) reaction led to the formation of a new isotope of uranium, $A = 239$, which decays by beta-ray emission with a half-life of 23.5 min resulting in the formation of neptunium. The following are the nuclear reactions:

$$^{238}_{92}U + ^1_0n \rightarrow (^{239}_{92}U^*) \rightarrow ^{239}_{92}U + \gamma \tag{16-6}$$

followed by

$$^{239}_{92}U \rightarrow ^{239}_{93}Np + \beta^- + \bar{\nu} \qquad T = 23.5 \text{ min} \tag{16-7}$$

McMillan and Abelson performed chemical experiments with the minute quantities, so-called *tracer* amounts, of the element formed in the above process and showed that its oxidation states differed from those of uranium. This was the first time that the existence of a new element was established by chemical experiments on a tracer scale of investigation.

The neptunium formed in the above reaction is also radioactive, emitting a beta particle to form a new transuranic element of atomic number 94, now known as *plutonium*. The following is the nuclear reaction for this process:

$$^{239}_{93}Np \rightarrow ^{239}_{94}Pu + \beta^- + \bar{\nu} \qquad T = 2.35 \text{ days} \tag{16-8}$$

This isotope of plutonium is radioactive with a long half-life and emits an alpha particle according to the following reaction:

$$^{239}_{94}Pu \rightarrow ^{235}_{92}U + ^4_2He \qquad T = 24{,}360 \text{ yr} \tag{16-9}$$

This isotope of plutonium plays an important part in an atomic bomb project because it is fissionable with both slow neutrons and fast neutrons.

A different isotope of plutonium was discovered shortly after the discovery of neptunium. This transuranic element was discovered by Seaborg, McMillan, Wahl, and Kennedy late in 1940. They bombarded uranium with deuterons and showed that the following reactions occurred:

$$^{238}_{92}U + ^2_1H \rightarrow (^{240}_{93}Np^*) \rightarrow ^{238}_{93}Np + n + n \tag{16-10}$$

followed by

$$^{238}_{93}Np \rightarrow ^{238}_{94}Pu + \beta^- + \bar{\nu} \qquad T = 2.1 \text{ days} \tag{16-11}$$

and then

$$^{238}_{94}Pu \rightarrow ^{234}_{92}U + ^4_2He \qquad T = 89 \text{ yr} \tag{16-12}$$

The chemical properties of plutonium determined with the minute amounts available from the above reactions formed the basis for setting up the so-called *chain-reacting* units at Oak Ridge, Tennessee, and Hanford, Washington, for the production of plutonium on a large scale for use in the atomic bomb. Once the chemical properties of plutonium were known it seemed reasonable to hunt for these elements in nature. Pitchblende ore, one of the sources of uranium, was examined by Seaborg and Perlman; they made a chemical separation of neptunium and plutonium and were able to show the presence of a small quantity of alpha activity in the transuranium fraction which they attributed to plutonium, $A = 239$. The amount of plutonium in the pitchblende corresponds to about 1 part in 10^{14}, an amount that could not possibly be found had the chemical properties not been known. It probably is being formed continuously as a result of the radiative capture of neutrons by some of the uranium present in the ore.

Another isotope of neptunium, $A = 237$, was discovered by Wahl and Seaborg in 1942. This is produced in the following reactions:

$$^{238}_{92}U + n \rightarrow (^{239}_{92}U^*) \rightarrow {}^{237}_{92}U + n + n \qquad (16\text{-}13)$$

followed by

$$^{237}_{92}U \rightarrow {}^{237}_{93}Np + \beta^- + \bar{\nu} \qquad T = 6.75 \text{ days} \qquad (16\text{-}14)$$

and

$$^{237}_{93}Np \rightarrow {}^{233}_{91}Pa + {}^4_2He \qquad T = 2.14 \times 10^6 \text{ yr} \qquad (16\text{-}15)$$

The fact that this isotope is comparatively stable makes it particularly suitable for the chemical investigations of the properties of neptunium.

A large number of isotopes of neptunium, ranging in mass number from 231 to 241, have since been produced. Similarly, a large number of isotopes of plutonium have since been produced; they range in mass number from 232 to 246.

Other transuranic elements, with atomic numbers up to 105, have also been produced. Some of these will be discussed in Section 16-17.

16-9 Photofission of Nuclei

Almost any method that will make a nucleus sufficiently unstable can be used to produce the fission of uranium and thorium. One method is to bombard these nuclei with high-energy photons. Haxby, Shoupp, Stephens, and Wells (1941) were the first to produce such fission with gamma rays. Gamma-ray photons of 6.3 MeV energy, obtained by bombarding fluorite, CaF_2, with high-energy protons, were used to irradiate a 12 cm² piece of uranium metal placed on the high-voltage plate of an ionization chamber. The fission products were measured by the pulses of ionization they produced in this chamber. Baldwin and Koch (1945) used the high-energy x-ray photons produced by the betatron to induce the fission of uranium. Uranium oxide coated on a cylinder was irradiated with x-ray photons of energies from 8 to 16 MeV. The fission fragments were collected on a paper cylinder held over this sample. The beta decay of the fission fragments was then examined with a counter. Fission products were observed for all values of the x-ray energies used in irradiating the

TABLE 16-3

Nuclide	Photofission Threshold (MeV)
^{230}Th	5.40
^{233}U	5.18
^{235}U	5.31
^{238}U	5.08
^{239}Pu	5.31

uranium down to about 8 MeV. They estimate that the threshold for the photofission of uranium is less than 7 MeV.

Baldwin and Koch also tried to produce photofission in lead but were unsuccessful, even with x-ray photons of 16 MeV energy.

Koch, McElhinney, and Gasteiger (1950) were able to make more accurate determinations of the photofission thresholds by allowing the x-rays from a 20-MeV betatron, operated at suitable energies, to be incident on samples of separated isotopes of uranium, plutonium, and thorium. The values obtained by them are given in Table 16-3. It will be noted that these values do not vary very much from nuclide to nuclide in this range of mass numbers.

The photofission of uranium and thorium was investigated by E. W. Titterton and co-workers (1950), using the continuous x-ray beam from the Harwell synchrotron with a maximum energy of 24 MeV. They used a photographic plate technique; for studying the photofission of uranium the emulsion of an Ilford D_1 plate of 100μ thickness was impregnated with uranium from a saturated solution of uranium nitrate. When dry, it was exposed to the x-ray beam and then developed. For the study of the photofission of thorium a similar plate was impregnated with thorium from a thorium nitrate solution. In each case they observed the typical tracks in the emulsion produced by the fission fragments moving in opposite directions. They measured the ranges of these particles and compared the kinetic energies of the fission fragments from thorium with those produced in slow neutron fission of ^{235}U. They found that the average kinetic energy released in the photofission of ^{232}Th was about 0.8 of that released in the slow-neutron fission of ^{235}U.

16-10 Ternary Fission

Shortly after nuclear fission was discovered, R. Present (1941), using the Bohr liquid drop model of the nucleus, predicted the possibility of tripartition, or ternary fission—that is, the division of an excited heavy nucleus disintegrating by division into three nuclei of comparable masses. This process is comparatively rare and the evidence for it is meager. L. Rosen and A. M. Hudson (1949), in a preliminary report on the tripartition of ^{235}U bombarded with slow neutrons, found about 4.3 ternary fissions per 10^6 binary fissions.

In other work on ternary fission two massive fragments have been observed with the third particle, usually a long-range alpha particle, and in a few cases a somewhat heavier particle; for example, K. W. Allen and J. T. Dewan (1950), in their investigation of the ternary fission of ^{233}U, ^{235}U, and ^{239}Pu bombarded by slow neutrons, found that a long-range alpha particle accompanied the fission of these nuclei in about 0.2 percent of the fissions. The average kinetic energy of the alpha particles was found to be about 15 MeV with a maximum energy of about 26 MeV. This is in agreement with the results reported for the ternary fission in the photofission of uranium.

In an investigation of the ternary fission of ^{235}U by slow neutrons E. W. Titterton, using the photographic emulsion technique, observed that most of the long-range alpha particles were emitted at right angles to the line of motion of the two massive fission fragments. The distribution of energy among these alpha particles showed a maximum at about 29 MeV, with the greatest number having energies in the neighborhood of 15 MeV. The observed maximum alpha-particle energy suggests the idea that the alpha particle is left between the massive fission fragments at the instant of fission and acquires its kinetic energy from the electrostatic fields of these nuclei. He found that the frequency of occurrence of ternary fission is about 1 in 400 binary fissions, in approximate agreement with the results of Allen and Dewan. He also found a few very short tracks that seemed to indicate the formation of nuclei with charges $Z > 2$, but found no evidence of ternary fission into three approximately equal masses. Allen and Dewan (1951) also investigated these short-range (< 1-cm air equivalent) particles accompanying nuclear fission of uranium; from the initial specific ionization along their tracks the masses were estimated to be of the order of $A = 13 \pm 4$. Their frequency of occurrence was comparatively high, about 1 in 72 fissions.

16-11   Spontaneous Fission

The occurrence of the spontaneous fission of a heavy nucleus was first observed by Petrzhak and Flerov (1940), who detected the spontaneous fission of natural uranium. E. Segrè (1952) reported on the work done on the spontaneous fission of heavy nuclei at Los Alamos up to January 1946 by himself and co-workers. The substance under investigation was deposited in a thin layer on a platinum disk and placed inside an ionization chamber connected to a linear amplifier; the ionization pulses produced by the fission fragments were then counted. They detected spontaneous fission in several of the heavy nuclei ranging from thorium, $Z = 90$, $A = 230$, to americium, $Z = 95$, $A = 241$, and measured their disintegration constants. They also measured the number of neutrons produced per spontaneous fission of uranium, $A = 238$, and obtained the value 2.2 ± 0.3 neutrons per fission.

The spontaneous fission of curium, $Z = 96$, $A = 242$, was first observed and studied by Hanna, Harvey, Moss, and Tunnicliffe (1951). They compared the fission rate with the rate of alpha-particle disintegration of curium and found 6.2 fissions per 10^8 alpha particles. They also measured the distribution in energies of the fission fragments and found that the most probable energy of the lighter fragment was about 95 MeV, whereas that of the heavier fragment was about 65 MeV, results very similar to those obtained in the slow-neutron fission of uranium and plutonium. This is in accord with the idea that the energy of the fission fragment is due entirely to the electrostatic field between them and is independent of the method of inducing fission.

16-12 Fission of Heavy Nuclei

Calculations based on the liquid drop model of the nucleus indicate that it should be possible to produce the fission of any heavy nucleus by providing sufficient energy of excitation. Probably the first definitely successful experiments on the fission of normally stable heavy nuclei were those performed by Perlman, Goeckermann, Templeton, and Howland (1947) in which high-energy helium ions, deuterons, and neutrons from the 184-in. Berkeley cyclotron were used to bombard tantalum, $Z = 73$, platinum, $Z = 78$, thallium, $Z = 81$, and bismuth, $Z = 83$. The occurrence of fission for each of these nuclei was verified by chemical identification of radioactive fission products. The energies of the bombarding particles were these: 200-MeV deuterons, 400-MeV helium ions or alpha particles, and 100-MeV neutrons. The fission yield curve for each of these elements differed from that of the slow-neutron fission of uranium in that it did not show the pronounced asymmetrical division of the fission products. They also found that in some cases the lighter isotopes of an element were produced and that the most probable type of fission was one in which the sum of the mass numbers of the fission products was much smaller than the target element mass number. This indicates that several neutrons are emitted in this process, probably before the actual occurrence of fission.

A similar result was observed by O'Connor and Seaborg (1948) in the fission of ^{238}U by 380-MeV helium ions. The fission-yield curve showed a maximum in the neighborhood of $A = 115$, extending to a low value of about $A = 27$, much lower than that of a slow-neutron fission of ^{235}U. On the high mass number side it extended up to about $A = 180$. They also found nuclei in the mass number range of 180 to 200 that were produced in the process of *spallation* rather than fission. Spallation is the name given to a nuclear reaction in which several light particles are emitted. The fission of thorium by alpha particles of 37.5-MeV average energy was investigated by A. S. Newton (1948), who found that the fission yield curve had a broad plateau extending from mass number 80 to about 150, with a small trough in the neighborhood of mass number 117. It is thus apparent that the fission yields depend on the excitation energy. Tewes and James (1951) investigated the fission of thorium when bombarded with protons of energies ranging from 6.7 to 21.1 MeV. They observed the trough in the fission-yield curve at the position corresponding to symmetrical fission and noted that the trough became shallower with increasing proton energy. Thus the probability of symmetrical fission increases with increasing energy of the bombarding particle.

16-13   Fission of Lighter Nuclei

The concept of nuclear fission as a type of nuclear disintegration in which the nucleus splits into two nuclei of comparable masses can be extended to the entire range of complex atomic nuclei. The fission of nuclei of elements of medium atomic weight has been investigated by Batzel and Seaborg (1951), using the high-energy protons from the 184-in. Berkeley cyclotron; for example, copper, $A = 63$, was bombarded by protons and one of the products looked for was chlorine, $A = 38$. ^{38}Cl has a half-life of 37.3 min. The production of ^{38}Cl was detected with proton energies as low as about 60 MeV, with the cross section for its production increasing with increasing bombarding energy, as shown in Figure 16-10. Of the several reactions by which ^{38}Cl could be produced, calculations can be made of the energy requirements for the following fission reaction:

$$^{63}\text{Cu} + p \rightarrow {}^{38}\text{Cl} + {}^{25}\text{Al} + n \qquad (16\text{-}16)$$

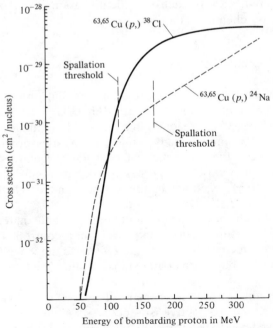

Figure 16-10   Cross section for the formation of ^{38}Cl and ^{24}Na from copper as a function of energy of the bombarding proton. [After Batzel and Seaborg, *Phys. Rev.*, **82**, 609 (1952).]

and the spallation reaction

$$^{63}\text{Cu} + p \rightarrow ^{38}\text{Cl} + p + n + 6\alpha \qquad (16\text{-}17)$$

in which several lighter particles are also emitted (see Section 17-14). The spallation reaction would require a threshold value of about 110 MeV, whereas the fission process would require a threshold energy of about 50 MeV. Of the latter amount about 30 MeV is needed for the excitation energy for passage over the potential barrier and the remainder for the mass difference between the reactants and the products. As shown in Figure 16-10, since ^{38}Cl is formed at energies less than the spallation threshold, ^{63}Cu is fissionable.

Another fission product of ^{63}Cu investigated by Batzel and Seaborg was ^{24}Na, which would be formed in the fission reaction

$$^{63}\text{Cu} + p \rightarrow ^{24}\text{Na} + ^{39}\text{K} + n \qquad (16\text{-}18)$$

with a threshold of about 50 MeV. The observed cross section for this reaction is also shown in Figure 16-10. The spallation reaction in which ^{24}Na could be formed is

$$^{63}\text{Cu} + p \rightarrow ^{24}\text{Na} + p + 3n + 9\alpha \qquad (16\text{-}19)$$

and would have a threshold of about 170 MeV.

Other nuclides of medium atomic weight in which fission was produced are ^{79}Br, ^{107}Ag, and ^{118}Sn. In each case identification of one of the fission products was made by chemical analysis.

16-14   Fission Chain Reaction

The fact that, on the average, more than one neutron is emitted per fission in the neutron fission of such isotopes as ^{232}Th, ^{233}U, ^{235}U, ^{238}U, and ^{239}Pu leads to the possibility of a chain reaction in a mass of fissionable material. Whether the chain reaction remains steady, builds up, or dies down depends on the competition between the production of neutrons through fission and the loss of neutrons through a variety of processes such as nonfission capture of neutrons, primarily (n,γ) reactions in the system, and the leakage of neutrons through the surface of the system.

A system in which the fissionable and nonfissionable materials are arranged so that the fission chain reaction can proceed in a controlled manner is called a *nuclear reactor*. In contrast, an *atomic bomb* is a device designed to produce a chain reaction that builds up at an explosive rate. The latter will not concern us here.

A nuclear reactor is a source of the products of the fission process, namely, energy, neutrons, and radioactive isotopes. As we have shown (Section 16-2), energy is released at the rate of 200 MeV per fission of one atom or about 23×10^6 kW-hr in the fission of 1 kg of fissionable material of mass number 235. As a source of neutrons, a nuclear reactor can provide a large number of neutrons per unit time with a wide range of energies (Section 16-6). Of the ν neutrons emitted per fission, only one neutron is required to cause further fission in order to maintain the chain reaction at a steady rate. The remaining $(\nu - 1)$ neutrons per fission are thus available for other purposes; these purposes are always taken into consideration in the design of the reactor. For example, a reactor may be designed for neutron experiments such as those discussed in earlier chapters; or the reactor may be designed to produce desired isotopes by neutron capture, such as the production of plutonium from ^{238}U, the production of ^{14}C from ^{14}N, or the production of fissionable ^{233}U from ^{232}Th through the reaction

$$^{232}_{90}\text{Th} + n \rightarrow {}^{233}_{90}\text{Th} + \gamma$$

followed by

$$^{233}_{90}\text{Th} \rightarrow {}^{233}_{91}\text{Pa} + \beta^- + \bar{\nu} \qquad T = 22.1 \text{ min} \tag{16-20}$$

and

$$^{233}_{91}\text{Pa} \rightarrow {}^{233}_{92}\text{U} + \beta^- + \bar{\nu} \qquad T = 27.4 \text{ days}$$

Once a fission chain reaction has started, the *effective multiplication factor* k_e will determine whether the chain reaction will continue at a steady rate, increase, or decrease. The effective multiplication factor is defined as *the ratio of the rate of production of neutrons P to the combined rate of absorption A and rate of leakage L of neutrons, or*

$$k_e = \frac{P}{A + L} \tag{16-21}$$

The term *absorption* includes all types of absorption, such as those that produce fission and those that produce (n, γ) processes in the material of the reactor. The fission chain reaction will be *critical* or steady when $k_e = 1$, it will be building up or *supercritical* when $k_e > 1$, and it will be dying down or *subcritical* when $k_e < 1$.

If F is the rate at which fission processes occur and ν is the average number of neutrons emitted per fission, then

$$P = \nu F \tag{16-22}$$

Equation (16-21) may then be written as

$$k_e = \frac{\nu F}{A + L}$$

from which

$$k_e = \nu \frac{F}{A} \frac{1}{1 + (L/A)} \tag{16-23}$$

The ratio F/A depends on the amount of fissionable and nonfissionable material and on their cross sections for fission and neutron capture. The ratio

L/A depends on the ability of the reactor to contain and absorb neutrons before they can escape through the surface. As the size of a reactor decreases, the rate of neutron leakage through the surface increases and the rate of neutron absorption decreases, so that L/A increases and approaches infinity, hence in the limit k_e approaches zero. As the size of the reactor increases, L/A decreases toward zero and k_e increases toward the limiting value $\nu F/A$. Hence, if the composition of the reactor is such that

$$\frac{\nu F}{A} > 1 \qquad (16\text{-}24)$$

there is some size of this reactor for which $k_e = 1$; for this size the reactor is critical. This size is called the *critical size* and the mass of fissionable material at this size is called the *critical mass*. The region containing the fissionable material is called the *reactor core*. The core may be surrounded by nonfissionable material capable of reflecting neutrons back into the core; in such a case both the critical size and the critical mass are reduced. On the other hand, if there is an insufficient amount of fissionable material or an excess of absorbing material in the reactor core so that $\nu F/A < 1$, there is no size for which a steady chain reaction can occur, irrespective of whether a reflector is used.

16-15 Processes within a Reactor

The power level at which a reactor operates is proportional to the number of fissions occurring per unit time and this, in turn, is proportional to the number of neutrons in the reactor. Hence the power level of the reactor can be controlled by controlling the number of neutrons in it. A common method of doing this is to introduce a neutron-absorbing material, usually in the form of a steel rod containing boron, and to adjust the position of this rod in the core. This essentially changes the value of k_e in the desired direction. As the reactor continues in operation, fissionable material is used up and the ratio F/A decreases. To keep the power level constant the control rods containing the neutron-absorbing material should be moved out of the reactor at a suitable rate. When the limit of such compensation is reached, all the control rods are out; the reactor will then become subcritical and die down unless new fissionable material is added.

It has already been shown that the neutron is unstable and decays with a half-life of about 11 min. However, this has practically no effect on the operation of a nuclear reactor because the lifetime of the neutron is very long in comparison with the time interval between the emission of a neutron in a fission process and its subsequent absorption or leakage. This time interval, sometimes called the time of a neutron generation, is usually smaller than 10^{-3} sec. Thus only negligible fractions of the neutrons produced are lost by the decay process.

The determination of the effective multiplication constant for a given dis-

TABLE 16-4   THERMAL NEUTRON CROSS SECTIONS FOR URANIUM

Process	Cross Sections in Barns		Natural U
	^{235}U	^{238}U	
Fission	515	0	3.7
Capture	99	2.80	3.5
Scattering	8.2	8.2	8.2

tribution of fissionable material in a nuclear reactor is a very difficult problem because of the fact that the neutron-fission cross sections and absorption cross sections are complicated functions of the neutron energy. The latter, in turn, are governed by the neutron-fission spectrum (Section 16-6), by the elastic and inelastic scattering cross sections of neutrons, and to some extent by the size of the reactor. When these calculations are carried out, it can be shown that pure natural uranium, no matter how large the amount, cannot support a chain reaction—that is, $\nu F/A < 1$. However, natural uranium, suitably arranged with either graphite (carbon) or heavy water, can support a chain reaction. The graphite or the heavy water acts as a *moderator* to slow down the highly energetic neutrons produced in the fission of uranium. The moderator has a very small capture cross section, so that most of the collisions between neutrons and moderator nuclei result in the scattering of neutrons at reduced energies. With a sufficient amount of moderator in the reactor, the neutrons are reduced to thermal energies. The values of the thermal cross sections are such that $\nu F/A > 1$, and thus a chain reaction can be produced. Table 16-4 lists the different cross sections of uranium for thermal neutrons.

A glance at Table 16-4 shows that thermal neutron fission in natural uranium is due entirely to the presence of the isotope ^{235}U, which has an abundance of only 0.71 percent in natural uranium. A small amount of fission will take place in ^{238}U with some of the fast neutrons being released in the fission process before they are reduced to thermal energies. The first reactor ever built, the so-called uranium-graphite pile, used natural uranium with graphite as the moderator.

It is interesting to consider the processes that occur during one generation of neutrons in a natural uranium graphite-moderated reactor operated at a constant power level. In one example, starting with 100 neutrons captured by uranium, mostly thermal neutrons captured by ^{235}U and the remainder captured in fast fission process by both ^{235}U and ^{238}U, 256 new neutrons are produced. Of these

100 neutrons will be used to carry on new fissions;
 90 neutrons will undergo radiative capture by ^{238}U;
 20 neutrons will undergo radiative capture by ^{235}U;
 30 neutrons will be absorbed by the moderator;
  5 neutrons will be absorbed by structural material;
  9 neutrons will escape from core; and

2 neutrons will be in excess, normally absorbed by control rods but otherwise available for increasing the power level.

16-16   Types of Nuclear Reactor

The design, construction, and operation of a nuclear reactor are part of a large and rapidly expanding field of nuclear engineering. (There are many nuclear reactors now in operation throughout the world, many for the production of power for conversion into electric power, some for the propulsion of ships, and many for experimental purposes and for the training of personnel.) Some of the nuclear reactors use natural uranium for the production of plutonium, ^{239}Pu, which is fissionable by both fast and slow neutrons. A reactor that could produce useful power from a natural fuel such as uranium and in the process produce additional fissionable nuclides equal to or exceeding the amount of the fuel used is called a *breeder* reactor.

Uranium ores are plentiful and widely dispersed over the surface of the earth, and even though the concentration of uranium in these ores is small, the amount of uranium in existence is probably very large. In constructing nuclear reactors, the fuel may be natural uranium or it may be uranium enriched with the lighter isotope ^{235}U. The enrichment may be produced in a number of ways, of which gaseous diffusion through a porous barrier is the most common, although separation by electromagnetic methods has also been used.

Reactors may be classified in a wide variety of ways; for example, they may be classified by the manner in which the fuel and the moderator are mixed. A *homogeneous reactor* is one in which the fuel and the moderator form a mixture that has uniform composition. One of the common types of homogeneous reactor consists of a solution of uranyl nitrate in water, with the uranium enriched with the lighter isotope by as much as one part of ^{235}U to six parts of ^{238}U, as compared with natural uranium, where the ratio is 1 to 140. The solution is put in a steel sphere and this is surrounded by a neutron reflector such as beryllium oxide and graphite. The entire reactor is shielded by lead, cadmium, and concrete. The operation of the reactor can be controlled by means of cadmium rods that penetrate the beryllium oxide reflector. These control rods can be positioned accurately to control the power level of the reactor, and one or more may be used as safety rods to be dropped into position should the power level get too high.

As shown previously, the critical mass of a given fissionable isotope depends on the shape and size of the core, the nature of the moderator, and the neutron reflector. The minimum critical mass of ^{235}U, in a homogeneous reactor consisting of a light water (H_2O) solution of a salt of uranium in a spherical container, is about 800 grams. Data on critical masses of ^{235}U in various concentrations in solutions of light and heavy water, and in spherical and cylindrical vessels, are given in a report that was presented by D. Callihan at the International Conference for the Peaceful Uses of Atomic Energy (see References: Charpie

Figure 16-11   Top of the reactor core of Argonne National Laboratory's heavy water reactor. The shield has been removed to show the ends of the uranium rods which are suspended in a tank of heavy water. The large hole in the center is the thimble which extends nearly to the bottom of the tank. Materials to be made radioactive are lowered into this hole. (Courtesy of Argonne National Laboratory.)

et al.); for example, the critical mass of ^{235}U in an aqueous solution of UO_2F_2 in a sphere 32 cm in diameter, without a reflector, is 2.13 kg. When the sphere is surrounded by a large quantity of water as a neutron reflector (infinite reflector), the critical mass of ^{235}U is reduced to about 855 grams at 25°C. Solutions with a concentration of less than about 12 grams/liter can never become critical no matter how large the mass of ^{235}U.

The other important isotopes fissionable by slow neutrons are ^{233}U and ^{239}Pu; the minimum critical masses of these isotopes are 588 and 509 grams, respectively.

Most of the nuclear reactors, especially the larger ones, are of the *heterogeneous* type; that is, the fissionable material is concentrated in containers suitably distributed throughout the moderator. In many cases the fissionable materials, in the form of cylinders of uranium or uranium oxide enclosed in aluminum tubes, are spaced in a lattice work in the moderator, usually graphite but sometimes heavy water, as shown in Figure 16-11. A heterogeneous reactor suitable for research work, designed and built at the Oak Ridge National Laboratory, consists of a set of fuel elements immersed in a large tank of

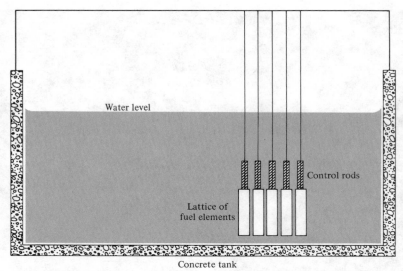

Figure 16-12 Simplified diagram of "swimming pool" type of nuclear reactor.

Figure 16-13 View of the Gulf-General Atomic TRIGA reactor core. The reactor, is designed for research, training and isotope production; it rests at the bottom of a tank 21 ft. deep, which is inside a concrete-lined pit. The surrounding earth and 16 ft. of demineralized water provide the required shielding. (Courtesy of Gulf Energy and Environmental Systems.)

ordinary water. This set of fuel elements consists of plates of uranium-aluminum alloy inside an aluminum container.

A set of these fuel elements containing sufficient uranium to become critical when immersed in water is mounted on a frame with suitable control rods of boron carbide. This fuel is lowered into a large concrete tank of water which acts as the moderator, as shown in Figure 16-12. About 3 kg of ^{235}U are sufficient for the critical amount of fuel, but if beryllium oxide is used as a reflector a smaller amount, about 2.4 kg of ^{235}U, will produce criticality. This reactor will develop approximately 100 kW and provide a thermal neutron flux of about 10^{12} neutrons/cm^2/sec. Samples to be irradiated by thermal neutrons can be placed conveniently anywhere in the water. Neutrons above thermal energy are also available for irradiating materials by placing the substances between the fuel elements or close to them.

Figure 16-13 is a photograph of a research reactor whose core consists of a set of fuel elements each of which is a homogeneous mixture of enriched uranium and zirconium hydride in the form of an alloy. The enriched uranium contains 20 percent of uranium 235. The control rods are made of boron carbide. The reactor can be operated at a steady power level at about 100 kW or it may be operated in short pulses, each lasting about ten thousandths of a second, with maximum power during the pulse of more than a million kilowatts. It is thus possible to study the effects of very intense radiation on various materials and organisms.

16-17 Transuranic Elements

We have already discussed the discovery and production of the first two transuranic elements, neptunium and plutonium, as a result of the resonance capture of neutrons by uranium. In the period since 1945, 11 additional transuranic elements have been produced so that there is now a total of 105 elements. The 13 transuranic elements are listed in Table 16-5.

These transuranic elements have been produced by bombarding uranium or heavier elements with high-energy helium ions (alpha particles) or with high-energy carbon ions produced in cyclotrons. Some of the transuranic elements are the products of radioactive decay. We shall mention only a few examples here.

Americium, for example, is the product of the beta decay of plutonium, which is formed in an (α, n) reaction of uranium, thus

$$^{238}_{92}\text{U} + {}^{4}_{2}\text{He} \rightarrow {}^{241}_{94}\text{Pu} + n \tag{16-25}$$

followed by

$$^{241}_{94}\text{Pu} \rightarrow {}^{241}_{95}\text{Am} + \beta^- + \bar{\nu} \qquad T = 13 \text{ yr} \tag{16-26}$$

^{241}Pu also disintegrates by alpha-particle emission to ^{237}U. The branching ratio of alpha-particle decay to beta decay is about 4×10^{-3}.

TABLE 16-5   TRANSURANIC ELEMENTS

Atomic Number	Name	Chemical Symbol	Range of Mass Numbers
93	Neptunium	Np	231–241
94	Plutonium	Pu	232–246
95	Americium	Am	237–246
96	Curium	Cm	238–250
97	Berkelium	Bk	243–250
98	Californium	Cf	244–254
99	Einsteinium	Es	245–256
100	Fermium	Fm	248–257
101	Mendelevium	Md	255–256
102	Nobelium	No	253–256
103	Lawrencium	Lw	255–257
104	Rutherfordium		257–260
105	Hahnium	Ha	260–261

Several different isotopes of curium are formed when ^{239}Pu is bombarded with alpha particles of 30 to 40 MeV. Two of the reactions are

$$^{239}_{94}\text{Pu} + ^4_2\text{He} \rightarrow ^{241}_{96}\text{Cm} + 2n \tag{16-27}$$

and

$$^{239}_{94}\text{Pu} + ^4_2\text{He} \rightarrow ^{240}_{96}\text{Cm} + 3n \tag{16-28}$$

These are followed by the radioactive disintegration of curium as follows:

$$^{240}_{96}\text{Cm} \rightarrow ^{236}_{94}\text{Pu} + ^4_2\text{He} \qquad T = 26.8 \text{ days} \tag{16-29}$$

$$^{241}_{96}\text{Cm} \rightarrow ^{237}_{94}\text{Pu} + ^4_2\text{He}$$

and

$$^{241}_{96}\text{Cm} + e^- \rightarrow ^{241}_{95}\text{Am} + \nu \qquad T = 35 \text{ days} \tag{16-30}$$

In addition, ^{240}Cm disintegrates by spontaneous fission.

Thompson, Ghiorso, and Seaborg (1950) produced berkelium by bombarding americium with 35-MeV alpha particles, the nuclear reaction being

$$^{241}_{95}\text{Am} + ^4_2\text{He} \rightarrow ^{243}_{97}\text{Bk} + 2n \tag{16-31}$$

This isotope of berkelium disintegrates mostly by electron capture and about 0.1 percent by alpha-particle emission, the reactions being

$$^{243}_{97}\text{Bk} + e^- \rightarrow ^{243}_{96}\text{Cm} + \nu$$

$$^{243}_{97}\text{Bk} \rightarrow ^{239}_{95}\text{Am} + ^4_2\text{He} \qquad T = 4.5 \text{ hr} \tag{16-32}$$

Thompson, Ghiorso, Seaborg, and Street (1950) produced californium by bombarding curium with 35-MeV alpha particles, the reaction being

$$^{242}_{96}Cm + ^{4}_{2}He \rightarrow ^{244}_{98}Cf + 2n \tag{16-33}$$

followed by

$$^{244}_{98}Cf \rightarrow ^{240}_{96}Cm + ^{4}_{2}He \qquad T = 25 \text{ min} \tag{16-34}$$

The same group of investigators (1951) bombarded natural uranium with carbon nuclei, charge $+6e$, with an energy of approximately 120 MeV. They separated the elements of $Z > 94$ chemically and investigated alpha-particle activity. They found one with a half-life of 45 min and an alpha-particle energy of 7.15 MeV, which is characteristic of ^{244}Cf. The reaction in which ^{244}Cf was formed was most likely

$$^{238}_{92}U + ^{12}_{6}C \rightarrow ^{244}_{98}Cf + 6n \tag{16-35}$$

They also found an alpha-particle activity with a longer half-life of 35 hr after the decay of ^{244}Cf. They ascribed this activity to the formation of a new isotope of californium, $A = 246$, the reaction being

$$^{238}_{92}U + ^{12}_{6}C \rightarrow ^{246}_{98}Cf + 4n \tag{16-36}$$

This isotope can also be produced in an (α,n) reaction with ^{243}Cm. Californium, $A = 246$, disintegrates by alpha-particle emission to ^{242}Cm.

The existence of element No. 102, called nobelium, was definitely established by the work of Ghiorso, Sikkleand, Walton, and Seaborg at Berkeley, California, in 1958. One reaction was

$$^{246}_{96}Cm + ^{12}_{6}C \rightarrow ^{254}_{102}No + 4n \tag{16-37}$$

followed by

$$^{254}_{102}No \rightarrow ^{250}_{100}Fm + ^{4}_{2}He \qquad T = 3 \text{ sec} \tag{16-38}$$

Because of its short half-life of 3 sec, its existence was confirmed by collecting the decay product, fermium, $A = 250$, which has a half-life of 30 min.

All of the transuranic elements are radioactive; all of them also undergo spontaneous fission. The probability for spontaneous fission increases rapidly with the quantity Z^2/A. Seaborg (1952) found that if Z^2/A is plotted against the fission half-life, it approaches the value 47 for a half-life of 10^{-20} sec, which may be considered the rate of instantaneous fission. The latter process may thus put a limit on the number of new elements which can be produced.

As a result of the intensive analysis of the chemical properties of the heavier elements, Seaborg suggested that the elements of atomic numbers greater than 88 probably form another transition group analogous to the rare-earth group of atomic numbers 58 to 71. In this latter group (see Table 9-2), each succeeding element is formed by the addition of an electron to the $4f$ shell until this shell is completed with 14 electrons. Seaborg suggested that in this new transition group, electrons are added to the $5f$ shell, the first $5f$ electron probably appearing in thorium. Just as the rare-earth group of elements is called the *lanthanide series,* so the new group of heavy elements is called the *actinide series.*

16-18   Stellar Energy of Nuclear Origin

Ever since the discovery of the transmutation of elements it has been thought that the conversion of mass into other forms of energy, which occurs during such transmutation, would provide a possible explanation of the origin of a great part of the energy radiated by the stars, but it was not until sufficient evidence had been accumulated in the laboratory concerning the probabilities of various types of reaction and their energy releases that it became possible to develop a fairly quantitative explanation. Astrophysical evidence shows that the most abundant type of nucleus present in the stars classified in the *main sequence* is the proton. Bethe (1939) developed the theory for the production of stellar energy in which protons, by suitable nuclear reactions, are transmuted into helium nuclei, thereby releasing energy which is transformed into radiation. One possible series of reactions, sometimes referred to as the *proton-proton* chain, which may occur in a star, starts with the reaction between a proton and a proton to form a deuteron—that is,

$$p + p \rightarrow d + e^+ + \nu \tag{16-39}$$

followed by

$$p + d \rightarrow {}^3\text{He} + \gamma \tag{16-40}$$

The chain ends with the reaction

$$^3\text{He} + {}^3\text{He} \rightarrow {}^4\text{He} + p + p \tag{16-41}$$

or with the less probable reaction

$$p + {}^3\text{He} \rightarrow {}^4\text{He} + e^+ + \nu \tag{16-42}$$

Either reaction leads to the net effect which is expressed by

$$4 \, {}^1\text{H} \rightarrow {}^4\text{He} + 2e^+ + 2\nu + \gamma \tag{16-43}$$

In this process an energy of about 24.7 MeV is changed from rest mass into kinetic energy. The student should verify this fact, keeping in mind that an extra amount of energy equal to 2×0.511 MeV must be included in the calculation for each positron formed if the masses used for ^{1}H and ^{4}He refer to neutral atoms. If these reactions occur inside a star, the positrons will readily find electrons and annihilate into gamma rays, which in turn give up their energy to other particles in the star. In effect, the energy liberated would be about 26.7 MeV.

Another possible series of reactions in which protons are converted into helium nuclei is one in which carbon and nitrogen act as catalysts; this series is usually referred to as the carbon-nitrogen cycle and consists of the following series of reactions:

$$p + {}^{12}\text{C} \rightarrow {}^{13}\text{N} + \gamma$$

$$^{13}\text{N} \rightarrow \, ^{13}\text{C} + e^+ + \nu \qquad T = 9.96 \text{ min}$$

$$p + \, ^{13}\text{C} \rightarrow \, ^{14}\text{N} + \gamma$$

$$p + \, ^{14}\text{N} \rightarrow \, ^{15}\text{O} + \gamma \qquad\qquad\qquad (16\text{-}44)$$

$$^{15}\text{O} \rightarrow \, ^{15}\text{N} + e^+ + \nu \qquad T = 124 \text{ sec}$$

$$p + \, ^{15}\text{N} \rightarrow \, ^{12}\text{C} + \, ^4\text{He}$$

It will be noticed that this chain of reactions can start with either nitrogen or carbon, since each one is reproduced in the reaction, except that in about one case in 10^5 the last reaction leads to the formation of ^{16}O and the emission of a gamma ray. After deducting catalysts from reactions in this cycle, it becomes clear that the net effect is the same as that of reaction (16-43), and the energy released by the carbon-nitrogen cycle is exactly the same as that released by the proton-proton cycle, namely 26.7 MeV or about 43×10^{-13} joule. A small amount of this energy is carried away by the neutrinos that are emitted along with the positrons. Bethe estimated this loss at about 3×10^{-13} joule, leaving about 40×10^{-13} joule for each alpha particle or approximately 10×10^{-13} joule for each proton consumed. In the particular case of the sun it has been estimated that a gram of its mass contains about 2×10^{23} protons; hence, if all the protons were consumed, the energy released would be 2×10^{11} joules. If the sun were to continue to radiate at its present rate, it would take about 30 billion years to exhaust its supply of protons, since its mass is about 2×10^{30} kg, and it radiates energy at a rate of about 4×10^{26} watts.

16-19 Stellar Evolution

It is interesting to speculate concerning the evolution of the stars. It is generally agreed that the first stage in the process of evolution is the contraction of a large mass of very tenuous matter from a large cloud of gas to a more compact star under the action of the gravitational attraction of the particles of the cloud. This produces a higher temperature at the center of the star and causes a flow of heat toward the surface of the star and thus radiation of energy from the surface. A type of equilibrium may be reached in which the rate of radiation of energy from the surface is just equal to the rate of change in gravitational energy. The time scale for this event is rather short in terms of the estimated age of our galaxy; the latter is of the order of a few billion years.

At some stage in the process of contraction the temperature at the center of the star must become sufficiently high that collisions between nuclei take place with sufficient relative kinetic energy of the particles so that nuclear reactions occur with the subsequent release of energy. Furthermore, such collisions must occur with sufficient frequency to account for the rate at which energy is known to be radiated by different classes of star. With the data on capture cross sections that are now available for several of the processes in the proton-proton chain and in the carbon-nitrogen cycle, it is possible to estimate the rate at which energy can be radiated from a star under a variety of conditions of temperature and density. H. Bondi and E. E. Salpeter (1952) developed empirical equations for the rate at which energy is liberated in each of the above chains of reaction. For the case of the sun they estimate that the rate of generation of energy in the proton-proton chain is about the same as that in the carbon-nitrogen cycle. For much more luminous stars the carbon-nitrogen cycle predominates as a source of energy, and for the less luminous stars the proton-proton chain supplies the greater fraction of the energy radiated.

When all of the hydrogen is used up in the above thermonuclear reactions, the star will consist mostly of helium. The temperature of the star is probably not sufficiently high for nuclear reactions to take place among the helium nuclei. At this stage gravitational contraction will occur once again until a temperature of about 10^8 K is reached with the density of the star about 10^4 grams/cm^3. Under these conditions it will be possible for three helium nuclei to combine to form ^{12}C with the release of about 7.3 MeV. F. Hoyle estimates that such processes can occur sufficiently often to provide for the radiation of energy for an additional 10^7 yr. When all the helium is used up, further gravitational contraction of the star will occur, producing a further rise in the temperature of the star and making conditions right for the formation of atoms of medium atomic weights. A glance at the results of Problem 15-20 will show that the region of mass number 60 is the most stable one and that any appreciable combination of these atoms to form heavier ones will lead to endothermic rather than exothermic reactions. Bondi and Salpeter, tracing this life history of a star, suggest that the endothermic reactions may account for the sudden collapse of a star, identified as the sudden appearance of a supernova.

16-20   Fusion of Light Nuclei

Nuclear fusion may be considered as the reverse of nuclear fission; that is, at least one of the products of the nuclear reaction will be more massive than any of the initial reactants. Nuclear fusion will lead to a release of energy in those cases in which the total mass of the product nuclei is less than the total mass of the reactants. This condition is almost always satisfied for light nuclei of mass numbers A_1 and A_2 such that $A_1 + A_2 < 60$, as can be seen in the table of Appendix IV by noticing that the minimum mass excess per nucleon occurs around $A = 60$.

The reactions considered in Section 16-18 in the proton-proton chain are examples of nuclear fusion. These reactions, however, take place at a comparatively slow rate. The deuteron-deuteron reaction discussed in Section 15-11 occurs at a much faster rate. This reaction leads to the following products:

$$^2_1H + ^2_1H \rightarrow ^3_1H + ^1_1H + 4.03 \text{ MeV} \tag{16-45}$$

or

$$^2_1H + ^2_1H \rightarrow ^3_2He + n + 3.27 \text{ MeV} \tag{16-46}$$

The cross sections for these two reactions are almost equal. The reaction between deuterium and tritium has a much larger cross section and releases a much greater amount of energy per atomic mass unit taking part in the reaction; thus

$$\ce{^2_1H + ^3_1H \rightarrow ^4_2He} + n + 17.6 \text{ MeV} \qquad (16\text{-}47)$$

The reaction between deuterium and ^{3}He also yields a much larger amount of energy but with a small cross section. This reaction is

$$\ce{^2_1H + ^3_2He \rightarrow ^4_2He + ^1_1H} + 18.4 \text{ MeV} \qquad (16\text{-}48)$$

It will be noted that both ^{3}H and ^{3}He are products of the deuteron-deuteron reactions. Each of these reactions can be produced in a suitable particle accelerator. Although all of these reactions are exothermic, the amount of energy released is a minute fraction of the energy supplied to the accelerator. Other methods must therefore be devised in order to utilize the above reactions as sources of energy. It is obvious from the above discussion that one method would be to start with a fairly large amount of deuterium, or deuterium mixed with tritium, and raise the mixture to a very high temperature. An idea of the order of magnitude of the temperature can be obtained from the fact that the above reactions take place when the deuterons are accelerated to energies greater than 10 keV. It will be recalled that so-called thermal neutrons—that is, neutrons in equilibrium with matter at room temperature, about 300 K— have an average energy of about 1/40 eV. Hence

$$1 \text{ eV} \approx 12{,}000 \text{ K}$$

$$1 \text{ keV} \approx 12{,}000{,}000 \text{ K} \qquad (16\text{-}49)$$

$$10 \text{ keV} \approx 120 \times 10^6 \text{ K}$$

Temperatures of this order of magnitude can be obtained by exploding an atomic bomb—that is, a fission reaction. Hence it is not surprising that the first successful thermonuclear fusion reactions were produced as military weapons. The actual methods used are, of course, military secrets.

Attempts to produce controlled thermonuclear fusion reactions have so far been unsuccessful, so much so that all of the work done on this problem was completely declassified in 1958. This makes it possible to attack this problem by the more conventional methods of science—that is, a careful study of the properties of a completely ionized gas, called a *plasma*, the gas in nearly all cases being deuterium. A whole new branch of physics known as *plasma physics* is being developed as a result of this work. It is interesting to note that the subject of atomic physics got its start from a study of the phenomena produced and observed in the gas discharge tube. Physicists are studying this once more with more sophisticated and refined methods and with the newer ideas of wave mechanics and quantum electrodynamics. It may be several decades before a controlled thermonuclear fusion is achieved.

Problems

16-1.   Calculate the electrostatic potential energy of two nuclei of atomic numbers $Z_1 = 36$ and $Z_2 = 56$ when their centers are 1.5×10^{-12} cm apart.

16-2.   Show that the ratio of the kinetic energies of two fission fragments that have equal and opposite momenta is M/m, where M and m are their respective masses. Which fragment has the greater kinetic energy?

16-3.   (a) Determine the potential, in volts, at a distance of 10^{-12} cm from a nucleus of charge $Ze = 45e$, where e is the electronic charge. (b) Determine the potential energy in MeV of two such nuclei when the distance between their centers is 10^{-12} cm.

16-4.   Calculate the energy released in the fission of uranium $A = 235$ by a slow neutron if the fission products have mass numbers 72 and 162, respectively.

16-5.   (a) Calculate the energy released in the fission of uranium $A = 235$ by a slow neutron if the two fission products have equal mass numbers and two neutrons are emitted promptly. (b) Calculate the kinetic energies of the two fission particles, assuming that they are acquired as a result of the Coulomb repulsion at the instant of fission. (c) Account for any difference between (a) and (b).

16-6.   Assuming that the most probable kinetic energy of a nucleus in thermal equilibrium at absolute temperature T is kT, where k is the Boltzmann constant ($k = 1.38 \times 10^{-16}$ erg/deg), calculate the energy in MeV of a neutron at 300 K, of a proton at 14×16^6 K, and of a helium nucleus at 10^8 K.

16-7.   Assume that the excitation energy required for passage over the nuclear potential barrier is equal to the Coulomb potential energy of the two nuclei formed in the fission process. (a) Calculate the excitation energy for the disintegration of ^{8}Be into two alpha particles. (b) Compare this value with the energy equivalent of the mass differences of the nuclei involved.

16-8.   The measured value of the total kinetic energy of the fission fragments from the thermal neutron fission of ^{235}U is 196 MeV. Assuming that the respective values of Z and A of the fission fragments are (35, 72) and (57, 162), calculate the distance r between the fragments at the instant of separation. Compare this value with the sum of the radii of the two fragments.

16-9.   In the fission reaction ^{63}Cu $+ p \rightarrow {}^{24}$Na $+ {}^{39}$K $+ n$ determine (a) the

excitation energy necessary for passage over the potential barrier;
(b) the energy equivalent of the mass difference between reactants
and products; and (c) the threshold energy for the reaction.

16-10.   In the fission reaction $^{118}_{50}Sn + p \rightarrow {}^{24}_{11}Na + {}^{94}_{40}Zr + n$ determine (a) the
excitation energy necessary for passage over the potential barrier,
(b) the energy equivalent of the mass difference between reactants
and products, and (c) the threshold energy for the reaction.

16-11.   In the fission reaction $^{118}_{50}Sn + p \rightarrow {}^{66}_{31}Ga + {}^{49}_{20}Ca + 4n$ determine (a) the
excitation energy necessary for passage over the potential barrier,
(b) the energy equivalent of the mass difference between reactants
and products, and (c) the threshold value for the reaction.

16-12.   In the liquid drop model of the nucleus the probability of spontaneous
fission increases rapidly with Z^2/A. G. T. Seaborg found that if Z^2/A
is plotted against the half-life, it approaches the value 47 for a half-life
of 10^{-20} sec, which may be considered the rate of instantaneous fission.
Use this value to determine an upper limit to the number of new elements that may be produced by nuclear processes.

16-13.   Determine the kinetic energy that an alpha particle can acquire from
the Coulomb field of a fission fragment of $Z = 56$ and $A = 140$, assuming that the alpha particle is released at rest at the instant of fission.

16-14.   (a) Show that the kinetic energy $\mathcal{E}_k$ of the fission fragments in MeV
for the slow-neutron fission of ^{235}U can be expressed in terms of the
mass excesses of the particles involved in the fission process by the
equation $\mathcal{E}_k = [m(^{235}U) - m(L) - m(H) - (\nu - 1)m(n)] - \nu\mathcal{E}_n - \mathcal{E}_\gamma$
where $m(L)$ is the mass excess of the light fragment, $m(H)$, the mass
excess of the heavy fragment, $m(n)$ the mass excess of the neutron,
ν the average number of neutrons emitted per fission, $\mathcal{E}_n$ the average
kinetic energy of these neutrons, and $\mathcal{E}_\gamma$ the average energy of the
prompt gamma rays. (b) Using the data from the table of mass excesses
and taking $\nu = 2.5$, $\mathcal{E}_n = 2$ MeV, and $\mathcal{E}_\gamma = 4.6$ MeV, calculate the
kinetic energy of the fission fragments for the most probable fission
mode.

16-15.   Assuming that the energy released per fission of ^{235}U is 200 MeV,
calculate the rate at which fission should occur in a nuclear reactor
in order to operate at a power level of 1 watt.

16-16.   (a) Show that if α is the ratio of the neutron capture cross section to
the neutron fission cross section, then $1/(1 + \alpha)$ is the fraction of fission
reactions and $\alpha/(1 + \alpha)$ is the fraction of capture reactions for a given
fissionable isotope. (b) Show that η, the average number of fission
neutrons produced per neutron absorbed, is $\nu/(1 + \alpha)$. (c) Using the
data of Table 16-4, calculate α and η for the thermal neutron fission
of ^{235}U.

16-17.   Low-energy neutrons are incident on a steel sheet which is 1 cm thick. Assume that the neutron capture cross section is 2.5 barns. What fraction of the incident neutrons will be captured?

16-18.   Kinetic energies of slow neutrons are often measured by their time of flight. Two detectors, separated by a distance L in meters, yield signals that differ in time by an amount t in microseconds. Show that $\mathscr{E}_K = 52 \times 10^2 \, (L^2/t^2)$ eV.

16-19.   Assume that neutron energies are measured by the time-of-flight method, with $L = 10$ meters. (a) What time interval is obtained for thermal neutrons ($\mathscr{E}_K = 1/40$ eV)? (b) What time interval is needed if $\mathscr{E}_K = 1$ MeV?

16-20.   The kinetic energy of neutrons at about 1 MeV is to be measured by the time-of-flight method. The time resolution of the circuitry is about 10^{-8} sec. How long a path is necessary to measure the energy to 1 percent?

16-21.   Watt's formula, Equation (16-5), expresses the energy distribution of neutrons from the fission of ^{235}U. Find the form of the velocity distribution.

16-22.   By differentiation of Watt's formula find the most probable energy of a neutron from fission.

16-23.   Assuming that $N(\mathscr{E})$ is normalized to unity, evaluate the constant C in the Watt formula.

16-24.   Expand the Watt formula and estimate the fraction of neutrons for which the kinetic energy is less than 0.1 MeV.

16-25.   (a) What is the electrostatic potential energy in electron volts associated with two protons separated by a distance of 1.2 fermis? (b) At what temperature does this energy coincide with the mean kinetic energy of a proton, assuming that the protons form an ideal gas?

16-26.   (a) Estimate the radius of the nucleus of ^{12}C. (b) How much potential energy is associated with the proton that gets within one radius of the center of the ^{12}C nucleus? (c) To what temperature does this energy correspond?

16-27.   Calculate the Q value for each reaction in the proton-proton cycle.

16-28.   Calculate the Q value for each reaction in the carbon-nitrogen cycle.

References

Nuclear Fission

Charpie, R. A., J. Horowitz, D. J. Hughes, D. J. Littler, Eds., *Progress in Nuclear Energy*, Series 1, *Physics and Mathematics*, Vol. 1. New York: McGraw-Hill Book Company, 1956.

Glasstone, S., and M. C. Edlund, *The Elements of Nuclear Reactor Theory*. New York: D. Van Nostrand Company, 1952.

Halpern, I., "Nuclear Fission," *Ann. Rev. Nucl. Sci., 9*, 245–342 (1959).

Soodak, H., and E. C. Campbell, *Elementary Pile Theory*. New York: John Wiley & Sons, 1950.

Weinberg, A. M., and E. P. Wigner, *The Physical Theory of Neutron Chain Reactors*. Chicago: The University Press of Chicago, 1958, Chapters I–V.

Nuclear Fusion

Burbidge, E. M., G. R. Burbidge, W. A. Fowler, and F. Hoyle, "Synthesis of the Elements in Stars." *Rev. Mod. Phys., 29*, 547 (1957).

Post, R. F., "High-Temperature Plasma Research and Controlled Fusion," *Ann. Rev. Nucl. Sci., 9*, 367 (1959).

Rose, D. J., and M. Clark, *Plasmas and Controlled Fusion*. Cambridge, Mass.: The M.I.T. Press, 1961.

17 | Nuclear Processes

17-1 Stability of Nuclei

More than 1400 different isotopes are now known, but only about 20 percent of them are stable. It can thus be inferred that stable configurations of nucleons are the exception rather than the rule. Figure 17-1 is a graph of the neutron number N plotted against the proton number Z of the stable isotopes. The region of stability on this neutron-proton diagram is rather narrow. For mass numbers below 40, N and Z are nearly equal. For heavier stable nuclides, N is always greater than Z. Lines of constant A can be drawn at angles of 135° with the Z axis passing through isobars. If A is an odd number, the line corresponding to that value of A passes through only one stable nuclide. If A is even, there will usually be either one or two stable isobars. In four cases the lines of constant A pass through three isobars at $A = 96$, 124, 130, and 136. These cases are shown on the graph.

A significant point is that for more than half the stable nuclides both Z and N are even numbers; these nuclides are referred to as *even-even*. About 20 percent have even Z and odd N (even-odd nuclides) and an almost equal number have odd Z and even N. Stable odd-odd nuclides are few in number: $^2_1$H, $^6_3$Li, $^{10}_5$B, $^{14}_7$N, and $^{180}_{73}$Ta. At one time $^{50}_{23}$V was listed as a stable odd-odd nuclide, but recent work has shown that it is radioactive with a half-life of about 6×10^{15} yr. Even numbers for Z and N are favored not only in the types of nuclide that are stable but also in the abundance of these nuclides; for example, it has been estimated that 87 percent of all nuclei on the earth have even Z.

Points to the right of the stability region of Figure 17-1 represent unstable nuclei which have an excess of protons or a deficiency of neutrons; their most probable mode of decay will be by positron emission or electron capture. Points to the left of the stability region represent unstable nuclei with an excess of neutrons or a deficiency of protons; their most probable mode of decay will

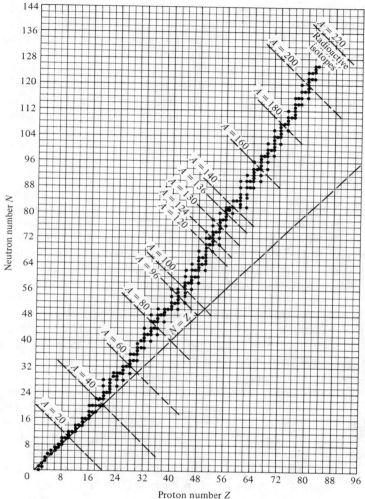

Figure 17-1 Neutron-proton diagram of stable nuclei.

be by negative beta-particle emission. However, heavy particle emission is a competing process. As we have seen, compound nuclei formed in an excited state may disintegrate by emitting protons, neutrons, or alpha particles in going toward more stable forms. Among the fission products (Section 16-4), neutron emission competes with beta decay to form stable nuclei.

Stability is, in a sense, a relative property depending on our ability to measure long lifetimes, that is, very weak activities. With improvements in experimental techniques, some of the isotopes now considered stable may be found to be radioactive with long lifetimes.

For a given assemblage of nucleons forming a nucleus of mass number A

and proton number Z, there is one configuration for which the energy is a minimum. This configuration is called the *ground state* of the nucleus; every other configuration is called an *excited state*. The difference in energy between an excited state of a nucleus and its ground state is called the *excitation energy* of that state. Several different nuclear models have been proposed to account for the behavior of nuclei, but at present there is no single nuclear model that is adequate for the wide range of nuclear energies of a given nucleus or for the entire range of mass numbers. Each of these models has a special usefulness for a limited range of energies or a limited range of mass numbers.

17-2 Nuclear Spins

In Chapter 3 we considered some basic properties of nuclei, including their size (Section 3-9) and mass (Section 3-13). In Chapter 9 we considered in some detail the properties of electrons in atoms which follow from the assumption that the electron has an intrinsic angular momentum known as spin. In this section we consider the angular momentum possessed by nuclei.

The original indication that nuclei have spin came from atomic spectroscopy. Many lines in atomic spectra, when examined with spectroscopes of very high resolving power, are found to consist of several lines very close together. Such lines are said to exhibit *hyperfine structure* (frequently abbreviated hfs in the periodical literature). One obvious source of such structure is the isotope effect, already considered in Section 8-3.

The second type of hyperfine structure cannot be explained as due to the presence of isotopes because it has been found in the spectral lines of elements, such as bismuth, which consist of single isotopes only. In these cases the number of components is different for different spectral lines, and their relative displacements are such that they cannot be explained on the basis of the existence of other isotopes. The explanation of this type of hyperfine structure of spectral lines, suggested by Pauli, is that the nucleus of the atom also spins about an axis and possesses a nuclear angular momentum due to spin of amount

$$I\hbar$$

where I is the *nuclear spin quantum number*.

The total angular momentum of the atom will now be the vector sum of the nuclear angular momentum and the total electronic angular momentum and is denoted by $F\hbar$, where

$$\mathbf{F} = \mathbf{I} + \mathbf{J} \tag{17-1}$$

$\mathbf{J}$ is the total angular momentum of the atomic electrons. $\mathbf{F}$, the vector sum of $\mathbf{I}$ and $\mathbf{J}$, is called the *hyperfine quantum number* and is restricted to integral or odd half-integral values. It is beyond the scope of this book to give an extended discussion of the analysis of hyperfine structure. Some of the results of this analysis, however, are of interest, since they have been used in predicting the types of particle that probably exist in the nucleus. For this reason some of the values of the *nuclear* spin quantum number I are listed in Appendix IV.

An examination of Appendix IV shows that the spin quantum number I is zero for even-even isotopes—that is, the spectral lines from these isotopes exhibit no hyperfine structure. The isotopes whose spectral lines exhibit hyperfine structure can be classified into three groups: (a) for isotopes of even atomic number and odd mass number, I is an odd half-integer; (b) for isotopes of odd atomic number and odd mass number, I is also an odd half-integer, and (c) for isotopes of odd atomic number and even mass number, I is an integer.

The argument that follows is based on the assumption that the spin angular momentum of the nucleus is the vector sum of the spin angular momenta of the particles within the nucleus and that these spins are aligned with their axes either parallel or antiparallel; that is, we are carrying over into the nucleus the same type of hypothesis that was found to work for the electrons outside the nucleus. Before the discovery of the neutron the nucleus was assumed to consist of A protons and $A-Z$ electrons; the resultant nuclear charge was therefore equivalent to Z protons. The total number of particles in the nucleus was thus $2A-Z$. Since $2A$ is always an even number, every element of odd atomic number would possess an odd number of particles and its spin quantum number I should be an odd half-integer. This is not always found to be so experimentally; for example, nitrogen, $^{14}_{7}N$, has a spin quantum number $I = 1$. Similarly, the lithium isotope, $^{6}_{3}Li$, has a spin quantum number $I = 1$. Consider the odd isotopes of mercury of mass numbers 199 and 201. Since $Z = 80$, $2A - Z$ is an even number for each of these isotopes, but I is found to be an odd half-integer in each case. After the discovery of the neutron, Heisenberg suggested that the nucleus should consist of protons and neutrons only and that each particle should have a spin angular momentum of $\frac{1}{2}\hbar$. On this basis the number of protons in the nucleus is equal to the atomic number Z and the number N of neutrons in the nucleus is $A-Z$. The total number of particles in the nucleus is equal to the mass number A. The isotopes of any one element therefore differ only in the number of neutrons in the nucleus. The discrepancies in the values of the spin quantum numbers mentioned above now disappear. On Heisenberg's hypothesis, I should be an integer for isotopes of even atomic mass number and should be an odd half-integer for isotopes of odd atomic mass number. These predictions are in agreement with experimental results.

17-3 Nuclear Magnetic Moments—Molecular Beam Method

We have shown that a nucleus has an angular momentum due to its spin. In addition, the nucleus also possesses a magnetic moment. Accurate data on the magnetic moments of atomic nuclei should provide additional information on the nature of nuclear forces and should also help in selecting an appropriate nuclear model. Very precise methods have been developed for the determination of nuclear magnetic moments. One of these, known as the *magnetic resonance method,* was developed by Rabi and his co-workers, and is a direct outgrowth of the Stern-Gerlach type of experiment. This experiment depends essentially on resonance between the precession frequency of the nuclear magnet about a constant magnetic field direction and the frequency of an impressed high-frequency magnetic field.

Just as the magnetic moment of an electron is expressed in terms of a Bohr magneton so the nuclear magnetic moment is expressed in terms of a *nuclear magneton M_n* defined by the equation

$$M_n = \frac{e\hbar}{2M} \tag{17-2}$$

in which M is the mass of the proton. The nuclear magneton is thus only about 1/1840 of a Bohr magneton. If I is the nuclear spin quantum number, the angular momentum of the nucleus due to its spin is $I\hbar$. Just as we introduced the Landé g factor to relate the magnetic moment of the electrons of an atom to their total angular momentum, so we can introduce a *nuclear g factor* to relate the magnetic moment μ of a nucleus to its spin angular momentum. *The nuclear g factor is defined as the ratio of the nuclear magnetic moment, expressed in units of nuclear magnetons, to the spin angular momentum, expressed in units of $\hbar$.* Thus

$$g = \frac{\mu}{IM_n} \tag{17-3}$$

Hence

$$\mu = gIM_n = gI\frac{e\hbar}{2M} \tag{17-4}$$

When a nucleus of magnetic moment μ is in a constant magnetic field of induction B, it will precess about the direction of B with a frequency ν given by Larmor's theorem

$$\nu = \frac{\mu B}{Ih} \tag{17-5}$$

The magnetic moment μ of a nucleus can thus be found by determining the Larmor frequency ν which the nucleus of spin quantum number I acquires in a known constant magnetic induction B. Instead of working with nuclei alone, Rabi and his co-workers used beams of molecules whose total electronic angular momentum is zero. Figure 17-2 shows the paths of typical molecules in the different magnetic fields used in the magnetic resonance experiment for measuring nuclear magnetic moments; Figure 17-3 is a schematic diagram of the apparatus. A narrow stream of molecules issues from the source at O. A very small fraction of these molecules will pass through the collimating slit

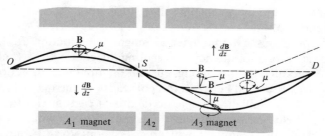

Figure 17-2 Paths of molecules in the molecular beam resonance experiment. The two solid curves indicate the paths of two molecules which have different magnetic moments and velocities and whose moments are not changed during passage through the apparatus. This is indicated by the small gyroscopes drawn on one side of the paths, in which the projection of the magnetic moment along the field remains fixed. The two dotted curves in the region of the A_3 magnet indicate the paths of the two molecules, the projection of whose nuclear magnetic moments along the field has been changed in the region of the A_2 magnet. This is indicated by means of the two gyroscopes drawn on the dotted curves, for one of which the projection of the magnetic moment along the field has been increased, and for the other of which the projection has been decreased.

S and reach the detector at D. In the absence of any inhomogeneous magnetic deflecting fields the molecules traverse straight-line paths OSD and form the so-called *direct* beam.

The magnets A_1 and A_3 are specially designed to produce inhomogeneous magnetic fields. The magnetic fields are in the same direction, but their gradients, dB/dz, are in opposite directions as shown in Figure 17-2. A molecule with magnetic moment μ will be deflected in the direction of the gradient if μ_z, the projection of μ in the direction of the field, is positive and will be deflected in the opposite direction if μ_z is negative. Molecules which left the source along the line OSD will be deflected to one side. Other molecules which leave O at some angle to the line OSD will follow paths indicated by the solid lines and reach the detector D. The force experienced by any such molecule in the inhomogeneous field due to the A_1 magnet is

$$F = \mu_z \left(\frac{\partial B}{\partial z}\right)_{A1} \tag{17-6}$$

A similar expression holds for the force due to the A_3 magnet. The actual deflection produced by each magnetic field can be established from a knowledge of the velocity of the molecule, which is determined by the temperature of the source, and from the geometry of the apparatus. If no change occurs in μ_z as the molecule goes from the A_1 field to the A_3 field, the deflections in these fields will be in opposite directions. It is a simple matter to adjust the two magnetic field gradients to make these deflections equal in magnitude and thus bring the molecules to the detector—that is, to "refocus" the beam. When the two magnetic fields are properly adjusted, the number of molecules reaching the detector D in any given time interval is about the same whether the magnets are on or off.

Magnet A_2 produces a homogeneous magnetic induction B. In the same region there is a high-frequency alternating magnetic field (not shown in Figure 17-2) produced by sending current in opposite directions through two parallel wires R placed between the pole faces of the magnet A_2, as shown in Figure 17-3. The oscillating magnetic field is at right angles to the homogeneous magnetic field produced by the A_2 magnet. When a molecule of magnetic moment μ enters this region, it will precess about B with the Larmor frequency ν. The interaction with the oscillating magnetic field will produce a torque which may either increase or decrease the angle between μ and B; in general, if the frequency f of the alternating magnetic field is different from the Larmor frequency of precession ν, the net effect produced will be small, since the torque produced by the alternating magnetic field will rapidly get out of phase with the precessional motion; but, when $f = \nu$, the increase or decrease produced in the angle between μ and B will be cumulative and this change in angle will become quite large. The molecule will then follow one of the dotted paths when it gets into the region of the A_3 magnet and will not enter the detector at D. In some of the experiments the frequency of the alternating magnetic field is kept at a constant value, and the value of B produced by the A_2 magnet is varied. Figure 17-4 is a typical curve which shows the beam intensity plotted as a function of B while the frequency of the alternating field is kept constant. It will be observed that resonance occurs for a definite value of B; the resonance value is the minimum value of the curve. The resonance curve of the ^{7}Li nucleus shown in Figure 17-4 was obtained with a beam of LiCl molecules. In other experiments with the ^{7}Li nucleus, molecular beams of LiF and Li$_2$ were used.

Solving Equations (17-4) and (17-5) for g and substituting the resonance frequency f for the Larmor frequency ν, we get

$$g = \frac{4\pi M}{e} \frac{f}{B}$$

(17-7)

Since the values of the constants M and e are accurately known, the substitution of the measured values of the resonance frequency f and the intensity of the homogeneous magnetic induction B in Equation (17-7) will yield the

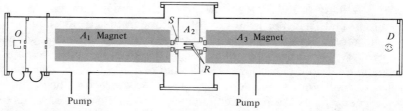

Figure 17-3　Schematic diagram of the apparatus used in the molecular beam experiment.

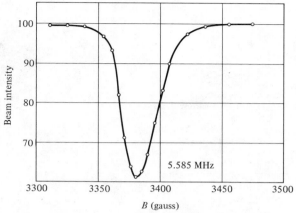

Figure 17-4   Resonance curve of the ^{7}Li nucleus observed in LiCl.

g factor for the particular nucleus under investigation. If its nuclear spin quantum number I is known, then

$$\mu = gI \qquad (17\text{-}8)$$

will give the magnetic moment of the nucleus in nuclear magnetons, whereas substitution of the values of g and I in Equation (17-4) will give the nuclear magnetic moment in mks units. For example, in the case of the ^{7}Li nucleus, $I = \frac{3}{2}$ and the measured value of g is 2.1688; hence its nuclear magnetic moment is 3.2532 nuclear magnetons.

Of very great importance in nuclear physics are the magnetic moments of the proton, deuteron, and neutron. Millman and Kusch (1941) made a precise measurement of the magnetic moment of the proton, and Kellogg, Rabi, Ramsey, and Zacharias (1939) made a precise determination of the ratio of the magnetic moments of the proton and deuteron. From these measurements the magnetic moment of the proton was found to be 2.7896 nuclear magnetons and that of the deuteron was found to be 0.8565 nuclear magneton. If we assume that a deuteron consists of a proton and a neutron and that the magnetic moment of the deuteron is the sum of the magnetic moments of the proton and the neutron, the magnetic moment of the neutron is $\mu_n = -1.933$ nuclear magnetons. Alvarez and Bloch (1940) made an independent determination of magnetic moment of the neutron by sending a beam of slow neutrons through a modified type of molecular beam magnetic resonance apparatus. Using the value $I = \frac{1}{2}$ for the spin of the neutron, they obtained $\mu_n = -1.935$ nuclear magnetons for the magnetic moment of free neutrons.

17-4   Nuclear Induction and Resonance Absorption

A significant modification of the magnetic resonance principle was proposed and developed by F. Bloch (1946). In the molecular beam experiments

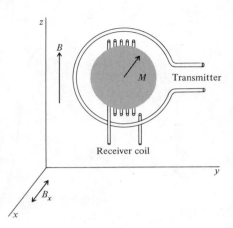

Figure 17-5 Schematic diagram of the arrangement of the transmitter and receiver coils in the nuclear-induction experiment. The cross section in the *y-z* plane of the spherical sample is shown shaded. *M* is the direction of the magnetic polarization, *B* is a constant magnetic field in the *z* direction produced by a magnet (not shown) with its poles above and below the sample. [After Bloch, Hansen, and Packard, *Phys. Rev.*, **70**, 475, (1946).]

the determination of the value of the nuclear moment consists in measuring the frequency of the alternating magnetic field at which the intensity of the molecular beam reached a minimum; this frequency is equal to the Larmor precession frequency of the nuclei in the constant magnetic field. The modification introduced by Bloch consists essentially in measuring the induced emf, or some effect due to this induced emf, produced by a change in the orientation of the nuclear magnetic moments in a sample of a substance. This sample is frequently in the form of a liquid or a solid.

The experimental arrangement used by Bloch and his co-workers is shown in outline in Figure 17-5. The sample under investigation, spherical in shape, is placed between the poles of an electromagnet that provides a constant magnetic field B in the z direction. A flat coil of several turns, called the transmitter coil, in the *y-z* plane, surrounds the spherical sample. Current of frequency f from a high-frequency generator is supplied to this coil, producing an alternating magnetic field B_x in the x direction.

A second coil, called the receiver coil, surrounds the sample; this coil has its axis in the y direction. An emf will be induced in the receiver coil whenever the magnetic flux in the y direction is changed. The receiver coil is connected to an appropriate circuit to measure the emf induced in it.

The effect of the constant magnetic field B is to cause the nuclear magnets to precess about the z axis with the Larmor precessional frequency; each nuclear magnet will have a component of its magnetic moment in the direction of B. Since we are dealing with matter in bulk, the effect of this alignment is to produce a paramagnetic substance having a magnetic moment M per unit volume. Some time will be required to establish this magnetic moment after the application of the constant magnetic field; this time is called the *relaxation time* and may vary from a fraction of a second to thousands of seconds. The relaxation time depends on the interactions between the nuclear magnetic moments and the electronic motions and configurations and thus depends on the temperature.

The magnetization of the sample will produce a magnetic flux in the system. Referring again to the figure, the receiving coil will record an emf only when the magnetic flux through it, in the y direction, is changed. A change in the magnetic flux will be produced by the action of the alternating magnetic field; the closer its frequency is to the Larmor precessional frequency of the nuclei in the constant magnetic field, the greater will be the change in flux produced.

In one of the earliest magnetic induction experiments, Bloch, Hansen, and Packard (1946), using water as the sample, observed a relaxation time of a few seconds. When a concentrated solution of $Fe(NO_3)_3$ in water was used, the relaxation time was found to be of the order of 10^{-4} to 10^{-5} sec. With this sample they found that for a resonance frequency $f = \nu = 7.765 \times 10^6$ Hz, the value of the constant magnetic field $B = 1826$ gauss, yielding a nuclear g value for the proton in agreement with the results obtained with the molecular beam experiments described previously. With improvement in the design of the apparatus, Bloch and his co-workers were able to make very accurate determinations of the ratios of the magnetic moment of the proton to that of the deuteron, the triton, and the neutron. Of great importance is the fact, demonstrated by Rogers and Staub (1949), using a rotating high frequency field, that the sign of the magnetic moment of the neutron is opposite to that of the proton.

The evaluation of the magnetic moment of any one type of particle depends on an accurate knowledge of the value of the constant magnetic field at the resonance frequency. Since convenient material samples containing hydrogen, such as water, paraffin, and oil, can be used in the nuclear induction experiments, Bloch pointed out that the procedure can be reversed to determine the value of the magnetic field in which nuclear resonance has been induced in such a sample. This method is now being widely used, for example, in measuring the magnetic fields in cyclotron magnets.

A method, parallel to the above, for measuring nuclear moments but using radiofrequency techniques was developed by Purcell, Torrey, and Pound (1946). This is called the *magnetic resonance absorption method*. In their first experiment they put a piece of paraffin (850 cm³) into a resonant cavity and placed the resonator in a strong magnetic field. Power was fed through a loop into the cavity at about 30×10^6 Hz. The output of the resonator was balanced against a portion of the output of the generator. When properly balanced, the magnetic field was varied slowly; at one particular value of B, 7100 gauss, a sharp resonance absorption was observed. The frequency was 29.8×10^6 Hz. The values of the proton magnetic moment calculated from these values of ν and B are in agreement with previously determined values within the limits of experimental error. They also suggested that this method could be used for accurate magnetic field determinations and for the determination of the sign of the moment by using radiofrequency fields with rotating components.

The molecular beam magnetic resonance method, the nuclear induction methods, and the magnetic resonance absorption methods have been developed and refined to be methods of very great precision. It would take us too far afield to discuss the various refinements; we shall merely quote the results wherever needed.

The present values of the nuclear magnetic moments of the proton, neutron, and deuteron, in units of the nuclear magneton, are

$$\mu_p = 2.792763 \pm 0.000030$$

$$\mu_n = -1.913148 \pm 0.000066 \qquad (17\text{-}9)$$

$$\mu_d = 0.857411 \pm 0.000020$$

17-5   Orbital Angular Momentum of the Deuteron

It will be noticed that the sum of the magnetic moments of the neutron and proton differ from the deuteron magnetic moment by about 0.022 nuclear magneton. This discrepancy is well outside experimental error. An analysis of this effect can provide insight into the nature of the wave function of the deuteron.

From studies of the hyperfine structure in the spectrum of the deuterium atom it is well established that the angular momentum of the deuteron is $I = 1$. Since the spins of the neutron and proton are both equal to $\frac{1}{2}$, the spin contribution is either zero or one.

$$\mathbf{S} = \mathbf{s}_p + \mathbf{s}_n = 0 \text{ or } 1 \qquad (17\text{-}10)$$

The two nucleons have an orbital angular momentum $\mathbf{L}$ about their center of mass, and following the usual rules for addition of angular momenta,

$$\mathbf{L} + \mathbf{S} = \mathbf{I} \qquad (17\text{-}11)$$

In the notation for Russell-Saunders coupling, introduced in Chapter 9, we observe that the deuteron has $^1P_1$, $^3S_1$, $^3P_1$, or $^3D_1$ as the only possibilities for its ground state. The principle of parity conservation (see Section 17-10) restricts the possibilities further; one of the following three conditions must be true: (a) the ground state is a pure state consisting of *one* of the four states listed, (b) it is a mixture of $^1P_1$ and $^3P_1$, or (c) it is a mixture of $^3S_1$ and $^3D_1$. Theoretical predictions can be made for the magnetic dipole moment of the deuteron, based on each of the four terms given above. We omit the details of the calculation here, and summarize the results in Table 17-1. The numerical values are to be compared with the experimental result, 0.8574 nuclear magnetons. It is clear from this comparison that the deuteron is a mixture of $^3S_1$ and $^3D_1$, with much higher statistical weight belonging to $^3S_1$ (see Problem 17-19).

The significance of this small admixture of D wave can be better appreciated if one is aware of the fact that a central force implies a pure state of orbital angular momentum for the ground state. The hydrogen atom is an example of a two-body system with a central force (i.e., the force acts along the line joining the proton and the electron and depends only on the distance of separation)

TABLE 17-1   THEORETICAL PREDICTIONS OF THE MAGNETIC MOMENT OF THE DEUTERON

Term	μ_d Algebraic Form	μ_d (Nuclear Magnetons) Numerical Value
$^1P_1$	$\frac{1}{2}$	0.5000
$^3S_1$	$\mu_p + \mu_n$	0.8796
$^3P_1$	$\frac{1}{2}(\mu_p + \mu_n + \frac{1}{2})$	0.6898
$^3D_1$	$\frac{1}{2}(\mu_p + \mu_n + \frac{1}{2}) - (\mu_p + \mu_n - \frac{1}{2})$	0.3102

and the ground state of the hydrogen atom is pure $^2S_{1/2}$. For the deuteron the presence of a mixed state ensures that the nuclear force must be more complicated than the more familiar Coulomb force.

17-6   Nuclear Models

In order to understand the processes of atomic electrons we began the development by considering the Coulomb force between pairs of charged particles and proceeded to consideration of the energy levels which are associated with electrons in an atom. Great progress in this study has been possible because the mathematical form of Coulomb's law is simple and well known and because the spectroscopy of atoms is rich with energy levels.

The situation in nuclear physics is quite different. The mathematical form of the nuclear force is not yet known, and from the previous section we know that a simple form (such as a potential function dependent only on the separation of two nucleons) is not a possible solution to the problem. It might be thought possible to guess at the answer as a result of studying the excited states of the deuteron, since it is the two-body system of nuclear physics—analogous to the hydrogen atom in atomic physics. Unfortunately for this purpose the deuteron has no bound states that decay into the ground state by photon emission. Nearly all other nuclides have such excited states, but they also consist of more than two nucleons.

From the above information it will be understandable that we cannot discuss the nuclear force in detail. Yet there are some qualitative observations that can be made. The nuclear force has a short range, probably no greater than nuclear dimensions, of the order of 10^{-13} or 10^{-14} cm. This specifically nuclear force must be strongly attractive in order to overcome the electrostatic repulsion between protons, but it may also be repulsive to some extent to prevent the collapse of the nucleus. Another attribute the nuclear force possesses is *charge independence:* that the *nuclear* force between two protons is the same as that between two neutrons. We know, of course, that the total force is not the same in these two cases, but the evidence indicates that after making allowance for Coulomb effects the forces would be equal. This topic is considered further in Section 18-15.

In the absence of detailed information about the nuclear force, theoretical descriptions of nuclei have been centered around models for nuclear structure which are admittedly inadequate but which represent to some extent the observed phenomena.

Among the models currently in use in nuclear physics are (a) the *uniform particle model*, (b) the *liquid-drop model*, (c) the *cluster model*, (d) the *shell model*, (e) the *collective model*, (f) the *optical model*, and (g) the *direct reaction model*.

The *uniform particle model* is sometimes called the *statistical model* or the *Fermi-gas model*. It was proposed by E. Wigner (1937), and it assumes that as a result of the very strong interactions between nucleons the motions of individual nucleons cannot be considered in detail but must be treated statistically. The mathematical treatment of nuclei in this model resembles the discussion of free electrons in a solid (see Section 11-6).

The *liquid-drop model* proposed by N. Bohr (1937) also ignores the motion of individual nucleons. The observation is made that nuclear matter is essentially incompressible, since the radii of nuclei are proportional to $A^{1/3}$ (see Section 3-9). By analogy with a drop of water the nucleus is assumed to have a definite surface tension, with nucleons behaving in a manner similar to that of molecules in a liquid. The decay of nuclei by emission of particles is analogous to the evaporation of molecules from the surface of a liquid. Nuclear fission is analogous to the division of a large drop of liquid into two smaller ones. We have been using a form of the liquid-drop model in previous chapters by listing reactions in which an intermediate *compound nucleus* is formed. This aspect of the model will be discussed more fully in Sections 17-7 and 17-8. Another use of the liquid-drop model is the semiempirical mass formula of Weizsaecker (1935) in which the mass of a nucleus is expressed as the sum of a number of terms. We list the content of each term: (a) Z times the mass of a proton, (b) $(A - Z)$ times the mass of a neutron, (c) a term proportional to A, (d) a term proportional to $A^{2/3}$ to allow for the surface tension of the droplet, (e) a term proportional to $(A/2 - Z)^2/A$ to express the tendency for the number of protons to be nearly equal to the number of neutrons, (f) a term proportional to $Z^2/A^{1/3}$ expressing the energy from the Coulomb repulsion of the protons, and (g) a term that adjusts the mass to lower values for even-even nuclei (and higher values for odd-odd nuclei). The constants of proportionality have been adjusted to give the best agreement with measured nuclear masses; the agreement obtained is quite good, but there are systematic deviations that can be explained by the shell model.

The *cluster model* or the *alpha-particle model* is based on the assumption that alpha particles form subgroups inside a nucleus. These alpha particles need not have a permanent existence but may exchange particles with one another. This model has usefulness that is limited mostly to elements with low mass numbers for which the mass number A can be expressed as $4n$, where n is an integer; for example, ^{12}C may be considered as a group of three alpha particles, with small interaction between them. Even in this limited range, however, there is a serious defect in that ^8Be is very unstable, although according to this model it should be stable. The cluster model has been moderately successful in treating nuclei for which $A = 4n \pm 1$ by considering them as n closed structures with either an additional nucleon or one missing from this closed structure. However, the model fails completely for nuclei of mass numbers $A = 4n + 2$.

The *shell model* is one of the more important nuclear models and will be discussed separately in Section 17-9. It is sometimes referred to as the *independent particle model* because it assumes that each nucleon moves independently of all the other nucleons and is acted on by an average nuclear field produced by the action of all the other nucleons.

The *collective model* of A. Bohr and B. Mottelson (1953) is an outgrowth of and an improvement on the shell model. In the shell model excited states of nuclei are predicted, but the detailed predictions often do not agree with experiment. The collective model considers effects such as the following: A nucleon near the surface of a nucleus moves in an orbit and, as it moves, draws with it a bulge in the ensemble of the other nucleons. This effect has been compared with the tides in the ocean that result from the gravitational attraction of the moon as it orbits the earth. Another analogy is that of a ripple in the surface of a liquid drop. The success of the collective model is that it can be used to calculate nuclear deformations, that is, departures from a spherical shape. Deformed nuclei give rise to additional energy levels by virtue of their rotation and vibration, just as in the case of molecular spectra (see Section 11-2).

The *optical model* is specifically designed to describe nuclear scattering processes. It helps to overcome some of the inadequacies of the compound-nucleus model, in which the projectile is considered as being absorbed in the nucleus and re-emitted at a later time in a direction independent of the initial state. Especially for increased bombarding energy, the compound-nucleus idea fails to describe the experimental results. In the optical model the scattering is treated by using wave mechanics in a way that is similar to the wave theory of physical optics. The projectile is analogous to a light wave and the nucleus is analogous to a translucent glass sphere. Elastic scattering from the nucleus corresponds to light that passes through the sphere or diffracts around it; inelastic processes correspond to the light absorbed in the sphere. This model is often called the *cloudy crystal ball* model.

The *direct reaction model* is a generic term for models that describe aspects of nuclear scattering that cannot be described by the compound-nucleus model. Sometimes the optical model is included under this category. We mention two other specific models for direct interactions: the *stripping* process, which we shall discuss in Section 17-14 and the *surface* direct interaction in which projectiles are treated as undergoing interactions with single nucleons near the surface of a nucleus.

17-7 Properties of the Compound Nucleus

Most of the nuclear processes discussed in the previous chapter in which a nucleus was bombarded by particles or photons to form a compound nucleus involved particle or photon energies of less than about 40 MeV. The existence of a compound nucleus was amply demonstrated in the (n,γ) and (p,γ) types of reaction; the product nucleus of one of these reactions is simply a lower energy state of the compound nucleus. We shall consider reactions as low-energy reactions if the energies of the bombarding particles are less than 40 MeV. The compound nucleus is always formed in an excited state. It may go to a state of lower energy by the emission of a gamma-ray photon or it may be unstable and disintegrate by particle emission, the latter process being the more probable one.

A given compound nucleus may be formed in several different ways by using different projectiles directed against suitable targets. The compound nucleus exists for a time that is long in comparison with the time of transit of a nucleon across the nuclear diameter. An idea of the time of transit of a nucleon can be obtained from the fact that its energy is of the order of a few MeV. Hence its velocity is of the order of 10^9 cm/sec. Since the nuclear diameter is of the order of 10^{-12} cm, the time of transit of a nucleon across a diameter is of the order of 10^{-21} sec. The lifetime of the compound nucleus, although short, of the order of 10^{-15} or 10^{-16} sec, is much longer than the time of transit.

In Bohr's theory of the compound nucleus the assumption is made that a projectile captured by a nucleus gives up its energy to a few nucleons and, as a result of the interaction of these nucleons with all the others, the energy is quickly distributed among all the nucleons of the compound nucleus. Hence, when a compound nucleus disintegrates, its mode of disintegration is independent of the mode of formation and depends only on the particular state of the nucleus thus formed. The disintegration of a compound nucleus by particle emission implies that energy exchanges take place among the nucleons until a particle or a group of particles acquires sufficient energy to leave the nucleus. This energy must be in excess of the binding energy of the emitted particle or particles in the compound nucleus.

17-8 Formation of a Compound Nucleus

When a compound nucleus is formed by bombarding a target with fast neutrons, the capture cross section σ_c of the target nuclei is

$$\sigma_c = \pi R^2 \qquad (17\text{-}12)$$

where R is the nuclear radius, given by

$$R = r_0 A^{1/3} \qquad (17\text{-}13)$$

with r_0 the radius parameter. The value of r_0 depends on the type of experiment used for its determination; these values range from about 1.2 to 1.5 fermis, where

$$1 \text{ fermi } (1f) = 10^{-13} \text{ cm} \qquad (17\text{-}14)$$

Unless otherwise stated, the value $1.2f$ will be used in all calculations.

There is almost no interaction between the neutron and the nucleus until the neutron gets within a distance R from its center. We can conclude from this that the forces acting on the neutron are *specifically nuclear forces* which have a very *short range of action*, probably of the order of nuclear dimensions. For slow neutrons the capture cross section varies inversely with the velocity v of the neutron; thus

$$\sigma_c = \frac{K}{v} \qquad (17\text{-}15)$$

where the constant K depends on the nature of the target nucleus. This dependence on velocity may be explained by the fact that the distance between the neutron and the target nucleus is determined by the De Broglie wavelength, $\lambda = h/mv$; hence the interaction between them may take place at distances larger than the nuclear radius R.

The problem is somewhat different when a charged particle such as a proton or an alpha particle is used as a projectile because of the Coulomb force between the projectile and the nucleus. This is a long-range force, proportional to r^{-2}, where r is the distance between them. Since the potential in a Coulomb field varies as $1/r$, the potential energy $V(r)$ of a system consisting of a nucleus of charge Ze and a proton of charge e is Ze^2/r. This will be the potential energy of the system at distances $r > R$, as shown in Figure 17-6. At distances $r < R$ the specifically nuclear forces come into play and the potential energy of the

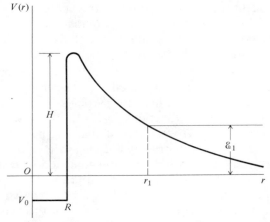

Figure 17-6   Assumed form of the potential energy curve in the neighborhood of a nucleus. R is the radius of the nucleus.

system drops sharply. Without adequate knowledge of the nature of these nuclear forces, it is impossible to draw a graph to represent accurately the potential energy for distances less than R from the center of the nucleus. For simplicity a constant value V_0 has been assumed for the potential energy of this system. The two regions have been joined at $r = R$ by a smooth curve.

A positively charged particle approaching a nucleus would encounter a *barrier* due to the Coulomb force of repulsion between them; this is sometimes called a *Coulomb barrier.* The maximum height H of this curve is called the *barrier height;* the magnitude of the barrier height for a Coulomb potential barrier is kZe^2/R. For a nucleus of low atomic number such as magnesium, $^{27}_{12}\text{Mg}$, for example, the barrier height H for protons as projectiles is approximately 4 MeV, whereas for a heavy nucleus such as $^{238}_{92}\text{U}$ it is about 15 MeV. On the basis of classical physics, a proton should have a kinetic energy in excess of the Coulomb barrier in order to be able to penetrate the nucleus. Protons with kinetic energies less than the barrier energy H are deflected before they get within the range of the specifically nuclear forces. We know from experiment, however, that some of the protons do penetrate the potential barrier. A qualitative explanation of this barrier penetration can be given on the basis of wave mechanics. Referring to Figure 17-6, suppose that a proton of kinetic energy $\mathcal{E}_1 < H$ is fired at a nucleus. There is a definite probability that this proton will penetrate the potential barrier. This probability is calculated by replacing the particle by its De Broglie wave and setting up the Schroedinger equation for this problem with suitable boundary conditions. The solution of this equation shows that when the wave approaches the potential wall it is partly reflected and partly transmitted, just as in any problem of wave motion. The *probability of penetration,* also called the *transparency* of the barrier, is defined as the ratio of the intensity of the transmitted wave to the intensity of the incident wave. The penetration probability is thus obtained from the ratio of the square of the amplitude of the transmitted wave to the square of the amplitude of the incident wave. The solution of the Schroedinger equation shows that the probability of penetration of the potential barrier is small for proton energies that are small fractions of the barrier height and increases to unity when the energy of the particle equals the height of the Coulomb barrier.

The probability of penetration of a potential barrier by a positively charged projectile can be determined experimentally by measuring the yield of a given reaction in which protons are the incident particles. Experiments already discussed in which alpha particles (Sections 15-2, -4, -9), protons (Section 15-10), and deuterons (Section 15-11) were the bombarding particles, show that the transparency of the Coulomb barrier is not always a smooth function of the energy of the projectile but exhibits the phenomenon of *resonance.* A resonance energy is shown by a sharp increase in the yield, as shown in Figure 15-8, hence in the probability of penetration, for a given value of the energy of the projectile. A resonance energy is evidence of the existence of a sharp energy level in the compound nucleus. Since the state of a compound nucleus is independent of the manner in which the compound nucleus is formed, a par-

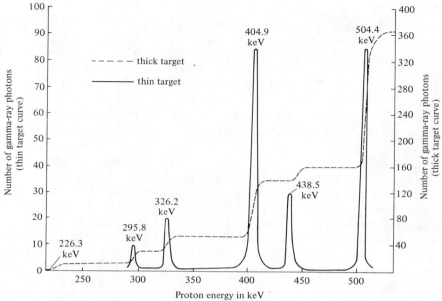

Figure 17-7   Yield curve showing the resonance energies as a function of the incident proton energy in the ^{27}Al (p,γ) ^{28}Si reaction. [After Hunt and Jones, *Phys. Rev.*, **89**, 1287 (1953).]

ticular energy level may be formed by one of several different reactions. For example, when Hunt and Jones (1953) investigated the $^{27}_{13}$Al (p,γ) $^{28}_{14}$Sr reaction, they found five resonances for proton energies less than 0.5 MeV, as shown in Figure 17-7, whereas the barrier height is about 4 MeV. Here the yield was obtained by counting the gamma-ray photons emitted by the compound nucleus $^{28}_{14}$Sr. It will also be noted that the resonances appear as sharp maxima when the target is very thin, but appear as sudden increases in the yield when the target is thick.

17-9   Nuclear Shell Structure

It is to be expected that the ideas and concepts that proved so effective in determining the electronic structure of atoms should be carried over into nuclear physics. One of these ideas is that of *shell structure* or *level structure* with certain shells *closed* because of the stability of the system with the given number of particles. The idea of closed nuclear shells was first put forward by W. Elsasser (1934); later Maria G. Mayer (1948) summarized the experimental facts to show that nuclei containing 20, 50, or 82 protons or 20, 50, 82, or 126 neutrons form very stable configurations. Among the types of evidence used by Mayer was the isotopic abundance of the elements, particularly those with $Z > 33$; for example, those isotopes of an element with an isotopic abundance greater than 60 percent are $^{88}_{38}$Sr, with $N = 50$, $^{138}_{56}$Ba, and $^{140}_{58}$Ce with $N = 82$. Another type of evidence is the number of stable isotopes of a given element; for example, tin, $Z = 50$, has the largest number of stable isotopes (10) of any element. Lead, $Z = 82$, has four stable isotopes; lead is also the end product of the naturally occurring radioactive series. There are seven stable *isotones*—that is, nuclides with the same number of neutrons—for $N = 82$; these are ^{136}Xe, ^{139}Ba, ^{140}Ce, ^{141}Pr, ^{142}Nd, and ^{144}Sm. On the other hand, there is only one stable isotope with $N = 81$ and only one stable isotope with $N = 83$. Also, there are six isotones with $N = 50$, but only one each for $N = 49$ and $N = 51$.

Evidence for the existence of a closed shell for 126 neutrons comes from several facts. Two of these are (a) the heaviest isotope of lead, $A = 208$, has $N = 126$; (b) when alpha-decay energies are plotted against the neutron number of the product nucleus, there is a sharp dip in energy when N drops below 126; this indicates a larger binding energy for the 126th neutron.

Other evidence for the existence of closed shells at $N = 50$, 82, and 126 may be obtained from the fact that the so-called *delayed neutron emitters* among the fission products of uranium 235 are $^{87}_{36}$Kr, with $N = 51$, and $^{137}_{54}$Xe, with $N = 83$; the binding energy of the last neutron in each case is so small that a neutron can be readily evaporated from each one to form ^{86}Kr, $N = 50$, and ^{136}Xe, $N = 82$, respectively. Also, the absorption cross section for neutrons as a function of neutron number is very small for nuclei with neutron numbers $N = 50$, 82, and 126.

The nuclear shell model is limited in its application to the ground states and low-lying excited states of nuclei. The present experimental evidence is that there are closed nuclear shells at neutron and proton numbers 2, 8, 20, 28, 50, and 82 and neutron number 126. The problem is to determine the order in which these shells are filled as nucleons are added to make heavier nuclei. This is similar to the problem encountered in the assignment of electrons to the different electronic level configurations of the atom. Quantum mechanical calculations have been made to determine this order by using the idea of a strong coupling between the orbital angular momentum of a nucleon and its spin angular momentum, particularly by Mayer (1949) and by Haxel, Jensen, and Suess (1949). In one calculation a single nucleon is assumed to move in a field due to the other nucleons whose potential has the shape of a square well. The results obtained from the square-well potential are not in complete agreement with experimental results obtained from analyses of beta-ray spectra, from measurements of magnetic moments, and from experiments on isomeric transitions. The results from such experiments have been used to modify the theoretical results. In addition, since each nucleon has a spin of $\frac{1}{2}\hbar$, the *Pauli exclusion principle* must be applied; this principle, as applied to the nucleus, states that *no two identical particles can be in the same quantum state*. A nuclear quantum state is usually specified by its total quantum number and by its parity.

A single nucleon is assumed to move in an orbit with an angular momentum l, in units of $\hbar$, in a nuclear field due to all the other nucleons. Each nucleon also has an angular momentum due to spin of $\frac{1}{2}\hbar$. The spectroscopic notation developed for electron configurations is taken over to describe nucleon configurations. Thus the values for the orbital angular momentum l have letters associated with them as indicated:

l value:	0	1	2	3	4	5	6
notation:	s	p	d	f	g	h	i

The assumption of strong spin-orbit coupling means that in determining the total angular momentum of the nucleus, the total angular momentum j of each nucleon is first determined; then these j values are added, subject to the usual quantum conditions. This type of coupling is referred to as the j-j coupling, in contrast to the Russell-Saunders or L-S coupling. The total angular momentum j for a single nucleon is

$$j = l + \tfrac{1}{2} \quad \text{or} \quad l - \tfrac{1}{2} \tag{17-16}$$

except in the case of $l = 0$. The value of j is always an odd-half integer. The j value is frequently written as a subscript in the lower right corner of the letter representing the l value. It can be seen that there are two possible energy levels for the same l value except for $l = 0$. In agreement with previous practice (Section 17-2), the total angular momentum quantum number of the nucleus is designated by the letter I. It was previously called the spin angular momentum quantum number of the nucleus; the term *spin* has been enlarged to include the angular momentum due to the nucleonic orbital motion as well.

In some notations the quantum number n bears the same relationship to l that it did in the electronic scheme. The more common notation today assigns the number unity to n for the lowest energy level of a particular value of l—that is, $1s$, $1p$, $1d$, and so forth. The latter notation will be used in this book. In our notation the number n is equal to one more than the number of nodes (or zeros) in the radial part of the wave function, not including a node that may occur at $r = 0$.

Let us examine a few nuclides to see how the shell model is used. The nucleus of deuterium has $N = 1$, $Z = 1$; hence the ground state is an s state in which each nucleon has a j value of $\frac{1}{2}$. Experiment shows that $I = 1$; hence the spin of the neutron is parallel to that of the proton. There are two nuclides with $A = 3$; they are $^3_1$H and $^3_2$He. In each case the value of $I = \frac{1}{2}$. For $^3_1$H, $N = 2$, $Z = 1$, so that the $s_{1/2}$ shell has its full complement of two neutrons with spins in opposite directions and the proton is in an $s_{1/2}$ shell with $j = \frac{1}{2}$. On the other hand, $N = 1$, $Z = 2$, for $^3_2$He, so that the two protons complete the $s_{1/2}$ proton shell with their spins in opposite directions and the neutron is in its $s_{1/2}$ shell with $j = \frac{1}{2}$. Two such nuclides, having the same number of nucleons A but with the protons and neutrons interchanged, are called *mirror nuclides*. For $A = 4$ there is only one known nuclide, the stable alpha-particle configuration of $^4_2$He. Here $N = 2$ and $Z = 2$ and the $1s_{1/2}$ shell is filled for both neutrons and protons. The spins of the protons are opposed, as are those of the neutrons. Experimentally, $I = 0$, in agreement with the above results. Also the binding energy of the alpha particle is very large—approximately 28 MeV.

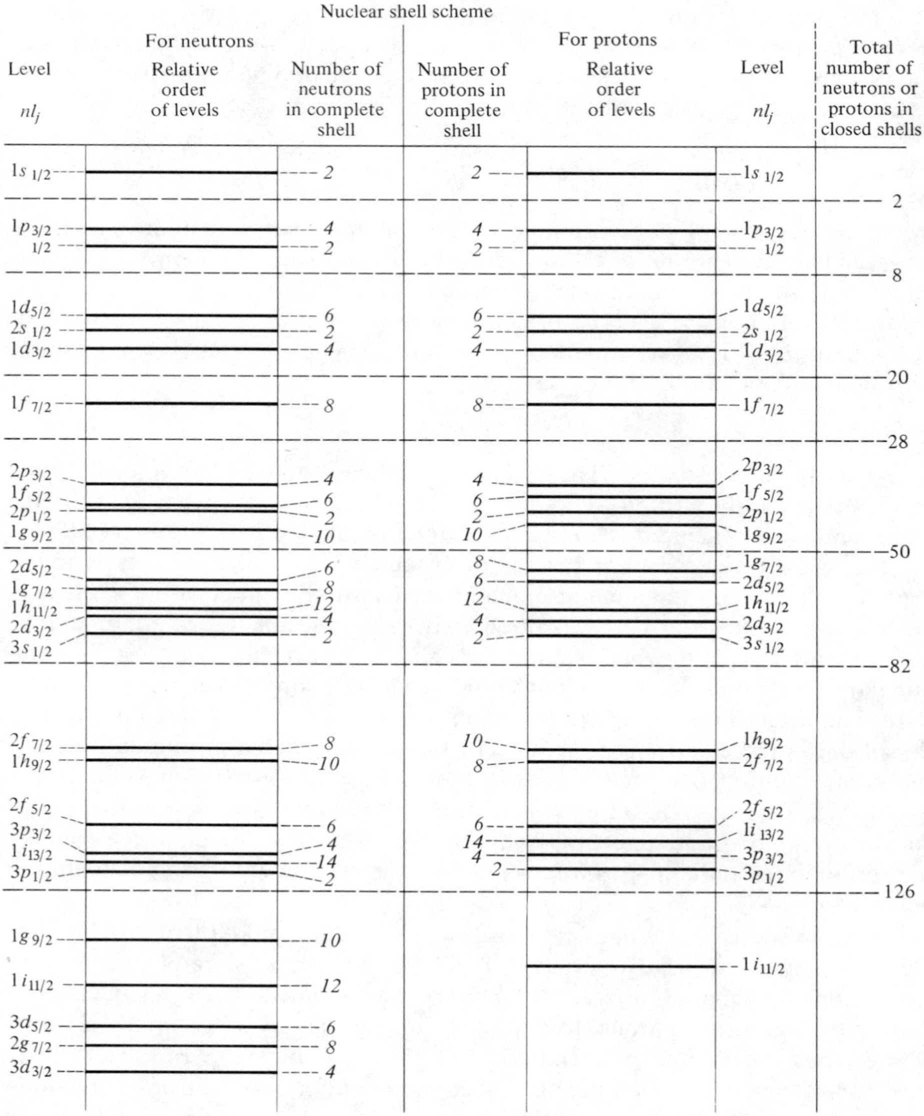

Nuclear shell scheme

	For neutrons			For protons		
Level nl_j	Relative order of levels	Number of neutrons in complete shell	Number of protons in complete shell	Relative order of levels	Level nl_j	Total number of neutrons or protons in closed shells
$1s_{1/2}$		2	2		$1s_{1/2}$	2
$1p_{3/2}$		4	4		$1p_{3/2}$	
$1p_{1/2}$		2	2		$1p_{1/2}$	8
$1d_{5/2}$		6	6		$1d_{5/2}$	
$2s_{1/2}$		2	2		$2s_{1/2}$	
$1d_{3/2}$		4	4		$1d_{3/2}$	20
$1f_{7/2}$		8	8		$1f_{7/2}$	28
$2p_{3/2}$		4	4		$2p_{3/2}$	
$1f_{5/2}$		6	6		$1f_{5/2}$	
$2p_{1/2}$		2	2		$2p_{1/2}$	
$1g_{9/2}$		10	10		$1g_{9/2}$	50
$2d_{5/2}$		6	8		$1g_{7/2}$	
$1g_{7/2}$		8	6		$2d_{5/2}$	
$1h_{11/2}$		12	12		$1h_{11/2}$	
$2d_{3/2}$		4	4		$2d_{3/2}$	
$3s_{1/2}$		2	2		$3s_{1/2}$	82
$2f_{7/2}$		8	10		$1h_{9/2}$	
$1h_{9/2}$		10	8		$2f_{7/2}$	
$2f_{5/2}$		6	6		$2f_{5/2}$	
$3p_{3/2}$		4	14		$1i_{13/2}$	
$1i_{13/2}$		14	4		$3p_{3/2}$	
$3p_{1/2}$		2	2		$3p_{1/2}$	126
$1g_{9/2}$		10				
$1i_{11/2}$		12			$1i_{11/2}$	
$3d_{5/2}$		6				
$2g_{7/2}$		8				
$3d_{3/2}$		4				

Figure 17-8   Order of level assignments in nuclei on the basis of the shell model of the nucleus. [After Klinkenberg, *Revs. Modern Phys.*, **24**, 65 (1952).]

There is no nucleus known with $A = 5$ that exists for a measurable length of time. There are two nuclides with $A = 6$; the stable nucleus $^6_3$Li, and the radioactive nucleus $^6_2$He. Since the first shell is closed with two protons and two neutrons, the two additional nucleons must go in a shell of higher energy. The spin-orbit coupling model predicts that the next shell is not a $2s_{1/2}$ shell but a $1p_{3/2}$ shell. The latter may contain up to four nucleons of one kind. Thus for $^8_3$Li, with $N = 3$ and $Z = 3$, the extra neutron and proton go into the $1p_{3/2}$ shells. Since $I = 1$, the spins of these nucleons must be parallel. There is at present no measured value of I for the ground state of $^8_2$He, but the shell model predicts that the two extra neutrons go into the $1p_{3/2}$ shell and their spins must be opposed so that I should be zero.

The assignment of nucleons can be continued on this shell model, checking with experimental results when available. Figure 17-8 shows the order in which the different levels appear and the maximum possible number of protons or neutrons in a given level; for example, the magic number 8 fits $^{16}_8$O, with $N = Z = 8$. This is a very stable nucleus—the measured value of $I = 0$. A great deal of information for determining the order in which shells are occupied by nucleons is obtained from a study of nuclides of odd mass numbers. The total quantum number I of such nuclides must be an odd half integer. Its value can often be obtained from the measured value of its magnetic moment determined by either one of the magnetic resonance methods, or from its hyperfine structure. The ground state may then be determined, and the odd nucleon, either a proton or a neutron, can then be assigned to the proper shell.

It will be observed from Figure 17-8 that the order of filling the nuclear levels is the same for protons as for neutrons up to $N = Z = 50$. Above this value there are a few differences; for example, the $2d_{5/2}$ level for neutrons is filled before the $1g_{7/2}$ level, whereas the opposite is the case for the protons.

17-10   Beta Decay

In the two preceding chapters we considered in some detail the types of radioactive beta-particle disintegration both in naturally occurring isotopes and in artificially produced isotopes. It was found that beta decay is a common type of disintegration throughout the entire range of mass numbers. Beta decay involves two isobaric nuclides of mass number A and produces a change in nuclear charge from Ze to $(Z + 1)e$ if an electron is emitted, and a change to $(Z - 1)e$ if a positron is emitted or an electron is captured. We have discussed the energy changes accompanying these disintegrations and the spectral distributions of the β^-- and β^+-particles. We have also given sufficient evidence that electrons do not exist as free particles in the nucleus, and hence must be created at the instant of nuclear disintegration. For β^--emission it was suggested that a neutron decays into a proton, an electron, and an antineutrino; thus

$$n \rightarrow p + e^- + \bar{\nu} \tag{17-17}$$

Since the mass of the neutron exceeds that of the proton and electron, this type of disintegration is energetically possible for free neutrons. This has actually been observed; the end-point energy was found to be 785 keV, nearly identical with the mass difference $(_0n^1 - _1H^1)$ of 783 KeV. The half-life of the free neutron in space was found to be about 10.5 min.

A positron was assumed to be produced by the disintegration in the nucleus of a proton into a neutron that stays in the nucleus and a positron and a neutrino that are emitted; thus

$$p \text{ (in nucleus)} \rightarrow n \text{ (in nucleus)} + e^+ + \nu \qquad (17\text{-}18)$$

Since the mass of the proton is less than that of the neutron and positron, however, the proton remains a stable particle in free space. The difference in energy, in this case $783 \text{ keV} + 2m_e c^2 = 1.805 \text{ MeV}$, must be supplied by the system in which β^+-emission occurs.

Electron capture can be written as

$$p + e^- \rightarrow n + \nu \qquad (17\text{-}19)$$

Here also the sum of the masses of the proton and electron is less than that of the neutron by 783 keV; hence this reaction cannot occur with free protons and electrons but can occur only in an atomic system in which the deficiency in energy may be supplied by other particles.

A satisfactory theory of beta disintegration should be able to predict the spectral distribution of the beta particles, which is found experimentally, and should relate beta disintegration to the energy states of the isobars involved in the process. Such a theory was developed by E. Fermi (1934); as experimental procedures became more refined, their results were found to be in better agreement with the predictions of the Fermi theory; for example, the Fermi theory gives the distribution in energy and in momentum of the beta particles, which is in agreement with those found experimentally. For allowed transitions the spectral distribution is of the form given by

$$N(p) \, dp = C p^2 (\mathscr{E}_m - \mathscr{E})^2 F(Z,p) \, dp \qquad (17\text{-}20)$$

where $N(p)$ is the number of beta particles emitted with momenta lying between p and $p + dp$, C is a constant for the nuclei involved in the beta decay, $\mathscr{E}_m$ is the maximum energy of the emitted beta particles and the end-point energy of the spectrum, $\mathscr{E}$ is the kinetic energy of the beta particle, $F(Z, p)$ is a Coulomb correction factor that takes into account the Coulomb interaction between the nuclear particles and the beta particle, and Z is the atomic number of the product nucleus.

A more convenient way of plotting the experimental results and checking with the theory was suggested by Kurie (1936). Writing the equation in the form

$$\sqrt{\frac{N(p)}{p^2 F(Z,p)}} = k(\mathscr{E}_m - \mathscr{E}) \qquad (17\text{-}21)$$

we see that if $\sqrt{N(p)/[p^2 F(Z,p)]}$ is plotted against the energy a straight line should be obtained with an energy intercept at $\mathscr{E}_m$. Figure 17-9a is a Kurie plot of the negative beta-particle spectrum of ^{64}Cu, and Figure 17-9b is a Kurie plot of the positron spectrum of ^{64}Cu obtained by Owen and Cook (1949) using very thin uniform sources prepared by evaporating the activated copper onto a thin aluminum foil. In making a Kurie plot, it is common to express the

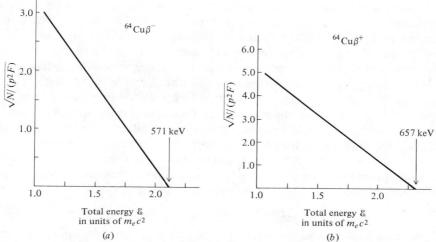

Figure 17-9 (a) Kurie plot of the negative beta-ray spectrum of ⁶⁴Cu; (b) Kurie plot of the positive beta-ray spectrum of ⁶⁴Cu. The value of the end-point energy is indicated by the arrow in each plot. [After Owen and Cook, *Phys. Rev.*, **76**, 1727 (1949).]

energy in units of $m_e c^2$. These spectra give straight lines down to very low energies in agreement with the Fermi theory of beta decay.

17-11 Comparative Half-Lives for Beta Decay

The number of beta particles $N(p)$ emitted per unit time per unit momentum interval is proportional to the disintegration probability per unit time; λ_β is the beta disintegration constant. It is to be expected that the probability of disintegration should depend on the energy available for the transition and on the characteristics of the initial nuclear state and the final nuclear state. A nuclear state is characterized by its energy, by its angular momentum I in units of $\hbar$, and by the arrangement of the nuclear particles. The state of the system due to arrangement of the particles will be given by a function of the coordinates such as $\psi(x, y, z)$ which is obtained from the solution of the wave equation for the system. Such a function will be either *symmetric* or *antisymmetric* or, more commonly stated, either *even* or *odd*. The function is an even one if, by changing the signs of all the coordinates, the function remains unchanged—that is, if

$$\psi(x, y, z) = \psi(-x, -y, -z)$$

The function is odd if, by changing the signs of all the coordinates, the sign of the function is changed—that is, if

$$\psi(x, y, z) = -\psi(-x, -y, -z)$$

Another way of designating the character of the wave function is to use the term *parity;* a symmetric function has *even parity* and an antisymmetric function has *odd parity*. When a system undergoes some change, the parity of the system may change or it may remain unchanged. Thus parity can be considered as a quantum number to represent one aspect of the state of a system.

Just as in optical and x-ray spectra, the transitions between nuclear states giving rise to beta-ray spectra are governed by selection rules, some transitions being more favored than others. One set of selection rules applies to the case in which the beta particle and the neutrino are emitted with their spins antiparallel. We may, using spectroscopic notation, say that they are emitted in a spin state for which $\mathbf{S} = 0$, that is, a singlet state. The resultant change in angular momentum for this process will be zero. The selection rule for this allowed transition, usually called the Fermi selection rule, is

$$\Delta I = 0, \text{ no change of parity} \tag{17-22}$$

If the beta particle and the neutrino are emitted with their spins parallel to each other, their spin state may be designated by $\mathbf{S} = 1$. The selection rules for this case, known as the Gamow-Teller selection rules, are

$$\Delta I = 0, \pm 1, \text{ no change of parity} \tag{17-23}$$

the transitions

$$I_i = 0 \text{ to } I_f = 0 \text{ are not allowed.}$$

The last may be understood from the fact that $\mathbf{S} = 1$ corresponds to a triplet state for the total angular momentum of the final system. It will be recalled that the total angular momentum $\mathbf{J} = \mathbf{L} + \mathbf{S}$ of an atomic system had three values when $\mathbf{S} = 1$: $J = L + 1$, $J = L - 1$, and $J = L + 0$. Similarly the final angular momentum $I_f = I_i + 1$, I_i, or $I_i - 1$ when $\mathbf{S} = 1$. However, I_f and I_i cannot both be zero when $\mathbf{S} = 1$ if angular momentum is to be conserved in this process; hence the last restriction is a necessary part of the Gamow-Teller selection rule.

Other changes of state than those given by the above selection rules do occur but are classed as *forbidden transitions,* there being different degrees of forbiddenness, such as first forbidden (with $\Delta I = \pm 1, 0$, change of parity), second forbidden, and higher orders of forbiddenness. It is beyond the scope of this book to consider beta-ray spectroscopy in greater detail. However, there is another interesting approach to this subject in terms of the *comparative half-lives* of the beta-particle disintegrations.

The probability per second λ_β for beta disintegration can be computed from Equation (17-20) by integrating over all values of the electron momentum from 0 to p_m, where p_m is the maximum momentum, obtaining

$$\lambda_\beta = Cf \tag{17-24}$$

where

$$f = \int_0^{p_m} p^2 (\mathscr{E}_m - \mathscr{E})^2 F(Z, p) \, dp \tag{17-25}$$

Values of the function f are available in tables. Substituting for λ_β its value

$$\lambda_\beta = \frac{\ln 2}{T} = Cf$$

we get

$$fT = \text{constant} \tag{17-26}$$

The product fT is called the *comparative half-life*. The numbers obtained for the comparative half-lives vary over such a wide range, from about 10^3 to 10^{18}, that it is more common to use values of $\log fT$ in discussing comparative half-lives. Log fT values of about 3-5 correspond to allowed transitions, with higher values for the unfavored or forbidden transitions.

An idea of the usefulness of the concept of comparative half-lives can be obtained by considering the two transitions

$$^3_1\text{H} \rightarrow \, ^3_2\text{He} + \beta^- + \bar{\nu} \qquad T = 12.26 \text{ yr} \tag{17-27}$$

and

$$n \rightarrow p + \beta^- + \bar{\nu} \qquad T = 10.8 \text{ min} \tag{17-28}$$

Langer and Moffat (1952) made an accurate determination of the end-point energy of the beta-ray spectrum of tritium and found it to be 17.95 keV. Using the previously determined half-life and tables of values of the function f, they obtained the value

$$fT = 1014 \text{ sec}$$

from which

$$\log fT = 3.006$$

Langer and Moffat suggested that if we assume that the comparative half-life of the beta decay of the neutron is the same as that for tritium, then, using the end-point energy of the beta-ray spectrum of the neutron as 783 keV, the half-life of the neutron should be 10.4 min. This is in agreement, within the limits of experimental error, with the results of the experiment listed in Equation (17-28).

17-12 Evidence for the Neutrino

The existence of the neutrino has been postulated to account for beta decay and also for several types of meson decay (see Chapter 18). The properties assigned to the neutrino make it very difficult to detect it by a direct experiment, and the earlier experiments on neutrinos were of an indirect type. The neutrino as postulated is a particle that has no charge, has a spin angular momentum of $\frac{1}{2}\hbar$, may have zero rest mass or a rest mass small in comparison with that of the electron, and has energy and momentum.

One of the earliest attempts to show experimentally that the neutrino hypothesis was correct was a cloud-chamber experiment by Crane and Halpern (1938) on the negative beta disintegration of ^{38}Cl. From the principle of conservation of linear momentum it is obvious that if only an electron is emitted the product nucleus should recoil with a momentum equal and opposite to that of the electron. If a neutrino is emitted simultaneously with the electron, then some of the momentum will be carried away by it. The distribution of momentum between the two light particles is such that when the energy of the electron is large its momentum is also large and that of the neutrino is negligible. When the energy of the electron is small, its momentum is also small and that of the neutrino should become significant experimentally; that is, the recoil momentum of the product nucleus should be greater than that of the electron. In the cloud-chamber experiment mentioned above, Crane and Halpern measured the recoil momenta of product nuclei, some of which were associated with high-energy electrons and others with low-energy electrons. They found that for high-energy beta-particle emission the momentum of the product nucleus was equal to that of the electron, but for low-energy beta-particle emission, the momentum of the nucleus was greater than that of the electron. The accuracy of the experiment was not very great, one of the difficulties being due to the fact that the velocity of the heavy nucleus was small, and instead of measuring the length of a track the number of droplets produced was counted. The assumption was made that the energy required to produce an ion pair was the same for a heavy nucleus as for an alpha particle.

Additional experiments on the recoil of nuclei produced in beta disintegrations have been performed with improved techniques; these have confirmed the above results. Some of the experiments involved electron capture; for example, Rodeback and Allen (1952) investigated the disintegration of ^{37}A by electron capture with the reaction

$$^{37}_{18}A + e^- \rightarrow {}^{37}_{17}Cl + \nu + Q \qquad (17\text{-}29)$$

The energy Q released in this reaction is simply equal to the difference in mass ^{37}A $- {}^{37}$Cl which is 814 keV, assuming that the neutrino rest mass is negligible. It has been found that about 93 percent of electrons captured come from the K shell of argon; the subsequent readjustment of the outer electrons in the

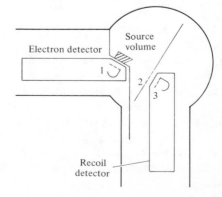

Figure 17-10 Schematic diagram of the apparatus for measuring the time of flight of recoil nuclei ^{37}Cl formed in the electron-capture process of ^{37}A. Shaded region indicates the effective source volume. [After Rodeback and Allen, *Phys. Rev.*, **86**, 447 (1952).]

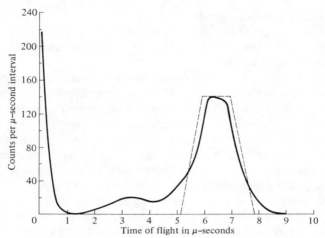

Figure 17-11 Time-of-flight distribution of recoil nuclei. Dashed curve is the distribution expected from monoenergetic recoils from the source volume. [After Rodeback and Allen, *Phys. Rev.*, **86**, 449 (1952).]

product nucleus ^{37}Cl results in the emission of Auger electrons in 90 percent of the transitions, with x-rays being emitted in the other 3 percent.

The experiment consisted in measuring the velocity of the recoiling nucleus, using the apparatus sketched in Figure 17-10. The chamber was kept at a low pressure of 10^{-5} mm Hg, the effective volume of the argon being the small shaded region in the figure. The Auger electrons traveled to grid 1 into the electron detector, and the recoil ^{37}Cl ions traveled in a field-free region to grid 2 and then into the recoil detector. Electron multipliers were used as detectors. The time of flight of the ^{37}Cl ions was measured by a 20-channel delayed coincidence circuit. The results are shown in the solid curve of the graph of Figure 17-11, with the dashed curve giving the distribution expected from monoenergetic recoil ions. This was calculated on the basis of the conservation of linear momentum between the neutrino and the recoil nucleus, with the reaction energy Q carried away by both of them. The agreement with experiment is thus good.

The first direct experiment to detect the neutrino (or antineutrino) by its interaction with matter, specifically by its interaction with protons, was performed by F. Reines and C. L. Cowan, Jr. (1953). They utilized the very great supply of antineutrinos being continuously emitted in the beta decay of the fission fragments which are produced during the operation of a large nuclear reactor. The reaction expected between an antineutrino $\bar{\nu}$ and a proton p is the formation of a neutron n and the emission of a positron, the reaction equation being

$$\bar{\nu} + p \rightarrow n + \beta^+ \tag{17-30}$$

To detect this reaction the antineutrinos are allowed to enter a large liquid scintillation counter; the organic liquid of this counter contains hydrogen sufficient to produce a proton density of the order of 5×10^{22} proton/cm^3. The estimated cross section of the above reaction is of the order of 6×10^{-20} barn, and with the large volume (10 ft^3) of liquid used, the number of these events should be about 10 to 30 hr.

In order to be certain that the above reaction was taking place, effects due to both the neutron and the positron had to be detected. The positrons, in their passage through the liquid, give rise to annihilation radiation and this, in turn, produces the scintillations that are detected by a suitable number of photomultiplier tubes placed around the tank containing the scintillation liquid. To detect the neutrons some cadmium was added to the liquid; the neutrons were captured by the cadmium in an (n,γ) reaction. The scintillations produced by these capture gamma rays occurred about 5 μsec after those of the annihilation radiation and were also detected by photomultiplier tubes. The annihilation radiation and the capture gamma rays could also be distinguished by the slightly different characteristics of the pulses produced by them in the detectors. The two pulses were fed into a delayed coincidence circuit and yielded 0.41 ± 0.20 delayed counts/min, in good agreement with the expected value.

The accuracy of the above experiment was improved greatly by Reines and Cowan in 1956; the effect of each term of Equation (17-30) was investigated separately. A schematic diagram of their experimental arrangement is shown in Figure 17-12a. The apparatus was placed near a larger nuclear reactor. Antineutrinos from this reactor passed through it. The target consisted of a tank of water to provide the protons and some CdCl$_2$ dissolved in it. This target was placed between two tanks containing the liquid scintillation detectors; the light from the scintillations was detected by photomultiplier tubes at the ends of the tanks and fed to an oscilloscope. As shown schematically in Figure 17-12a, if an antineutrino is captured by a proton in the target, resulting in the emission of a positron and a neutron, the positron is captured by an electron in the water, producing two annihilation gamma rays of 0.51 MeV each; these are detected by the two liquid scintillation detectors and a small pulse is recorded on the oscilloscope trace for that detector, as shown schematically in Figure 17-12b. The neutron released in this reaction diffuses through the water, losing speed by collision with protons, and is finally captured by a cadmium nucleus, resulting in the emission of several gamma rays whose total energy is 9 MeV. These gamma rays are now detected by the scintillation detectors and produce higher pulses on the oscilloscope. The time interval between the pulses from the β^+ annihilation radiation and the gamma rays from neutron capture is 5.5 μsec.

The results of this experiment showed 2.88 ± 0.22 counts/hr, in good agreement with the predicted cross section of 6×10^{-44} cm^2 or 6×10^{-20} barn.

Among the checks made during the course of the experiment was a check on the dependence of the experiment on the number of protons in the target. When the light water H$_2$O was replaced by heavy water D$_2$O, the number of counts decreased by a factor of 2; this is due principally to the fact that the protons are bound to neutrons in deuterium. The character of the annihilation pulses was checked by using positrons from ^{64}Cu to produce this type of radia-

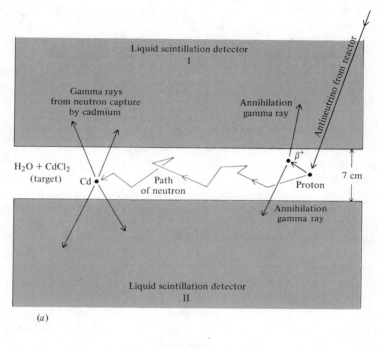

(a)

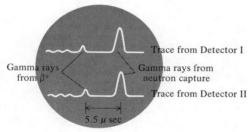

(b)

Figure 17-12 (a) Schematic diagram illustrating the processes occurring in the experiment on the detection of the antineutrino. [After Cowan and Reines, *Phys. Rev.*, **107**, 528 (1957).] (b) Oscillograph record of pulses.

tion. The assignment of the second pulse to gamma rays from neutron capture by cadmium was verified both by calculation of capture time and also by removing the cadmium from the solution; when the cadmium was removed, the neutron gamma-ray pulse disappeared. Thus there is no longer any doubt of the existence of the neutrino and antineutrino.

17-13 Rest Mass of the Neutrino

In the original formulation of the theory of beta decay it was assumed that the rest mass of the neutrino was very small and probably zero. Attempts were made to determine this value by studying the shape of the beta-ray spectrum on a Kurie plot near the end-point energy. For zero neutrino rest mass the Kurie plot will be a staight line; for a finite rest mass the curve will deviate from a straight near the end-point energy toward lower values. The best that has been done with this type of experiment is to place an upper limit on the value of the rest mass; this upper limit is about 0.2 percent of the rest mass of an electron. Using the results of experiments which show that the neutrino is a polarized particle—that is, that its axis of spin is always antiparallel to its direction of motion, or parallel in the case of the antineutrino—it is very easy to show that the neutrino rest mass must be zero (see Section 18-11).

Let us assume that a neutrino is moving in the positive x direction with a finite momentum p; its axis of spin is directed in the negative x direction. If its rest mass $m \neq 0$, its velocity v is less than the velocity of light c. We can now imagine an observer in a reference frame S' moving with a velocity v' in the x direction such that $v' > v$. The neutrino will appear to be moving in the negative x direction with respect to this observer; the direction of its spin, however, will remain unchanged. This implies that a simple Lorentz transformation of coordinates has changed a particle, the neutrino, into its antiparticle, the antineutrino; this is impossible, therefore the velocity v must equal the velocity of light c. Hence the rest mass of the neutrino must be zero.

Both the neutrino and the photon move with the speed of light and have zero rest mass. One important distinction between them is that the angular momentum of the neutrino is $\frac{1}{2}\hbar$, whereas that of the photon is $1\hbar$.

17-14 Intermediate-Energy Nuclear Reactions

In recent years the investigation of nuclear reactions has proceeded into regions of higher beam energy. An early result of this effort was the discovery of many new types of particles, which we discuss in Chapter 18. Particle physics, or high-energy physics, is principally concerned with the energy region above 1 GeV. The field of intermediate-energy physics covers energies from about 50 MeV up to about 1 GeV, with the emphasis on nuclear structure rather than on the properties of any single particle. The energy limits given here are to be taken as being approximate, since any limits would have to be chosen arbitrarily.

The more important processes in this intermediate region are not the same as those in low-energy nuclear physics. In particular, the concept of the formation of a compound nucleus has to be modified considerably by combining it with various other processes.

One of these processes involves the production of neutrons using high-energy deuterons incident on a target. In a typical (d,n) reaction with low-energy deuterons, a compound nucleus is formed and neutrons are then emitted from the target in all directions; but when high-energy deuterons are used as projectiles, the neutrons come out predominantly in the forward direction with very high energies. In one of these experiments Helmholtz, McMillan, and

Sewell (1947) used 190-MeV deuterons to bombard thin targets such as Be, Al, Cu, and others up to U; in each case they obtained a very narrow beam of neutrons proceeding in a forward direction with high energy. The explanation of this phenomenon given by R. Serber (1947) is that the deuteron is not captured by the nucleus but passes close by it, nearly at grazing incidence; the proton is *stripped* off and the neutron continues with about the same velocity as the deuteron. Because of the high velocity of the deuteron, the duration of the interaction between it and the nucleus is very small so that very little change is produced in the motion of the neutron in this *stripping* process. The final velocity of the neutron will be the vector sum of the velocity of the center of mass of the deuteron plus the velocity of the neutron relative to the center of mass. For a high-energy deuteron the latter is comparatively small so that the final velocity of the neutron will have a very large component in the forward direction; its energy will be approximately half that of the deuteron.

A phenomenon allied to that of stripping, observed at lower energies but resulting in a (d,p) process, occurs when a high-energy deuteron passes close to the nucleus, say at a distance between R and $3R$, where R is the nuclear radius. The Coulomb force between the target nucleus and the proton in the deuteron may be sufficient to break the bond between the proton and the neutron, the proton being repelled and the neutron being captured by the nucleus. This process, which is a special type of stripping, is sometimes called the *Oppenheimer-Phillips process.*

The inverse of stripping is a process in which a projectile, such as a proton, a neutron, or a deuteron, picks up another particle as it passes close to the nucleus. Among these processes, called *pickup,* are (p,d), (n,d), and (d,t) reactions. These reactions differ from similar reactions in which a compound nucleus is formed by the fact that the particle formed in pickup has a large forward momentum, whereas particles emitted by compound nuclei show almost isotropic distributions of momenta in the center of mass system. Examples of reactions in which pickup occurs are

$$^9\text{Be}(d,t)^8\text{Be} \quad \text{and} \quad ^{13}\text{C}(d,t)^{12}\text{C}$$

Another important process, observed when high-energy particles bombard a target, involves the emission of several nuclear fragments such as protons, neutrons, and alpha particles. This process is known as *spallation.* When observed in a cloud chamber, or in a photographic research nuclear emulsion, the result of spallation is the production of a *star;* several examples of stars are given in Chapter 18. When a target of medium atomic weight is bombarded by high-energy particles, the spallation products usually cover a large range of mass numbers and atomic numbers; for example, in one experiment Lindner and Perlman (1950) bombarded thin strips of antimony metal with high-energy deuterons and alpha particles; the isotopes formed in the target were separated by chemical means and identified by measurements of the half-lives and absorption characteristics of their radiations with Geiger counters. Targets bombarded with 190-MeV deuterons yielded a large number of nuclides with atomic numbers ranging from $Z = 52$ down to $Z = 39$, and mass numbers ranging from $A = 124$ down to $A = 87$. Similar results were obtained with 380-MeV alpha particles as projectiles. Many of the isotopes formed were found to be neutron-deficient and decayed by electron capture.

The production of such a large number of different nuclides by bombarding a target with one kind of particle of high energy must involve several different types of nuclear reaction, with the formation of a compound nucleus being only one type. Serber (1947) examined the various possibilities and indicated that among them are the transfer of only a part of the projectile energy to a nucleus in a single collision; the projectile with reduced energy makes collisions with another nucleus, transferring a different amount of energy; some of the particles emitted in these collisions act as projectiles in bombarding other nuclei. Low-energy particles may be captured by other nuclei resulting in the formation of a compound nucleus with ensuing reactions that are already known.

Spallation may also be produced with high-energy neutrons as projectiles. Marquez (1952) bombarded a copper target with neutrons with a range of energies of 300 to 440 MeV and a peak at 370 MeV. Chemical separation of the product nuclides, and subsequent analyses of these nuclides, showed a series of products extending from $^{45}_{22}\text{Ti}$ to $^{64}_{29}\text{Cu}$.

The method of indicating a spallation reaction by a concise notation depends on the knowledge that is available concerning the reactants and the products. For example, if $^{52}_{26}\text{Fe}$, which is one of the spallation products, is produced by the bombardment of $^{63}_{29}\text{Cu}$ by a neutron, we may write the reaction as

$$^{63}_{29}\text{Cu} + n \rightarrow \, ^{52}_{26}\text{Fe} + \text{nuclear fragments} \qquad (17\text{-}31)$$

In this reaction it is apparent that the charge of $^{52}_{26}\text{Fe}$ is three units less, and its mass number is 12 units less than those of the reactants. This may be represented schematically by writing the reaction as

$$^{63}_{29}\text{Cu}(n, 3z12a)^{52}_{26}\text{Fe} \qquad (17\text{-}32)$$

This group of fragments may consist of one alpha particle, one proton, and seven neutrons, in which case the reaction may be written as

$$^{63}_{29}\text{Cu}(n, \alpha p 7n)^{52}_{26}\text{Fe} \qquad (17\text{-}33)$$

17-15 Charge Distribution in Nuclei

The radii of nuclei have been measured by several methods, some of which were described earlier, such as Rutherford scattering experiments and alpha-particle decay. Experiments analogous to Rutherford scattering but utilizing high-energy electrons from the Stanford linear accelerator have been performed over a period of years by Robert Hofstadter and his co-workers. Equations can be derived for the scattering cross section of high-energy electrons, assuming the nuclei to be point charges; a comparison of the actual scattering cross section with the theoretical cross section will show how the charge distribu-

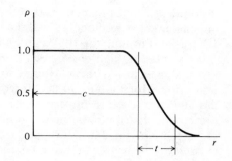

Figure 17-13 Fermi charge distribution in nuclei.

tion must be modified. *Form factors* have been developed to fit the experimental results. Good results are obtained by a charge density distribution such as that sketched in Figure 17-13. The nucleus appears to consist of a core of approximately uniform charge distribution and a "skin" of thickness t in which the charge density ρ drops rapidly. The actual shape of the charge distribution can be approximated by several different analytical expressions. One that seems to give a good fit, known as the Fermi distribution, is given by

$$\rho = \frac{\rho_1}{\exp\ [(r - c)/z_1] + 1} \qquad (17\text{-}34)$$

where c is the distance from the center to the point at which ρ drops to half the constant value and is equal to

$$c = 1.07A^{1/3} \times 10^{-13}\ \text{cm}$$

and the skin thickness t is given by

$$t = 2{\cdot}4 \times 10^{-13}\ \text{cm} = 4.4z_1$$

The skin thickness t is measured from $\rho = 0.9$ to $\rho = 0.1$ of the core distribution and z_1 is proportional to it.

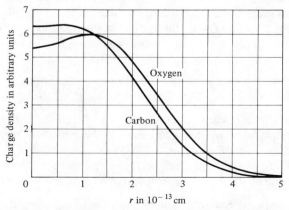

Figure 17-14 Charge distribution for ^{12}C and ^{16}O nuclei as determined for the shell model with parabolic potential well. [After Ehrenberg, Hofstadter, Meyer-Berkhout, Ravenhall, and Sobottka, *Phys. Rev.*, **113**, 666 (1959).]

Another interesting point is that the charge density of the core decreases with increasing mass number. It is about 14 for the proton, 1.4 for magnesium, and 1.1 for gold in units of 10^{-19} coulomb cm^{-3}.

By scattering high-energy electrons from the lighter nuclei such as ^{12}C and ^{16}O and measuring the elastic scattering cross section at various angles of scattering, Hofstadter and his co-workers were able to determine the effect of the nuclear structure on the nuclear charge distribution. Figure 17-14 shows the charge distributions for these nuclei obtained from data on the scattering cross sections for 240-, 360-, and 420-MeV electrons; these curves were calculated on the basis of the shell model using a parabolic shape for the potential well for the nucleons in the $1p$ shell. The rms values of the nuclear radii are 2.50 and 2.70 f for ^{12}C and ^{16}O, respectively.

Problems

17-1.   (a) Calculate the difference between the binding energy of a nucleus of ^{12}C and the sum of the binding energies of three alpha particles. (b) Assuming that these alpha particles from a triangular structure with three "alpha-particle bonds" between them, calculate the binding energy provided by each alpha-particle bond.

17-2.   What target isotope must be used to form the compound nucleus ($^{24}_{11}Na^*$) when the projectile is (a) a neutron, (b) a proton, and (c) an alpha particle?

17-3.   Determine the De Broglie wavelength of (a) a thermal neutron whose energy is 0.025 eV; (b) a neutron whose energy is 200 keV.

17-4.   Calculate the Coulomb barrier of $^{108}_{47}Ag$ (a) for protons and (b) for alpha particles.

17-5.   (a) Calculate the distance of closest approach r_1 of an alpha particle with a kinetic energy of 4.2 MeV to a nucleus of charge $Ze = 90e$. (b) Using the value of R as the radius of the nucleus of uranium, calculate the width w of the potential barrier through which the alpha particle must pass in the disintegration of ^{238}U.

17-6.   Calculate the energy of the recoil nucleus for the EC reaction with (a) ^7Be and (b) ^{37}A. (c) Calculate the velocity of the ion in each case.

17-7. ^{14}C disintegrates by β^--emission with an end-point energy of 0.155 MeV. (a) Calculate the recoil energy of the product nucleus. (b) A beta particle with an energy of 0.025 MeV is emitted in a direction at 135 degrees to the direction of motion of the recoil nucleus. Determine the momenta of the three particles involved in this disintegration.

17-8.   The end-point energy of the negative beta-ray spectrum of ^6He is 3.50 MeV. (a) Determine the recoil energy of the product nucleus. (b) A beta particle is ejected with a kinetic energy of 1.5 MeV at an angle of

90 degrees to the direction of motion of the recoil nucleus. Determine the momenta of the three particles involved in this disintegration.

17-9.   (a) Write the reaction equation for electron capture by ^{7}Be. (b) Determine the Q value of this reaction. (c) Determine the recoil energy of the product nucleus.

17-10. (a) Write the equation for the β^--decay of ^{32}P. (b) Calculate the Q value for this decay. (c) When the angle between the recoil nucleus and the beta particle is 130 degrees, the recoil momentum of the nucleus yields a value of $Br = 5750$ gauss cm and that of the beta particle a value of 2150 gauss cm. Using a vector diagram, determine the momentum of the neutrino.

17-11. (a) Using the known values of the nuclear magnetic moment and the value of I for the ground state of iron 57, calculate the nuclear g value. (b) Determine the splitting of this energy level, in electron volts, if the field at the nucleus is 3.33×10^5 oersteds. (c) Calculate the value of the velocity of the source that will produce the same energy shift of the gamma-ray line.

17-12. The value of the magnetic moment of the excited state of ^{57}Fe was found to be $\mu = -0.153$ nuclear magneton. (a) What is the significance of the negative sign for the magnetic moment? (b) Determine the nuclear g value for this state. (c) Calculate the splitting of this energy level in a field of 3.33×10^5 oersteds. (d) Calculate the values of the velocities of the source that will produce the same energy changes of the gamma rays.

17-13. On a neutron-proton plot (such as Fig. 17-1) draw transitions corresponding to (a) alpha decay, (b) e^- decay, (c) e^+ decay, (d) electron capture.

17-14. On a neutron-proton plot display the transitions corresponding to the following types of nuclear reaction:

$$(\alpha,n) \qquad (d,p)$$

$$(\gamma,p) \qquad (n,p)$$

$$(\gamma,n) \qquad (n,\alpha)$$

17-15. Obtain data on the absolute abundances of elements on the earth and defend the assertion in Section 17-1 that 87 percent of all nuclei on earth have even Z.

17-16. (a) Estimate the energy in electron volts associated with an electron that is confined inside a nucleus. (b) Do the same for a proton confined in a nucleus. The answers should indicate an additional argument against the presence of electrons in nuclei.

17-17. Compare the electrostatic force with the gravitational force for two protons separated by a distance of one fermi.

17-18. If the angular momenta of the proton and the neutron in the deuteron can be coupled with the Russell-Saunders scheme, the magnetic dipole moment of the deuteron can be expressed in the form

$$\mu_d = \frac{I}{2}\left[(\mu_p + \mu_n + \tfrac{1}{2}) + (\mu_p + \mu_n - \tfrac{1}{2})\,\frac{S(S+1) - L(L+1)}{I(I+1)}\right]$$

From this formula verify the information given in Table 17-1.

17-19. From the experimental value of the magnetic moment of the deuteron, estimate the fraction of the ground state which is a D-wave.

17-20. Devise three nuclear reactions that lead to ^{24}Mg* as a compound nucleus, starting with stable nuclides as both targets and projectiles.

17-21. Why are these processes forbidden?

$$^{23}\text{Na}(n,\alpha)^{19}\text{F}$$

$$^{3}\text{H} \rightarrow {}^{3}\text{He} + e^+ + \nu$$

$$^{238}\text{U} \rightarrow {}^{235}\text{Th} + \alpha$$

$$^{3}\text{H}(d,p)^{4}\text{He}$$

17-22. (a) When ^{27}Al is bombarded with protons, what compound nucleus can be formed? (b) List five possible decay modes of this compound nucleus. You may assume that adequate bombarding energy exists for the modes that you choose.

References

Arya, A. P., *Fundamentals of Nuclear Physics*. Boston: Allyn and Bacon, 1966.

Bethe, H. A., and P. Morrison, *Elementary Nuclear Theory*, 2d ed. New York. John Wiley & Sons, 1956.

Blatt, J. M., and V. F. Weisskopf, *Theoretical Nuclear Physics*. New York: John Wiley & Sons, 1952.

Elton, L. R. B., *Introductory Nuclear Theory*. New York: Interscience Publishers, 1959.

Feingold, A. M., "Table of *ft* Values in Beta Decay." *Rev. Mod Phys.*, **23**, 11 (1951).

Fermi, E., *Nuclear Physics*. Notes compiled by J. Orear, A. H. Rosenfeld and R. A. Schluter from a course given by E. Fermi. Chicago: The University of Chicago Press, 1950.

Harvey, B. G., *Nuclear Physics and Chemistry*. Englewood Cliffs, N. J.: Prentice-Hall, 1969.

Kopfermann, H., *Nuclear Moments*. New York: Academic Press, 1958.

Landolt, H., *Energy Levels of Nuclei: A = 5 to A = 257*. Berlin: Springer-Verlag, 1961.

Leighton, R. B., *Principles of Modern Physics*. New York: McGraw-Hill Book Company, 1959, Chapters 16-19.

Lipkin, H. M., *Beta Decay for Pedestrians*. Amsterdam: North-Holland Publishing Company, 1962.

Paul, E. B., *Nuclear and Particle Physics*. Amsterdam: North-Holland Publishing Company, 1969.

18 | Fundamental Particles

18-1 The Four Forces

In a study of Newtonian mechanics a physicist is likely to get the impression that nature provides all sorts of forces to be inserted into Newton's second law. The problems of classical mechanics tend to exemplify a great variety of mathematical expressions which describe forces between and among objects. However, a more detailed picture, on the atomic and nuclear scale, discloses that in reality only a small number of fundamental forces seem to exist. At present, the number is set at four. Further research may change this number, but the forces listed in Table 18-1 seem to account for the general features of known physical processes.

Somewhat paradoxically, the weakest of these forces is the most familiar and was the first to be described mathematically. The gravitational force affects our daily lives in a way that does not seem especially feeble. Yet it is only because the mass of the earth is great that we find it so easy to experience gravity. It is well known, that the law of force is $F = Gm_1m_2/r^2$, the form discovered by Newton. The force is always attractive. For all its familiarity the gravitational force still has some mysteries associated with its details. The general theory of relativity, for instance, is extremely difficult to test because the gravitational effects are so small. As a result gravitational concepts are usually excluded from discussions of fundamental particles and from atomic and nuclear physics generally, although a full treatment must eventually face this problem. We shall yield to tradition and mention gravitation no more.

Historically, the next force is the electromagnetic force. Coulomb's law and the law of Biot and Savart are the appropriate force relations, and they are also inverse-square laws. Maxwell's equations provide the information that the electric force and the magnetic force are only two manifestations of what is really one general phenomenon. Light constitutes an obvious example of an electromagnetic phenomenon. As our treatment of atomic physics has shown, the electromagnetic forces govern the behavior of electrons in atoms and

TABLE 18-1   THE FOUR FORCES

Interaction	Relative Strength
Strong	~1
Electromagnetic	1/137
Weak	~10^{-13}
Gravitational	~10^{-38}

molecules, implying that all of chemistry is included in a really complete study of electromagnetic interactions. A result is that matter in bulk is held together electromagnetically, and the everyday forces of things like baseballs and bats are in reality complicated combinations of many electromagnetic forces. Nuclear processes sometimes involve electromagnetism; gamma decay is an outstanding example. It is certainly fair to say that electromagnetism is the most pervasive and the best understood of the interactions.

The strong interaction provides the force that holds nuclei together. It must be very strong, since it must overcome the Coulomb force which ought to render all nuclei except hydrogen unstable. This strong (or nuclear) force must have a short range, of the order of 10^{-13} cm, since it is of no importance in the electronic structure of atoms. Processes that are governed by this force are alpha decay, nuclear fission, nuclear fusion, and scattering of nucleons by nuclei at high energy. Despite much effort, the mathematical form of the strong force is not yet known.

The weak interaction is also a nuclear force. A typical manifestation is the beta decay of a nucleus. We know that it is not the same as the strong nuclear force because the decay rate is so slow for beta decay; if it were a strong process, beta emission would imply half-lives that would be many orders of magnitude shorter than those observed experimentally.

Our description of fundamental particles will involve frequent reference to these few types of force. They serve as a sort of framework for our understanding of the processes which occur at the shortest distances available to modern experimental techniques.

18-2   The Discovery of the Meson

We have already learned that electromagnetic waves are quantized and that the photon is the particle that embodies this quantization. Around 1930 the theory was developed that the electromagnetic force arises from the exchange of photons between two charged particles. The success of this idea led H. Yukawa (1935) to try the same sort of scheme for the nuclear force. In this case one needs to exchange a massive particle in order to account for the short range of the force. A result of much calculation is that the range of a force is of the same order of magnitude as the Compton wavelength of the exchanged particle; for example, a photon has no rest mass; therefore its Compton wave-

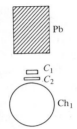

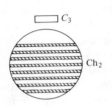

Figure 18-1 A schematic drawing showing the two cloud chambers Ch$_1$ and Ch$_2$ and Geiger counters C_1, C_2, and C_3 which are connected in triple coincidence. The passage of a charged particle through the three counters actuated the simultaneous expansion of the two chambers. The upper cloud chamber was in a magnetic field directed at right angles to the plane of the paper. [After Fretter, *Phys. Rev.*, **70**, 626 (1946).]

length is infinite and the Coulomb force ($\sim 1/r^2$) has infinite range. By analogy the nuclear force has a range of about 1.4×10^{-13} cm. A particle for which $\hbar/(mc) = 1.4 \times 10^{-13}$ cm will have its rest energy equal to 140 MeV, or about 275 times the mass of an electron. The name *mesotron* was given to this exchanged particle whose mass is intermediate between that of the electron and the proton. The modern name is *meson*.

In 1937 a particle believed to be of this type was discovered by S. H. Neddermeyer and C. D. Anderson and independently by J. C. Street and E. C. Stevenson in cloud-chamber studies of cosmic rays. Estimates of the mass of this "meson" were made from measurements of the curvature of its track in a magnetic field, the density of ionization along this track as observed in the Wilson cloud-chamber photographs, or sometimes its range in the gas of the chamber. Such estimates yielded values for the mass of the meson in the neighborhood of 200 electron masses. Both positive and negative particles were observed. It was also observed that they decay into electrons. The symbol μ (mu) was used to designate this particle.

After World War II interest in mesons was resumed. W. B. Fretter (1946) made some very careful measurements of the masses of the mu particles, using two cloud chambers, one above the other. They were expanded simultaneously whenever a penetrating particle passed through them. This was accomplished by placing Geiger counters above each chamber, the two sets of counters actuating the expansion mechanism whenever an ionizing particle passed them, as shown in Figure 18-1. The upper cloud chamber was placed in a magnetic induction of 5300 gauss so that the momentum of the particle could be measured. The lower cloud chamber had a set of lead plates 0.5 in. thick and placed 1.5 in. apart, so that the range in lead of the particles could be measured. Of the 2100 particle tracks observed, 26 were suitable for measurement. Their mass determinations yielded a value of 202 m_e. The present accepted value is 207 m_e.

This mass value is in mild disagreement with the predicted value, but there is another discrepancy that is more troublesome. Of special interest in nuclear physics is whether these mu particles (or muons) are the mesons of the Yukawa theory. One of the most decisive experiments which settled the question in the negative was performed by Conversi, Pancini, and Piccioni (1947), who investigated the difference in behavior between positive and negative muons stopped in iron and carbon. They used a magnetic field of 15,000 gauss between pole faces 20 cm in diameter to sort out positive from negative muons. They found that only the positive muons disintegrated in the iron; the negative muons were probably captured by the nuclei of iron; but in the case of the passage of muons through carbon both the positive and the negative muon, together with disintegration electrons, both positive and negative, were observed. This result was interpreted by Fermi, Teller, and Weisskopf (1947) as showing a much weaker interaction between muons and nuclei than that required by Yukawa particles. They calculated that as the negative muon slows down in its passage through matter, losing energy by electron collision and radiation, it gets to a distance from a nucleus corresponding to a K orbit of the Bohr type in a time of the order of 10^{-12} sec (see Section 18-3). The radius of the K orbit of a muon is smaller than the K orbit of an electron about the same nucleus in the ratio of the masses of the two particles; that is, about 200:1. In the case of carbon the radius of this K orbit is about 10 times the nuclear radius. The muon apparently moves in this orbit for about 2.5×10^{-6} sec and then decays. On the basis of the Yukawa theory a nuclear meson should be captured in a time of the order of 10^{-18} sec. The disagreement between the two is of the order of 10^{12}, showing a much weaker interaction between muon and nucleus than that needed for a Yukawa particle.

It was suggested by R. E. Marshak and H. Bethe (1947) that the fact that mesons are produced in the atmosphere by nuclear interactions but that muons have very small interactions with nuclei could be explained only by assuming that the mesons produced in the primary nuclear interactions are heavy mesons with a short lifetime of about 10^{-8} sec and that these heavy mesons decay into muons. At about this time Occhialini and Powell and D. H. Perkins, using special nuclear emulsion photographic plates exposed at high altitudes, observed that some of the mesons stopped in the photographic emulsions and produced so-called "stars"—that is, nuclear disintegrations with the emission of slow protons or alpha particles. The photographs in Figure 18-2 shows the star observed in the photographic emulsion of an Ilford C-2 plate by Perkins. The noticeably curved track is that of the heavy meson, now known as the *π meson;* when captured by a nucleus in the emulsion, the resulting nuclear disintegration produces a star in which three charged particles are emitted. Shortly after that Lattes, Muirhead, Occhialini, and Powell (1947), using similar nuclear photographic emulsions exposed at high altitudes, found tracks of some mesons, each of which when brought to rest in the emulsion, decayed with the emission of a muon. This was interpreted as the decay of a π^+ meson

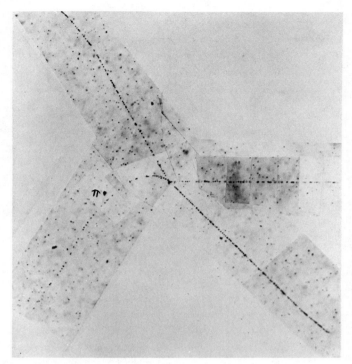

Figure 18-2   Capture of a π^-meson by a nucleus producing a star in an Ilford C-2 emulsion. [Photograph by Perkins, *Nature*, **159**, 126 (1947).]

into a μ^+. The Ilford C-2 plate was not sensitive enough to show the further disintegration of the muon into a positive electron. With improved photographic emulsions the complete $\pi^+ \rightarrow \mu^+ \rightarrow e^+$ decay scheme has been successfully recorded many times. Figure 18-3 is a photograph which shows the tracks in a more sensitive nuclear emulsion of a π^+ meson that, when stopped, decays into a μ^+, the latter then continuing until its characteristic kinetic energy has been expended in the emulsion and then decaying into an electron. The kinetic energy of the muon emitted in the decay of a pi meson is always the same and is equal to 4 MeV. To conserve energy and also momentum in this process, an additional particle, most likely a neutrino, must be emitted simultaneously. Thus the disintegration scheme for a positive pi meson is

$$\pi^+ \rightarrow \mu^+ + \nu \tag{18-1}$$

Examples have also been found of the decay of a negative pi meson *in flight,* as shown in Figure 18-4; its mode of decay is similar to that of the positive pi meson; thus

$$\pi^- \rightarrow \mu^- + \bar{\nu} \tag{18-2}$$

Early estimates of the rest mass of the charged π meson yielded values of about $330m_e$, or about 165 MeV; this value is now known to be too high. At

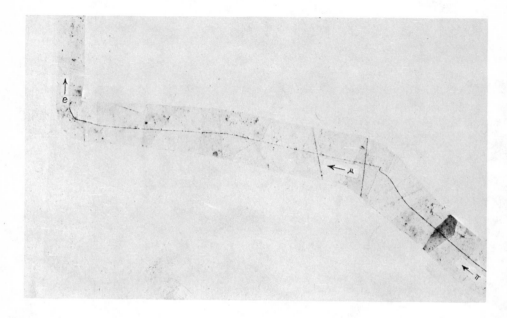

Figure 18-3 π-μ-e decay. A positive π meson comes to rest and decays into a μ^+ meson and a neutrino. The muon track shows the rapid increase in grain density (ionization) and multiple coulomb scattering characteristic of slow mesons. Upon coming to rest, the muon disintegrates into an electron (thin track at minimum ionization) and two neutrinos. (Photomosaic courtesy of Maurice M. Shapiro and Nathan Seeman, Naval Research Laboratory.)

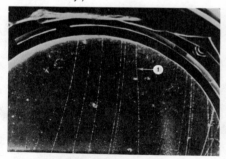

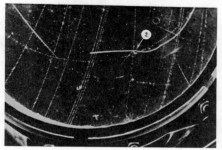

Figure 18-4 Photograph of π^--meson tracks and associated events in a cloud chamber placed in a magnetic field of 4500 gauss. At (1) a π^- meson decays in flight into a μ^-. At (2), a meson, which lost some energy in a brass bolt, was slowed sufficiently so that it came to rest in the gas and was captured by a nucleus, resulting in a nuclear explosion forming a *star*. (Photograph courtesy of L. Lederman, Nevis Cyclotron Laboratory, Columbia University.)

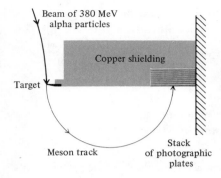

Beam of 380 MeV
alpha particles

Copper shielding

Target

Meson track

Stack
of photographic
plates

Figure 18-5   Schematic diagram of the apparatus in the first experiment on the production of π mesons, using 380 MeV alpha particles from the 184-in. synchrocyclotron to bombard a target; π^- mesons travel in semicircular paths from place of origin in the target to stack of photographic plates. The synchrocyclotron is located at the Lawrence Radiation Laboratory in Berkeley, California.

the time of the discovery of the pi meson the new large synchrocyclotron at the University of California was already producing alpha particles of about 380 MeV. Gardner and Lattes (1948) performed experiments in which these high-energy alpha particles bombarded a thin target made of a material such as carbon, beryllium, or uranium to determine whether π mesons can be produced by the bombardment of nuclei by alpha particles. A stack of photographic plates (Ilford C-2 emulsions) was placed at an appropriate position so that any π^- mesons produced by this bombardment would be deflected by the magnetic field of the cyclotron to reach the plates, as shown in Figure 18-5. Examination of the tracks in the emulsion showed them to be of the same type as the cosmic-ray meson tracks. In later experiments on the artificial production of π mesons high-energy protons were used as the bombarding particles, with a variety of substances as targets. A plentiful supply of *pions* (shortened version of pi meson) thus became readily available wherever there were high-energy particle accelerators.

18-3   Properties of Muons and Pions

Muons are unstable particles. At the end of their paths they disintegrate by emitting an electron. The mean life is 2.198×10^{-6} sec—much longer than the lifetime predicted by Yukawa—providing additional evidence that the muon does not mediate the nuclear force. The electrons that are emitted from stopped muons have no unique energy. In one series of experiments (1949) Leighton, Anderson, and Seriff measured the energy spectrum of these electrons by using a Geiger-counter-controlled cloud chamber placed in a magnetic field of 7200 gauss. They found a range of values of 9 to 55 MeV for the decay electrons. The shape of the spectrum indicated that two neutral particles of very small mass, now known to be a neutrino and an antineutrino, are emitted in the decay process. The decay of the muon can thus be represented by the reaction

$$\mu^{\pm} \rightarrow e^{\pm} + \nu + \bar{\nu} \tag{18-3}$$

An immediate consequence is that the muon is a Fermi particle, since its three decay products all have spin $\frac{1}{2}$. In fact, the muon also has spin $\frac{1}{2}$. Here is another piece of evidence against the muon's identification with the meson. Yukawa's particle must be a boson.

It should be noted that the muon is often incorrectly called the "mu meson." The modern definition of a meson is as follows.

> *A meson is a particle that is*
>   (1)   *believed to be fundamental, not composite,*
>   (2)   *capable of participating in strong interactions,*
>   (3)   *a boson.*

The muon fails the last two of these criteria. Notice that the definition says nothing about the mass being intermediate between that of the electron and that of the proton. Mesons of many varieties have been identified and some of them are much heavier than protons. Notice also that the deuteron is not a meson; it satisfies the last two criteria, but fails the first one, since it is clearly a composite of a neutron and a proton.

The muon is now classified in the category of particles known as *leptons*. A lepton is any Fermi particle lighter than the proton. Electrons, muons, and neutrinos are all leptons. They all have spin $\frac{1}{2}$; none of them is capable of participating in strong interactions.

The pions, on the other hand, have all the properties that one might expect from the original prediction of Yukawa. A detailed study of their properties is lengthened by the fact that there are three kinds of pi meson, denoted by π^+, π^0, and π^-. The charged pions are very similar to each other and somewhat different from the π^0. Either charged pion possesses a mass of 139.6 MeV; that of the neutral pion is 135.0 MeV. All pions have zero spin and participate in strong interactions. The method of production of pions is similar for all types. Typical reactions are

$$p + p \rightarrow \pi^+ + n + p$$
$$p + p \rightarrow \pi^0 + p + p \qquad (18\text{-}4)$$
$$p + n \rightarrow \pi^- + p + p$$

The decay properties of pions mark the greatest difference between the charged and the neutral types. The charged pions, as previously mentioned, decay into muons:

$$\pi^\pm \rightarrow \mu^\pm + \nu \qquad (18\text{-}5)$$

This is a weak process; the mean life is 2.60×10^{-8} sec, readily measurable by counter techniques. The neutral pion decays in quite a different way; the process is

$$\pi^0 \rightarrow \gamma + \gamma \qquad (18\text{-}6)$$

The presence of photons in the final state leads us to expect that the process is electromagnetic in nature and therefore a more vigorous reaction than was the case for charged pions. Indeed, the photons from the decay always seem to come from the spot at which the π^0 was produced in some bombardment process. The measurement of the lifetime of such a short-lived object is not easy; but emulsion techniques provide enough spatial resolution so that in the case of the rare decay modes

$$\pi^0 \to \gamma + e^+ + e^-$$
$$\pi^0 \to 2e^+ + 2e^-$$

(18-7)

it is just barely possible to measure the separation of the electrons from the place at which the π^0 was produced. Recently, counter methods have been developed which permit measurement of the π^0 mean life; they yield the result 0.89×10^{-16} sec, one of the shortest time intervals that can be measured directly.

We have already hinted at a property possessed by negative pions and muons. They can be slowed down in their passage through matter and then stopped and captured in orbits about nuclei. Since the masses are several hundred times that of an electron, the radius of the Bohr orbit (which varies inversely as the mass) will be quite small. The energy of a Bohr orbit varies directly as the mass of the orbiting particle. So the energies can get rather large, especially for elements in which Z is large. The charge on the nucleus is not well screened from the negative particle because the electron orbits are so much larger than the muonic orbits. Also, a muon or a pion can easily tumble into the $1S$ state; a muon is not identical to any of the electrons in the atom, and there is nothing to exclude it from the ground state. Pions are not subject to the Pauli exclusion principle, and they too, fall into the lowest state. Observations indicate that the transition rates are fast enough so that there is often time enough for a π^- to get to the $1S$ state before it decays. Once it is there it might decay or it might be captured by the nucleus.

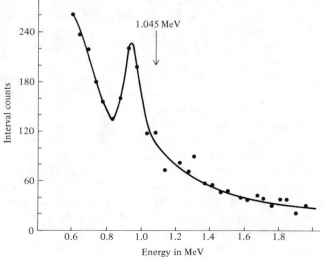

Figure 18-6   X-ray spectrum from muonic atom of titanium. [After Fitch and Rainwater, *Phys. Rev.*, **92**, 796 (1953).]

Transitions in muonic atoms give rise to a discrete x-ray spectrum which can be measured by the usual x-ray techniques. Figure 18-6 shows a typical x-ray spectrum obtained by stopping muons in titanium. There is a characteristic line at 0.955 MeV, an energy that is in reasonable agreement with a theoretical calculation based on the nucleus as a point charge. For heavier atoms, such as Pb ($Z = 82$) and Bi ($Z = 83$), fine structure has been observed in the muonic x-ray spectrum, resulting from the spin of the muon. The transitions

$$2P_{3/2} \rightarrow 1S$$
$$2P_{1/2} \rightarrow 1S \qquad\qquad (18\text{-}8)$$

will yield x-rays of slightly different energy. Experiments with muonic x-rays have become an important tool for studying nuclear charge distributions.

18-4   Cosmic Rays

Early measurements on radioactivity were always troubled by background particles, which showed up in detectors even when no radioactive sources were near. One possible source of this radiation was thought to be residual radioactivity in the earth's crust, but balloon experiments by W. Kolhörster and V. F. Hess (1912) showed that the radiation decreases at first and then increases with altitude, indicating that the source is outside the atmosphere. A number of empirical facts were discovered about this cosmic radiation.

First of all, the radiation is crudely constant in time. From this fact it follows that the sun is not a major source of cosmic rays. However, there are periodic changes in cosmic-ray intensities which are well correlated with sun spots, flares, and other solar disturbances. Among the particles emitted by the sun are streams of low energy protons and electrons which have escaped from the solar atmosphere. These low-energy charged particles are profoundly affected by the earth's magnetic field, many of them being trapped in oscillatory orbits north-to-south giving rise to such well-known effects as the Van Allen belts, the aurora borealis, and the aurora australis.

Another effect which the earth's magnetic field imposes on cosmic rays is an apparent lack of isotropy. There is a north-south effect: the latitudes near the poles receive more cosmic rays (at sea level) than the tropics. There is also an east-west effect which does not change the intensity but does change the direction of incidence of cosmic rays. Both effects can be understood completely by simply considering the earth's magnetic field and its effect on charged particles.

A third effect, which is most pertinent to the subject of nuclear particles and interactions, is as follows. At the surface of the earth there seem to be two components to the cosmic radiation: *hard* and *soft*. The distinction is not based on energy because both components are the secondary results of the interaction of primary cosmic rays with nuclei in the upper atmosphere. Both components necessarily involve high energies, since they must penetrate the atmosphere. The difference lies in their ability to penetrate matter at the surface of the earth. The hard penetrating component consists of muons. The

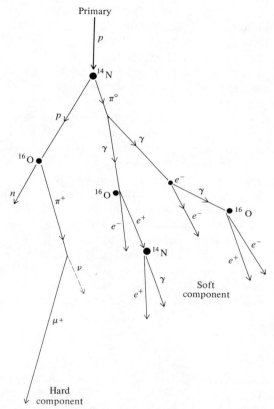

Figure 18-7   Origins of cosmic-ray secondaries.

soft component consists of photons, electrons, and positrons. These particles owe their origin to a strong interaction between a primary particle and a nucleus, leading to the production of pions of all charges. The charged pions decay into muons (hard component); the π^0 particles decay into photons, the result of which is a profusion of particles from pair production, Compton scattering and bremsstrahlung. The situation is illustrated in Figure 18-7.

18-5   Primary Cosmic-Ray Particles

Although the earth's atmosphere may seem sparse to us, nevertheless it is so dense that a cosmic-ray primary has no chance of getting very far into it. Therefore any study of these primaries requires high-altitude devices such as balloons, rockets, and satellites. Numerous studies have determined that the primary cosmic radiation consists of positive ions, mostly protons, but sometimes with Z as large as 40. These particles have no favored direction in space. They have really impressive energies—10^{12} to 10^{18} eV would not be unusual.

It is interesting to speculate how the cosmic-ray primaries can attain such high energies. The isotropy of the particles argues against any stellar origin. Also, it is hard to imagine any mechanism that can occur inside a star which is capable of imparting such extremely high energies to the primaries.

In 1949 Fermi suggested a mechanism to account for the general features of the cosmic-ray primaries. His hypothesis began with the assumption that interstellar space is filled with an ionized gas (mostly hydrogen). The ions and electrons tend to stream, causing magnetic fields that are very weak in magnitude but very far-reaching. A single charged particle can be considered with regard to its interaction with these fields. If a particle (such as a proton) is deflected by a static magnetic field, its energy will be unchanged, but if it collides with a moving region of inhomogeneity in the magnetic field it can either gain or lose energy, depending on the nature of the collision. If the proton overtakes the inhomogeneity (a kink or constriction in the field), it will lose energy. If the collision is a head-on type, the proton will gain energy. The head-on type is more probable, as can be seen by a one-dimensional analogy: a car on a highway encounters more cars coming toward it than cars it overtakes. The net result for the proton is that, on the average, it will gradually gain energy from a multitude of collisions. Given enough time, the sparse ionized gas of the interstellar regions will produce some protons with energies appropriate for primary cosmic rays.

18-6 Antiparticles

The production of pairs of oppositely charged electrons leads to an inquiry concerning the properties of the positron. We shall consider some characteristics of the production and annihilation of positrons.

Very soon after the discovery of the positron, cloud-chamber experiments showed examples of pair production, leading to the idea that a photon can give up its energy to materialize as two electrons of opposite charge. Certainly the photon must have an energy of at least $2m_ec^2$ in order to produce a pair. But one must be aware of the fact that a photon cannot produce a pair without external aid. No photon, regardless of its energy, can produce a pair in a perfect vacuum. The proof of this statement is not difficult.

To prove the impossibility of the reaction

$$\gamma \rightarrow e^+ + e^- \qquad (18\text{-}9)$$

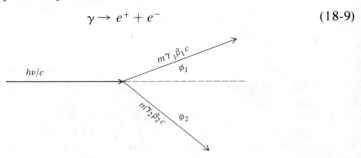

Figure 18-8

we employ the method of *reductio ad absurdum,* assuming that the process occurs and then showing that it leads to inconsistency. Using the diagram in Figure 18-8, we can write down the three equations that express conservation of the longitudinal component of momentum, the transverse component of momentum, and the energy.

$$\frac{h\nu}{c} = m\gamma_1\beta_1 c \cos \phi_1 + m\gamma_2\beta_2 c \cos \phi_2$$

$$0 = m\gamma_1\beta_1 c \sin \phi_1 - m\gamma_2\beta_2 c \sin \phi_2 \qquad (18\text{-}10)$$

$$h\nu = m\gamma_1 c^2 + m\gamma_2 c^2$$

In these equations $\beta = v/c$, $\gamma = (1 - \beta^2)^{-1/2}$. Next we (a) divide the momentum equations by c, (b) divide the energy equation by c^2, (c) square and add the momentum equations, and (d) square the energy equation. The result is

$$\left(\frac{h\nu}{c^2}\right)^2 = m^2(\gamma_1{}^2\beta_1{}^2 + \gamma_2{}^2\beta_2{}^2 + 2\beta_1\beta_2\gamma_1\gamma_2 \cos (\phi_1 + \phi_2))$$

$$(18\text{-}11)$$

$$\left(\frac{h\nu}{c^2}\right)^2 = m^2(\gamma_1{}^2 + \gamma_2{}^2 + 2\gamma_1\gamma_2)$$

Since two things each equal to a third are equal to each other, we get

$$\gamma_1{}^2 + \gamma_2{}^2 + 2\gamma_1\gamma_2 = \gamma_1{}^2\beta_1{}^2 + \gamma_2{}^2\beta_2{}^2 + 2\beta_1\beta_2\gamma_1\gamma_2 \cos (\phi_1 + \phi_2) \quad (18\text{-}12)$$

Since $\beta < 1$ for any massive particle, it follows that each term on the right-hand side is less than the corresponding term on the left. Therefore we have proved the inconsistency of our original assumption.

The arguments of the preceding paragraph do not apply if an extra particle is present to serve as a target in the initial state and as a means of carrying away momentum and energy from the final state. Several processes that actually occur are

$$\gamma + p \;\rightarrow\; e^- + e^+ + p$$

$$\gamma + \text{Pb} \rightarrow e^- + e^+ + \text{Pb} \qquad (18\text{-}13)$$

$$\gamma + e^- \;\rightarrow\; e^- + e^- + e^+$$

Pair production is strictly an electromagnetic process. It seems to occur mostly in the intense electric field *near* the nucleus rather than *inside* the nucleus. As a result of the peripheral nature of the reaction very little energy gets transferred to the target particle. At high energies or with heavy targets it is typically reasonable to ignore the energy transferred to the target, so that nearly all the energy from the photon goes into the electron pair. The energy equation

$$h\nu = 2m_e c^2 + \mathcal{E}_1 + \mathcal{E}_2 \qquad (18\text{-}14)$$

holds approximately; $m_e c^2$ is the rest energy of each electron; $\mathcal{E}_1$ and $\mathcal{E}_2$ are the kinetic energies of the particles at the instant of production. The heavier the target used, the more nearly Equation (18-14) is satisfied. Therefore we

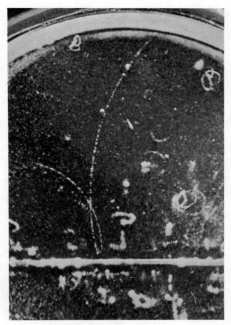

Figure 18-9 Cloud-chamber photograph of the paths of a pair of oppositely charged particles, an electron and a positron, formed by a 5.7 MeV gamma-ray photon in its passage through a sheet of lead 0.002 inch thick. Magnetic field of 1680 gauss is directed into the paper. (Photograph by H. R. Crane.)

might expect that the results for lead would agree well with the case in which all the energy of the photon goes into the pair of electrons. An experiment by H. R. Crane tested this hypothesis by using gamma rays from fluorine bombarded by protons. The photon energy was 5.7 MeV. For this case the total kinetic energy of the electrons should be 4.7 MeV. Crane placed a lead foil, 0.002 in. thick, in a cloud chamber and observed events such as the one in Figure 18-9. The measured energies of the electrons add up to give good agreement with the predicted value.

Pair production can occur in the vicinity of an electron. An example of this phenomenon is seen in Figure 18-10, in which a photon of energy 270 MeV was incident in a diffusion cloud chamber. Measurement of the momenta of the visible electrons indicates that both momentum and total energy are conserved in this process.

The process of pair production provides a method for the detection of photons for which the energy exceeds 1.02 MeV. The most common method is to send the beam of gamma rays through a sheet of dense metal with large atomic number and then to detect the electron pairs directly. Lead and tantalum are metals frequently used in this way. Another idea is to use a bubble chamber filled with a type of freon that includes bromine or iodine to provide a large value of Z. The cross section for pair production increases approximately as Z^2.

To continue with the subject of the properties of positrons, we examine their ultimate fate. Positrons are stable, so there are no decay processes to discuss. However, they eventually terminate their existence by meeting electrons and annihilating by means of the process

$$e^+ + e^- \rightarrow 2\gamma \qquad (18\text{-}15)$$

The same line of reasoning that forbids the process $\gamma \rightarrow e^+ + e^-$ can be used to prove that an electron-positron pair cannot annihilate into just one photon. Annihilation into three or more photons is possible but less likely. Each extra photon tends to suppress the rate of annihilation by a factor of the order of magnitude of the fine-structure constant (1/137).

It is instructive to trace the history of a positron. A positron, perhaps from a pair or perhaps from a nuclear beta decay, moves through matter and forms ion pairs, giving up energy in the process. There is about a 2 percent chance that the positron will hit an electron and annihilate in flight, but the more likely outcome is that the positron will stop and become attracted to an electron. The "atom" formed by these two particles is called *positronium,* and its properties

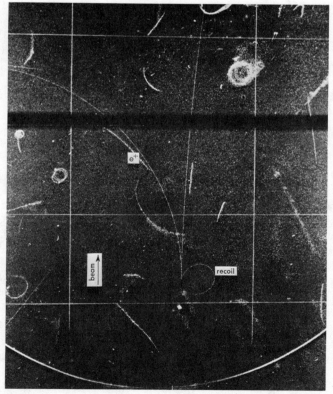

Figure 18-10   Photograph of a triplet produced by a 270 MeV photon in a diffusion cloud chamber placed in a magnetic field. The momentum of the recoil electron is 1.2 MeV/*c*. [Photograph by E. L. Hart, G. Cocconi, V. T. Cocconi, and J. M. Sellin, *Phys. Rev.,* **115**, 678 (1959).]

are readily calculated by using the same techniques that were so successful in describing the hydrogen atom. (We must not make the mistake of ignoring the reduced-mass correction for positronium, since we do not have the situation in which the two masses are very different.) The positron-electron system drops into successively lower energy states, emitting (low-energy) photons, until it arrives in the ground state.

The lowest Bohr orbit of positronium is one for which $n = 1$ and $l = 0$, so that the lowest state is an S state. This state, however, has fine structure due to the spins of the particles; when the two spins are oppositely directed, the atom is in a ^1S state; when the two spins are parallel, it is in a ^3S state and has the higher energy. The triplet state is a metastable state and has an appreciably longer lifetime than the singlet state. Theoretical calculations predicting the existence of these states were made by J. Pirenne (1944) and J. A. Wheeler (1946), who showed that the lifetime of the singlet state should be of the order of 10^{-10} sec. Furthermore, the annihilation radiation emitted by the combining of a positron-electron pair in the ^1S state should consist of two gamma-ray photons emitted simultaneously; but the radiation from the ^3S state of this system should consist of three gamma-ray photons emitted simultaneously. Theoretical calculations by Ore and Powell (1949) showed that the lifetime of the ^3S state of positronium decaying and producing this three-photon annihilation should be about 1.4×10^{-7} sec.

The first experimental evidence for the formation of positronium was obtained by M. Deutsch (1951), who observed a time delay between the emission of a positron from ^{22}Na and the appearance of the annihilation photon from the substance in which the positrons were absorbed. Several different gases, such as N_2, O_2, or methane (CH_4), were used as absorbers of the positrons. He also observed that the delay time increased with an increase in pressure of the gas. He concluded that the time delay was due to the formation of positronium. To show the existence of the two states ^1S and ^3S, and to distinguish them, Deutsch made use of the fact that the triplet state, being a metastable state with an energy of only 0.013 eV above the ^1S state, could be converted to the singlet state by interaction with some substance such as NO, which has an odd number of electrons. He found that by adding a small amount, about 5 percent, of NO to the gas in which positronium was being formed, the triplet state was quickly converted to the singlet state. The evidence for this was the rapid decrease in the number of delayed counts which would otherwise have come from the decay of the longer lived triplet state. An examination of the gamma-ray spectrum with an NaI scintillation spectrometer showed an increase in intensity of the 0.51-MeV line when NO was added to the gas, such as N_2, and a decrease in intensity at the lower-energy region corresponding to the three-photon annihilation radiation.

Deutsch (1951) measured the lifetime of the ^3S state of the positronium formed in O_2 and in freon (CCl_2F_2). The latter seems to have a special affinity for positrons. Deutsch found the disintegration constant λ, extrapolated to zero pressure of the gas, to be 6.8×10^6 sec^{-1}, so that the mean life of the ^3S state becomes 1.5×10^{-7} sec, in good agreement with the theoretical prediction.

The simultaneity of emission of the three photons in the decay of the ^3S state of positronium was verified by De Benedetti and Siegel (1952), using three coplanar scintillation counters set at angles of 120 degrees around a circle with the source at the center. The latter consisted of ^{22}Na in a small vessel containing freon at a pressure of 6 atmospheres. The three counters were connected in coincidence; when one of the counters was moved out of the plane of the circle, the number of coincidences obtained was reduced by a factor of 10. This showed that the radiation reaching the three counters simultaneously came from the positronium in the source. They also measured the energy of the radiation in one of the counters and found it to consist of a single line with its maximum point at an energy of $2/3m_ec^2$, which is to be expected, since for this special arrangement the three annihilation photons should have equal energies. They also used an arrangement with only two counters in coincidence to detect the two-photon annihilation radiation from the ^1S state of positronium and measured its energy in one of the counters, obtaining only one line with its maximum at an energy of m_ec^2, in agreement with other experiments.

The idea that the positron is the antiparticle of the electron proved so successful that it has been extended to other particles. Among those that we have considered, we note that the muon occurs in positive and negative varieties; the μ^+ is considered as being the antiparticle of the μ^-. A particle and its antiparticle need not be distinct. The photon is its own antiparticle, and the same is true of the π^0 meson. The antiparticle of the π^- meson is the π^+ meson.

It is convenient to define a quantum-mechanical operator C (called the charge-conjugation operator), which converts a particle into its antiparticle. Thus, if a wave function ψ_- describes an electron, then $\psi_+ = C\psi_-$ would be the wave function that describes a positron under similar conditions. Since the equations of electricity and magnetism remain the same, even if all charges and currents are given the opposite sign, it was thought for many years that *all* the laws of nature would be unchanged if a system were acted on by the operator C. In the following sections we shall see that this idea had to be revised, since some of the laws of nature happen *not* to be invariant under charge conjugation.

18-7 The Concepts of Parity and Time-Reversal

It will be recalled that a wave function can be either odd or even under the operation of changing the sign of all its coordinates. We can express this fact in operator form, defining P as the parity operator.

$$P\psi(x, y, z) = \psi(-x, -y, -z) \tag{18-16}$$

Specifically, we have the two cases

$$P\psi = +\psi \quad \text{(even parity)}$$
$$P\psi = -\psi \quad \text{(odd parity)} \tag{18-17}$$

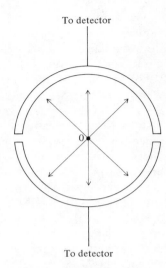

To detector

To detector

Figure 18-11 Schematic diagram for testing the law of conservation of parity. A radioactive sample is placed at O. Conservation of parity implies an up-down symmetry in the decay.

Some examples of systems that possess a definite parity are atoms, nuclei, and pions, always provided that they are referred to a system of coordinates having its origin at the center of mass.

The law of conservation of parity states that an isolated system with a definite parity will continue to have the same parity. In particular, a nucleus with a certain parity may undergo a radioactive decay, and the product of the parities of the final particles will be the same as the parity of the initial nucleus. In enumerating the final particles, it is important to note that any orbital angular momentum l will contribute a factor of $(-1)^l$ (see Problem 8-18).

A consequence of the law of conservation of parity is that any process that occurs in nature can also occur as a mirror-image process. Put in other words, neither left-handedness nor right-handedness is favored by nature. This notion has far-reaching consequences in unexpected fields, such as the chemistry of sugars, tartaric acid, and 2-butanol. Synthesis of these asymmetric molecules from symmetric ingredients always yields equal amounts of the left- and right-handed forms. (Of course, it must be recognized that such biological phenomena as snail shells, flounders, and baseball are governed by laws that are not invariant under parity.)

A more appropriate result of the law of conservation of parity concerns nuclear decay. If a small sample of a radioactive element is surrounded symmetrically by two hemispherical detectors, the average number of counts in the two detectors will be the same. The situation is shown in Figure 18-11. Initially, a nucleus at the origin is in a definite state of parity. If parity is *not* conserved, then the final-state wave function ψ (which describes the system after decay occurs) will consist of an even and an odd part:

$$\psi = \psi_{\text{even}} + \psi_{\text{odd}} \tag{18-18}$$

The chances of seeing a decay product at a given place will be proportional to $\psi^*\psi$, evaluated at that place. It is easily verified that

$$\psi^*\psi = |\psi_{even}|^2 + |\psi_{odd}|^2 + 2\,\mathrm{Re}\,\psi_{even}^*\psi_{odd} \qquad (18\text{-}19)$$

The expression Re means "take the real part of the complex quantity." Then the difference in the counting rates of the two hemispheres will be

$$N_{up} - N_{down} = \int_{UH} \psi^*\psi \, dV - \int_{LH} \psi^*\psi \, dV \qquad (18\text{-}20)$$

where the first integral is taken over the upper hemisphere (UH) and the second integral over the lower hemisphere (LH); dV is the element of volume. A simple sketch of an even and an odd function, along with their squares, will convince the reader that the square of either an even or an odd function is even. Therefore the first two terms in Equation (18-19) will not contribute to the up-down asymmetry in question. The third term will appear twice, so

$$N_{up} - N_{down} = 4 \int_{UH} \mathrm{Re}\,\psi_{even}^* \, \psi_{odd} \, dV$$

Now, if parity is conserved, one or the other of ψ_{even} or ψ_{odd} will be zero, since the initial state had a definite parity. Therefore any up-down asymmetry in the decay of a collection of nuclei will provide evidence for nonconservation of parity.

A third concept, closely allied to parity and charge conjugation, is the operation of time-reversal. We denote the operator by T and define it as the operator that makes clocks run backward. The basic laws of physics—such as Newton's laws, Maxwell's equations and the Schroedinger equation—are invariant under time reversal.

A movie of single-particle motion will look reasonable whether it is run forward or backward. Admittedly, there is a problem with the second law of thermodynamics, but that law pertains to collections of particles and is really not under discussion here. It was maintained for a long time that nature was invariant under C, P, and T separately. Pauli was able to arrive at a proof that the product CPT was conserved, based on the principle of causality. The first serious doubt about any of the separate conservation laws was suggested by T. D. Lee and C. N. Yang (1956). They agreed that C, P, and T are separately conserved for strong and electromagnetic interactions, but they pointed out that it would be good to obtain an experimental test that involved weak interactions. Within a few months after they put forth their ideas several experiments were performed which showed that parity is not always conserved in weak interactions. Some of these experiments will be described in the following sections.

18-8 Nonconservation of Parity in Beta Decay

The first observation of parity failure in weak interactions was an experiment performed by Wu, Ambler, Hayward, Hoppes, and Hudson (1957) in which they measured the asymmetry in the emission of beta particles by aligned

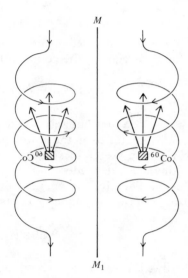

Figure 18-12 Mirror image of an experiment in the β decay of cobalt 60.

nuclei of cobalt 60. We have already seen in the preceding section that the law of conservation of parity predicts an up-down symmetry for such a process. Figure 18-12 is a schematic diagram showing a sample containing aligned cobalt nuclei placed inside a coil which carries an electric current. The image of this coil in the mirror MM_1 shows the current reversed. If parity is to be conserved, the beta particles should be emitted symmetrically, in which case it would be impossible to distinguish between the two cases. But if the beta particles are emitted upward, for example, the mirror image of this motion will show the beta particles moving upward even though the current and the associated magnetic field have been reversed. In this case parity is not conserved, since the phenomenon shown in the mirror image is not realizable in actual experiment.

Cobalt 60 decays by beta emission from a state for which $I_i = 5^+$ (even parity) to a state of nickel 60 for which $I_f = 4^+$; that is $|\Delta I| = 1$ and no change of parity.

The nucleus of cobalt 60 has a nonzero magnetic moment, since I is greater than zero, so that it is possible to align the nuclei by placing them in a very strong magnetic field. To maintain this alignment or polarization, the temperature must be extremely low, 1K or less. The degree of polarization of the nuclei can be determined by measuring the anisotropy of the gamma rays that are emitted following beta decay. The method of polarizing the nuclei was to form a very thin layer of cobalt 60 on the surface of a paramagnetic substance—in this case a single crystal of cerium magnesium nitrate—and place it in a cryostat containing liquid helium, thus reducing the temperature (see Figure 18-13). A strong magnetic field is then applied to this system, polarizing the paramagnetic ions; the heat developed during magnetization is transferred to the liquid helium. The latter is then pumped off, creating a very good vacuum, and the magnetic field is reduced slowly—that is, adiabatically. During this adiabatic demagnetization the temperature is reduced still further without changing the polarization. The magnetic field now acting on the cobalt nuclei is that due to the paramagnetic ions and is much greater than the original external field. Two sodium iodide scintillation counters placed outside the cryostat measure the intensity of the gamma rays emitted horizontally and almost vertically. Figure 18-14(a) shows the graphs obtained with these counters. It will be noted that the intensity of the gamma rays was initially much greater in the horizontal direction than in the vertical direction, but as time went on the two intensities became equal. The reason for this is that after all the helium has been removed the temperature of the system begins to increase and the thermal motion of the ions and nuclei destroys the polarization.

The beta-ray emission of the cobalt is measured by an anthracene crystal placed in the cryostat a short distance above the sample; the space between them is a vacuum. The light of the scintillations produced by the beta rays is transmitted through a lucite rod to a photomultiplier tube and to a detecting system. A solenoid is raised up around the cryostat shortly after the adiabatic demagnetization; current through the solenoid produces a magnetic field H in the vertical direction. The beta rays are counted simultaneously with the gamma rays. The results are shown in the graphs of Figure 18-14b. It will be observed that the emission of beta rays is asymmetrical during the time that the cobalt nuclei are polarized. The asymmetry disappears at the same time that the polarization of the nuclei is destroyed.

The mirror-image experiment consists of reversing the current in the solenoid so that the field H is now reversed. Again there is asymmetry in beta-particle emission of polarized nuclei, this time in the opposite direction. The results of this experiment also show that the beta particles are emitted in a direction opposite that of the direction of spin of the nuclei.

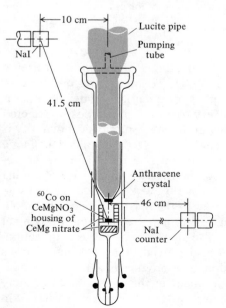

Figure 18-13 Schematic diagram of lower part of cryostat used by Wu, Ambler, Hayward, Hoppes, and Hudson. [*Phys. Rev.*,**105**, 1413 (1957).]

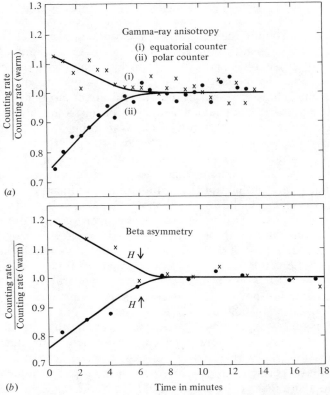

Figure 18-14 (a) Gamma-ray anisotropy; (b) beta asymmetry for polarizing field pointing up and pointing down. [After C. S. Wu, E. Ambler, R. W. Hayward, D. D. Hoppes, and R. P. Hudson, *Phys. Rev.*,**105**, 1414 (1957).]

The fact that ^{60}Co decay is capable of violating P conservation raises a question concerning the conservation of C and T. To answer this question we need to consider the behavior of various physical quantities under the operators C, P, and T. Table 18-2 contains a resumé of the properties of selected physical entities. In the cobalt 60 experiment the relevant physical quantity is $\mathbf{p} \cdot \mathbf{H}$, where $\mathbf{p}$ is the momentum of a beta particle and $\mathbf{H}$ is the applied magnetic field. From Table 18-2 we see that $\mathbf{p}$ is odd under parity and $\mathbf{H}$ is even. Therefore the scalar product is odd under parity. The average value $<\mathbf{p} \cdot \mathbf{H}>$ must vanish (if parity is conserved); but experimentally we know that $<\mathbf{p} \cdot \mathbf{H}> \neq 0$ because the electrons prefer to go opposite to the magnetic field. The implication, as we have already learned, is that parity is not conserved. Looking at Table 18-2 again, we find that $\mathbf{p} \cdot \mathbf{H}$ is also odd under charge conjugation. Therefore the cobalt 60 experiment has also proved that weak interactions are capable of violating charge conjugation. The product $\mathbf{p} \cdot \mathbf{H}$ is *even* under time reversal, so that the cobalt 60 experiment in the form presented here sheds no light on the question whether time reversal invariance is a valid sym-

TABLE 18-2   PROPERTIES OF CERTAIN PHYSICAL QUANTITES WHICH ARE ACTED ON
BY THE OPERATORS C, P, AND T (CHARGE CONJUGATION, PARITY, AND
TIME REVERSAL).

Physical Quantity	Behavior under C	Behavior under P	Behavior under T
Mass, energy	+	+	+
Time, angular momentum	+	+	−
Displacement, acceleration, force	+	−	+
Velocity, momentum	+	−	−
Electric charge	−	+	+
Magnetic field, electric current	−	+	−
Electric field	−	−	+
Electric current density	−	−	−

metry property. Notice also that $\mathbf{p} \cdot \mathbf{H}$ is even under CPT, and so this experiment is not useful as a test of the CPT theorem.

It is possible in principle to test invariance under time reversal in a cobalt 60
experiment. If, in addition, the direction of motion of the recoiling nucleus can
be measured, we can study the quantity $\mathbf{H} \cdot \mathbf{p} \times \mathbf{q}$, where $\mathbf{q}$ is the momentum
of the recoil. This triple product is even under P and odd under both C and T.
Therefore, if the weak interaction is invariant under the operation of time-
reversal, the quantity $<\mathbf{H} \cdot \mathbf{p} \times \mathbf{q}>$ must vanish. Recoil measurements of this
type are difficult to perform experimentally; so far (1971) no beta-decay experi-
ments have yielded any indication of a violation of time-reversal invariance.

18-9   Nonconservation of Parity in Meson Decays

The nonconservation of parity in the reactions

$$\pi^+ \rightarrow \mu^+ + \nu$$

and

$$\mu^+ \rightarrow e^+ + \nu + \bar{\nu} \tag{18-21}$$

was investigated by Garwin, Lederman, and Weinrich (1957) following the
suggestion by Lee and Yang. They pointed out that nonconservation of parity
in π-μ decay implies that the spin of the muon emitted by a stopped pion must
be polarized along the direction of motion and that the electrons emitted by
the muons following the π-μ decay should have an angular distribution given by

$$1 + a \cos \theta \qquad\qquad (18\text{-}22)$$

where θ is the angle between the velocity of the electron and the velocity of the muon. The factor a should have the value $-\frac{1}{3}$ if the spin of the muon is $\frac{1}{2}$.

The experimental arrangement used is shown schematically in Figure 18-15. The positive meson beam coming from the Nevis cyclotron contains about 10 percent muons, which originate from the decay of pions in flight near the target in the cyclotron. The muons are separated from the pions by allowing the meson beam to enter a carbon absorber 8 in. thick; the pions are stopped by 5 in. of carbon. The muons emitted by the stopped pions continue through the absorber to a carbon target in which they are stopped. The electrons emitted in the μ-e decay are deflected by a magnetic field and are detected by counters 3 and 4; counters 1 and 2, in coincidence, are used to indicate when the muons are stopped. The pulse from counters 1–2 operates a delay circuit so that electrons emitted between 0.75 and 2.00 μsec after the muon has come to rest are counted by 3–4.

Assuming that the muons emitted by the stopped pions are polarized parallel to their direction of motion and that they maintain this polarization until they are stopped, the electrons coming from them should have the angular distribution given by

$$1 - \tfrac{1}{3} \cos \theta$$

The counters are fixed in place to measure the distribution at $\theta = 100$ degrees.

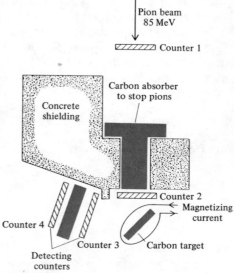

Figure 18-15 Experimental arrangement to detect polarization of muons. Magnetizing coil wound closely on carbon target to provide uniform vertical field. [After R. L. Garwin, L. M. Lederman, and M. Weinrich, *Phys. Rev.*,**105**, 1415 (1957).]

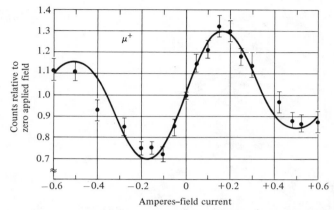

Figure 18-16 Variation of counting rate with magnetizing current. [*Phys. Rev.*,**105**, 1416 (1957).]

Instead of moving the counters, a small vertical magnetic field is applied to the target which causes the muons to precess at a rate of

$$\frac{\mu}{s\hbar} B$$

where μ is the magnetic moment of the muon and s is its spin, assumed equal to $\frac{1}{2}$. The distribution in the angle θ is carried with the precession. By varying the current in the coil, both in magnitude and direction, the value of H (and therefore B) was changed so that, effectively, the angular distribution of the emitted electrons was measured over a range of 2π radians. Figure 18-16 shows the experimental values, and the solid curve is a theoretical fit to a $1 - \frac{1}{3} \cos \theta$ distribution; this shows good agreement between experiment and theory. The conclusion is that the muons emitted from stopped pions are polarized parallel to their direction of motion; this is in violation of the principle of the conservation of parity. The violation is again easily understood in terms of the information in Table 18-2. This experiment in essence involves a measurement of a magnetic field and a momentum vector (for the emitted electron). Just as in the ^{60}Co experiment, the relevant quantity is $\mathbf{p} \cdot \mathbf{H}$; it is odd under both P and C.

An interesting additional result obtained in this experiment is the value of the gyromagnetic ratio g of the muon. This value is $g = 2$, showing that the spin of the muon is $\frac{1}{2}$.

Results similar to the above were obtained for the negative muon; its magnetic moment is negative and equal to that of the positive muon within experimental error.

18-10 Longitudinal Polarization of Beta Particles

One of the results that follow from the theory of beta decay is that if parity is not conserved in the process, then the negative beta particles should be

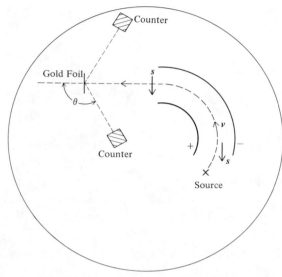

Figure 18-17 Schematic representation of Frauenfelder experiment for detecting polarization of β particles.

emitted with their spins antiparallel to their velocities; that is, the electrons should be polarized longitudinally. It is not necessary to have the nuclei aligned in an experiment to detect this polarization, since the line of reference is simply the direction of motion of the beta particles. On the basis of present theory, the degree of polarization is simply v/c, where v is the velocity of the beta particles.

The first experiment to show that beta particles are longitudinally polarized was performed by Frauenfelder and his co-workers (1957) using the beta rays from a thin source of cobalt 60 as shown in Figure 18-17. It is very difficult to detect longitudinal polarization, so it was changed to a transverse polarization by sending the beta particles through a cylindrical capacitor which deflected the path through 108 degrees. The radial electrostatic field between the plates has no effect on the direction of spin; it simply changes the direction of motion of the particles. The determination of the degree of polarization was made by scattering the electrons from gold foils and determining the asymmetry between those scattered to the left and those scattered to the right as measured by the Geiger counters.

The reason for the asymmetrical scattering of transversely polarized electrons is that in heavy atoms such as gold there is considerable *spin-orbit* coupling due to the strong electric and magnetic fields of the nuclei. Figure 18-18 shows two paths of transversely polarized electrons; in (a) the spin of the electron s is oppositely directed to the angular momentum l of the orbit and in (b) s and l are parallel. These vectors are all at right angles to the planes of the orbits. The magnetic field **H** due to the relative motion of the electron and nucleus is parallel to the magnetic moment μ of the electron in (a) but oppositely directed to it in (b). The interaction energy in (a) is less than that in

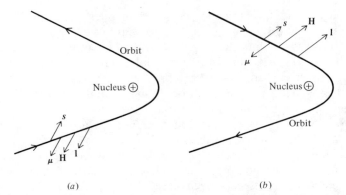

Figure 18-18   Orbit of transversely polarized electrons scattered by heavy nucleus. In (a) μ and **H** are parallel; in (b) they are opposed. All vectors are perpendicular to the planes of the orbit.

(b); hence more electrons with s and l opposed will be scattered than those with s and l parallel. (*Note:* it will be recalled that the $^2P_{3/2}$ energy level of sodium is slightly higher than the $^2P_{1/2}$ level.)

One set of results of this experiment is shown in Table 18-3. They show a definite left-right asymmetry; the scattering was measured over an angular interval of 95 to 140 degrees by two symmetrically placed Geiger counters, one to the left of the beam, the other to the right. The negative sign of the longitudinal polarization shows that when they are created the beta particles are polarized with their spins antiparallel to their velocities.

TABLE 18-3

Electron Energy in KeV	v/c	Thickness of Gold Scatterer in mg/cm²	Left-Right Asymmetry	Longitudinal Polarization
50	0.41	0.15	1.03 ± 0.03	−0.04
68	0.47	0.15	1.13 ± 0.02	−0.16
77	0.49	0.05	1.35 ± 0.06	−0.40
77	0.49	0.15	1.30 ± 0.09	−0.35

Many other experiments have been performed since this first one, with both positive and negative beta particles and with other methods of analyzing the direction of polarization. All of these experiments have led to the same results: the negative beta particles are emitted longitudinally polarized with their spins antiparallel to their velocities or with "left-handed" polarization and the positive beta particles are emitted with their spins parallel to their velocities or with "right-handed" polarization. The degree of polarization is $\pm v/c$ for $e^{\pm}$.

18-11   Neutrinos and Their Interactions

We have already mentioned the fact that neutrinos are distinct from anti-neutrinos, that each type of particle has spin $\frac{1}{2}$, that the neutrino (ν) is left-handed, and that the antineutrino ($\bar{\nu}$) is right-handed. The spin vector for the neutrino always points opposite to its momentum vector and the spin of the antineutrino always points along its direction of motion. This idea was first proposed by Weyl (1929), but because it violates the principle of conservation of parity it was discarded. Since 1956 the generally accepted line of reasoning is that neutrinos are not capable of interacting strongly or electromagnetically; they participate in weak interactions, but because parity is not conserved in weak interactions there is no objection to a *two-component* theory of the neutrino such as we have just described.

The fact that neutrinos have a definite helicity distinguishes them from the other leptons (electrons and muons). We have seen in the Frauenfelder experiment that electrons from beta decay tend toward having their spins aligned along their directions of motion but only to an extent $\pm v/c$, with the positive sign (or right-handed polarization) for the positron and the negative sign (or left-handed polarization) for the electron.

In a very interesting experiment Maurice Goldhaber, Grodzins, and Sunyar (1958) were able to show that the spin of the neutrino does indeed point in the direction opposite to its momentum. The process they used was electron capture by a nucleus followed by the gamma decay of the recoiling nucleus. The specific sequence used began with the nucleus $^{152}\mathrm{Eu}^m$; the superscript m refers to the fact that the original nucleus is in an isomeric state. The ground state of $^{152}\mathrm{Eu}$ has spin and parity 3^-; it, too, undergoes electron capture, but the analysis associated with its decay is more complicated and so the simpler 0^- isomer was used. The sequence is

$$^{152}_{63}\mathrm{Eu}^m + e^- \rightarrow {}^{152}_{62}\mathrm{Sm}^* + \nu$$

$$^{152}_{62}\mathrm{Sm}^* \rightarrow {}^{152}_{62}\mathrm{Sm} + \gamma$$

$$(18\text{-}23)$$

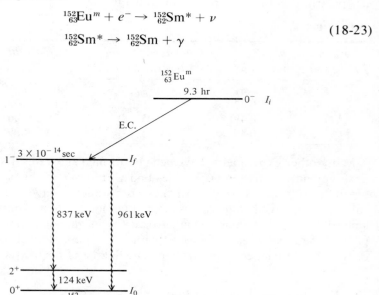

Figure 18-19 Levels involved in β decay by e^- capture of $^{152}\mathrm{Eu}^m$ to $^{152}\mathrm{Sm}$.

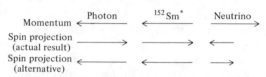

Momentum Photon ←—————— ←———— ^{152}Sm* Neutrino ——————→

Spin projection ——————→ ——————→ ←——
(actual result)

Spin projection ←—————— ←———— ——→
(alternative)

Figure 18-20   Momenta and spins involved in the experiment which determined the helicity
of the neutrino. The arrows in the second row illustrate the true situation. The
arrows in the third row illustrate the arrangement that would be correct if the
neutrino's spin pointed along its direction of motion.

The energy levels associated with these processes are shown in Figure 18-19.
Notice that there are two ways for the transition from the 1⁻ level of ^{152}Sm to
the ground state to occur. The transition can be direct, with emission of a photon of 961 keV, or it can cascade through the 2⁺ level, emitting photons of 837
and 124 keV. In the emission of either the 961 or the 837 keV photon the nu-
cleus undergoes a change of angular momentum $|\Delta I| = 1$; this entire angular
momentum is carried away by the emitted photon in the form of the spin of the
photon (which is unity). A photon can be either left or right circularly
polarized. The nature of the polarization can be determined by a Compton
type of experiment—that is, by scattering the photons from aligned electrons,
for example, from the electrons in magnetized iron. Those electrons whose
spins are directed parallel to the spins of the photons will have less effect on
the photon beam than those that are spinning antiparallel. The latter can absorb
energy and angular momentum from the photons to change their direction of
spin or perform a "spin flip." By placing the iron in a magnetic field the direction of spin of the majority of electrons can be controlled by controlling the
direction of this field.

The analysis of the angular momenta involved in this process begins with
the assumption that the metastable Eu nucleus captures an electron from a K
orbit (for which $l = 0$). Therefore the total angular momentum results entirely
from the spin of the electron and is equal to one-half. The axis of quantization
of angular momentum is taken to be the direction of motion of the decay products (neutrino, ^{152}Sm*) and the situation is summarized in Figure 18-20. Any
relative orbital angular momentum between ^{152}Sm* and the neutrino can contribute nothing to the components of angular momentum along the direction
of motion; therefore the figure does not need to display this orbital angular
momentum. When the excited state of ^{152}Sm decays, the photon spin points
in the same direction *if the photon's direction of motion is the same as that of*
^{152}Sm. In the experiment, Goldhaber and his co-workers used only those gamma rays that were emitted in this favored direction. There were two reasons for

this choice: first, the orbital angular momentum component is thereby zero in the direction of motion; second, the photons were detected by means of resonance absorption. It will be recalled (see Section 17-14) that resonance absorption of gamma-ray photons is a rather rare event because the energy of the radiation is usually less than the energy difference between the two levels involved because some of the energy goes to the recoil nucleus. This deficiency may be supplied by adding energy from some other source. In this case the extra energy comes from the Doppler shift provided by the motion of the emitting excited state, with the photons emitted in the forward direction. The method of selecting these forward gamma-ray photons was to have them resonantly absorbed by a sample of ^{152}Sm. Following this resonance absorption, the nucleus returns to the ground state by emitting either 961 keV gamma-ray photons or by emitting two photons successively of 837 and 124 keV. The 961 and 837 keV photons were readily observed in the detector.

The direction of polarization of these forward photons was determined by first scattering them from magnetized iron and then allowing them to be resonantly absorbed by ^{152}Sm. It was found that the gamma-ray photons were transmitted more readily when the electron spins in the iron were parallel to the angular momentum of the left circularly polarized photons. Hence the forward gamma-ray photons are left circularly polarized photons.

It is apparent from Figure 18-20 that the result obtained from this experiment implies that the spin of the neutrino is directed opposite to its momentum vector. Other ways of stating this fact are "the neutrino has negative helicity" or "the neutrino has left-handed polarization."

A recent development in our knowledge about neutrinos has led to an additional doubling in the number of types. The neutrino and antineutrino of nuclear beta decay processes are two distinct particles, as we have seen. In addition, a second neutrino and a second antineutrino exist in association with muons. The experiment that led to this result was performed at Brookhaven National Laboratory by a group of physicists from Columbia University: Danby, Gaillard, Goulianos, Lederman, Mistry, Schwartz, and Steinberger (1962). Their experiment investigated whether the neutrinos (and antineutrinos) in the decays

$$\pi^+ \rightarrow \mu^+ + \nu \quad \text{(a)}$$
$$\pi^- \rightarrow \mu^- + \bar{\nu} \quad \text{(b)}$$

(18-24)

are the same as those in the beta decay of the neutron (or a bound proton)

$$n \rightarrow p + e^- + \bar{\nu} \quad \text{(a)}$$
$$p \text{ (bound in a nucleus)} \rightarrow n + e^+ + \nu \quad \text{(b)}$$

(18-25)

The experimental procedure (see Fig. 18-21) was to use a beam of protons at the AGS (see Section 12-8) incident at 15 GeV on a beryllium target. A variety of particles was produced, including many pions of both charges. The decay of these secondary pions produced an intense beam of neutrinos and antineutrinos, with energies distributed continuously over a range of zero to

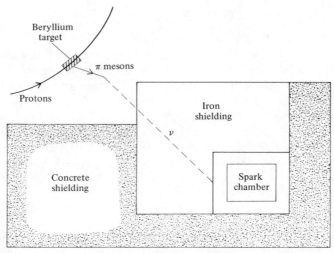

Figure 18-21 Experimental arrangement for distinguishing the electron-neutrino from the muon-neutrino.

about 2 GeV. The 13.5 meters of iron shielding were sufficient to eliminate all particles except the neutrinos and antineutrinos, some of which were then detected in a spark chamber with aluminum plates (2.54 cm thick). The reactions sought in the spark chamber were

$$\nu + n \rightarrow p + e^- \qquad \text{(a)}$$
$$\bar{\nu} + p \rightarrow n + e^+ \qquad \text{(b)}$$
$$\nu + n \rightarrow p + \mu^- \qquad \text{(c)} \qquad \qquad (18\text{-}26)$$
$$\bar{\nu} + p \rightarrow n + \mu^+ \qquad \text{(d)}$$

The apparatus was sensitive to all four processes, and if the neutrino of Process (18-26a) were the same as that of Process (18-26c), they should have occurred with equal probability. Instead, after using much accelerator time at maximum intensity, 29 examples of muon production were observed and no examples of electron production. It was concluded that the muon-neutrino (ν_μ) and its antiparticle ($\bar{\nu}_\mu$) are distinct from the neutrinos of beta decay (ν_e and $\bar{\nu}_e$). We should therefore rewrite all of the reactions that contain neutrinos to reflect this distinction. We choose several familiar process to serve as examples:

$$\pi^+ \rightarrow \mu^+ + \nu_\mu$$
$$\pi^- \rightarrow \mu^- + \bar{\nu}_\mu$$
$$n \rightarrow p + e^- + \bar{\nu}_e$$
$$\mu^- \rightarrow e^- + \bar{\nu}_e + \nu_\mu$$
$$\mu^+ \rightarrow e^+ + \nu_e + \bar{\nu}_\mu$$

18-12 K Mesons

Soon after the discovery of the pi meson, Rochester and Butler (1947), using a cloud chamber, discovered a neutral particle which decayed into two pions of opposite charge. This particle became known as the θ^0 meson. Its mass was measured to be between those of the pion and the nucleon. Measurement of the time of flight for many examples of these particles led to a mean lifetime of about 10^{-10} sec. A group of physicists in England—Brown, Camerini, Fowler, Muirhead, Powell, and Ritson (1949)—exposed a photographic emulsion to cosmic rays and discovered a particle that decayed into three charged π mesons. Subsequent experiments proved that these particles occur in both positively and negatively charged forms; it became known as the τ meson. The decay can be written as

$$\tau^{\pm} \to \pi^{\pm} + \pi^{+} + \pi^{-}$$

An example of τ decay in emulsion is shown in Figure 18-22. The tracks a, b, and c are the paths of the three pions in the emulsion. These three tracks are coplanar, consistent with the idea that neutrinos, photons, or other neutral particles are absent from the final state. The measured values of the kinetic energies of the particles forming these tracks are 17.1, 27.8, and 41.6 MeV, respectively, or a total of 86.5 MeV. If the mass of a pion is taken to be 140

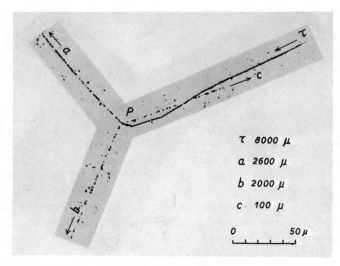

Figure 18-22 Photograph of tracks of a τ meson decaying at rest into three π mesons, a, b, and c. Tracks are in an Ilford G-5 nuclear emulsion, 1200 microns thick, which was exposed on Mt. Rosa, 4500 meters above sea level, surrounded by an aluminum absorber 5 cm thick. The numbers on the photograph indicate the respective lengths of the tracks in the emulsion. Tracks a and b end in the glass plate, track c goes into the air after only 100 microns in the emulsion. [From photograph supplied by Ceccarelli, Dallaporta, Merlin, and Rostagni, *Nature*,**170**, 454 (1952).]

MeV, the mass (rest energy, see Section 3-14) of the decaying τ meson is about 506 MeV.

During the years from 1947 to 1953 various other mesons were observed. A θ^+ meson was observed to decay by

$$\theta^+ \to \pi^+ + \pi^0$$

and similarly a negative form was discovered,

$$\theta^- \to \pi^- + \pi^0$$

A particle called the K meson was observed to decay by

$$K^+ \to \mu^+ + \nu_\mu$$

$$K^- \to \mu^- + \bar{\nu}_\mu$$

Of course, at the time of its discovery it was not realized that the neutrino from K decay was different from that of beta decay.

At first these new particles could be studied only by means of cosmic-ray experiments. The numbers of events were small and measurements were not always accurate, so that each of the decay processes mentioned so far was assigned to a different particle. With the development of high-energy accelerators, these new particles could be produced in larger quantities, allowing for better accuracy in the determination of their properties. As more data were accumulated it became clear that all of these particles had masses close to 500 MeV and that all of the charged θ, τ, and K mesons had the same lifetime (about 10^{-8} sec). It was tempting to assume that they were all essentially the same particle, just as was assumed for the π mesons. The difference in lifetime between the θ^0 and the θ^+ was not a barrier to this idea, since the lifetime of the π^0 is very different from that of the charged pions. In fact, the modern point of view is to include all the θ, τ, and K particles under the name "K meson."

The most formidable stumbling block which delayed the acceptance of the unity of the K-meson family came from the law of conservation of parity. The wave function of a π meson is odd under the operation of parity. Further, it was found that all angular momenta in both θ and τ decays are zero. Therefore

$$\text{parity}(\theta) = (-1)^2 = +1$$

$$\text{parity}\ (\tau) = (-1)^3 = -1$$

Clearly, the θ and the τ could not be the same particle because their parities were not the same, yet measurement showed their masses and lifetimes to be identical. This problem of the two particles being alike in all respects except for parity became known as the theta-tau puzzle. The resolution of this puzzle was a prominent consideration for Lee and Yang (1956) when they suggested that parity might not be conserved in weak interactions. As we have seen (Sections 18-7 through 18-10), parity is often not conserved in weak interactions. The resolution of the puzzle consists of recognition of the fact that all the K mesons (including both θ and τ) decay according to the weak interaction.

TABLE 18-4 DECAY PROPERTIES OF THE K MESONS

K^+	K^-	Percentage
$\mu^+ + \nu_\mu$	$\mu^- + \bar{\nu}_\mu$	63.5
$\pi^+ + \pi^0$	$\pi^- + \pi^0$	20.8
$\pi^+ + \pi^+ + \pi^-$	$\pi^- + \pi^- + \pi^+$	5.5
$\pi^+ + \pi^0 + \pi^0$	$\pi^- + \pi^0 + \pi^0$	1.7
$\mu^+ + \pi^0 + \nu_\mu$	$\mu^- + \pi^0 + \bar{\nu}_\mu$	3.4
$e^+ + \pi^0 + \nu_e$	$e^- + \pi^0 + \bar{\nu}_e$	5.0

K^0 (and $\overline{K}{}^0$)

$\pi^+ + \pi^-$

$\pi^0 + \pi^0$

$\pi^0 + \pi^0 + \pi^0$

$\pi^+ + \pi^- + \pi^0$

$\pi^+ + \mu^- + \bar{\nu}_\mu$

$\pi^- + \mu^+ + \nu_\mu$

$\pi^+ + e^- + \bar{\nu}$

$\pi^- + e^+ + \nu_e$

More recent experiments have shown that the K meson has odd parity. When it decays in the τ mode, parity is conserved. When it decays in the θ mode, parity is violated. In the process $K \rightarrow \mu + \nu$ the helicity of the neutrino and the consequent longitudinal polarization of the muon imply nonconservation of parity, just as in the decay $\pi \rightarrow \mu + \nu$ (see Section 18-9).

The conclusion is that all the various processes discussed represent competing decay modes of the K meson. There are four types of meson: K^+, K^-, K^0 and $\overline{K}{}^0$. The neutral K mesons are antiparticles of each other and they are distinct. In Section 18-15 we shall discuss the ways in which they differ. Table 18-4 contains a list of observed decay modes of the K mesons (or kaons). The list contains the most probable modes; other decay modes exist, but they are rare, accounting for a small fraction of 1 percent of all K-meson decays.

18-13 Hyperons

At about the time (1947) when K^0 particles were discovered similar cloud-chamber photographs showed V-shaped events in which the two tracks corresponded to a π^- meson and a proton (rather than a π^- and a π^+, as in K^0 decay). The events of this type were all consistent with an interpretation that a neutral particle exists with a lifetime of the order of 10^{-10} sec and with a rest energy of about 1115 MeV. This particle was given the name lambda (Λ). Its two principal modes of decay are

$$\Lambda \rightarrow p + \pi^- \tag{18-27a}$$

$$\Lambda \rightarrow n + \pi^0 \tag{18-27b}$$

On rare occasions the Λ particle has been observed to decay in the processes

$$\Lambda \rightarrow p + e^- + \bar{\nu}_e \tag{18-28a}$$

$$\Lambda \rightarrow p + \mu^- + \bar{\nu}_\mu \tag{18-28b}$$

The decay mode (18-28a) suggests the idea that the Λ particle may be very similar to the neutron because one of the decays of the Λ is identical to the only decay mode of the neutron. We shall see (in Section 18-18) that this idea has some merit.

If one adds the angular momenta on the right-hand side of Processes (18-27a and b), it becomes apparent that the Λ has half-integral spin. The results of a number of ingenious experiments have shown that when the Λ decays it does not conserve parity. Its decay into a pion and a nucleon proceeds by a mixture of S-waves and P-waves, with the angular momentum of the final state being partly $S_{1/2}$ and partly $P_{1/2}$. Thus the initial angular momentum (i.e., the spin of the Λ) is $\frac{1}{2}$. An immediate result of the half-integer spin assignment is that the Λ is not a meson. In Section 18-3 it was pointed out that a meson must have integer spin. Among fermions we have already learned that light fermi particles are called leptons and do not interact strongly. Fermions whose mass is at least as great as the mass of the proton, and which can interact strongly, are called *baryons*. The name *hyperon* is applied to certain baryons of which the lightest (and the first to be discovered) is the Λ. Specifically, a hyperon is a baryon that cannot decay strongly into a nucleon. The process of Λ decay is a weak interaction and therefore agrees with this definition. We know that Λ decay is a weak interaction because the lifetime is of the right order of magnitude and because parity is not conserved.

Although its decay is weak, the Λ particle is capable of participating in strong interactions. Figure 18-23 is an example of a strong process involving a Λ. The photograph was taken in a hydrogen bubble chamber using a beam of protons with energy 2.85 GeV. The protons are moving upward in the picture; the magnetic field was arranged so that positive particles go in a clockwise direction. The reaction shown is

$$p + p \rightarrow \Lambda + K^0 + p + \pi^+$$

The Λ and the K^0 appear as V's which point back to the primary interaction. The Λ is the one on the left. The momentum vectors of the π^- and the proton from Λ decay add to agree with the momentum of a neutral particle (mass = 1115 MeV) produced in the primary interaction. Thus we see that the Λ was produced strongly and decayed weakly.

Soon after the discovery of the Λ, other hyperons were found. Three particles with rest energies close to 1190 MeV were observed and given the name sigma (Σ). The three forms differ in their electric charge and are denoted Σ^+, Σ^0, and Σ^-. They have much in common with the Λ: they have spin $\frac{1}{2}$, they are produced strongly, and they do not decay strongly. The most common decay modes are the following:

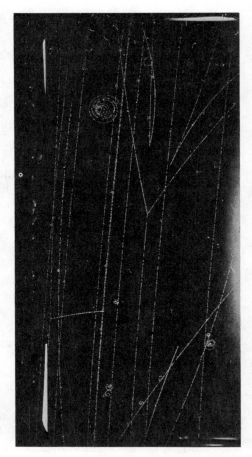

Figure 18-23 A 2.85-GeV proton from the cosmotron collides with a proton in the liquid hydrogen bubble chamber. (Courtesy of Brookhaven National Laboratory.)

$$\Sigma^+ \rightarrow p + \pi^0$$

$$\Sigma^+ \rightarrow n + \pi^+$$

$$\Sigma^0 \rightarrow \Lambda + \gamma$$

$$\Sigma^- \rightarrow n + \pi^-$$

The Σ^0 decay is an electromagnetic process; the other decays are weak. We therefore expect that the lifetimes will be very different, and indeed they are. The charged sigmas have a mean life around 10^{-10} sec, comparable to Λ decay. The lifetime of the Σ^0 has not been measured with precision for the same reason that it is very difficult to measure the lifetime of the π^0 meson, but attempts to do so have indicated that this lifetime is shorter than 10^{-14} sec.

The photograph in Figure 18-24 shows a strong interaction in which a Σ^+ hyperon was produced. The reaction is

$$\pi^+ + p \rightarrow \Sigma^+ + K^+$$

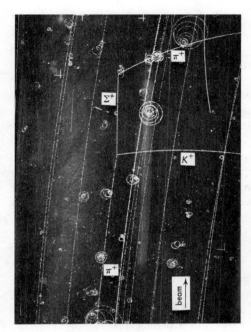

Figure 18-24 Photograph of the tracks of particles in a liquid-hydrogen bubble chamber showing the production of a Σ^+ particle by the interaction of a positive pion and a proton. The pion track starts at the bottom of the photograph. The initial momentum of the pion is 1.23 BeV/c; the bubble chamber is in a magnetic field of 17,000 gauss. [Photograph by C. Baltay *et al.*, *Rev. Mod. Phys.*,**33**, 374 (1961).]

followed by the decay

$$\Sigma^+ \to \pi^+ + n$$

In this picture, as in other bubble-chamber events, identification of the tracks is made possible by the application of the laws of conservation of energy and momentum at each vertex. A set of mass assignments is made for each track and a computer tests for the validity (within experimental error) of the conservation laws. This process is repeated for all reasonable permutations of masses for the various tracks. If one set of masses yields agreement with conservation of energy and momentum and all other sets yield disagreement, the experimenters feel justified in accepting the event with the track labels which gave best agreement.

By the early 1960s the list of hyperons had been completed by the discovery of the xi particles (Ξ). There are two types, the Ξ^- and the Ξ^0, with masses around 1320 and 1315 MeV, respectively. There is no evidence for a positively charged hyperon in this mass region, a fact that is in agreement with the theory to be presented in Section 18-15. The properties of the Ξ hyperons are similar to those of the Σ and the Λ. Their lifetime is of the order of 10^{-10} sec and they have spin $\frac{1}{2}$. They decay almost entirely by the weak processes

$$\Xi^0 \to \Lambda + \pi^0 \qquad \Xi^- \to \Lambda + \pi^-$$

In particular, the Ξ has never been observed to decay directly into a nucleon and a pion; rather it always undergoes a "cascade" to the nucleon by decaying first to a Λ, then to a nucleon, each time emitting a pion. For this reason the Ξ is often called the *cascade hyperon*.

18-14 Antibaryons

The original concept of a particle-antiparticle pair was developed by P. A. M. Dirac (1930) in his relativistic theory of the electron. The discovery of the positron provided striking confirmation of this prediction and led to notions of charge conjugation which were discussed in Section 18-6. Dirac's theory applies only to particles with spin $\frac{1}{2}$, but all the baryons discussed so far have this spin. There are difficulties in the extension of the Dirac theory to include strongly interacting particles, but the possibility is suggested that distinct antiparticles exist for the proton, the neutron, and the Λ, Σ, and Ξ hyperons. For each of these particles the antiparticle also has spin $\frac{1}{2}$, but the electric charge and the magnetic moment would have the opposite sign. The antiparticle is conventionally denoted by placing a bar over the symbol for the particle, as was done for neutrinos.

The fact that the proton is stable against gamma-decay into a positron led to the suggestion that each of the baryons mentioned so far should possess a quantum number, called the baryon number, which is conserved in all interactions and decays. For baryons this quantum number is given the value of $+1$; for the antiparticles, $\bar{p}$, $\bar{n}$, $\overline{\Lambda}$, $\overline{\Sigma}$, and $\overline{\Xi}$ it is -1. Mesons are assigned a baryon number of zero. From this conservation principle it is apparent that the following reactions would be typical of those in which antibaryons are produced:

$$p + p \rightarrow p + p + p + \bar{p}$$

$$\pi^- + p \rightarrow \pi^- + p + p + \bar{p}$$

$$\pi^+ + p \rightarrow p + p + \bar{n}$$

$$p + p \rightarrow p + p + \Lambda + \overline{\Lambda}$$

In each of these examples the net baryon number is the same in the final state as it was in the initial state.

Antibaryons decay with the same lifetimes as those of their respective baryons, and the decay processes exhibit symmetry with respect to charge conjugation; for example,

$$\bar{n} \rightarrow \bar{p} + e^+ + \nu_e$$

$$\overline{\Lambda} \rightarrow \bar{p} + \pi^+$$

$$\overline{\Sigma^+} \rightarrow \bar{p} + \pi^0$$

In the third example we have made explicit use of the fact that the π^0 meson is its own antiparticle; it has spin zero and is not described by Dirac's theory. The $\overline{\Sigma^+}$, like the $\bar{p}$, is a particle with negative charge.

The process $p + p \rightarrow 3p + \bar{p}$ requires less beam energy than any other reaction for producing antibaryons in the laboratory. The threshold is at about 5.6 GeV kinetic energy. With the completion of the 6-GeV bevatron at the University of California it was possible to perform an unambiguous experi-

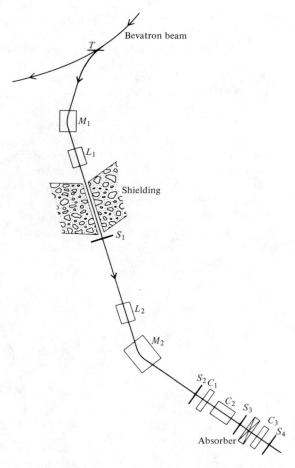

Figure 18-25 Schematic diagram of the original mass spectrograph for the detection of antiprotons.

ment for the production and detection of antiprotons. The first successful experiment was performed by Chamberlain, Segrè, Wiegand, and Ypsilantis (1955). A schematic diagram of their detecting apparatus, which is essentially a mass spectrograph, is shown in Figure 18-25. Protons of 6-GeV energy are incident on a copper target T; negatively charged particles are deflected out of the bevatron by its magnetic field. Most of the particles are negative pions; the estimate is 50,000 pions to 1 antiproton. Since they travel in the same trajectory in a magnetic field, these particles all have the same momentum; in this experiment the momentum was 1.19 GeV/c. Because of its larger mass the proton had a velocity of only 0.78c whereas the pion had a velocity of 0.99c. The trajectory took these particles through deflecting magnets M_1 and M_2, focusing magnets (quadrupole lenses) L_1 and L_2, scintillation counters S_1, S_2, and S_3, and Cerenkov counters C_1 and C_2. The Cerenkov counter C_2 was designed to respond only to particles whose velocities were between 0.75c

and $0.78c$; counter C_1 was designed to respond to particles with velocities greater than $0.79c$. The distance between S_1 and S_2 was 12 meters; the time of flight for protons was 51 msec and for pions 40 msec. Counters S_1, S_2, S_3, and C_2 were in coincidence and C_1 in anticoincidence, so that the signals that reached S_3 were those produced by antiprotons. Preliminary mass measurements showed that the mass of the antiprotons was equal to that of the proton to an accuracy of 5 percent. The accuracy has been improved in later experiments.

Any antibaryon that collides with a baryon is capable of annihilation, just as positrons are destroyed when they encounter electrons, but with an important difference. Electrons and positrons do not interact strongly and their annihilation products are always photons. It is possible for a baryon-antibaryon pair to annihilate into photons, but it is much more probable that they will produce mesons instead. The reason is that mesons and baryons (and therefore the antibaryons) interact strongly, leading to probabilities that are much enhanced in comparison with the electromagnetic processes involving photons. Typical examples are

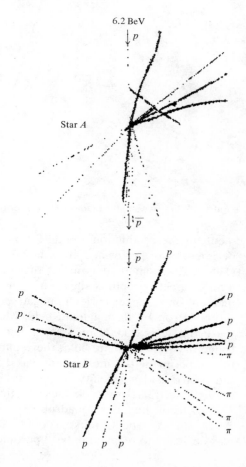

Figure 18-26 A 6.2-BeV proton p produces a star at A in the nuclear emulsion. An antiproton $\bar{p}$ from this star travels 1.4 mm in the emulsion and is annihilated at B, producing 14 visible tracks of protons (p) and mesons (π). [Photograph courtesy of R. D. Hill, S. D. Johansson, and F. T. Gardner, *Phys. Rev.*, **103**, 250 (1956).]

Figure 18-27   Photograph showing an antiproton entering a propane bubble chamber. Its track ends abruptly (near upper arrowhead). At this point it undergoes charge exchange with a proton forming a neutron-antineutron pair. The antineutron travels 9.5 cm and interacts with a nucleus forming an annihilation star (near lower arrowhead). (Photograph courtesy of Professor E. Segrè, Lawrence Radiation Laboratory, University of California, Berkeley, California.)

$$\bar{p} + p \rightarrow \pi^+ \ \pi^+ \ \pi^- \ \pi^-$$

$$\bar{p} + p \rightarrow \pi^+ \ \pi^- \ \pi^0$$

An example of both the production of an antiproton and its annihilation is shown in Figure 18-26. In this experiment by Hill, Johansson, and Gardner, the beam of 6.2 GeV protons was incident on a photographic emulsion. The proton p produced a "star" in a collision with a nucleus. One of the particles produced in this event is an antiproton $\bar{p}$ that proceeds in the emulsion and produces a second star. It is most likely annihilated by a proton in this nucleus; the evidence for this is a measurement of the total energy liberated as kinetic energy of the protons and mesons forming this star. The minimum energy that must be accounted for is the rest-mass energy of two protons plus the kinetic energy of the antiproton, estimated to be a total of 2600 MeV. The measured value of the energies of the protons and pions forming the star is 1410 MeV. In stars of this kind an additional amount of energy is carried away by neutral particles that do not leave tracks in the emulsion; these particles are probably neutrons and neutral mesons. It is thus highly probable that the second star was initiated by the annihilation of a proton-antiproton pair.

The detection of an antineutron is much more difficult, since it leaves no track until it is annihilated. Figure 18-27 is a photograph showing the track of an antiproton in a propane bubble chamber. The track ends suddenly and a star appears with its origin at a distance of 9.5 cm from the end of the antiproton track. The interpretation of this photograph is that the antiproton gave up its charge to a proton, thus forming a neutron-antineutron pair. The antineutron is then annihilated in a nucleus, giving rise to the star. The antiproton is a stable particle, whereas the antineutron should decay by positron emission with the same half-life as the neutron.

Examples of antihyperons have been observed experimentally. A beautiful

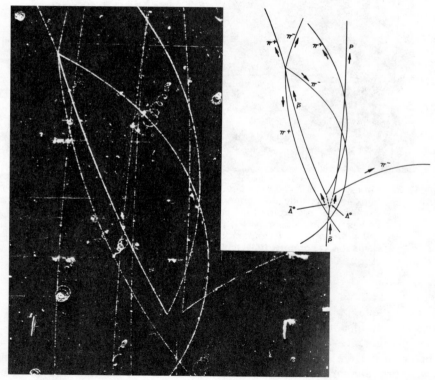

Figure 18-28 Photograph of the first production of an antilambda particle in the Alvarez 72-in. hydrogen bubble chamber. The incident antiproton has a momentum of 1.6 Gev/c and enters from the bottom of the photograph. The identification of the tracks of the particles is shown in the line drawing. The antiproton from the decay of the antilambda annihilated with a proton in the chamber to produce four charged pions. [G. R. Lynch, *Revs. Mod. Phys.*, **33**, 395 (1961). Photograph courtesy of Lawrence Radiation Laboratory, Berkeley, California.]

instance is illustrated in Figure 18-28 and the accompanying drawing. The antiproton enters the 72-in. hydrogen bubble chamber and interacts with a proton to form a lambda-antilambda pair—that is,

$$\bar{p} + p \rightarrow \Lambda + \bar{\Lambda}$$

Since the lambda particles are electrically neutral, they produce no visible tracks; they decay with a mean lifetime of 2.52×10^{-10} sec, yielding

$$\Lambda \rightarrow p + \pi^-$$

and

$$\bar{\Lambda} \rightarrow \bar{p} + \pi^+$$

The decay products form visible tracks; the sign of the charge and the momentum of each particle can be found from the curvature of its track and the known magnetic field in which the bubble chamber is situated. It will be observed that the antiproton formed in the decay of the antilambda is annihilated with a proton to produce four charged pions—that is,

$$\bar{p} + p \rightarrow \pi^+ + \pi^+ + \pi^- + \pi^-$$

It is interesting to speculate on the existence of antiatoms; that is, atoms whose nuclei are composed of antinucleons and whose electronic structure consists of shells of positrons. Such antimatter undoubtedly exists but must constitute a minute fraction of the mass of our universe. Antideuterons have been observed in experiments with proton synchrotrons.

18-15   The Classification of Particles

The particles that we have considered so far have been listed in Table 18-5, along with some of their properties. We have already developed the beginning of a classification scheme for these particles in that we have grouped them into leptons, mesons, and baryons. Within these large categories we have also recognized that some of the particles are related; our recognition consists of using the same Greek letter to designate, for example, the π^+, π^0, and π^- particles. The basic idea involved in grouping the pions together goes by various equivalent names: charge independence, isotopic spin, isobaric spin, and isospin or I-spin.

The concept of I-spin applies only to strongly interacting particles (i.e., to mesons and baryons). The assumption is made that if only the *strong* interaction were present, the pions would be completely alike. If the *electromagnetic* interaction is turned on, the mass of the neutral pion becomes different from that of the charged pions and the lifetimes become different as well. A similar statement applies to nucleons; in the absence of electromagnetic effects the neutron and the proton are indistinguishable. The addition of the electromagnetic interaction causes the mass difference of about 2 MeV, as well as differences in the lifetimes and magnetic moments. The situation is similar to Zee-

TABLE 18-5   PARTICLES WHICH DO NOT DECAY STRONGLY

Type	Particle	Spin in Units of $\hbar$	Mass (in MeV)	Mean Lifetime (in sec)
Photon	γ	1	0	Stable
Leptons	$\nu_e \bar{\nu}_e$	$\frac{1}{2}$	$0 (< 2 \times 10^{-4})$	Stable
	$\nu_\mu \bar{\nu}_\mu$	$\frac{1}{2}$	$0 (< 1.6)$	Stable
	$e^+ e^-$	$\frac{1}{2}$	0.511006	Stable ($> 2 \times 10^{21}$ yr)
	$\mu^+ \mu^-$	$\frac{1}{2}$	105.659	2.20×10^{-6}
Mesons	π^0	0	134.97	0.89×10^{-16}
	$\pi^+ \pi^-$	0	139.58	2.60×10^{-8}
	$K^+ K^-$	0	493.8	1.23×10^{-8}
	$K^0 \overline{K}^0$	0	497.8	K^0_{short}: 0.86×10^{-10} K^0_{long}: 5.3×10^{-8}
	η	0	549	$\sim 10^{-19}$
Baryons	p	$\frac{1}{2}$	938.26	Stable ($> 2 \times 10^{28}$ yr)
	n	$\frac{1}{2}$	939.55	0.96×10^3
	Λ	$\frac{1}{2}$	1115.6	2.5×10^{-10}
	Σ^+	$\frac{1}{2}$	1189.4	0.81×10^{-10}
	Σ^0	$\frac{1}{2}$	1192.6	$< 1.0 \times 10^{-14}$
	Σ^-	$\frac{1}{2}$	1197.4	1.6×10^{-10}
	Ξ^0	$\frac{1}{2}$	1315	3.0×10^{-10}
	Ξ^-	$\frac{1}{2}$	1321.3	1.7×10^{-10}
	Ω^-	$\frac{3}{2}$?	1672	1.3×10^{-10}

man splitting of a level in atomic spectra—in the absence of a magnetic field, several magnetic states have the same energy; application of the magnetic field causes a splitting of the level. Perhaps an even better analogy is provided by the fine structure in atomic spectra in which we cannot control experimentally the splitting of the levels. We must be content with surmising what the energy-level diagram might be if we could eliminate the spin-orbit effect that causes the splitting. In the study of fundamental particles the electromagnetic force cannot be turned off, and we must deal with multiplets of particles in which the masses (i.e., rest energies) differ. We therefore assume that the variously charged states are really substates of a single level which has been split.

To describe the charge splitting, a new pair of quantum numbers is introduced: the I-spin T and its projection onto a special axis T_3. These quantum numbers are quite analogous to the angular momentum J and its projection m_J, respectively. Each set of particles (or multiplet) of I-spin T possesses $(2T + 1)$ charged states. Thus we consider the nucleon to be an I-spin doublet, with $T = \frac{1}{2}$. The proton has $T_3 = \frac{1}{2}$, the neutron has $T_3 = -\frac{1}{2}$. Table 18-6 contains the I-spin assignments for those particles that were listed in Table 18-5.

TABLE 18-6   QUANTUM NUMBERS OF MESONS AND BARYONS

Particle	I-spin T	T_3	Baryon Number B	Strangeness S	Charge Q
π^+	1	$+1$	0	0	$+1$
π^0	1	0	0	0	0
π^-	1	-1	0	0	-1
K^+	$\frac{1}{2}$	$+\frac{1}{2}$	0	$+1$	$+1$
K^0	$\frac{1}{2}$	$-\frac{1}{2}$	0	$+1$	0
$\overline{K^0}$	$\frac{1}{2}$	$+\frac{1}{2}$	0	-1	0
K^-	$\frac{1}{2}$	$-\frac{1}{2}$	0	-1	-1
η	0	0	0	0	0
p	$\frac{1}{2}$	$+\frac{1}{2}$	$+1$	0	$+1$
n	$\frac{1}{2}$	$-\frac{1}{2}$	$+1$	0	0
$\bar{p}$	$\frac{1}{2}$	$-\frac{1}{2}$	-1	0	-1
$\bar{n}$	$\frac{1}{2}$	$+\frac{1}{2}$	-1	0	0
Λ	0	0	$+1$	-1	0
$\bar{\Lambda}$	0	0	-1	$+1$	0
Σ^+	1	$+1$	$+1$	-1	$+1$
Σ^0	1	0	$+1$	-1	0
Σ^-	1	-1	$+1$	-1	-1
Ξ^0	$\frac{1}{2}$	$+\frac{1}{2}$	$+1$	-2	0
Ξ^-	$\frac{1}{2}$	$-\frac{1}{2}$	$+1$	-2	-1
Ω^-	0	0	$+1$	-3	-1

Soon after the K mesons and the hyperons were discovered it became apparent that these new particles were produced in pairs; for instance, no examples of the following processes were observed:

$$p + p \not\rightarrow K^+ + n + p$$
$$\pi^- + p \not\rightarrow \Lambda + \pi^0 \qquad (18\text{-}29)$$
$$\pi^+ + p \not\rightarrow \Sigma^+ + \pi^+$$

Another peculiarity of K mesons and hyperons is that although they can decay into strongly interacting particles, the decay processes are weak. In order to explain these results Gell-Mann and Nishijima (1953) introduced an additional quantum number, known as strangeness (S). This quantity is conserved for strong and electromagnetic interactions but can change by one unit in weak interactions. Table 18-6 includes the values of S for various mesons and baryons. It is instructive to review the strong processes listed in Sections

18-12 and 18-13 for production of K mesons and hyperons; in each case it will be seen that the total strangeness before the scattering is equal to the total strangeness afterward.

Included in the strangeness scheme is the Gell-Mann-Nishijima formula

$$Q = T_3 + \frac{B + S}{2} \tag{18-30}$$

which connects the quantum numbers listed in Table 18-6; Q is the electric charge of the particle in units of e, the charge of the electron. This generalization has several important consequences which agree with experiment, which are now accepted as correct but which were new when the strangeness scheme was first proposed.

The K^+ meson, with $Q = +1$ and $B = 0$, must have $S = +1$ in order to be produced along with Λ and Σ hyperons. The formula (18-30) then requires $T_3 = +\frac{1}{2}$, so that the K^+ is part of a doublet, rather than a triplet, in I-spin. The temptation to form a triplet from K^+, K^0, and K^- must be rejected in favor of a pair of doublets, as shown in the table. Therefore the K^0 must be distinct from the $\overline{K^0}$; this prediction of the existence of two neutral K mesons marked one of the early triumphs of the strangeness scheme.

A second example involves the Ξ hyperon. It is produced from nonstrange particles in association with two K mesons, indicating that its strangeness is -2. Application of the Gell-Mann-Nishijima formula (18-30) shows that Ξ^0 and Ξ^- are members of a doublet rather than a triplet. This result agrees with the experimentally proved absence of the Ξ^+.

18-16 Selection Rules

The interactions of fundamental particles, like nuclear reactions or electronic transitions in atoms, obey various selection rules, which in many cases are simply conservation laws. We are already quite familiar with the laws expressing conservation of energy and momentum; these laws are valid, even at the highest energies and shortest distances at which measurements have been made. Electrical charge is another quantity that is conserved for all interactions. The absolute stability of the proton suggests that the baryon number is a conserved quantity.

Other physical quantities exist which are conserved in some processes but not in others, as shown in Table 18-7. We have already discussed at some length the fact that parity is not conserved in weak interactions. Similarly, I-spin projections and strangeness are changed by weak interactions. The total I-spin is conserved only in strong interactions. The fact that masses differ within an I-spin multiplet is due to the fact that electromagnetic interactions do not conserve I-spin.

Some physical quantities are not conserved by any of the interactions. An example is the magnetic dipole moment.

Certain selection rules describe in more detail the way in which a conserva-

TABLE 18-7   VALIDITY OF CONSERVATION LAWS

Quantity	Type of Interaction		
	Strong	Electromagnetic	Weak
Energy	Yes	Yes	Yes
Linear momentum	Yes	Yes	Yes
Angular momentum			
(**J**)	Yes	Yes	Yes
Parity	Yes	Yes	No
Baryon number (B)	Yes	Yes	Yes
Total I-spin (**T**)	Yes	No	No
T_3	Yes	Yes	No
Strangeness (S)	Yes	Yes	No
Charge (Q)	Yes	Yes	Yes
Magnetic moment	No	No	No

tion law breaks down; for instance, strangeness is not conserved in weak interactions, but it never seems to change by more than one unit in a single weak process, leading to the selection rule $|\Delta S| = 0, 1$ for weak interactions. For this reason the Ξ hyperon does not decay into a nucleon and a pion; rather it undergoes a cascade (see Section 18-13), changing strangeness by one unit each time.

18-17   Resonances

In the physics of fundamental particles resonant scattering can occur in a fashion similar to that which has already been discussed for alpha particles and for gamma rays at low energy. If the cross section for a scattering process is plotted as a function of energy, a resonance will typically show up as a peak, with the maximum occurring near the resonant energy.

The earliest example of a high-energy resonance was discovered by Fermi and his co-workers at the University of Chicago (1952). They used a synchro-cyclotron to produce a beam of protons which was directed onto a target to produce a secondary beam of pions. The pions passed into a target of liquid hydrogen, where they could collide with protons. Scintillation counters were used to detect both the incident beam and the scattered pions. The ratio of scattered pions to incident pions was measured as a function of the energy of the incident pions. This experiment and others in the same energy region yielded results similar to those shown in Figure 18-29. Measurements were performed on both positive and negative pions, and in both cases a pronounced peaking was observed at about 195 MeV kinetic energy.

The best interpretation of the concept of "resonance" in a scattering process is not immediately clear. The name implies a connection with oscillatory phenomena, as in the case of resonance in alternating-current circuits or in a driven harmonic oscillator. Indeed, the idea of resonance in scattering is best defined

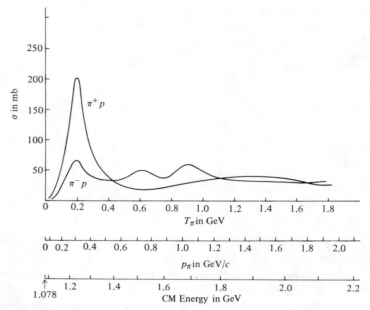

Figure 18-29   Cross sections for π^+p and π^-p scattering as a function of energy. The scale is linear in the laboratory kinetic energy of the incident pion (T_π).

with reference to the wave theory of matter (since a wave is an oscillatory phenomenon), but a particle model is still of value. A fruitful way of thinking about the 195-MeV resonance is to use the compound nucleus idea. When a pion with kinetic energy equal to about 195 MeV collides with a nucleon, it tends to form an intermediate compound state, which we shall call Δ. The Δ exists for a short time as a particle and then decays into a pion and a nucleon. The justification for using particlelike language to describe a wave phenomenon lies in the fact that a resonance such as the Δ has many properties in common with particles.

1. It has a definite mass. In the case of the Δ this mass is 1236 MeV and the resonance is often referred to as the Δ(1236). It might seem from looking at Figure 18-29 that the mass is not very definite, since the peak is rather broad. The width is about 120 MeV and is a result of the relation

$$\Delta \mathcal{E} \, \Delta t \gtrsim \hbar \qquad (18\text{-}31)$$

The short lifetime of the Δ(1236) leads to an uncertainty in a single measurement of its energy.

2. It has a definite lifetime. The Δ never lives long enough to travel a distance that can be measured directly, but the lifetime can be inferred from Equation (18-31); if we choose $\Delta \mathcal{E} \approx 120$ MeV, the lifetime Δt is of the order of 10^{-23} sec, a typical time interval associated with the strong interactions.

3. It has a definite set of quantum numbers. These are listed in Table 18-8, both for the Δ(1236) and for various other resonances that have been studied in

TABLE 18-8 PROPERTIES OF SELECTED RESONANCES IN HIGH-ENERGY PARTICLE PHYSICS

Meson Resonances	Mass (MeV)	Width (MeV)	I-Spin T	Baryon Number B	Strange-ness S	Spin J	Parity
$\rho(765)$	765	~120	1	0	0	1	−
$\omega(783)$	783	12	0	0	0	1	−
$f(1260)$	1264	145	0	0	0	2	+
$K^*(890)$	891	50	$\frac{1}{2}$	0	+1	1	−
$\overline{K}^*(890)$	891	50	$\frac{1}{2}$	0	−1	1	−
Baryon Resonances							
$\Delta(1236)$	1236	120	$\frac{3}{2}$	1	0	$\frac{3}{2}$	+
$N(1470)$	1470	210	$\frac{1}{2}$	1	0	$\frac{1}{2}$	+
$N(1518)$	1525	115	$\frac{1}{2}$	1	0	$\frac{3}{2}$	−
$N(1550)$	1550	130	$\frac{1}{2}$	1	0	$\frac{1}{2}$	−
$\Delta(1640)$	1640	180	$\frac{3}{2}$	1	0	$\frac{1}{2}$	−
$N(1680)$	1680	170	$\frac{1}{2}$	1	0	$\frac{5}{2}$	−
$N(1688)$	1688	130	$\frac{1}{2}$	1	0	$\frac{5}{2}$	+
$N(1710)$	1710	300	$\frac{1}{2}$	1	0	$\frac{1}{2}$	−
$\Delta(1950)$	1950	220	$\frac{3}{2}$	1	0	$\frac{7}{2}$	+
$\Lambda(1405)$	1405	40	0	1	−1	$\frac{1}{2}$	−
$\Lambda(1520)$	1519	16	0	1	−1	$\frac{3}{2}$	−
$\Sigma(1385)$	1382	36	1	1	−1	$\frac{3}{2}$	+
$\Xi(1530)$	1529	7	$\frac{1}{2}$	1	−2	$\frac{3}{2}$	+

detail. It will be noticed that the list is long; even so, it is not exhaustive. New resonances have been discovered in abundance in recent years; many of them are not well established and some of those whose existence can no longer be doubted are still controversial with regard to their correct masses and quantum numbers. The table attempts to present only those results that are not likely to change drastically as additional research is reported.

During the late 1950s the measurement of cross sections for pion-nucleon scattering was extended to energies beyond 200 MeV. Resonant peaks were found in the π^- cross section at energies of 600 and 900 MeV, corresponding to masses of 1520 and 1680 MeV, respectively. No such peaks were observed in the π^+ cross sections at those energies. The conclusion can be drawn that these resonances correspond to I-spin $\frac{1}{2}$. Since the nucleon also has I-spin $\frac{1}{2}$, the symbol N is used to denote these resonances, with the masses in parentheses: $N(1518)$ and $N(1688)$. Soon after the discovery of the $N(1688)$ it was suspected that the peak in the cross section consists of more than one resonance. This conjecture was later proved to be correct; the peak corresponds to several resonances that are rather broad and cannot be resolved without recourse to rather sophisticated analysis. At still higher energy there is a reso-

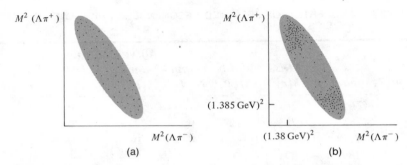

Figure 18-30 Dalitz plots for the reaction $K^- + p \rightarrow \Lambda + \pi^+ + \pi^-$: (a) appearance if no resonances are present; (b) actual results.

nance in the π^+-proton system. The third component of I-spin is $+\frac{3}{2}$ for this system, implying that $T \neq \frac{1}{2}$. Since the pion has $T = 1$ and the nucleon has $T = \frac{1}{2}$, the only possibilities are $\frac{1}{2}$ and $\frac{3}{2}$ for the I-spin of a pion-nucleon system. Therefore this resonance has $T = \frac{3}{2}$; it is called $\Delta(1950)$. The symbol N^* is sometimes used to denote these resonances in the pion-nucleon system, the star denoting an excited state.

In 1960 an experiment performed by the Alvarez group at Berkeley (using a hydrogen bubble chamber) led to the discovery of the first example of a resonance that involves a hyperon. The reaction studied was

$$K^- + p \rightarrow \Lambda + \pi^+ + \pi^-$$

For each event the effective mass of the Λ combined with one of the pions can be calculated from

$$M^2(\Lambda\pi) = (\mathcal{E}_\Lambda + \mathcal{E}_\pi)^2 - (\mathbf{p}_\Lambda + \mathbf{p}_\pi)^2 \qquad (18\text{-}32)$$

Since there are two pions, two such masses can be calculated for each event. An analysis by Dalitz shows that in the absence of resonances a scatter plot of $M^2(\Lambda\pi^+)$ versus $M^2(\Lambda\pi^-)$ ought to be uniformly populated inside an oval region, as shown in Figure 18-30a. The experimental result is that nearly every event has one or the other combined mass close to 1385 MeV, as seen in Figure 18-30b. The conclusion is that there is a resonance in the $\Lambda\pi$ system at a mass of about 1385 MeV. The symbol Y was often used for hyperons (Λ, Σ, and Ξ), so the resonance was originally called the Y^*. The modern name is $\Sigma(1385)$; the symbol Σ reminds us that $T = 1$ for this resonance; the Λ has $T = 0$ and the pion has $T = 1$; just as in the case of adding ordinary angular momentum, the only possible result of combining zero with one is $T = 1$.

The year 1961 saw a large increase in the number of resonances. In a study of the processes

$$K^- + p \rightarrow \Sigma^\pm + \pi^\mp + \pi^+ + \pi^- \qquad (18\text{-}33)$$

the Alvarez group reported that the Σ often resonates with one of the pions in a way analogous to the $\Lambda\pi$ resonance. This resonance was found to have $T = 0$ and was called the Y_0^* to distinguish it from the previously discovered Y^*

(which was then called Y_1^*). The modern name for the $\Sigma\pi$ resonance is $\Lambda(1405)$. In the same year the same group of physicists studied the reaction

$$K^- + p \rightarrow \overline{K}^0 + \pi^- + p \qquad (18\text{-}34)$$

using a Dalitz-plot analysis similar to the one used in the discovery of the Σ (1385). It was found that the effective mass of the $\overline{K}^0\pi^-$ system was near 890 MeV far more often than expected. The interpretation of this clustering is that the reaction proceeds in two steps:

$$K^- + p \rightarrow \overline{K}^{*-} + p, \qquad \overline{K}^{*-} \rightarrow \overline{K}^0 + \pi^- \qquad (18\text{-}35)$$

Both the production and the decay are strong processes. The idea of I-spin suggests that the decay

$$\overline{K}^{*-} \rightarrow K^- + \pi^0 \qquad (18\text{-}36)$$

should also be observed. Subsequent experiments have proved that it does occur.

As early as 1956 R. Hofstadter and others at Stanford studied the scattering of electrons by protons and neutrons. For proton targets at small scattering angles the results agree with those of Rutherford scattering; but for large angles (corresponding to small impact parameters and to large values of the momentum transferred to the nucleon) the results depart from the Rutherford formula. This fact is not surprising because the Rutherford formula is derived with the assumption that the target is a point charge. So the deviation from Rutherford scattering can be interpreted as being due to an extended charge distribution or, equivalently, that the nucleon has structure. Detailed analysis of the Stanford results led to theoretical speculation that there should exist vector mesons with masses somewhat greater than that of the pion (vector particles have spin 1 and odd parity). The prediction included vector mesons with $T = 0$ and $T = 1$. In 1961 both kinds were discovered experimentally. Walker and his co-workers at Wisconsin studied the reactions

$$\pi^- + p \rightarrow \pi^- + \pi^0 + p \qquad (18\text{-}37)$$

$$\pi^- + p \rightarrow \pi^- + \pi^+ + n \qquad (18\text{-}38)$$

which occurred in a hydrogen bubble chamber at Brookhaven National Laboratory. For both reactions the number of events, plotted as a function of the effective mass of the two-pion system, showed a large peak at about 765 MeV. This effect is known as the ρ meson. The decay

$$\rho \rightarrow \pi + \pi \qquad (18\text{-}39)$$

has been studied extensively and experiment has shown that the decay proceeds as a P-wave ($l = 1$). Since the pions have no intrinsic spin, the conclusion is that the spin of the ρ is 1 and that it has odd parity. The I-spin of the ρ is also equal to 1. The vector meson with $T = 0$ was discovered by the Alvarez group at Berkeley. They used a beam of antiprotons incident on the protons in the 72-in. hydrogen bubble chamber. They measured the events for which four charged particles left the interaction. Using a computer, they isolated those measurements which were examples of the five-pion annihilation process

$$\bar{p} + p \rightarrow \pi^+ + \pi^+ + \pi^- + \pi^- + \pi^0 \tag{18-40}$$

They calculated effective masses for the various combinations of pions and plotted histograms for each combination. For most of the graphs the results showed no surprising features, but the plot showing the number of events observed as a function of $M(\pi^+\pi^-\pi^0)$ yielded a peak at about 785 MeV. This effect was named the omega (ω) meson. Subsequent analysis (consisting of an extension of the Dalitz-plot technique) showed that the ω, like the ρ, has spin 1 and odd parity. It is a resonant state that decays strongly.

Late in 1961 Pevsner and his co-workers at Johns Hopkins reported on their experiment which studied the interactions of positive pions in a deuterium bubble chamber. The reaction was

$$\pi^+ + d \rightarrow p + p + \pi^+ + \pi^- + \pi^0 \tag{18-41}$$

The spectrum of the quantity $M(\pi^+ \pi^- \pi^0)$ was plotted. As expected, the ω meson appeared in the form of a peak near 785 MeV. A smaller peak was also observed at 550 MeV and was named the eta (η) meson. At first it was believed that the η is a resonance, similar to the ρ and the ω; several later findings combined to change this idea. As experiments with good resolution and high statistics examined the η, it was found that the peak is much narrower than those belonging to the various resonances. The conclusion to be drawn from a narrow peak is long lifetime, suggesting that the decay of the η is not a strong process. Further evidence was added by the discovery of the process

$$\eta \rightarrow \gamma + \gamma \tag{18-42}$$

This decay mode is actually more probable than the mode in which the η was discovered:

$$\eta \rightarrow \pi^+ + \pi^- + \pi^0 \tag{18-43}$$

The two-photon decay, reminiscent of π^0 decay, is a typical electromagnetic process and is evidence that the η is not a resonance, since photons cannot interact strongly.

In the year 1962 many additional resonances were discovered. The methods used were similar to those that we have already described, so we shall omit the details and simply list a few of the strong decay processes that were observed:

$$\Lambda\,(1520) \rightarrow \Sigma + \pi$$

$$\phi \rightarrow K + \overline{K}$$

$$f \rightarrow \pi^+ + \pi^-$$

$$\Xi\,(1530) \rightarrow \Xi + \pi$$

The years since 1962 have led to the discovery of resonant states in such large numbers that their classification has become one of the most important problems in high-energy nuclear physics. The following section describes a scheme for classifying these states (resonances and/or particles).

18-18   Unitary Symmetry

The notion of I-spin, discussed in Section 18-15, has been useful in classifying particles and resonances. Gell-Mann and Ne'eman (1961) independently obtained an extension of the I-spin theory which has been successful in that it points to further regularities among the many states that we have listed in Tables 18-6 and 18-8. This new theory is referred to as SU(3).

To understand how SU(3) applies to particles it is necessary to re-examine the concept of I-spin. The mathematics of I-spin is identical to that which describes angular momentum in quantum mechanics. The algebraic structure is defined by the properties of 2×2 matrices which are unitary and which have a determinant equal to $+1$. This group of matrices is called SU(2), an abbreviation for *special unitary*. It is not necessary to understand the details of the theory in order to appreciate the results, so we shall omit the mathematics. In SU(2), multiplets exist which can be portrayed by weight diagrams like those in Figure 18-31. Each multiplet is characterized by a quantum number J, the total angular momentum. The multiplet consists of $2J + 1$ substates which correspond to equally spaced values of J_z, going from $J_z = -J$ to $J_z = +J$ with no states missing. We have had numerous examples of this symmetric arrangement, both in atomic spectra and in I-spin.

The SU(3) theory extends the algebraic structure to 3×3 matrices which are unitary and which have determinants equal to $+1$. Again, multiplets exist. They are often called *supermultiplets* because they contain SU(2) multiplets. In SU(2) only one number was needed to specify a multiplet; in SU(3) two such numbers are needed to serve the purpose which J served for angular momentum. We shall use λ_1 and λ_2 to denote these numbers. The number of substates in a multiplet is given by the expression

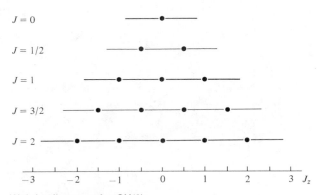

Figure 18-31   Weight diagrams for SU(2).

$$(\lambda_1 + 1)\,(\lambda_2 + 1)\left(\frac{\lambda_1 + \lambda_2}{2} + 1\right) \tag{18-44}$$

Weight diagrams for SU(3) are two-dimensional; the coordinates refer to two quantum numbers, one of which is T_3 (the third component of the I-spin) and the other is Y (the *hypercharge* defined by the sum of the strangeness and the baryon number, $Y = S + B$). Points on these weight diagrams must lie on a triangular lattice, as in Chinese checkers. The permissible diagrams are hexagons, triangles, and single points. Examples appear in Figure 18-32. The quantity $(\lambda_1 + 1)$ is the number of points in the topmost side of a hexagon and $(\lambda_2 + 1)$ is the number of points in the bottom side. Each figure is symmetric with respect to rotations through 120 degrees and with respect to reflection about the line $T_3 = 0$. An important difference between SU(3) weight diagrams and those of SU(2) is that the SU(3) multiplets may contain lattice sites with more than one state. This fact appears in two of the multiplets of Figure 18-32, in which a dot within a circle denotes two states at the same values of T_3 and Y. This doubling (or tripling or more) occurs in a definite way: the lattice points on the margin of a hexagon are single points; the points next within are double, and so on, increasing until the hexagon becomes a triangle or a point; within a triangular figure, no increase occurs. In two respects the SU(3) weight diagrams resemble those of SU(2): first, no interior points can be absent; second, the "center-of-mass" of the diagram is located at the origin.

A subsidiary condition seems to be necessary in addition to the rules already specified for weight diagrams. Only those SU(3) multiplets can exist in nature for which a state is located at the origin. The corresponding rule for SU(2)

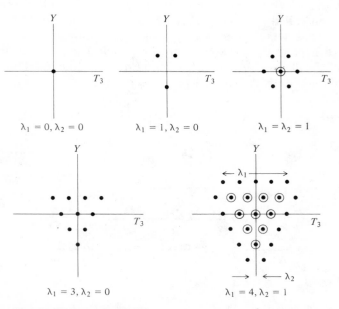

Figure 18-32 Weight diagrams for SU(3).

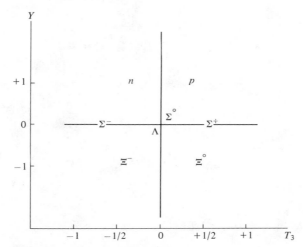

Figure 18-33 The octet of baryons with spin 1/2.

would have the effect of requiring all angular momenta to be integers, and, indeed, the *orbital* angular momentum is an example of a physical quality which is described by SU(2) with this extra restriction. Introduction of spin $\frac{1}{2}$ particles leads to the realization of all possible multiplets of SU(2) in nature. Various theories have been suggested which include all multiplets of SU(3), but at this writing no evidence has been found for the existence in nature of particles that must be described by a model lacking a state at the origin.

So far the SU(3) idea may appear rather abstract. Its utility in physics results from the fact that an SU(3) multiplet can be interpreted as a single type of particle, or a single level, which splits when interactions are applied. Large numbers of particles can be grouped together in these multiplets. The first example to be recognized was that of the baryons with spin $\frac{1}{2}$ and even parity: the nucleon, the Λ, Σ, and Ξ particles. Figure 18-33 shows how these particles fit into an octet of states in SU(3). In this grouping both the Λ and the Σ^0 are situated at the origin. The interpretation in terms of splitting of levels is that in the absence of all interactions all eight baryons would be identical (same spin, same mass, etc.). The strong interactions split this baryonic level into four sublevels: N, Λ, Σ, and Ξ. The electromagnetic interaction causes these four levels to split further, as we have already seen in our study of I-spin.

Groupings of other particles are possible. The mesons with zero spin and odd parity (π, K, $\overline{K}$, η) form an octet, as seen in Figure 18-34. It was one of the early successes of SU(3) that this grouping was recognized before the η meson was observed experimentally. Another grouping of mesons involves those with spin 1 and odd parity (ρ^+, ρ^0, ρ^-, K^{*+}, K^{*0}, $\overline{K}^{*0}$, K^{*-}, ω, ϕ). These nine resonant vector mesons are grouped into an octet and a singlet.

The multiplet which represents the most important success of the SU(3) theory is the one that contains the baryons with spin $\frac{3}{2}$ and even parity. Experiments showed that the $\Delta(1236)$, the $\Sigma(1385)$, and the $\Xi(1530)$ all have spin-parity $\frac{3}{2}^+$. They could not be included in an octet because the I-spin of the

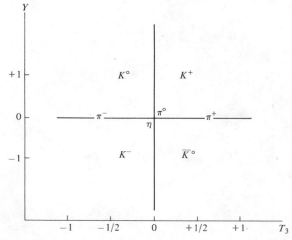

Figure 18-34 The octet of mesons with spin zero.

$\Delta(1236)$ is too large. However, the multiplet with $\lambda_1 = 3$ and $\lambda_2 = 0$, a decuplet, could accommodate these particles if in addition a particle existed for which $T_3 = 0$ and $Y = 2$. The situation is shown in Figure 18-35, in which the Ω^- is the particle that was lacking. The Ω^- has some rather interesting properties. Its baryon number is $+1$ and its hypercharge is -2, implying that its strangeness is -3. Gell-Mann and Okubo predicted its mass by showing that the masses of members of a decuplet should be linear in the hypercharge. The value of this mass, about 1670 MeV, is small enough that the obvious strong decay processes

$$\Omega^- \nrightarrow \Xi^0 + K^-$$

or

$$\Omega^- \nrightarrow \Xi^- + \overline{K}^0$$

cannot occur because of conservation of energy and momentum. In fact, the only way that the Ω^- can decay is by way of the weak interaction:

$$\Omega^- \rightarrow \Xi^- + \pi^0$$

$$\Omega^- \rightarrow \Xi^0 + \pi^-$$

$$\Omega^- \rightarrow \Lambda + K^-$$

It is remarkable that the Ω^- is not a resonance, although it is classified with strongly decaying resonant states in the SU(3) scheme. The distinction between the concepts of "particle" and "resonance" is not so clear when considered in this light.

After a large experimental effort the Ω^- was observed (1964) by a group of physicists at Brookhaven National Laboratory. The experiment consisted of exposing the 80-in. hydrogen bubble chamber to a beam of K^- mesons at a momentum of about 5 GeV/c. Figure 18-36 shows a photograph of the first

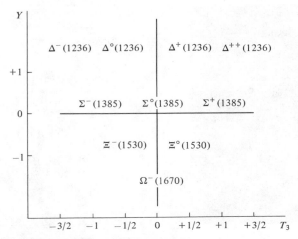

Figure 18-35   The decuplet of baryons with spin 3/2.

Ω^- event. The reactions are

$$K^- + p \to \Omega^- + K^+ + K^0$$

$$\Omega^- \to \Xi^0 + \pi^-$$

$$\Xi^0 \to \Lambda + \pi^0$$

$$\pi^0 \to \gamma + \gamma$$

$$\Lambda \to p + \pi^-$$

Both photons from the π^0 decay underwent pair production in the bubble chamber, aiding considerably in the identification of the event. In subsequent

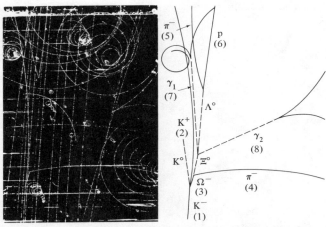

Figure 18-36   The first experimental evidence for the existence of the Ω^-. (Courtesy of Brookhaven National Laboratory.)

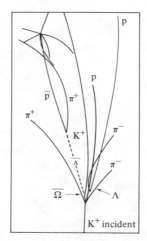

Figure 18-37 Production of the $\overline{\Omega}$ in a deuterium bubble chamber. [Photograph by A. Firestone, G. Goldhaber, D. Lissauer, B. Sheldon, and G. Trilling, *Phys. Rev. Letters*, **26**, 410 (1971).]

months other examples of the Ω^- were found, which provided ample proof of its existence and lending credence to SU(3), the theory that had predicted the particle and its properties.

The concepts of classification using SU(3) are readily extended to include antibaryons (see Problem 18-30). We have already noted that $\overline{\Lambda}$ particles have been observed. The other antibaryons that are stable against strong decay have been observed also, culminating with the discovery (1971) of the $\overline{\Omega}$. The experiment was performed by A. Firestone, Gerson Goldhaber, D. Lissauer, B. M. Sheldon, and G. H. Trilling from the University of California at Berkeley. They used a beam of K^+ mesons at a momentum of 12 GeV/c to bombard deuterons. The target-detector was the 82-inch bubble chamber located at the Stanford Linear Accelerator Center. The $\overline{\Omega}$ event is shown in Figure 18-37. The reactions are

$$K^+ + d \rightarrow \overline{\Omega} + \Lambda + \Lambda + \pi^+ + \pi^- + p$$

$$\overline{\Omega} \rightarrow \overline{\Lambda} + K^+$$

$$\overline{\Lambda} \rightarrow \overline{p} + \pi^+$$

$$\Lambda \rightarrow p + \pi^-$$

The $\overline{p}$ from the $\overline{\Lambda}$ decay encounters a deuteron and forms an annihilation star.

18-19 Frontiers

In this section we shall attempt to point out a few areas in which new developments can be expected. The progress made so far in nuclear and high-energy physics has left unanswered some basic questions, which research workers are trying to answer. One reason for posing these questions here is to alert the student to the circumstance that our understanding of nature is far from being complete, in spite of the volume of facts already learned. As the coming years bring solutions to some of the puzzles listed here, it is hoped that the difficulty of the problems will be appreciated.

What is the correct classification scheme for particles? We have already discussed I-spin and SU(3) as classification techniques, and we have seen that I-spin ideas are completely contained in SU(3). Perhaps there exist other and better symmetry techniques that will contain or even replace SU(3). Efforts made so far along this line have met with only partial success.

How many fundamental particles are there? The list of particles (if resonances are included) has grown to a great extent. Will more new particles and resonances be discovered or are we near the end of the list? Do particles exist with properties radically different from those possessed by particles already discovered? Searches have been made for particles with magnetic monopole moments (i.e., isolated magnetic poles), with speeds greater than c, or with charges not equal to an integer multiple of e. The results to date have set upper limits on the formation rates and lower limits on the masses for such particles. The completion of accelerators with energies higher than those now available may see the discovery of new types of particle or they may lead to changes in the limits on masses and rates.

Which particles (if any) are really fundamental? The list of particles is already so lengthy that it seems unlikely that all of them can be truly fundamental. Perhaps only a small subset of particles is fundamental and the rest of them are composite. If this idea is true, then which particles are fundamental and which are composite? It has been conjectured that all particles are composites of all the rest; this idea is called "nuclear democracy." It is difficult to think of an experimental test that can prove or disprove this hypothesis that "all particles are equal."

What is the purpose of leptons? The classification schemes considered so far apply only to strongly interacting particles. A similar question is: what is the basic difference between the electron and the muon? They have different masses; all the rest of their properties that differ seem to be consequences of this mass difference. But is there a deeper, more fundamental difference?

Why do the selection rules for strong, electromagnetic, and weak interactions have the form that we observe? It would be desirable to possess a theory that is capable of deriving selection rules, for example, the law of conservation of strangeness, from mathematical assumptions rather than inferring them from observation. In fact a dynamical theory of strong interactions is what is needed. Four forces were listed at the beginning of this chapter. Are there others? If not, why not?

In the field of nuclear structure there are almost certain to be some new results that come from bombardment of nuclei with π mesons and K mesons. This branch of nuclear physics may provide clues to a better understanding of nuclei. Certain theories of nuclear structure predict regions of stable nuclei for values of Z and A which are much larger than those observed so far. Do stable "superheavy" nuclei exist? Experiments in which uranium nuclei are used as both target and projectile might yield an affirmative answer to this question and raise other questions in turn.

Toward answering these and other questions, new equipment and new ideas are needed. The development of new apparatus will extend the ideas presented in Chapters 12 and 13. One technique that is being investigated is the use of superconducting coils for magnets used in accelerators, beam transport, and detectors. Are there other devices for detection and identification of particles? Can neutral particles be detected with better efficiency? Can accelerators be built at higher energy without exorbitant cost?

A clear-thinking scientist must be alert to new approaches and must therefore pose one additional query: Have we been asking the right questions?

Problems

18-1. (a) Calculate the momentum of a proton whose kinetic energy is 400 MeV. (b) Determine the velocity of this proton.

18-2. Determine (a) the velocity and (b) the momentum of a muon whose kinetic energy is 4 MeV. (c) Determine the radius of curvature of the path of this particle in a magnetic induction of 5 kg.

18-3. Calculate (a) the velocity and (b) the momentum of a charged pion whose kinetic energy is 10 MeV. (c) Calculate the radius of curvature of its path in a magnetic induction of 6 kg.

18-4. A gamma-ray photon incident on a hydrogen target produces a π^+ meson that moves off with an initial kinetic energy of 70 MeV in a direction at 90° to the direction of the incident photon. (a) Determine the momentum of the meson. (b) Determine the energy of the incident photon.

18-5. (a) Show that the kinetic energy $\mathscr{E}_k$ of a charged particle of mass m moving in a magnetic induction of intensity B in a circular path of radius R satisfies the equation

$$\mathscr{E}_k \left(1 + \frac{\mathscr{E}_k}{2mc^2} \right) = \frac{B^2 e^2 R^2}{2m}$$

where all quantities are in mks units. (b) Calculate the radius of curvature of the path of a π^+ meson in a magnetic induction of 15 kg when its kinetic energy is 14 MeV. (c) Calculate the radius of curvature of a muon in the same magnetic field when its kinetic energy is 4 MeV.

18-6. Suppose that a neutral particle decays into two charged particles which make an angle of 40° with each other. The measured kinetic energy of one particle is 191 MeV and its momentum is 300 MeV/c; the kinetic energy of the second particle is 870 MeV and its momentum is 1000 MeV/c. Identify all three particles.

18-7.   Assuming the nucleus to be a point charge, (a) calculate the radii of the muonic Bohr orbits of titanium ($Z = 22$) for quantum numbers $n = 1$ and $n = 2$. (b) Compute the energy radiated from such an atom in transitions between these orbits.

18-8.   Assuming the nucleus to be a point charge, (a) calculate the radii of the π-mesic Bohr orbits of beryllium ($Z = 4$) for quantum numbers $n = 1$ and 2. (b) Compute the energy radiated from this atom in a transition between these orbits.

18-9.   Perform the calculations of the preceding problem for K-mesic beryllium.

18-10. Assuming that the nucleus is a point charge, calculate the radius of the muonic $1s$ orbit for ^{208}Pb. Compare this value with that of the radius of the ^{208}Pb nucleus.

18-11. For what nuclide does the nuclear radius coincide with the radius of the $1s$ muonic Bohr orbit?

18-12. In about 0.012 percent of the cases a charged pion decays directly into an electron. (a) Write the reaction equation. (b) Determine the energy released in this reaction.

18-13. Calculate the Q value for the decay of a K^+ meson into (a) two pions, (b) three pions, and (c) a muon and a neutrino.

18-14. Determine the laboratory threshold values for the production of antiprotons (a) in $p - p$ and (b) in $\pi - p$ interactions.

18-15. The cross section for neutrino interactions in lead is about 10^{-40} cm². What is the mean free path for neutrinos in lead? Express your answer in centimeters and in light years.

18-16. Calculate the momentum and kinetic energy of a muon which is produced by a pion that decays at rest.

18-17. Use conservation of momentum and energy to show that the minimum and maximum energies of electrons from muons (decaying at rest) are 0 and 53 MeV.

18-18. Find the minimum energy that can be carried away by the neutrino-antineutrino system in muon decay.

18-19. Show that the minimum photon energy for pair production near an electron is $4m_e c^2$.

18-20. What is the ionization energy for the positronium atom?

18-21. Prove that the wavelength of each photon produced from the annihilation of ^1S positronium is equal to the Compton wavelength of the electron.

18-22. Negative pions with momentum of 1 GeV/c enter a hydrogen bubble chamber. (a) What is the kinetic energy of a pion with this momentum? (b) At this momentum a large number of scatterings in hydrogen produce two charged secondaries. Some of these correspond to elastic scattering. What are some inelastic processes that can occur?

18-23. Which upper-case letters of the Latin alphabet remain unchanged under the operation of parity?

18-24. Which upper-case letters of the Greek alphabet are invariant under parity?

18-25. Given the following table, derive the rest of the information in Table 18-2.

Quantity	C	P	T
Mass	+	+	+
Time	+	+	−
Displacement	+	−	+
Electric current	−	+	−

18-26. How does a magnetic dipole moment transform under C, P, and T?

18-27. In the cobalt 60 experiment let us assume that **p** (the momentum of the decay electron), **H** (the applied magnetic field), **q** (the momentum of the recoiling nucleus), and μ (the magnetic dipole moment of the recoil) are all measurable. Which of the symmetries C, P, T, and CPT can be tested by a measurement of (a) $\mu \cdot \mathbf{p} \times \mathbf{q}$; (b) $\mu \cdot \mathbf{H} \times \mathbf{p}$?

18-28. Which of the symmetries C, P, T, and CPT can be tested by making an energy measurement?

18-29. How does helicity (see Section 14-13) behave under C, P, and T, respectively?

18-30. Evaluate the quantum numbers T, T_3, B, S, and Q for the $\overline{\Sigma}$ particles, for the $\overline{\Xi}$, and the $\overline{\Omega}$.

18-31. The first pion-nucleon resonance is observed for pions with a laboratory kinetic energy of 195 MeV. (a) Calculate the momentum of pions with this energy. (b) Calculate the total energy in the center-of-mass system for a pion ($\mathcal{E}_k = 195$ MeV) incident on a proton (at rest).

18-32. A resonance appears as a peak in a plot of cross section versus energy ($\mathcal{E}$). The peak is centered at a value $\mathcal{E}_0$ The energy dependence can be approximated by a form obtained by G. Breit and E. Wigner in which the cross section is proportional to $[(\mathcal{E} - \mathcal{E}_0)^2 + \Gamma^2/4]^{-1}$ For the rho-meson resonant state, $\mathcal{E}_0$ is about 765 MeV and Γ is about 120 MeV. (a) Sketch the energy dependence for this resonance. (b) Show that Γ is the full width at half-maximum for the peak. (c) Show that half the counts lie within $\Gamma/2$ of the value $\mathcal{E}_0$ for a Breit-Wigner shape, provided that $\mathcal{E}_0$ is much larger than $\mathcal{E}_t$, where $\mathcal{E}_t$ is the threshold.

18-33. What orbital wave is involved in each of the following decays: (a) $\Delta(1236) \rightarrow N\pi$, (b) $\rho \rightarrow \pi\pi$, and (c) $\Lambda(1405) \rightarrow \Sigma\pi$?

18-34. Construct the weight diagram for the multiplet of SU(3) having $\lambda_1 = \lambda_2 = 2$.

18-35. Construct the weight diagram for the multiplet of SU(3) having $\lambda_1 = 6$ and $\lambda_2 = 0$.

18-36. Gell-Mann and Zweig have proposed that the baryons are composed of three particles called "quarks," which have baryon number $\frac{1}{3}$ and which belong to a multiplet of SU(3) with $\lambda_1 = 1$ and $\lambda_2 = 0$. The three types of quark are called p, n, and λ. The p and n quarks have zero strangeness and the λ has strangeness -1. (a) What are the electric charges of the quarks? (b) What combination of them yields a proton? (c) What combination of them yields a Ξ^-?

18-37. In the quark model described in the preceding problem we can construct mesons from quark-antiquark pairs. (a) What kind of meson results from $p\bar{\lambda}$? (b) How would a π^+ meson be constructed in this model?

18-38. (a) What wavelength is necessary in order that a photon may be capable of performing pair production? (b) What is the ratio of this result to the Compton wavelength of the electron?

18-39. List all of the strong interactions that can occur as a result of the capture of a K^- meson by a proton, both particles being at rest before the capture occurs.

18-40. What selection rule forbids each of the following processes?

$$\Lambda \rightarrow p + e^-$$

$$\gamma \rightarrow \mu^+ + \mu^- \quad \text{in a vacuum}$$

$$\Sigma^+ \rightarrow n + e^- + \bar{\nu}$$

$$\pi^0 \rightarrow \mu^+ + e^- + \bar{\nu}$$

$$\Sigma^+ \rightarrow \pi^+ + \pi^0$$

18-41. (a) What is the energy of the photon emitted in Σ^0 decay? (b) What is the wavelength of the above photon?

References

Adair, R. K., and E. C. Fowler, *Strange Particles*. New York: Wiley-Interscience, 1963.

Chew, G., M. Gell-Mann, and A. H. Rosenfeld, "Strongly Interacting Particles." *Scientific American* (February 1964).

Lichtenberg, D. B., *Meson and Baryon Spectroscopy*. New York: Springer-Verlag, 1965.

Segrè, E., *Nuclei and Particles*. New York: W. A. Benjamin, 1964.

Swartz, C. E., *The Fundamental Particles*. Reading, Mass.: Addison-Wesley Publishing Company, 1965.

Swartz, C. E., "Resource Letter on Subatomic Particles." *Am. J. Phys.* **34**, No. 12 (December 1966).

Williams, W. S. C., *An Introduction to Elementary Particles*. New York: Academic Press, 1961.

Yang, C. N., *Elementary Particles*. Princeton, N.J.: Princeton University Press, 1962.

Appendixes

Appendix I

Physical Constants and Conversion Factors

Numerical values for the constants are from B. N. Taylor, W. H. Parker, and D. N. Langenberg, *Revs. Mod. Phys.*, **41**, 375 (1969). Units are quoted in mksa (Système International) and in cgs.

Physical Quantity	*Symbol*	*Value*	*Units (SI)*	*Units (cgs)*
1. Speed of light	c	2.9979250	10^8 meters sec^{-1}	10^{10} cm sec^{-1}
2. Electron charge	e	1.6021918	10^{-19} C	10^{-20} emu
		4.803250		10^{-10} esu
3. Planck's constant	h	6.6261965	10^{-34} joule sec	10^{-27} erg sec
	$\hbar = h/2\pi$	1.05459198	10^{-34} joule sec	10^{-27} erg sec
4. Avogadro's number	N_0	6.022169	10^{26} kmole^{-1}	10^{23} mole^{-1}
5. Electron rest mass	m_e	9.109559	10^{-31} kg	10^{-28} gm
		0.5110041	MeV	
6. Proton rest mass	m_p	1.672614	10^{-27} kg	10^{-24} gm
		938.2592	MeV	
7. Neutron rest mass	m_n	1.674920	10^{-27} kg	10^{-24} gm
		939.5528	MeV	
8. Faraday's constant	F	9.648671	10^7 coul kmole^{-1}	10^3 emu mole^{-1}
		2.892599		10^{14} esu mole^{-1}
9. Gas constant	R	8.31434	10^3 joule kmole^{-1} K^{-1}	10^7 erg mole^{-1} K^{-1}
10. Boltzmann's constant	k	1.380623	10^{-23} joule K^{-1}	10^{-16} erg K^{-1}
11. Gravitational constant	G	6.6732	10^{-11} newt meter^2 kgm^{-2}	10^{-8} dyne cm^2 gm^{-2}

Clusters of Constants

In many atomic and nuclear calculations the above constants appear in clusters, the evaluation of which can become tedious. We provide here a list of some of the more frequently encountered combinations. The serious student is urged to become familiar with the clusters listed here so that he will recognize them when they occur in a problem and use them in preference to looking up the constants separately, followed by several multiplications or divisions.

In these clusters the symbols in brackets apply for calculations in mksa units. For cgs the symbols in parentheses are correct if e is in esu.

Physical Quantity	Symbol	Value	Units (SI)	Units (cgs)
1. Fine-structure constant	$\alpha = [\mu_0 c^2/4\pi]\ (e^2/\hbar c)$ α^{-1}	7.297351 137.03602	10^{-3}	10^{-3}
2. Ratio of proton mass to electron mass	m_p/m_e	1836.109		
3. Electron charge-to-mass ratio	e/m_e	1.7588029	10^{11} coul kg^{-1}	10^7 emu gm^{-1}
4. Rydberg's constant	$[\mu_0 c^2/4\pi]^2\ (m_e e^4/4\pi\hbar^3 c)$	1.09737312	10^7 meter^{-1}	10^5 cm^{-1}
5. $\hbar c$		1.9732891	10^{-11} MeV cm	
6. Bohr radius a_0	$[\mu_0 c^2/4\pi]^{-1}\ (\hbar^2/m_e e^2)$	5.2917716	10^{-11} meter	10^{-9} cm
7. Classical electron radius	$[\mu_0 c^2/4\pi]\ (e^2/m_e c^2)$	2.817939	10^{-15} meter	10^{-13} cm
8. Compton wavelength of the electron	$h/m_e c$	2.4263097	10^{-12} meter	10^{-10} cm
9. Compton wavelength of the proton	$h/m_p c$	1.3214410	10^{-15} meter	10^{-13} cm
10. Compton wavelength of the neutron	$h/m_n c$	1.3196218	10^{-15} meter	10^{-13} cm
11. Bohr magneton	$[c]\ (e\hbar/2m_e c)$	9.274097	10^{-24} joule/tesla	10^{-21} erg/gauss
12. Nuclear magneton	$[c]\ (e\hbar/2m_p c)$	5.050952	10^{-27} joule/tesla	10^{-24} erg/gauss

Conversion Factors

1 atomic mass unit ($^{12}C = 12$) $= 1.660531 \times 10^{-27}$ kg
$= 1.660531 \times 10^{-24}$ gm
$= 931.4812$ MeV
1 hertz $= 1$ cycle sec^{-1}
1 tesla $= 10^4$ gauss
1 eV $= 1.6021918 \times 10^{-19}$ joule
$= 1.6021918 \times 10^{-12}$ erg
1 kg $= 5.609538 \times 10^{29}$ MeV
1 coulomb $= 10^{-1}$ emu
$= c \times 10^{-1}$ esu
1 volt $= 10^8$ emu
$= 10^8\ c^{-1}$ esu
$1/4\pi\epsilon_0 = \mu_0 c^2/4\pi$
$\mu_0/4\pi = 10^{-7}$ henry meter^{-1}
300K $\approx \frac{1}{40}$ eV
1 year $\approx \pi \times 10^7$ sec
1 torr $= 1$ mm of Hg pressure

The Greek Alphabet

Lower-Case Letter	Capital Letter	Name of Letter
α	A	alpha
β	B	beta
γ	Γ	gamma
δ	Δ	delta
ϵ	E	epsilon
ζ	Z	zeta
η	H	eta
θ	Θ	theta
ι	I	iota
κ	K	kappa
λ	Λ	lambda
μ	M	mu
ν	N	nu
ξ	Ξ	xi
o	O	omicron
π	Π	pi
ρ	P	rho
σ, ς	Σ	sigma
τ	T	tau
υ	Υ	upsilon
ϕ	Φ	phi
χ	X	chi
ψ	Ψ	psi
ω	Ω	omega

Appendix II

Atomic Weights of the Elements

	Symbol	Atomic Number	Atomic Weight (amu)		Symbol	Atomic Number	Atomic Weight (amu)
Actinium	Ac	89		Germanium	Ge	32	72.5_9
Aluminum	Al	13	26.9815^a	Gold	Au	79	196.9665^a
Americium	Am	95		Hafnium	Hf	72	178.4_9
Antimony	Sb	51	121.7_5	Helium	He	2	4.00260^b
Argon	Ar	18	$39.94_8^{b,c}$	Holmium	Ho	67	164.9303^a
Arsenic	As	33	74.9216^a	Hydrogen	H	1	$1.008_0^{b,c}$
Astatine	At	85		Indium	In	49	114.82
Barium	Ba	56	137.3_4	Iodine	I	53	126.9045^a
Berkelium	Bk	97		Iridium	Ir	77	192.2_2
Beryllium	Be	4	9.01218^a	Iron	Fe	26	55.84_7
Bismuth	Bi	83	208.9806^a	Krypton	Kr	36	83.80
Boron	B	5	10.81^c	Lanthanum	La	57	138.905_5^b
Bromine	Br	35	79.904	Lawrencium	Lr	103	
Cadmium	Cd	48	112.40	Lead	Pb	82	207.2^c
Calcium	Ca	20	40.08	Lithium	Li	3	6.94_1^c
Californium	Cf	98		Lutetium	Lu	71	174.97
Carbon	C	6	$12.011^{b,c}$	Magnesium	Mg	12	24.305
Cerium	Ce	58	140.12	Manganese	Mn	25	54.9380^a
Cesium	Cs	55	132.9055^a	Mendelevium	Md	101	
Chlorine	Cl	17	35.453	Mercury	Hg	80	200.5_9
Chromium	Cr	24	51.996	Molybdenum	Mo	42	95.9_4
Cobalt	Co	27	58.9332^a	Neodymium	Nd	60	144.2_4
Copper	Cu	29	63.54_6^c	Neon	Ne	10	20.17_9
Curium	Cm	96		Neptunium	Np	93	237.0482^a
Dysprosium	Dy	66	162.5_0	Nickel	Ni	28	58.7_1
Einsteinium	Es	99		Niobium	Nb	41	92.9064^a
Erbium	Er	68	167.2_6	Nitrogen	N	7	14.0067^b
Europium	Eu	63	151.96	Nobelium	No	102	
Fermium	Fm	100		Osmium	Os	76	190.2
Fluorine	F	9	18.9984^a	Oxygen	O	8	$15.999_4^{b,c}$
Francium	Fr	87		Palladium	Pd	46	106.4
Gadolinium	Gd	64	157.2_5	Phosphorus	P	15	30.9738^a
Gallium	Ga	31	69.72	Platinum	Pt	78	195.0_9

	Symbol	Atomic Number	Atomic Weight (amu)		Symbol	Atomic Number	Atomic Weight (amu)
Plutonium	Pu	94		Sulfur	S	16	32.06_c
Polonium	Po	84		Tantalum	Ta	73	180.947_9^b
Potassium	K	19	39.10_2	Technetium	Tc	43	98.9062^d
Praseodymium	Pr	59	140.0977^a	Tellurium	Te	52	127.6_0
Promethium	Pm	61		Terbium	Tb	65	158.9254^a
Protactinium	Pa	91	231.0359^a	Thallium	Tl	81	204.3_7
Radium	Ra	88	$226.0254^{a,d}$	Thorium	Th	90	232.0381^a
Radon	Rn	86		Thulium	Tm	69	168.9342^a
Rhenium	Re	75	186.2	Tin	Sn	50	118.6_9
Rhodium	Rh	45	102.9055^a	Titanium	Ti	22	47.9_0
Rubidium	Rb	37	85.467_8	Tungsten	W	74	183.8_5
Rutheium	Ru	44	101.0_7	Uranium	U	92	238.029^b
Samarium	Sm	62	150.4	Vanadium	V	23	50.941_4^b
Scandium	Sc	21	44.9559^a	Xenon	Xe	54	131.30
Selenium	Se	34	78.9_6	Ytterbium	Yb	70	173.0_4
Silicon	Si	14	28.08_6^c	Yttrium	Y	39	88.9059^a
Silver	Ag	47	107.868	Zinc	Zn	30	65.3_7
Sodium	Na	11	22.9898^a	Zirconium	Zr	40	91.22
Strontium	Sr	38	87.62				

[a] Elements with no isotopes. These elements exist as one type of atom only.
[b] Elements with one predominant isotope (around 99 to 100% one isotope).
[c] Elements for which the variation in isotopic abundance in earth samples limits the precision of the atomic weight.
[d] Most commonly available isotope (radioactive).

Appendix III

Metalloids and Non-metals

METALS

Transition Metals

I	II	III	IV	V	VI	VII	VIII
1 H Hydrogen							2 He Helium
3 Li Lithium	4 Be Beryllium	5 B Boron	6 C Carbon	7 N Nitrogen	8 O Oxygen	9 F Fluorine	10 Ne Neon
11 Na Sodium	12 Mg Magnesium	13 Al Aluminum	14 Si Silicon	15 P Phosphorus	16 S Sulfur	17 Cl Chlorine	18 Ar Argon

Transition Metals (Groups between II and III):

21 Sc Scandium	22 Ti Titanium	23 V Vanadium	24 Cr Chromium	25 Mn Manganese	26 Fe Iron	27 Co Cobalt	28 Ni Nickel	29 Cu Copper	30 Zn Zinc
39 Y Yttrium	40 Zr Zirconium	41 Nb Niobium	42 Mo Molybdenum	43 Tc Technetium	44 Ru Ruthenium	45 Rh Rhodium	46 Pd Palladium	47 Ag Silver	48 Cd Cadmium
57-71* Lanthanides	72 Hf Hafnium	73 Ta Tantalum	74 W Tungsten	75 Re Rhenium	76 Os Osmium	77 Ir Iridium	78 Pt Platinum	79 Au Gold	80 Hg Mercury
89-103* Actinides									

Period 4 (Group I–II): 19 K Potassium, 20 Ca Calcium
Period 5 (Group I–II): 37 Rb Rubidium, 38 Sr Strontium
Period 6 (Group I–II): 55 Cs Cesium, 56 Ba Barium
Period 7 (Group I–II): 87 Fr Francium, 88 Ra Radium

Period 4 (Group III–VIII): 31 Ga Gallium, 32 Ge Germanium, 33 As Arsenic, 34 Se Selenium, 35 Br Bromine, 36 Kr Krypton
Period 5 (Group III–VIII): 49 In Indium, 50 Sn Tin, 51 Sb Antimony, 52 Te Tellurium, 53 I Iodine, 54 Xe Xenon
Period 6 (Group III–VIII): 81 Tl Thallium, 82 Pb Lead, 83 Bi Bismuth, 84 Po Polonium, 85 At Astatine, 86 Rn Radon

*Lanthanides (Rare Earth Metals)

57 La Lanthanum	58 Ce Cerium	59 Pr Praseodymium	60 Nd Neodymium	61 Pm Promethium	62 Sm Samarium	63 Eu Europium	64 Gd Gadolinium	65 Tb Terbium	66 Dy Dysprosium	67 Ho Holmium	68 Er Erbium	69 Tm Thulium	70 Yb Ytterbium	71 Lu Lutetium

*Actinides

89 Ac Actinium	90 Th Thorium	91 Pa Protactinium	92 U Uranium	93 Np Neptunium	94 Pu Plutonium	95 Am Americium	96 Cm Curium	97 Bk Berkelium	98 Cf Californium	99 Es Einsteinium	100 Fm Fermium	101 Md Mendelevium	102 No Nobelium	103 Lw Lawrencium

Appendix IV

Table of Nuclear Properties

We provide here a listing of a few of the properties of some of the more important nuclides. We list Z (the atomic number), the chemical symbol for each element, and the mass number A.

The *mass excess* is the amount of energy by which the mass of an atom exceeds an integer multiple of atomic mass units. In symbols

$$\text{(mass excess)} = \text{(atomic mass)}c^2 - 931.4812\,A$$

The numbers quoted here are from J. H. E. Mattauch, W. Thiele, and A. H. Wapstra, *Nucl. Phys.*, **67**, 1 (1965) and are based on the mass of ^{12}C being exactly 12 u. Note that all masses listed refer to the *neutral atom*. They may be used directly for most nuclear calculations without regard for the masses and binding energies of the associated atomic electrons. Positron decay, however, is an example of a process for which it is necessary to allow explicitly for electron masses (see Section 15-8).

The spin and parity indicated here refer to the gound state of the nucleus. Percentage abundances are quoted for those that occur in nature. Half-lives are quoted for unstable isotopes. Data mentioned in this paragraph are from *Chart of the Nuclides*, General Electric Company (1965).

Z	Element	Atomic Mass No.	Mass Excess (MeV)	Spin and Parity	Percent Abundance	Lifetime
0	n	1	8.072	$\frac{1}{2}^+$		10.8 min (β^-)
1	H	1	7.289	$\frac{1}{2}^+$	99.985	
		2	13.136	1^+	0.015	
		3	14.950	$\frac{1}{2}^+$		12.26 yr (β^-)
2	He	3	14.931	$\frac{1}{2}^+$	0.00013	
		4	2.425	0^+	~ 100	
		5	11.454	$\frac{3}{2}^-$		2×10^{-21} sec $(n + \alpha)$
		6	17.598	0^+		0.81 sec (β^-)

Z	Element	Atomic Mass No.	Mass Excess (MeV)	Spin and Parity	Percent Abundance	Lifetime
3	Li	5	11.679	$\frac{3}{2}^-$		$\sim 10^{-21}$ sec (p)
		6	14.088	1^+	7.42	
		7	14.907	$\frac{3}{2}^-$	92.58	
		8	20.946	2^+		0.85 sec (β^-)
4	Be	7	15.769	$\frac{3}{2}^-$		53 days (e capt.)
		8	4.944	0^+		$\sim 3 \times 10^{-16}$ sec (2α)
		9	11.350	$\frac{3}{2}^-$	100	
		10	12.607	0^+		2.7×10^6 yr (β^-)
5	B	8	22.923			0.78 sec (β^+)
		9	12.419			$\geq 3 \times 10^{-19}$ sec ($p + 2\alpha$)
		10	12.052	3^+	19.78	
		11	8.668	$\frac{3}{2}^-$	80.22	
		12	13.370	1^+		0.020 sec (β^-)
6	C	10	15.658	0^+		19 sec (β^+)
		11	10.648	$\frac{3}{2}^-$		20.5 min (β^+)
		12	0.000	0^+	98.89	
		13	3.125	$\frac{1}{2}^-$	1.11	
		14	3.020	0^+		5730 yr (β^-)
		15	9.873	$\frac{1}{2}^+$		2.25 sec (β^-)
7	N	12	17.364	1^+		0.011 sec (β^+)
		13	5.345	$\frac{1}{2}^-$		9.96 min (β^+)
		14	2.864	1^+	99.63	
		15	0.100	$\frac{1}{2}^-$	0.37	
		16	5.685	2^-		7.35 sec (β^-)
		17	7.871			4.14 sec (β^-)
8	O	14	8.008	0^+		71 sec (β^+)
		15	2.860	$\frac{1}{2}^-$		124 sec (β^+)
		16	−4.737	0^+	99.759	
		17	−0.808	$\frac{5}{2}^+$	0.037	
		18	−0.782	0^+	0.204	
		19	3.333	$\frac{5}{2}^+$		29 sec (β^-)
9	F	17	1.952			66 sec (β^+)
		18	0.872	1^+		110 min (β^+, e capt.)
		19	−1.486	$\frac{1}{2}^+$	100	
		20	−0.012			11 sec (β^-)
		21	−0.046			4.4 sec (β^-)
10	Ne	18	5.319	0^+		1.46 sec (β^+)
		19	1.752	$\frac{1}{2}^+$		18 sec (β^+)
		20	−7.041	0^+	90.92	
		21	−5.730	$\frac{3}{2}^+$	0.257	
		22	−8.025	0^+	8.82	
		23	−5.148			38 sec (β^-)

Z	Element	Atomic Mass No.	Mass Excess (MeV)	Spin and Parity	Percent Abundance	Lifetime
11	Na	21	−2.185	$\frac{3}{2}^+$		23 sec (β^+)
		22	−5.182	3^+		2.58 yr (β^+, e capt.)
		23	−9.528	$\frac{3}{2}^+$	100	
		24	−8.418	4^+		15.0 hr (β^-)
12	Mg	23	−5.472	$\frac{3}{2}^+$		12 sec (β^+)
		24	−13.933	0^+	78.70	
		25	−13.191	$\frac{5}{2}^+$	10.13	
		26	−16.214	0^+	11.17	
		27	−14.583	$\frac{1}{2}^+$		9.5 min (β^-)
13	Al	27	−17.196	$\frac{5}{2}^+$	100	
		28	−16.855			2.30 min (β^-)
14	Si	28	−21.490	0^+	92.21	
		29	−21.894	$\frac{1}{2}^+$	4.70	
		30	−24.439	0^+	3.09	
		31	−22.962	$\frac{3}{2}^+$		2.62 hr (β^-)
15	P	30	−20.197	1^+		2.5 min (β^+)
		31	−24.438	$\frac{1}{2}^+$	100	
		32	−24.303	1^+		14.3 days (β^-)
		33	−26.335			25 days (β^-)
16	S	32	−26.013	0^+	95.0	
		33	−26.583	$\frac{3}{2}^+$	0.76	
		34	−29.933	0^+	4.22	
		35	−28.847	$\frac{3}{2}^+$		86.7 days (β^-)
		36	−30.655	0^+	0.014	
17	Cl	35	−29.014	$\frac{3}{2}^+$	75.53	
		36	−29.520	2^+		3×10^5 yr (β^-, e capt.)
		37	−31.765	$\frac{3}{2}^+$	24.47	
18	Ar	36	−30.232	0^+	0.337	
		37	−30.951	$\frac{3}{2}^+$		35.1 days (e capt.)
		38	−34.718	0^+	0.063	
		39	−33.238	$\frac{7}{2}^-$		270 yr (β^-)
		40	−35.038	0^+	99.60	
19	K	39	−33.803	$\frac{3}{2}^+$	93.10	
		40	−33.533	4^-	0.0118	1.3×10^9 yr (β^-, e capt.)
		41	−35.552	$\frac{3}{2}^+$	6.88	
20	Ca	40	−34.848	0^+	96.97	
		41	−35.140	$\frac{7}{2}^-$		7.7×10^4 yr (e capt.)
		42	−38.540	0^+	0.64	
		43	−38.396	$\frac{7}{2}^-$	0.145	
		44	−41.460	0^+	2.06	
		46	−43.138	0^+	0.0033	
		48	−44.216	0^+	0.18	

Z	Element	Atomic Mass No.	Mass Excess (MeV)	Spin and Parity	Percent Abundance	Lifetime
21	Sc	41	−28.645	$\frac{7}{2}^-$		0.55 sec (β^+)
		45	−41.061	$\frac{7}{2}^-$	100	
22	Ti	46	−44.123	0^+	7.93	
		47	−44.927	$\frac{5}{2}^-$	7.28	
		48	−48.483	0^+	73.94	
		49	−48.558	$\frac{7}{2}^-$	5.51	
		50	−51.431	0^+	5.34	
23	V	48	−44.470			16.1 days (β^+, e capt.)
		50	−49.216	6^+	0.24	$\sim 6 \times 10^{15}$ yr (β^-, e capt.)
		51	−52.199	$\frac{7}{2}^-$	99.76	
24	Cr	48	−43.070	0^+		23 hr (e capt.)
		50	−50.249	0^+	4.31	
		52	−55.411	0^+	83.76	
		53	−55.281	$\frac{3}{2}^-$	9.55	
		54	−56.930	0^+	2.38	
25	Mn	54	−55.552	3^+		303 days (e capt.)
		55	−57.705	$\frac{5}{2}^-$	100	
26	Fe	54	−56.245	0^+	5.82	
		56	−60.605	0^+	91.66	
		57	−60.175	$\frac{1}{2}^-$	2.19	
		58	−62.146	0^+	0.33	
27	Co	59	−62.233	$\frac{7}{2}^-$	100	
		60	−61.651	5^+		5.26 yr (β^-)
28	Ni	58	−60.228	0^+	67.88	
		60	−64.471	0^+	26.23	
		61	−64.220	$\frac{3}{2}^-$	1.19	
		62	−66.748	0^+	3.66	
		64	−67.106	0^+	1.08	
29	Cu	63	−65.583	$\frac{3}{2}^-$	69.09	
		65	−67.266	$\frac{3}{2}^-$	30.91	
30	Zn	64	−66.000	0^+	48.89	
		66	−68.881	0^+	27.81	
		67	−67.863	$\frac{5}{2}^-$	4.11	
		68	−69.994	0^+	18.57	
		70	−69.550	0^+	0.62	
31	Ga	69	−69.326	$\frac{3}{2}^-$	60.4	
		71	−70.135	$\frac{3}{2}^-$	39.6	
32	Ge	70	−70.558	0^+	20.52	

Z	Element	Atomic Mass No.	Mass Excess (MeV)	Spin and Parity	Percent Abundance	Lifetime
		72	−72.579	0^+	27.43	
		73	−71.293	$\frac{9}{2}^+$	7.76	
		74	−73.418	0^+	36.54	
		76	−73.209	0^+	7.76	
33	As	75	−73.031	$\frac{3}{2}^-$	100	
34	Se	74	−72.212	0^+	0.87	
		76	−75.257	0^+	9.02	
		77	−74.601	$\frac{1}{2}^-$	7.58	
		78	−77.020	0^+	23.52	
		80	−77.753	0^+	49.82	
		82	−77.586	0^+	9.19	
35	Br	79	−76.075	$\frac{3}{2}^-$	50.54	
		81	−77.972	$\frac{3}{2}^-$	49.46	
36	Kr	78	−74.143	0^+	0.35	
		80	−77.891	0^+	2.27	
		82	−80.589	0^+	11.56	
		83	−79.985	$\frac{9}{2}^+$	11.55	
		84	−82.433	0^+	56.90	
		86	−83.259	0^+	17.37	
37	Rb	85	−82.156	$\frac{3}{2}^-$	72.15	
		87	−84.591	$\frac{3}{2}^-$	27.85	4.7×10^{10} yr (β^-)
38	Sr	84	−80.638	0^+	0.56	
		86	−84.499	0^+	9.86	
		87	−84.865	$\frac{9}{2}^+$	7.02	
		88	−87.894	0^+	82.56	
39	Y	89	−87.678	$\frac{1}{2}^-$	100	
40	Zr	90	−88.770	0^+	51.46	
		91	−87.893	$\frac{5}{2}^+$	11.23	
		92	−88.462	0^+	17.11	
		94	−87.267	0^+	17.40	
		96	−85.430	0^+	2.80	
41	Nb	93	−87.203	$\frac{9}{2}^+$	100	
42	Mo	92	−86.804	0^+	15.84	
		94	−88.406	0^+	9.04	
		95	−87.709	$\frac{5}{2}^+$	15.72	
		96	−88.794	0^+	16.53	
		97	−87.539	$\frac{5}{2}^+$	9.46	
		98	−88.110	0^+	23.78	
		100	−86.185	0^+	9.63	

Z	Element	Atomic Mass No.	Mass Excess (MeV)	Spin and Parity	Percent Abundance	Lifetime
43	Tc	97	−87.240			2.6×10^6 yr (*e* capt.)
		98	−86.520			1.5×10^6 yr (β^-)
		99	−87.327	$\frac{9^+}{2}$		2.1×10^5 yr (β^-)
44	Ru	96	−86.071	0^+	5.51	
		98	−88.221	0^+	1.87	
		99	−87.619	$\frac{5^+}{2}$	12.72	
		100	−89.219	0^+	12.62	
		101	−87.953	$\frac{5^+}{2}$	17.07	
		102	−89.098	0^+	31.61	
		104	−88.090	0^+	18.58	
45	Rh	103	−88.014	$\frac{1^-}{2}$	100	
46	Pd	102	−87.923	0^+	0.96	
		104	−89.411	0^+	10.97	
		105	−88.431	$\frac{5^+}{2}$	22.23	
		106	−89.907	0^+	27.33	
		108	−89.524	0^+	26.71	
		110	−88.338	0^+	11.81	
47	Ag	107	−88.403	$\frac{1^-}{2}$	51.82	
		109	−88.717	$\frac{1^-}{2}$	48.18	
48	Cd	106	−87.128	0^+	1.22	
		108	−89.248	0^+	0.88	
		110	−90.342	0^+	12.39	
		111	−89.246	$\frac{1^+}{2}$	12.75	
		112	−90.575	0^+	24.07	
		113	−89.041	$\frac{1^+}{2}$	12.26	
		114	−90.018	0^+	28.86	
		116	−88.712	0^+	7.58	
49	In	113	−89.339	$\frac{9^+}{2}$	4.28	
		115	−89.542	$\frac{9^+}{2}$	95.72	
50	Sn	112	−88.644	0^+	0.96	
		114	−90.565	0^+	0.66	
		115	−90.031	$\frac{1^+}{2}$	0.35	
		116	−91.523	0^+	14.30	
		117	−90.392	$\frac{1^+}{2}$	7.61	
		118	−91.652	0^+	24.03	
		119	−90.062	$\frac{1^+}{2}$	8.58	
		120	−91.100	0^+	32.85	
		122	−89.942	0^+	4.72	
		124	−88.237	0^+	5.94	

Z	Element	Atomic Mass No.	Mass Excess (MeV)	Spin and Parity	Percent Abundance	Lifetime
51	Sb	121	−89.593	$\frac{5^+}{2}$	57.25	
		123	−89.224	$\frac{7^+}{2}$	42.75	
52	Te	120	−89.400	0^+	0.089	
		122	−90.291	0^+	2.46	
		123	−89.163	$\frac{1^+}{2}$	0.87	1.2×10^{13} yr (e capt.)
		124	−90.500	0^+	4.61	
		125	−89.032	$\frac{1^+}{2}$	6.99	
		126	−90.053	0^+	18.71	
		128	−88.978	0^+	31.79	
		130	−87.337	0^+	34.48	
53	I	127	−88.984	$\frac{5^+}{2}$	100	
54	Xe	124	−87.450	0^+	0.096	
		126	−89.154	0^+	0.090	
		128	−89.850	0^+	1.92	
		129	−88.692	$\frac{1^+}{2}$	26.44	
		130	−89.880	0^+	4.08	
		131	−88.411	$\frac{3^+}{2}$	21.18	
		132	−89.272	0^+	26.89	
		134	−88.120	0^+	10.44	
		136	−86.422	0^+	8.87	
55	Cs	133	−88.160	$\frac{7^+}{2}$	100	
56	Ba	130	−87.331	0^+	0.101	
		132	−88.380	0^+	0.097	
		134	−88.852	0^+	2.42	
		135	−87.980	$\frac{3^+}{2}$	6.59	
		136	−89.140	0^+	7.81	
		137	−88.020	$\frac{3^+}{2}$	11.32	
		138	−88.490	0^+	71.66	
57	La	138	−86.710	$\frac{5^-}{2}$	0.089	1.1×10^{11} yr (e capt., β^-)
		139	−87.428	$\frac{7^+}{2}$	99.911	
58	Ce	136	−86.550	0^+	0.193	
		138	−87.720	0^+	0.250	
		140	−88.125	0^+	88.48	
		142	−84.631	0^+	11.07	5×10^{15} yr (α)
59	Pr	141	−86.072	$\frac{5^+}{2}$	100	
60	Nd	142	−86.010	0^+	27.11	
		143	−84.039	$\frac{7^-}{2}$	12.17	
		144	−83.797	0^+	23.85	2.4×10^{15} yr (α)

Z	Element	Atomic Mass No.	Mass Excess (MeV)	Spin and Parity	Percent Abundance	Lifetime
		145	−81.469	$\frac{7^-}{2}$	8.30	
		146	−80.959	0^+	17.22	
		148	−77.435	0^+	5.73	
		150	−73.666	0^+	5.62	
61	Pm	143	−82.910			265 days (*e* capt.)
62	Sm	144	−81.980	0^+	3.09	
		147	−79.300	$\frac{7^-}{2}$	14.97	1.06×10^{11} yr (α)
		148	−79.371	0^+	11.24	1.2×10^{13} yr (α)
		149	−77.145	$\frac{7^-}{2}$	13.83	$\sim 4 \times 10^{14}$ yr? (α)
		150	−77.056	0^+	7.44	
		152	−74.746	0^+	26.72	
		154	−72.393	0^+	22.71	
63	Eu	151	−74.670	$\frac{5^+}{2}$	47.82	
		153	−73.361	$\frac{5^+}{2}$	52.18	
64	Gd	152	−74.710	0^+	0.20	1.1×10^{14} yr (α)
		154	−73.653	0^+	2.15	
		155	−72.037	$\frac{3^-}{2}$	14.73	
		156	−72.493	0^+	20.47	
		157	−70.769	$\frac{3^-}{2}$	15.68	
		158	−70.627	0^+	24.87	
		160	−67.891	0^+	21.90	
65	Tb	159	−69.534	$\frac{3^+}{2}$	100	
66	Dy	156	−70.860	0^+	0.052	2×10^{14} yr (α)
		158	−70.374	0^+	0.090	
		160	−69.673	0^+	2.29	
		161	−68.049	$\frac{5^+}{2}$	18.88	
		162	−68.182	0^+	25.53	
		163	−66.363	$\frac{5^-}{2}$	24.97	
		164	−65.949	0^+	28.18	
67	Ho	165	−64.811	$\frac{7^-}{2}$	100	
68	Er	162	−66.370	0^+	0.136	
		164	−65.867	0^+	1.56	
		166	−64.918	0^+	33.41	
		167	−63.285	$\frac{7^+}{2}$	22.94	
		168	−62.983	0^+	27.07	
		170	−60.020	0^+	14.88	
69	Tm	169	−61.249	$\frac{1^+}{2}$	100	
70	Yb	168	−61.330	0^+	0.135	
		170	−60.530	0^+	3.03	

Z	Element	Atomic Mass No.	Mass Excess (MeV)	Spin and Parity	Percent Abundance	Lifetime
		171	−59.220	$\frac{1}{2}^-$	14.31	
		172	−59.280	0^+	21.82	
		173	−57.690	$\frac{5}{2}^-$	16.13	
		174	−57.060	0^+	31.84	
		176	−53.390	0^+	12.73	
71	Lu	175	−55.290	$\frac{7}{2}^+$	97.41	
		176	−53.410	7^-	2.59	2.2×10^{10} yr (β^-)
72	Hf	174	−55.550	0^+	0.18	4.3×10^{15} yr (α)
		176	−54.430	0^+	5.20	
		177	−52.720	$\frac{7}{2}^-$	18.50	
		178	−52.270	0^+	27.14	
		179	−50.270	$\frac{9}{2}^+$	13.75	
		180	−49.530	0^+	35.24	
73	Ta	180	−48.862		0.0123	
		181	−48.430	$\frac{7}{2}^+$	99.988	
74	W	180	−49.365	0^+	0.14	
		182	−48.156	0^+	26.41	
		183	−46.272	$\frac{1}{2}^-$	14.40	
		184	−45.619	0^+	30.64	
		186	−42.438	0^+	28.41	
75	Re	185	−43.725	$\frac{5}{2}^+$	37.07	
		187	−41.140	$\frac{5}{2}^+$	62.93	4×10^{10} yr (β^-)
76	Os	184	−44.010	0^+	0.018	
		186	−42.970	0^+	1.59	
		187	−41.141	$\frac{1}{2}^-$	1.64	
		188	−40.909	0^+	13.3	
		189	−38.840	$\frac{3}{2}^-$	16.1	
		190	−38.540	0^+	26.4	
		192	−35.910	0^+	41.0	
77	Ir	191	−36.670	$\frac{3}{2}^+$	37.3	
		193	−34.454	$\frac{3}{2}^+$	62.7	
78	Pt	190	−37.300	0^+	0.0127	6×10^{11} yr (α)
		192	−36.190	0^+	0.78	
		194	−34.721	0^+	32.9	
		195	−32.776	$\frac{1}{2}^-$	33.8	
		196	−32.633	0^+	25.3	
		198	−29.905	0^+	7.21	
79	Au	197	−31.166	$\frac{3}{2}^+$	100	
80	Hg	196	−31.838	0^+	0.146	

Z	Element	Atomic Mass No.	Mass Excess (MeV)	Spin and Parity	Percent Abundance	Lifetime
		198	−30.966	0^+	10.02	
		199	−29.547	$\frac{1}{2}^-$	16.84	
		200	−29.503	0^+	23.13	
		201	−27.658	$\frac{3}{2}^-$	13.22	
		202	−27.346	0^+	29.80	
		204	−24.689	0^+	6.85	
81	Tl	203	−25.753	$\frac{1}{2}^+$	29.50	
		205	−23.807	$\frac{1}{2}^+$	70.50	
		206	−22.259			4.3 min (β^-)
		207	−21.005			4.78 min (β^-)
		208	−16.754	5^+		3.1 min (β^-)
		209	−13.697			2.2 min (β^-)
		210	−9.264			1.3 min (β^-)
82	Pb	204	−25.109	0^+	1.48	1.4×10^{17} yr (α)
		205	−23.772			3×10^7 yr $(e$ capt.$)$
		206	−23.783	0^+	23.6	
		207	−22.446	$\frac{1}{2}^-$	22.6	
		208	−21.750	0^+	52.3	
		209	−17.622			3.3 hr (β^-)
		210	−14.730	0^+		22 yr (β^-)
		211	−10.486			36.1 min (β^-)
		212	−7.541	0^+		10.64 hr (β^-)
		214	−0.218	0^+		26.8 min (β^-)
83	Bi	208	−18.880			3.7×10^5 yr $(e$ capt.$)$
		209	−18.262	$\frac{9}{2}^-$	100	
		210	−14.791	1^-		5.0 days (β^-)
		211	−11.830			2.15 min (β^-)
		212	−8.124	1^-		60.6 min (β^-)
		213	−5.294			4.7 min (β^-)
		214	−1.224			19.7 min (β^-)
84	Po	209	−16.370	$\frac{1}{2}^-$		103 yr $(\alpha, e$ capt.$)$
		210	−15.950	0^+		138.40 days (α)
		211	−12.429			0.52 sec (α)
		212	−10.371	0^+		0.30×10^{-6} sec (α)
		213	−6.683			4×10^{-6} sec (α)
		214	−4.470	0^+		0.164×10^{-3} sec (α)
		215	−0.537			0.0018 sec (α, β^-)
		216	1.790	0^+		0.15 sec (α)
		218	8.318	0^+		3.05 min (α, β^-)
85	At	212	−8.641			0.31 sec (α)
		214	−3.410			< 5 sec (α)
		215	−1.245			$\sim 10^{-4}$ sec (α)

Z	Element	Atomic Mass No.	Mass Excess (MeV)	Spin and Parity	Percent Abundance	Lifetime
		216	2.246			$\sim 3.0 \times 10^{-4}$ sec (α)
		217	4.329			0.032 sec (α)
		218	8.017			1.3 sec (α)
86	Rn	216	0.254	0^+		45×10^{-6} sec (α)
		217	3.629			500×10^{-6} sec (α)
		218	5.219	0^+		0.035 sec (α)
		219	8.832			4.0 sec (α)
		220	10.620	0^+		56 sec (α)
		222	16.329	0^+		3.823 days (α)
87	Fr	218	7.020			< 5 sec (α)
		219	8.622			0.02 sec (α)
		220	11.492			28 sec (α)
		221	13.211			4.8 min (α)
		223	18.383			22 min (β^-)
88	Ra	220	10.274	0^+		0.02 sec (α)
		221	12.940			30 sec (α)
		222	14.322	0^+		37 sec (α)
		223	17.233	$\frac{1}{2}^+$		11.7 days (α)
		224	18.832	0^+		3.64 days (α)
		225	21.916			14.8 days (β^-)
		226	23.623	0^+	100	1620 yr (α)
		227	27.161			41 min (β^-)
		228	29.005	0^+		5.7 yr (β^-)
89	Ac	222	16.540			5 sec (α)
		223	17.832			2.2 min (α, e capt.)
		224	20.200			2.9 hr (e capt., α)
		225	21.566			10.0 days (α)
		227	25.851	$\frac{3}{2}^+$		21.2 yr (β^-, α)
		228	28.950			6.13 hr (β^-)
90	Th	224	20.005	0^+		~ 1 sec (α)
		225	22.288			8 min (α, e capt.)
		226	23.195	0^+		31 min (α)
		227	25.807			18.17 days (α)
		228	26.780	0^+		1.91 yr (α)
		229	29.483	$\frac{5}{2}^+$		7300 yr (α)
		230	30.820	0^+		76000 yr (α)
		231	33.804	$\frac{5}{2}^+$		25.6 hr (β^-)
		232	35.512	0^+		1.41×10^{10} yr (α)
		233	38.628			22.1 min (β^-)
		234	40.596	0^+		24.10 days (β^-)
91	Pa	226	25.900			1.8 min (α)
		227	26.837	$\frac{5}{2}^-$		38.3 min (α, e capt.)

Z	Element	Atomic Mass No.	Mass Excess (MeV)	Spin and Parity	Percent Abundance	Lifetime
		228	28.880			22 hr (e capt., α)
		229	29.828	$\frac{5}{2}^-$		1.5 days (e capt., α)
		230	32.074	2^-		17 days (e capt., β^-, β^+, α)
		231	33.419	$\frac{3}{2}^-$		32480 yr (α)
		232	35.966			1.32 days (β^-)
		233	37.383	$\frac{3}{2}^-$		27.4 days (β^-)
		234	40.332			6.66 hr (β^-)
92	U	228	29.236	0^+		9.3 min (α, e capt.)
		229	31.187			58 min (e capt., α)
		230	31.612	0^+		20.8 days (α)
		232	34.621	0^+		72 yr (α)
		233	36.814	$\frac{5}{2}^+$		1.62×10^5 yr (α)
		234	38.102	0^+	0.0057	2.48×10^5 yr (α)
		235	40.906	$\frac{7}{2}^-$	0.72	7.13×10^8 yr (α)
		237	45.277			6.75 days (β^-)
		238	47.291	0^+	99.27	4.51×10^9 yr (α)
		239	50.579			23.5 min (β^-)
93	Np	231	35.660			$\sim$ 50 min (e capt., α)
		237	44.763	$\frac{5}{2}^+$		2.14×10^6 yr (α)
		238	47.408	2^+		2.10 days (β^-)
		239	49.297	$\frac{5}{2}^+$		2.35 days (β^-)
94	Pu	232	38.360	0^+		36 min (e capt.)
		234	40.347	0^+		9 hr (e capt.)
		236	42.914	0^+		2.85 yr (α)
		238	46.118	0^+		89 yr (α)
		239	48.573	$\frac{1}{2}^+$		24360 yr (α)
		241	52.849	$\frac{5}{2}^+$		13 yr (β^-)
95	Am	239	49.383			12 hr (e capt., α)
		241	52.828	$\frac{5}{2}^-$		458 yr (α)
		242	55.425	1^-		16 hr (β^-, e capt., α)
96	Cm	240	51.739	0^+		26.8 days (α)
		242	54.760	0^+		163 days (α)

Appendix V-1

Review of Vector Notation

In this appendix we review the notation that is employed in the manipulation of vector quantities. A vector quantity is printed in boldface type, as in the vector **r**. The same symbol in italic type, r, denotes the magnitude of the vector; by definition,

$$|\mathbf{r}| = r \tag{1}$$

The sum or difference of two vectors is readily obtained by the parallelogram method: the sum of two vectors is the diagonal of the parallelogram, as seen in Figure A-1.

Two kinds of product of vector are defined. The scalar product (also dot product, inner product) of two vectors **a** and **b** is defined by

$$\mathbf{a} \cdot \mathbf{b} = ab \cos \theta \tag{2}$$

where θ is the angle between the two vectors. As its name indicates, the result of **a** • **b** is a scalar quantity. The vector product (also cross product, outer product) is defined by

$$\mathbf{a} \times \mathbf{b} = ab \sin \theta \boldsymbol{\epsilon} \tag{3}$$

where θ is defined as before and $\boldsymbol{\epsilon}$ is a vector of unit magnitude which is perpendicular to both **a** and **b** and chosen according to a righthand rule. Clearly

$$\mathbf{b} \cdot \mathbf{a} = \mathbf{a} \cdot \mathbf{b}$$

$$\mathbf{b} \times \mathbf{a} = -\mathbf{a} \times \mathbf{b} \tag{4}$$

Products involving three vectors can also be defined. One of these is written

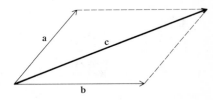

Figure A-1   The sum of two vectors, **c** = **a** + **b**.

a · **b** × **c**. The meaning of this expression is unambiguously **a** · (**b** × **c**); the result is a scalar. The interpretation (**a** · **b**) × **c** is without meaning, since the cross product of a scalar with a vector is not defined. The following identities are readily derived by use of components of the vectors involved:

$$\mathbf{a} \cdot \mathbf{b} \times \mathbf{c} = \mathbf{a} \times \mathbf{b} \cdot \mathbf{c}$$

$$= \mathbf{c} \times \mathbf{a} \cdot \mathbf{b}$$

$$= \mathbf{b} \times \mathbf{c} \cdot \mathbf{a}$$

$$= -\mathbf{b} \cdot \mathbf{a} \times \mathbf{c} \tag{5}$$

Expressed in words, the quantity **a** · **b** × **c** is invariant under interchange of the dot and cross but changes sign under an odd number of permutations of two of the vectors.

The triple product **a** × (**b** × **c**) can also be defined. In this instance the result is a vector. Its meaning is sensitive to the grouping and accounts for our inclusion of parentheses.

Again, the use of components leads to a useful result:

$$\mathbf{a} \times (\mathbf{b} \times \mathbf{c}) = \mathbf{b} \, (\mathbf{a} \cdot \mathbf{c}) - \mathbf{c} \, (\mathbf{a} \cdot \mathbf{b}) \tag{6}$$

Appendix V-2

Displacement Equation
for Brownian Motion

The second type of experiment performed by Perrin for the determination of Avogadro's number consisted in observing the displacement of a particle in a given time interval. The fundamental assumption used in the derivation of the equation for the displacement of such particles is that the particles suspended in the fluid have a mean kinetic energy equal to the mean kinetic energy of gas molecules at the same temperature. The equation of state for one mole of an ideal gas is

$$Pv = RT \tag{1}$$

and on the basis of the kinetic theory of gases it can be shown that this equation takes the form

$$Pv = \tfrac{1}{3} N m \overline{c^2} \tag{2}$$

where $\overline{c^2}$ is the average of the squares of the velocities of the molecules of the gas at temperature T. Thus the mean kinetic energy of a gas molecule is

$$\mathcal{E} = \tfrac{1}{2} m \overline{c^2} = \frac{3}{2} \frac{R}{N} T \tag{3}$$

In general, a particle has three degrees of freedom of translatory motion; but if the observations are confined to a single direction, say the x direction, the average kinetic energy due to the motion in this direction will be one third of its total kinetic energy, or

$$\tfrac{1}{2} m \overline{\left(\frac{dx}{dt}\right)^2} = \frac{\mathcal{E}}{3} = \frac{1}{2} \frac{R}{N} T \tag{4}$$

where dx/dt is the instantaneous velocity of the molecule in the x direction and $\overline{(dx/dt)^2}$ is the average of the squares of its velocity in the x direction.

The motion of a Brownian particle in a horizontal plane is determined by two forces: (a) that due to the bombardment of the molecules of the medium giving an unbalanced force whose component in the x direction is X and (b) the resistance to the motion due to the viscosity of the medium. From Stokes' law the resistance to the motion is proportional to the velocity; hence the force on the particle in the x direction at any instant is, from Newton's second law,

$$F_x = X - K\frac{dx}{dt} = m\frac{d^2x}{dt^2} \tag{5}$$

where K is a factor of proportionality determined by the viscosity of the medium and is given by

$$K = 6\pi\eta a \tag{6}$$

where η is the coefficient of viscosity of the fluid and a is radius of the particle. In studying Brownian motion it is easier to consider the magnitude only of the displacement; hence the equation of motion is modified to yield x^2 instead of x. Multiplying through by x, we get

$$Xx - Kx\frac{dx}{dt} = mx\frac{d^2x}{dt^2}$$

Now

$$x\frac{d^2x}{dt^2} = \frac{1}{2}\frac{d^2}{dt^2}(x^2) - \left(\frac{dx}{dt}\right)^2$$

and

$$x\frac{dx}{dt} = \frac{1}{2}\frac{d}{dt}(x^2)$$

so that

$$Xx - \tfrac{1}{2}K\frac{d(x^2)}{dt} = \frac{m}{2}\frac{d^2(x^2)}{dt^2} - m\left(\frac{dx}{dt}\right)^2 \tag{7}$$

We are interested only in average values of these quantities that can be observed in a time interval t. A bar over a symbol will represent its average value. Let $z = d(\overline{x^2})/dt$ and note that $\overline{Xx} = 0$ since Xx is probably positive as often as it is negative; then Equation (7) becomes

$$-\frac{Kz}{2} = \frac{m}{2}\frac{dz}{dt} - m\overline{\left(\frac{dx}{dt}\right)^2} \tag{8}$$

or

$$-\frac{Kz}{2} = \frac{m}{2}\frac{dz}{dt} - \frac{RT}{N} \tag{9}$$

from Equation (4).

Separating variables, we get

$$\frac{dz}{z - (2RT/KN)} = -\frac{K}{m}dt \tag{10}$$

and, integrating from 0 to t, this yields

$$z - \frac{2RT}{NK} = A\exp\left(-\frac{Kt}{m}\right) \tag{11}$$

where A is a constant of integration. For any reasonable time interval required for observation, the exponential term becomes negligible so that

$$z = \frac{2RT}{NK} = \frac{d(\overline{x^2})}{dt} \tag{12}$$

and

$$\overline{x^2} = \frac{2RT}{NK} t \tag{13}$$

where t is the time interval for observation. This equation states that the average square of the displacements of a particle depends on the time interval t and on the temperature of the medium. This equation was derived by Einstein (1905).

Appendix V-3

Path of an Alpha Particle in a Coulomb Field of Force

Consider a nucleus of charge Ze stationary at point C and an alpha particle of mass M and charge Q approaching it along the line AB (Fig. A-2). The original velocity of the alpha particle in the direction of AB is V. There will be a force of repulsion between the two charges given by Coulomb's law.

$$F = k \frac{ZeQ}{r^2} \tag{1}$$

where r is the distance of the alpha particle from the nucleus. The alpha particle will be deflected from its original direction by this force of repulsion and its path will be a conic section, since the motion is governed by an inverse square law of force. This conic section will be one branch of a hyperbola with the nucleus at the focus on the convex side of this branch.

To derive the expression for the path of the alpha particle let us choose polar coordinates with the nucleus at the pole. The acceleration of the particle may be resolved into two components, one along the radius, a_r, and the other transverse

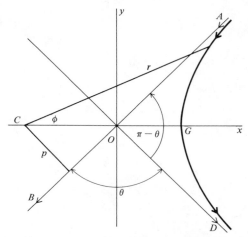

Figure A-2   The hyperbolic path of an alpha particle in the field of force of a nucleus.

to the radius, a_ϕ. In texts on mechanics it is shown that these components are given by the equations

$$a_r = \frac{d^2r}{dt^2} - r\left(\frac{d\phi}{dt}\right)^2 \tag{2}$$

$$a_\phi = \frac{1}{r}\frac{d}{dt}\left(r^2\frac{d\phi}{dt}\right) \tag{3}$$

Since the force of repulsion is along the radius vector r, the radial component of the acceleration is

$$a_r = \frac{F}{M} = \frac{kZeQ}{Mr^2} = \frac{J}{r^2} \tag{4}$$

where

$$J = \frac{kZeQ}{M}$$

Since there is no force transverse to the radius vector,

$$a_\phi = 0 = \frac{1}{r}\frac{d}{dt}\left(r^2\frac{d\phi}{dt}\right) \tag{5}$$

Integrating Equation (5) yields

$$r^2\frac{d\phi}{dt} = K \tag{6}$$

Equation (6) is a statement of Kepler's law of areas and also expresses the principle of conservation of angular momentum, for, multiplying both sides by M, we get

$$Mr^2\frac{d\phi}{dt} = MK = MVp \tag{7}$$

where Mr^2 is the moment of inertia of the alpha particle with respect to an axis through C, $d\phi/dt$ is the angular velocity of the particle, and MVp is the initial angular momentum of the particle. The distance p from the nucleus to AB is called the *impact parameter*.

The differential equation of the path of the particle may be obtained by combining Equations (2) and (4); this yields

$$\frac{d^2r}{dt^2} - r\left(\frac{d\phi}{dt}\right)^2 = \frac{J}{r^2} \tag{8}$$

Before integrating this equation, let us change the independent variable from t to ϕ, and also, for convenience, let $u = 1/r$. To carry out this transformation note that

$$\frac{dr}{dt} = \frac{dr}{d\phi}\frac{d\phi}{dt}$$

$$\frac{dr}{d\phi} = -\frac{1}{u^2}\frac{du}{d\phi}$$

and from Equation (6)

$$\frac{d\phi}{dt} = \frac{K}{r^2} = Ku^2$$

and

$$\frac{d^2r}{dt^2} = -K^2u^2 \frac{d^2u}{d\phi^2}$$

Equation (8) now becomes

$$\frac{d^2u}{d\phi^2} + u = -\frac{J}{K^2}$$

or

$$\frac{d^2}{d\phi^2}\left(u + \frac{J}{K^2}\right) = -\left(u + \frac{J}{K^2}\right) \tag{9}$$

This is a well-known differential equation and its solution is

$$u + \frac{J}{K^2} = A \cos(\phi - \delta) \tag{10}$$

where A and δ are constants of integration. To verify that this is the solution of the differential equation, we need only to differentiate Equation (10) twice.

By the proper choice of the x axis the phase angle δ can be set equal to zero. Further, by setting

$$\epsilon = \frac{AK^2}{J} \tag{11}$$

Equation (10) becomes

$$u + \frac{J}{K^2} = \frac{\epsilon J}{K^2} \cos \phi \;\cdot$$

from which

$$u = \frac{1}{r} = -\frac{J}{K^2}(1 - \epsilon \cos \phi) \tag{12}$$

Equation (12) is one form of the equation of a conic section in polar coordinates; ϵ is the eccentricity of this conic section. When ϵ is greater than unity, the conic section is a hyperbola. The eccentricity of this path can be determined with the aid of the principle of conservation of energy, which yields

$$\tfrac{1}{2}MV^2 = \tfrac{1}{2}Mv^2 + \frac{kZeQ}{r} = \tfrac{1}{2}Mv^2 + \frac{JM}{r} \tag{13}$$

where $kZeQ/r$ is the potential energy of the particle at any point in its path. Resolving the velocity v into two components, one along the radius and one transverse to the radius, we get

$$v^2 = \left(\frac{dr}{dt}\right)^2 + \left(r\frac{d\phi}{dt}\right)^2 = \left[\left(\frac{dr}{d\phi}\right)^2 + r^2\right]\left(\frac{d\phi}{dt}\right)^2 \tag{14}$$

From Equation (12) we get

$$\left(\frac{dr}{d\phi}\right)^2 = \frac{J^2\epsilon^2 r^4 \sin^2 \phi}{K^4}$$

and since

$$\left(\frac{d\phi}{dt}\right)^2 = \frac{K^2}{r^4}$$

Equation (14) becomes

$$v^2 = \frac{J^2\epsilon^2 \sin^2 \phi}{K^2} + \frac{J^2}{K^2}(1 - \epsilon \cos \phi)^2 \tag{15}$$

Substituting this value of v^2 in Equation (13) and the value for r from Equation (12) and simplifying, we get

$$V^2 = \frac{J^2\epsilon^2}{K^2} - \frac{J^2}{K^2}$$

from which

$$\epsilon^2 - 1 = \frac{K^2}{J^2} V^2$$

or

$$\epsilon^2 - 1 = \left(\frac{MV^2 p}{kZeQ}\right)^2 \tag{16}$$

The eccentricity of the orbit is greater than unity and is expressed in terms of the initial energy of the alpha particle, the impact parameter p, and the nuclear charge $Ze;$ thus

$$\epsilon = \sqrt{1 + \left(\frac{MV^2 p}{kZeQ}\right)^2} \tag{17}$$

The orbit is therefore a hyperbola with the nucleus at the focus outside the branch of the curve followed by the alpha particle. The asymptote AB represents the direction of the initial velocity of the alpha particle, and the second asymptote OD represents the direction of the final velocity of the alpha particle. The angle θ between the two asymptotes is the angle through which the particle has been deflected and is the angle of scattering. This angle can be expressed in terms of the eccentricity from a consideration of the properties of the hyperbola.

The equation of the hyperbola in rectangular coordinates is

$$\frac{x^2}{a^2} - \frac{y^2}{b^2} = 1 \tag{18}$$

where

$$a\epsilon = OC$$

and

$$b^2 = a^2(\epsilon^2 - 1)$$

The equations of the asymptotes are obtained by setting the left-hand side of

Equation (18) equal to zero, yielding

$$y = \pm \frac{b}{a} x$$

From the figure

$$\frac{b}{a} = \tan \frac{\pi - \theta}{2} = \cot \frac{\theta}{2}$$

hence

$$\epsilon^2 - 1 = \cot^2 \frac{\theta}{2} \tag{19}$$

Using Equation (16), we get

$$\cot \frac{\theta}{2} = \frac{MV^2}{kZeQ} p \tag{20}$$

Appendix V-4

Derivation of the Equations for the Compton Effect

The three equations derived in Section 5-19 are

$$h\nu = h\nu' + mc^2 (\gamma - 1) \tag{1}$$

$$\frac{h\nu}{c} = \frac{h\nu'}{c} \cos \phi + \gamma m \upsilon \cos \theta \tag{2}$$

$$0 = \frac{h\nu'}{c} \sin \phi - \gamma m \upsilon \sin \theta \tag{3}$$

where

$$\gamma = \frac{1}{(1 - \upsilon^2/c^2)^{1/2}}$$

To solve these equations, let

$$\alpha = \frac{h\nu}{mc^2} = \frac{h}{mc\lambda}$$

$$\alpha' = \frac{h\nu'}{mc^2} = \frac{h}{mc\lambda'}$$

$$b = \sqrt{\gamma^2 - 1} = \frac{\gamma\upsilon}{c}$$

or

$$\gamma = \sqrt{1 + b^2}$$

$$l_1 = \cos \phi$$

$$n_1 = \sin \phi$$

$$l_2 = \cos \theta$$

$$n_2 = \sin \theta$$

By dividing Equation (1) by mc^2 we get

$$\alpha = \alpha' + \sqrt{1 + b^2} - 1 \tag{4}$$

and by dividing Equations (2) and (3) by mc we get

$$\alpha = \alpha' l_1 + bl_2 \tag{5}$$

and

$$0 = \alpha' n_1 - bn_2 \tag{6}$$

From Equation (5)

$$b^2 l_2{}^2 = \alpha^2 - 2\alpha\alpha' l_1 + \alpha'^2 l_1{}^2 \tag{7}$$

and from Equation (6)

$$b^2 n_2{}^2 = \alpha'^2 n_1{}^2 \tag{8}$$

Adding Equations (7) and (8), we get

$$b^2(l_2{}^2 + n_2{}^2) = \alpha^2 - 2\alpha\alpha' l_1 + \alpha'^2(l_1{}^2 + n_1{}^2)$$

or

$$b^2 = \alpha^2 - 2\alpha\alpha' l_1 + \alpha'^2 \tag{9}$$

since

$$l_2{}^2 + n_2{}^2 = 1 = l_1{}^2 + n_1{}^2$$

From Equation (4),

$$b^2 = \alpha^2 - 2\alpha\alpha' + \alpha'^2 + 2\alpha - 2\alpha' \tag{10}$$

Subtracting Equation (10) from Equation (9), we get

$$0 = 2\alpha\alpha'(1 - l_1) - 2\alpha + 2\alpha'$$

from which

$$0 = \alpha(1 - l_1) - \left(\frac{\alpha}{\alpha'} - 1\right)$$

or

$$\frac{\alpha}{\alpha'} - 1 = \alpha(1 - l_1) \tag{11}$$

from which

$$\frac{\lambda'}{\lambda} - 1 = \frac{h}{mc\lambda}(1 - \cos\phi)$$

or

$$\lambda' - \lambda = \frac{h}{mc}(1 - \cos\phi) \tag{12}$$

which is the same as Equation (5-34) in the text. To get the expression for the kinetic energy of the recoil electron solve Equation (11) for α' to obtain

$$\alpha' = \frac{\alpha}{1 + \alpha(1 - l_1)} \tag{13}$$

Substitute this value in Equation (4) to obtain

$$\sqrt{1 + b^2} - 1 = \alpha - \frac{\alpha}{1 + \alpha(1 - l_1)}$$

$$= \frac{\alpha^2(1 - l_1)}{1 + \alpha(1 - l_1)}$$

Now the kinetic energy $\mathcal{E}$ is given by

$$\mathcal{E} = mc^2(\gamma - 1) = mc^2(\sqrt{1 + b^2} - 1)$$

$$= mc^2 \frac{\alpha^2(1 - \cos \phi)}{1 + \alpha(1 - \cos \phi)}$$

But

$$mc^2 = \frac{h\nu}{\alpha}$$

therefore

$$\mathcal{E} = h\nu \frac{\alpha(1 - \cos \phi)}{1 + \alpha(1 - \cos \phi)} \tag{14}$$

The energy of the recoil electron can also be put in terms of the angle θ by noting that, from Equations (5) and (6),

$$\frac{l_2}{n_2} = \frac{\alpha - \alpha' l_1}{\alpha' n_1} = \frac{1}{n_1}\left(\frac{\alpha}{\alpha'} - l_1\right) \tag{15}$$

and eliminating α' with the aid of Equation (11) we get

$$\frac{l_2}{n_2} = \frac{1}{n_1}(1 + \alpha)(1 - l_1) \tag{16}$$

Substituting the values of l_1, l_2, n_1, and n_2, we get

$$\cot \theta = (1 + \alpha)\frac{1 - \cos \phi}{\sin \phi} \tag{17}$$

Now

$$\frac{1 - \cos \phi}{\sin \phi} = \tan \frac{\phi}{2}$$

therefore

$$\cot \theta = (1 + \alpha) \tan \frac{\phi}{2} \tag{18}$$

also

$$\tan \theta = \frac{\cot (\phi/2)}{1 + \alpha} \tag{19}$$

which is Equation (5-35) of the text.
By dividing Equation (18) by Equation (19) we get

$$\frac{\cos^2 \theta}{\sin^2 \theta} = \frac{\cos^2 \theta}{1 - \cos^2 \theta} = (1 + \alpha)^2 \tan^2 \frac{\phi}{2}$$

Now

$$\tan^2 \frac{\phi}{2} = \frac{1 - \cos \phi}{1 + \cos \phi}$$

hence

$$\frac{\cos^2 \theta}{1 - \cos^2 \theta} = (1 + \alpha)^2 \frac{1 - \cos \phi}{1 + \cos \phi} \tag{20}$$

Solving Equation (20) for $\cos \phi$ and then for $(1 - \cos \phi)$, we get

$$1 - \cos \phi = \frac{2 \cos^2 \theta}{(1 + \alpha)^2 - 2\alpha \cos^2 \theta - \alpha^2 \cos^2 \theta} \tag{21}$$

Substituting this value in Equation (14) yields

$$\mathcal{E} = h\nu \frac{2\alpha \cos^2 \theta}{(1 + \alpha)^2 - \alpha^2 \cos^2 \theta} \tag{22}$$

which is Equation (5-36) of the text.

Appendix V-5

Evaluation of Integrals of the Form $\int_0^\infty v^n \exp(-\lambda v^2)dv$

Integrals of the form $I_n = \int_0^\infty v^n \exp(-\lambda v^2)\, dv$ arise in varied applications, most notably in the quantum-mechanical treatment of the harmonic oscillator and in the classical statistical description of an ideal gas. The easiest of these integrals is the one for which $n = 1$.

$$I_1(\lambda) = \int_0^\infty v \exp(-\lambda v^2)dv$$

$$= -\frac{1}{2\lambda}\left[\exp(-\lambda v^2)\right]_0^\infty$$

$$= \tfrac{1}{2}\lambda^{-1} \tag{1}$$

If we now take $\frac{\partial I_1}{\partial \lambda}$, differentiating both the first and third lines of (1), we get

$$-\int_0^\infty v^3 \exp(-\lambda v^2)\, dv = -\tfrac{1}{2}\lambda^{-2} \tag{2}$$

Clearly this process can be repeated to supply the result for all odd values of n. To evaluate I_0 it is necessary to use a trick.

$$I_0 = \int_0^\infty \exp(-\lambda v^2)\, dv \tag{3}$$

$$2I_0 = \int_{-\infty}^\infty \exp(-\lambda v^2)\, dv \tag{4}$$

We next square both sides, remembering that the variable of integration for a definite integral is a dummy variable for which the symbol may be changed without affecting the result.

$$4I_0{}^2 = \int_{-\infty}^{\infty} \exp{(-\lambda x^2)}\, dx \int_{-\infty}^{\infty} \exp{(-\lambda y^2)}\, dy \tag{5}$$

Next we write the product of integrals as a double integral over all $x - y$ space.

$$4I_0{}^2 = \int_{-\infty}^{\infty} \int_{-\infty}^{\infty} \exp{[-\lambda (x^2 + y^2)]}\, dx\, dy \tag{6}$$

The two-dimensional integral can be performed in polar coordinates, for which $r^2 = x^2 + y^2$, $\theta = \tan^{-1}(y/x)$, and the element of area is $r\, d\theta\, dr$.

$$4I_0{}^2 = \int_0^{\infty} \int_0^{2\pi} \exp{(-\lambda r^2)}\, r\, d\theta\, dr \tag{7}$$

Using (1), we get

$$4I_0{}^2 = 2\pi \tfrac{1}{2} \lambda^{-1}$$

or $\tag{8}$

$$I_0 = \tfrac{1}{2}\sqrt{\pi}\, \lambda^{-1/2}$$

By taking partial derivatives with respect to λ we can now generate the integrals corresponding to even values of n.

$$\int_0^{\infty} \exp{(-\lambda v^2)}\, dv = \tfrac{1}{2}\sqrt{\pi}\, \lambda^{-1/2} \tag{9}$$

$$-\int_0^{\infty} v^2 \exp{(-\lambda v^2)}\, dv = -\tfrac{1}{4}\sqrt{\pi}\, \lambda^{-3/2} \tag{10}$$

$$\int_0^{\infty} v^4 \exp{(-\lambda v^2)}\, dv = \tfrac{3}{8}\sqrt{\pi}\, \lambda^{-5/2} \tag{11}$$

and so on. These results are listed in tabular form.

n	$I_n = \int_0^{\infty} v^n \exp{(-\lambda v^2)}\, dv$
0	$\tfrac{1}{2}\sqrt{\pi}\, \lambda^{-1/2}$
1	$\tfrac{1}{2} \lambda^{-1}$
2	$\tfrac{1}{4}\sqrt{\pi}\, \lambda^{-3/2}$
3	$\tfrac{1}{2} \lambda^{-2}$
4	$\tfrac{3}{8}\sqrt{\pi}\, \lambda^{-5/2}$

Appendix V-6

Quantum Mechanical Solution of the Harmonic Oscillator

In this appendix we examine the solutions of the Hermite differential equation (7-66). We shall employ the method of factorization, a method used by Schroedinger. We begin by applying this method to the solution of the classical equation (7-59).

The classical oscillator is described by

$$\frac{d^2x}{dt^2} + \omega^2 x = 0 \tag{1}$$

If we use the symbol D to represent the operator d/dx, the equation becomes

$$(D^2 + \omega^2)x = 0$$

The quantity in parentheses is easily factored to yield

$$(D + i\omega)(D - i\omega)x = 0 \tag{2}$$

But any solution of $(D - i\omega)x = 0$ is also a solution of Equation (2), and it is easily verified that $x = A \exp(i\omega t)$ is a solution. Similarly, we can reverse the order of the factors in Equation (2) and obtain

$$(D - i\omega)(D + i\omega)x = 0$$

This reversal is possible only because ω is a constant. Any solution of $(D + i\omega)x = 0$ is now also a solution of Equation (2), and the result is $x = B \exp(-i\omega t)$. So the general solution of Equation (1) is

$$x = A \exp(i\omega t) + B \exp(-i\omega t)$$

where A and B are arbitrary constants.

We now proceed to work with the Hermite equation,

$$\frac{d^2\psi}{dy^2} + (\eta - y^2)\psi = 0 \tag{3}$$

Again, we define $D = d/dy$ to obtain

$$[D^2 + (\eta - y^2)]\psi = 0 \tag{4}$$

683

But the factorization is not so straightforward this time because the coefficient $(\eta - y^2)$ is not constant. To prepare for factoring we need to recall the rule for the differentiation of a product:

$$D(y\psi) = \psi + yD\psi = (1 + yD)\psi \tag{5}$$

By use of this rule we can derive three useful identities:

$$(D + y)(D - y)\psi = [D^2 - (1 + yD) + yD - y^2]\psi = (D^2 - y^2 - 1)\psi \tag{6a}$$

Similarly,

$$(D - y)(D + y)\psi = (D^2 - y^2 + 1)\psi \tag{6b}$$

Subtracting these two, we get

$$[(D - y)(D + y)]\psi - [(D + y)(D - y)]\psi = 2\psi \tag{6c}$$

It is worthwhile writing these three identities as operator equations:

$$(D + y)(D - y) = D^2 - y^2 - 1 \tag{7a}$$

$$(D - y)(D + y) = D^2 - y^2 + 1 \tag{7b}$$

$$(D - y)(D + y) = (D + y)(D - y) + 2 \tag{7c}$$

Using Identity (7b), we now rewrite the Hermite equation in the form

$$[(D - y)(D + y) - 1]\psi = -\eta\psi \tag{8}$$

It might seem that little has been gained, but it will be noticed that the equation has been factored. It is also apparent that if $\eta = 1$ we can solve the equation. Let ψ_0 represent the solution for which $\eta = 1$ and for which

$$(D - y)(D + y)\psi_0 = 0. \tag{9}$$

Then any solution of $(D + y)\psi_0 = 0$ will also be a solution. We separate variables as follows:

$$\frac{d\psi_0}{dy} + y\psi_0 = 0,$$

$$\frac{d\psi_0}{\psi_0} = -y\,dy,$$

$$\ln \psi_0 = -\frac{y^2}{2} + \ln C$$

$$\psi_0 = C \exp\left(-\frac{y^2}{2}\right) \tag{10}$$

As we shall see, this is the wave function for the ground state of the oscillator. The energy of the ground state is obtained by recalling from Equation (7-65) that $\eta = 2\mathcal{E}/\hbar\omega$. For the ground state $\eta_0 = 1$, so

$$\mathcal{E}_0 = \tfrac{1}{2}\hbar\omega \tag{11}$$

At this point it is reasonable to ask why we did not use Identity (7a) instead of

(7b). Then the Hermite equation would have read

$$[(D + y)(D - y) + 1]\psi = -\eta\psi$$

To obtain the ground state we would have chosen $\eta = -1$ and then we would have proceeded to solve the equation $(D - y)\psi_0 = 0$; but the solution to this equation is proportional to $\exp(+y^2/2)$, a result that would not allow us to calculate $\int_{-\infty}^{\infty} \psi^2 \, dy$.

The correct result $\psi_0 = C \exp(-y^2/2)$ can be used in integrals over all values of y. Furthermore, the choice of $\eta = -1$ implies a negative total energy for the oscillator; but both the kinetic energy $(p^2/2m)$ and the potential energy $(\frac{1}{2}kx^2)$ are positive, implying that the total energy must be positive. We conclude then that Identity (7a) does not lead to a useful solution.

Next we shall show that if ψ_0 is a solution of the equation

$$[(D - y)(D + y) - 1]\psi_0 = -\eta_0\psi_0 \qquad (8)$$

the function

$$\psi_1 = (D - y)\psi_0 \qquad (12)$$

is a solution of the equation

$$[(D - y)(D + y) - 1]\psi_1 = -\eta_1\psi_1 \qquad (13)$$

This fact is easily proved by starting with Equation (8) and operating from the left with the operator $(D - y)$.

$$[(D - y)(D - y)(D + y) - (D - y)]\psi_0 = -(D - y)\eta_0\psi_0$$

Next we refer to Identity (7c) for use in rewriting the first term on the left. We get

$$\{(D - y)[(D + y)(D - y) + 2] - (D - y)\}\psi_0 = -\eta_0(D - y)\psi_0$$

Regrouping terms, we get

$$[(D - y)(D + y) - 1]\,(D - y)\psi_0 = -(\eta_0 + 2)(D - y)\psi_0$$

This equation has the desired form if

$$\psi_1 = (D - y)\psi_0 \qquad (14)$$

and

$$\eta_1 = \eta_0 + 2$$

Notice that this result does not depend on the fact that η_0 refers to the ground state; so the proof that we used is general enough to enable us to write immediately

$$\psi_2 = (D - y)\psi_1$$

$$\eta_2 = \eta_1 + 2$$

or more generally

$$\psi_{n+1} = (D - y)\psi_n$$

$$\psi_{n+1} = \eta_n + 2 \qquad (15)$$

686

Since we know ψ_0 and η_0, it is now possible to generate an infinite number of solutions to the Hermite equation just by repeated application of the operator $(d/dy - y)$. The general form of η is

$$\eta_n = 2n + \eta_0 \tag{16}$$

Since $\eta_0 = 1$ and $\eta_n = 2\mathcal{E}_n/(\hbar\omega)$, we get

$$\mathcal{E}_n = (n + \tfrac{1}{2})\hbar\omega, \tag{17}$$

the general expression for the energy of the nth state of the harmonic oscillator.

A few concluding comments are appropriate. We have derived an infinite set of solutions for the harmonic oscillator, but we have not proved that we have found *all* of the possible solutions. In fact, there are many others but they all have the property that $\int_{-\infty}^{\infty} \psi^2 \, dy$ is infinite; those solutions that we have found are the only ones for which the integral remains finite, allowing us to normalize by appropriate choice of the constant C in Equation (10).

Instead of using mathematics to prove that we have found all the useful solutions, it is perhaps better to use a physical argument. If the curves for the infinite square well in Figure 7-3 are studied, it will be seen that the ground-state wave function has no zeros in the interior of the physical region. The first-excited-state wave function has one zero and so on. Comparison with Figure 7-7 shows that the same statements apply to the wave functions for the harmonic oscillator. For the infinite square well we proved mathematically that we found all the physically useful solutions. By analogy we infer that we have done the same for the oscillator, a fact that can be proved.

The approach we have used to solve this problem is not the one that is ordinarily used. The more conventional treatment can be studied in nearly any of the references at the end of Chapter 7. The method of factorization has the advantages of compactness and a certain amount of elegance. Another attribute is the fact that it is useful in the study of the quantum mechanical nature of angular momentum. Unfortunately the method is not so general as was hoped when it was first discovered. There are numerous problems in quantum mechanics for which factorization is not a useful technique.

Appendix V-7

Evaluation of $\oint p_r dr = n_r h$

We wish to evaluate

$$\oint P_r dr = n_r h. \tag{1}$$

Now

$$p_r = m \frac{dr}{dt} \tag{2}$$

is the momentum along the radius. Changing the independent variable from t to ϕ by the relationship

$$\frac{dr}{dt} = \frac{dr}{d\phi} \frac{d\phi}{dt} \tag{3}$$

we get

$$p_r = m \frac{d\phi}{dt} \frac{dr}{d\phi} \tag{4}$$

and noting that

$$p_\phi = mr^2 \frac{d\phi}{dt} \tag{5}$$

and

$$dr = \frac{dr}{d\phi} d\phi \tag{6}$$

we may write the integral as

$$\oint \frac{p_\phi}{r^2} \frac{dr}{d\phi} \cdot \frac{dr}{d\phi} \cdot d\phi = n_r h \tag{7}$$

or

$$p_\phi \oint \frac{1}{r^2} \left(\frac{dr}{d\phi} \right)^2 d\phi = n_r h \tag{8}$$

The equation of the ellipse may be written as

$$\frac{1}{r} = \frac{1 + \epsilon \cos \phi}{a(1 - \epsilon^2)} \tag{9}$$

where a is the semimajor axis and ϵ is the eccentricity. Therefore

$$\frac{dr}{d\phi} = \frac{a(1 - \epsilon^2)\epsilon \sin \phi}{(1 + \epsilon \cos \phi)^2} \tag{10}$$

and

$$\frac{1}{r}\frac{dr}{d\phi} = \frac{\epsilon \sin \phi}{1 + \epsilon \cos \phi} \tag{11}$$

The integral equation now becomes

$$p_\phi \int_0^{2\pi} \frac{\epsilon^2 \sin^2 \phi}{(1 + \epsilon \cos \phi)^2} \, d\phi = n_r h \tag{12}$$

The integral

$$I = \int_0^{2\pi} \frac{\epsilon^2 \sin^2 \phi}{(1 + \epsilon \cos \phi)^2} \, d\phi \tag{13}$$

can be integrated by parts. In the usual standard form

$$\int u \, dv = uv - \int v \, du \tag{14}$$

let

$$u = \epsilon \sin \phi$$

$$du = \epsilon \cos \phi \, d\phi$$

$$dv = \frac{\epsilon \sin \phi}{(1 + \epsilon \cos \phi)^2} \, d\phi$$

so that

$$v = \frac{1}{1 + \epsilon \cos \phi}$$

then

$$I = \frac{\epsilon \sin \phi}{1 + \epsilon \cos \phi}\Bigg]_0^{2\pi} - \int_0^{2\pi} \frac{\epsilon \cos \phi}{1 + \epsilon \cos \phi} \, d\phi \tag{15}$$

On substitution of the limits of integration, the first term on the right-hand side becomes zero. The value of the integral is

$$I = -\int_0^{2\pi} \frac{\epsilon \cos \phi}{1 + \epsilon \cos \phi} \, d\phi \tag{16}$$

which may be written in the form

$$I = \int_0^{2\pi} \left(\frac{1}{1 + \epsilon \cos \phi} - 1\right) d\phi \tag{17}$$

yielding

$$I = \frac{2\pi}{(1 - \epsilon^2)^{1/2}} - 2\pi \tag{18}$$

(see Peirce's *Table of Integrals,* page 41). Putting this back in the integral equation, we get

$$\frac{2\pi p_\phi}{(1 - \epsilon^2)^{1/2}} - 2\pi p_\phi = n_r h \tag{19}$$

or

$$\frac{n_\phi h}{(1 - \epsilon^2)^{1/2}} - n_\phi h = n_r h \tag{20}$$

since

$$p_\phi = n_\phi \hbar$$

Hence

$$(n_r + n_\phi) = \frac{n_\phi}{(1 - \epsilon^2)^{1/2}} \tag{21}$$

so that

$$\epsilon = \sqrt{1 - \left(\frac{n_\phi}{n_\phi + n_r}\right)^2} \tag{22}$$

which is the expression for the eccentricity of the ellipse in terms of the azimuthal and radial quantum numbers.

Of great interest is the expression for the total energy $\mathcal{E}$ of an elliptic orbit. The potential energy is $-kZe^2/r$. The kinetic energy can be written as

$$\tfrac{1}{2}m \left[\left(\frac{dr}{dt}\right)^2 + \left(r\frac{d\phi}{dt}\right)^2\right]$$

where dr/dt is the radial component of the velocity and $r(d\phi/dt)$ is the transverse component of the velocity. Now, using Equations (3) and (5), we get

$$\frac{dr}{dt} = \frac{p_\phi}{mr^2}\frac{dr}{d\phi} \tag{23}$$

and

$$\left(r\frac{d\phi}{dt}\right)^2 = \frac{p_\phi^2}{m^2 r^2} \tag{24}$$

The expression for the total energy then becomes

$$\begin{aligned}
\mathcal{E} &= \tfrac{1}{2}m\left[\frac{p_\phi^2}{m^2 r^4}\left(\frac{dr}{d\phi}\right)^2 + \frac{p_\phi^2}{m^2 r^2}\right] - \frac{kZe^2}{r} \\
&= \frac{p_\phi^2}{2mr^2}\left[\left(\frac{1}{r}\frac{dr}{d\phi}\right)^2 + 1\right] - \frac{kZe^2}{r} \tag{25}
\end{aligned}$$

Solving this for $\left(\dfrac{1}{r}\dfrac{dr}{d\phi}\right)^2$ yields

$$\left(\frac{1}{r}\frac{dr}{d\phi}\right)^2 = \frac{2m\mathcal{E}r^2}{p_\phi^2} + \frac{2mkZe^2r}{p_\phi^2} - 1 \tag{26}$$

From Equation (11) we get

$$\left(\frac{1}{r}\frac{dr}{d\phi}\right)^2 = \frac{\epsilon^2 \sin^2 \phi}{(1 + \epsilon \cos \phi)^2} \tag{27}$$

Eliminating the angle ϕ between this equation and the equation of the ellipse (9) yields

$$\left(\frac{1}{r}\frac{dr}{d\phi}\right)^2 = -\frac{r^2}{a^2(1 - \epsilon^2)} + \frac{2r}{a(1 - \epsilon^2)} - 1 \tag{28}$$

By equating coefficients of like powers of r in Equations (26) and (28), we get

$$\frac{2m\mathcal{E}}{p_\phi^2} = -\frac{1}{a^2(1 - \epsilon^2)} \tag{29}$$

and

$$\frac{mkZe^2}{p_\phi^2} = \frac{1}{a(1 - \epsilon^2)} \tag{30}$$

from which

$$\mathcal{E} = -\frac{kZe^2}{2a} \tag{31}$$

The total energy depends only on the major axis of the ellipse and is the same as that for a circle of radius a. Eliminating a from Equations (29) and (30) yields

$$\mathcal{E} = -\frac{mkZ^2e^4\,(1 - \epsilon^2)}{2p_\phi^2} \tag{32}$$

Appendix V-8

Derivation of the Fermi-Dirac and Bose-Einstein Distributions

We assume a collection of identical indistinguishable particles; we also assume that any forces of interaction between pairs (or among groups) of particles are not very strong. Suppose that each particle has a number of energy levels available to it, and that the energies of these levels are $\mathscr{E}_1, \mathscr{E}_2, \mathscr{E}_3, \ldots, \mathscr{E}_j, \ldots$ Further, suppose that the ground state contains n_1 particles, the state of energy $\mathscr{E}_j$ contains n_j particles, and so on.

The key assumption we make concerns $f(\mathscr{E}_j, 1)$, the probability of finding one particle in the jth state. A rigorous derivation of the form of f is beyond the scope of this work. Even a nonrigorous derivation is lengthy and will not be given here. Qualitatively, we expect that $f(\mathscr{E}_j, 1)$ should decrease as $\mathscr{E}_j$ increases, and one of the simplest forms that satisfies this condition happens to be correct; so we shall assume this form:

$$f(\mathscr{E}_j, 1) = C_1 \exp\left(\frac{\mu - \mathscr{E}_j}{kT}\right) \tag{1}$$

The quantity C_1 is a constant, k is Boltzmann's constant, T is the absolute temperature, and μ is a constant for which the interpretation depends on the specific problem.

Since the particles do not have much interaction with each other, the probability of finding two of them in the jth state will be the square of the one-body expression.

$$f(\mathscr{E}_j, 2) = C_2 \left[\exp\left(\frac{\mu - \mathscr{E}_j}{kT}\right)\right]^2 \tag{2}$$

Clearly, we can generalize to n_j particles in the jth state.

$$f(\mathscr{E}_j, n_j) = C \left[\exp\left(\frac{\mu - \mathscr{E}_j}{kT}\right)\right]^{n_j} \tag{3}$$

The next task is the evaluation of the constant C. The principle is simply that there must be *some* integer number of particles in the state, and so the sum of all the probabilities must add up to unity:

$$\sum_{n_j} f(\mathscr{E}_j, n_j) = 1 \tag{4}$$

691

Using the explicit form in Equation (3),

$$C \sum_{n_j} \left[\exp \left(\frac{\mu - \mathscr{E}_j}{kT} \right) \right]^{n_j} = 1 \tag{5}$$

It is convenient to denote the quantity in the brackets by y. Then

$$C = \frac{1}{\sum_{n_j} y^{n_j}} \tag{6}$$

The probability of finding n_j particles becomes

$$f(\mathscr{E}_j, n_j) = \frac{y^{n_j}}{\sum_{n_j} y^{n_j}} \tag{7}$$

The statistical distribution function is, by definition, the *average number of particles* in the state. This quantity can now be calculated.

$$f(\mathscr{E}_j) = \sum_{n_j} n_j \, f(\mathscr{E}_j, n_j)$$

$$= \frac{\sum_{n_j} n_j y^{n_j}}{\sum_{n_j} y^{n_j}} \tag{8}$$

The details of the summation depend on the kind of particle.

For Fermi particles the Pauli exclusion principle ensures that no two identical particles may occupy the same quantum state. Therefore n_j will assume only the values 0 and 1.

$$f_{FD}(\mathscr{E}_j) = \frac{y}{1+y} = \frac{1}{1/y + 1} \tag{9}$$

$$f_{FD}(\mathscr{E}_j) = \frac{1}{\exp \left[(\mathscr{E}_j - \mu)/(kT) \right] + 1}$$

For Bose particles n_j can be any integer from zero to infinity. So the distribution has the form

$$f_{BE}(\mathscr{E}_j) = \frac{\sum_{n=0}^{\infty} n_j y^{n_j}}{\sum_{n=0}^{\infty} y^{n_j}} \tag{10}$$

The infinite series in the numerator and denominator are most easily summed by long division. The results are

$$\sum_{n_j=0}^{\infty} y^{n_j} = 1 + y + y^2 + y^3 + \ldots = \frac{1}{1-y} \tag{11}$$

$$\sum_{n_j=0}^{\infty} n_j y^{n_j} = y + 2y^2 + 3y^3 + \cdots = \frac{y}{(1-y)^2}$$

Then the Bose-Einstein distribution becomes

$$f_{BE} = \frac{y(1-y)}{(1-y)^2} = \frac{y}{1-y} = \frac{1}{1/y - 1} \tag{12}$$

$$f_{BE} = \frac{1}{\exp \left[(\mathscr{E}_j - \mu)/(kT) \right] - 1}$$

Appendix V-9

Probability Density Functions

In both quantum mechanics and statistical mechanics the idea of a probability density function (or statistical distribution function) has great importance. If a variable x can take on any value in some range, the probability of measuring x and getting x_0 as an answer is zero, but if we ask for the probability that the result will lie between x_1 and x_2, the result is not zero in general. Typically there exists a function $f(x)$ such that

$$P\,(x_1 < x < x_2) = \int_{x_1}^{x_2} f(x)\ dx \tag{1}$$

where P means "the probability that . . . ."

Since competent measurement must result in some answer, it is clear that $f(x)$ satisfies the normalization condition

$$\int_{-\infty}^{\infty} f(x)\ dx = 1 \tag{2}$$

The situation often occurs when we know $f(x)$ but when x is not really the relevant physical quantity. Some other quantity y is the interesting parameter, and $y(x)$ is a known monotonic function. The problem is to determine $g(y)$, the probability density function for the new variable. The solution to this problem is contained in the integral property of a probability density, expressed in Equation (1). Since the

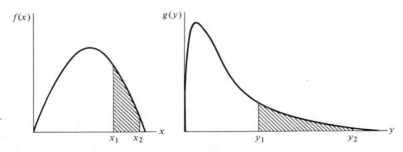

Figure A-3   Example of a transformed probability density function.

function $y(x)$ is monotonic, there exists a one-to-one correspondence between values of x and values of y. Further, the probability of a certain physical result must be independent of whether it was measured as x or y. Therefore we expect that

$$P(x_1 < x < x_2) = P(y_1 < y < y_2) \tag{3}$$

where $y_1 = y(x_1)$ and $y_2 = y(x_2)$. Using Equation (1), we get

$$\int_{x_1}^{x_2} f(x) \, dx = \int_{y_1}^{y_2} g(y) \, dy \tag{4}$$

This equation is illustrated in Figure A-3. The shaded areas are equal, as implied by the equality of the integrals in the equation. Since this equality must be true for every choice of x_1 and x_2, it follows that

$$f(x) \, dx = g(y) \, dy \tag{5}$$

This equation is the statement of the solution to our problem. The differentials dx and dy are essential to the result and must not be ignored.

Equation (5) is correct whenever y is a monotonically increasing function of x; however, if $y(x)$ is a monotonically decreasing function, then dx and dy will have opposite signs. Since f and g must be positive (to ensure that any probability is positive which results from integration), we see that an extra minus sign must be inserted into the equation when $y(x)$ is decreasing.

We illustrate the transformation of probability density functions by an example. Suppose x is the cosine of a scattering angle and it is observed that for many scatterings the distribution of x is uniform between $x = -1$ and $x = +1$; that is, $f(x) = \frac{1}{2}$. We may inquire about the distribution of y where y is the scattering angle itself, bounded by zero and π radians. The functional relation between x and y is

$$x = \cos y$$

from which we obtain

$$dx = -\sin y \, dy$$

Here y is a monotonically decreasing function of x over the interval under consideration, so we suppress the minus sign that appears above. Then

$$f(x) \sin y \, dy = g(y) \, dy$$

or

$$g(y) = \tfrac{1}{2} \sin y$$

Answers to Selected Problems

Chapter 1

1-1. 1.59×10^{-19} coul
1-2. (b) 1.84×10^{11} coul/kg; (c) 1.83×10^{11} coul/kg
1-3. (a) 4.8×10^{-10} erg; (b) 1.03×10^9 cm/sec
1-4. 1.2 cm
1-5. 2.64×10^7 m/sec
1-6. (a) Helical trajectories; axes parallel to B

$$R = \frac{mv \sin \theta}{Be}$$

Chapter 2

2-1. $0.986 \ c$
2-2. $0.94 \ c$
2-3. $\dot{x} = \dfrac{w_x + v}{1 + vw_x/c^2}$

$\dot{y} = \dfrac{w_y}{K(1 + vw_x/c^2)}$

$\dot{z} = \dfrac{w_z}{K(1 + vw_x/c^2)}$
2-5. (a) $2.3 \ mc^2$; (b) $1.3 \ mc^2$
2-8. $0.42 \ c$
2-9. $2 \times 10^3 \ c$
2-10. $u < 0.38 \ c$
2-11. $u < 0.995 \ c$
2-15. $\frac{5}{16}(mu^6/c^4)$
2-16. (a) $u = 0.0199 \ c$; (b) $u = 0.99 \ c$
2-17. $0.995 \ c$
2-18. $0.80 \ c$
2-19. 1.18 MeV

2-23. $\begin{pmatrix} K & -\beta K \\ -\beta K & K \end{pmatrix}$

Chapter 3

3-1. (c) 55 min
3-2. 8.5×10^{-6}
3-3. (a) 11.4×10^{-12} cm; (b) 95 r_p; (c) 11.4×10^{-12} cm
3-4. 2.3×10^{14} gm/cm³
3-5. (a) 8.80 MeV; (b) 2.06×10^9 cm/sec
3-6. (a) 13.27 cm; (b) 5.30 MeV
3-7. (a) 1.17 mm; (b) 6.059 MeV; 6.098 MeV
3-8. 8.82 cm; 9.53 cm
3-9. (a) 0.315 kilogauss; (b) 313 cm, 315 cm
3-10. 0.007781 u; 7.248 MeV
3-11. $v = E/B$
3-12. (a) $R = MV/qB$
3-13. 5.93×10^8 reactions/sec

Chapter 4

4-1. (a) 1.17×10^{23} cm/sec²; (b) 22.5×10^{-10} nt/coul; 7.5×10^{-14} stvolt/cm; (c) 2.24×10^{-29} joule/m³; 2.24×10^{-28} erg/cm³
4-2. 7.8×10^{-5} erg/sec
4-3. (a) 2668 Å; (b) 11.2 eV; (c) 15.83 eV
4-4. $\mathscr{E} \leq 1.5$ eV
4-5. (a) 1.8×10^{34} erg/sec; (b) $1.73 \times$

696 Answers to Selected Problems

10^{15} kg/day; (c) 8.95 cal/cm² min; 4.5 times solar constant

4-6. 4829 Å
4-7. 0.375×10^{-14} volt-sec
4-8. (a) 1.222 keV; (b) 0.069 c
4-9. 59°C
4-10. (a) 3.5×10^{-5} erg/cm³; (b) 2.96×10^{-2} dyne/stcoul; (c) 3.18×10^{17} photons/sec
4-11. (b) 1.9×10^{18} photons/sec

Chapter 5

5-1. (a) 6.81×10^{-21} cm²/atom; 1.97×10^{-21} cm²/atom
5-2. (a) 2.06×10^{-20} cm²/atom; 1.73×10^{-19} cm²/atom
5-3. (a) 0.27 cm²/gm; (b) 1.22×10^{-23} cm²/atom
5-4. Al: 0.051 cm; Cu: 0.00155 cm
5-5. (a) 4.07×10^{-15} joule sec/coul; (b) 6.6×10^{-34} joule sec
5-6. 3.035×10^{-8} cm
5-7. (a) 0.309 Å; (b) 2° 55.4'
5-8. (a) 5° 19'; (b) 10° 42'; (c) 10th order
5-9. (a) 1.363 Å; (b) $\delta = 1 - \mu = 4.9 \times 10^{-7}$; (c) 3.42 min
5-10. (b) 13.5 min
5-12. (a) 0.024 Å; (b) 28 min
5-13. (a) −43° 59'; (b) 579 eV
5-14. (a) 0.172 Å; (b) 2.79×10^4 eV
5-17. (b) $p = \dfrac{4(h\nu)^2 + 4h\nu mc^2}{4\,h\nu c + 2\,mc^3}$
5-18. 0.136

Chapter 6

6-1. 6.11×10^{-13} cm
6-3. (a) 1.025; (b) 29° 11'; (c) 9.56×10^{11} cm/sec; (d) 9.41×10^8 cm/sec
6-4. (a) 59.8 min; (b) 8.75×10^8 m/sec
6-5. (a) $\lambda = h\sqrt{2\pi/MkT}$; (b) 3.14 Å
6-8. (b) 0.9789
6-10. (e) $u = dv/dk$
6-11. 1.15×10^8 cm/sec
6-12. 65.8 MeV
6-14. Photon

Chapter 7

7-2. $a = n\lambda/2$
7-3. $a = n\lambda/2$
7-4. (a) $\epsilon/a - 1/2\pi n \sin(2\pi n\epsilon/a)$
  (b) ϵ/a
  (c) $\frac{2}{3}n^2\,\pi^2(\epsilon/a)^3$
7-5. (b) $\mathcal{E}_n = n^2\pi^2\hbar^2/2ma^2$
  $\psi_n = \sqrt{2/a}\,\sin(n\pi x/a)$
7-11. $R = 1$
7-12. $\mathcal{E}/V_0 \approx 1.03$
7-15. 1.64×10^{-34} joule
7-16. $n = 3.74 \times 10^{30}$
7-17. $<x> = 0$; $1x^2> = \frac{3}{2}^2$
7-18. $<p> = 0$; $<p^2> = \frac{3}{2}\hbar^2\alpha^2$
7-19. $\frac{3}{2}\hbar$
7-21. (a) ground state; (b) $A = \sqrt{30}/a^{5/2}$; (c) $<x> = a/2$; (d) $<x^2> = 2a^2/7$; (e) more localized, $\Delta x = 0.19\,a$
7-22. (a) $\overline{T} = \overline{V} = k\omega A^2/8\pi$; $\omega^2 = k/m$; (b) $<T> = <V> = \hbar^2\alpha^2/4m$, $\alpha^4 = mk/\hbar^2$
7-23. $\lim\limits_{n \to \infty}$ Prob $= 4\,\Delta x/a$
7-24. Transmission $= 3.22 \times 10^{-7}$ percent
7-25. 0.49, 3.7, 10.0, 18.8, 0.96, 7.5, 15
7-26. $a = 1.94 \times 10^{-13}$ cm
7-27. $ka = n\pi$; $a = n\lambda/2$
7-28. $\left|\dfrac{E}{A}\right|^2 + \left|\dfrac{B}{A}\right|^2 = 1$; therefore probability is conserved.
7-30. (a) $\frac{1}{8}$; (b) $<x> = \frac{3}{4}$
7-31. $a \to b$, allowed; $a \to c$, forbidden; $b \to c$, allowed
7-32. (a) $\left[\dfrac{\hbar^2}{2m_1}\dfrac{d^2}{dx_1{}^2} + \dfrac{\hbar^2}{2m_2}\dfrac{d^2}{dx_2{}^2} + (\mathcal{E}_1 + \mathcal{E}_2)\right]\psi = 0$; (b) $\psi_1\psi_2$; (c) $\mathcal{E}_1 + \mathcal{E}_2$
7-33. (a) $<x> = 0$; (b) $<p> = \hbar k$; (c) $A = -\hbar/i$
7-34. *Harmonic Oscillator*

nodes	$n + 2$
maxima	$\begin{cases} n/2 + \frac{1}{2} & n \text{ odd} \\ n/2 + 1 & n \text{ even} \end{cases}$
minima	$\begin{cases} n/2 + \frac{1}{2} & n \text{ odd} \\ n/2 & n \text{ even} \end{cases}$

points of inflection $n + 2$

Square Well

nodes $n + 1$

$$\text{maxima} \quad \begin{cases} n/2 + \frac{1}{2} & n \text{ odd} \\ n/2 & n \text{ even} \end{cases}$$

$$\text{minima} \quad \begin{cases} n/2 - \frac{1}{2} & n \text{ odd} \\ n/2 & n \text{ even} \end{cases}$$

points of inflection $n - 1$

Chapter 8

8-1. $\nu = cR \dfrac{2n - 1}{n^4 - 2n^3 + n^2}$

$\lim\limits_{n \to \infty} \nu = 2cR/n^3$

8-3. $\Delta\lambda = 2.38$ Å

8-5. 1.85 eV; 12.06 eV; 13.94 eV; 14.60 eV

8-6. 12,193 cm^{-1}; 8201 Å; 1.51 eV

8-7. $n_i = 5 \to n_f < 5$; $n_i = 4 \to n_f < 4$; etc.

8-8. $\epsilon = 0$

8-9. (a) 303.75 Å, 256.3 Å, ultraviolet; (b) 1640.25 Å, 1180.01 Å, ultraviolet

8-10. 22.6×10^{38}

8-11. (a) 40.8 volts; (b) 54.4 volts

8-12. 2.49×10^{-11} cm

8-13. (a) 7.29×10^{-3}; (b) 4.14×10^{14} rad/sec; (c) 1.26

8-14. 137

8-15. 0.114 Å

8-16. (c) k = wave number

8-19. $6\hbar^2$; $2\hbar$

8-20. 0; 0

8-21. Allowed

8-22. Forbidden

8-23. Forbidden

8-24. $3a_0/2$

8-25. (c) 8×10^{-10} sec

Chapter 9

9-1. $L = 5, 4, 3, 2, 1$

9-2. (a) $j = \frac{7}{2}, \frac{5}{2}$; (b) $j = \sqrt{63}/2, \sqrt{35}/2$; (c) $60°, 131° \, 12'$

9-3. (a) $J = 5, 4, 3, 2, 1$; (b) $J' = 5, 4, 3, 2, 1$

9-4. (a) $J = \frac{11}{2}, \frac{9}{2}, \frac{7}{2}, \frac{5}{2}, \frac{3}{2}, \frac{1}{2}$; (b) $J = \frac{9}{2}, \frac{7}{2}, \frac{5}{2}, \frac{3}{2}, \frac{1}{2}$

9-6. 2.65×10^{11} sec^{-1}

9-7. 2.77×10^{-16} erg

9-9. (a) 1.88 eV, 2.92 eV; (b) $4 \, {}^3P_1 \to 4 \, {}^1S_0$

9-10. (a) 2542.6 Å, 1852.6 Å; (b) 84405.6 cm^{-1}

9-11. (c) 5.37 eV; (e) 1.84 eV

9-12. (a) 2; (b) $\pm 4.635 \times 10^{-17}$ dyne; (c) 62.7 m

9-13. (a) $\pm 4.635 \times 10^{-17}$ dyne; (b) 590 m; (c) 1.74×10^{16} gauss/cm

9-15. (a) 4, 3, 2; (b) 1, 0; (c) 5, 4, 3, 2, 1

9-16. $2n^2$

9-17. (a) $L = 1, S = 1, J = 0$; (b) ${}^1S_0, {}^1D_2$ ${}^3P_1 \, {}^3P_2$

9-19. $\langle S_x \rangle = 0$, $\langle S_y \rangle = 0$, $\langle S_z \rangle = \frac{1}{2}$

9-20. $0, 0, -\frac{1}{2}$

9-24. iS_x

9-25. iS_y

9-26. 0, 0, 0

Chapter 10

10-1. 78.1 kV

10-3. 3.87 cm

10-4. (a) 5.37 keV, 5.93 keV, 7.26 keV; (b) 1.65 cm, 1.73 cm, 1.92 cm; (c) 0.16 cm, 0.38 cm

10-5. (a) 72.68 keV, 94.60 keV, 94.75 keV; (b) 315.6 cm, 361.34 cm, 361.43 cm; (c) 2.88×10^{10} cm/sec

10-6. (b) 16.33 keV

10-7. (b) 7.05 keV

10-8. 54.41 keV

Chapter 11

11-2. (a) 2.42×10^{-21} gm; (b) 3.08×10^{-37} gm cm^2; (c) 2.23×10^{-6} eV

11-3. (a) 7.4×10^{-39} gm cm^2; (b) 9.3×10^{-5} eV

11-5. 4×10^{-8} cm

11-7. $k = 2a^2D$

11-12. $n(v)dv = C\sqrt{m^3/2} \, v^2 \exp{(-\frac{1}{2}mv^2/kT)}dv$

11-13. $\sqrt{8kT/m\pi}$

11-14. $\sqrt{3kT/m}$

11-15. (a) 99 percent; (b) 125 eV

11-23. (a) 1 joule (approx); (b) 2×10^4 joule

11-25. 0.494

11-28. $\mathcal{E}_F/k$

11-32. (a) $v_g =$
$$2\sqrt{\frac{\mathcal{E}}{2m}} \frac{L\sin kL - (2k/\alpha^2)\sin\alpha L}{L\sin\alpha L + (2kL/\alpha)\cos\alpha L},$$
(b) $v_g = -\frac{2\pi}{5}\sqrt{\frac{\mathcal{E}}{2m}}\sin kL;$
(c) $v_g = \frac{2\pi}{5}\sqrt{\frac{\mathcal{E}}{2m}}\sin kL$

Chapter 12

12-1. (b) 17.2 MeV

12-4. (a) 412 MeV; (b) 970 MeV/c; (c) 0.72 c; (d) 191 cm

12-5. (a) 153 MeV; (b) 766 MeV/c; (c) 0.42 c; (d) 165 cm

12-6. (a) 1.07×10^9 rad/sec; (b) 1.71×10^8 Hz; (c) 8350 gauss

12-7. (a) 2.5×10^9 rad/sec; (b) 3.96×10^8 Hz; (c) 32.4 MeV

12-8. 15 eV/cycle

12-10. (a) 4.4×10^7 m²/weber sec; (b) 7×10^6 Hz; (c) 2.24 meters

12-11. (a) 12.1 GeV/sec; (b) 70.7 eV/cycle; (c) 0.865

12-12. (a) 34 GeV/sec; (b) 714 eV/cycle; (c) 0.566

12-13. (a) 17.5 eV/sec; (b) 1.06×10^{-7} eV/cycle; (c) $0.69c$, 1.035×10^8 rad/sec

12-14. (a) 8×10^{-3} amp; (b) 4.8×10^{-7} amp; (c) 8.0 megawatts

12-15. 6.25×10^{15} protons/sec

12-16. 5 percent

12-17. 16.5 cm

12-18. $R_1 = R_2/e$

12-19. (a) 22.9 MHz; (b) 15.7 MeV

12-20. 5.09×10^9 wavelengths

Chapter 13

13-1. (b) 290 m_e; (c) 0.915 c

13-2. (a) 874 MeV; (b) 0.355 c

13-3. (a) 15 percent; (b) 5 percent

13-4. (a) 0.66; (b) 0.29; (c) 0.052

13-8. (a) ~ 0.9 rad; (b) 0.09 rad

13-11. (a) 0.36×10^{23} protons/cm³; (b) 0.48×10^{23} protons/cm³;

(c) 7.3×10^{20} protons/cm³; (d) 3.2 protons/cm³

13-12. 2.22 meters

13-13. 8.25×10^{-31} cm²

13-14. $n \geqslant 1.371$

13-15. (a) 2×10^{10} cm/sec; (b) 0.174 MeV; (c) 0°

13-16. 39° 6′

13-17. 0.525 MeV

13-18. (a) 0.566 c; (b) 11.0°

13-19. (a) 0.557 c; (b) 191 MeV

13-20. 295 MeV

13-21. 7.47×10^6 Hz for deuterons; 7.37×10^6 Hz for protons

13-22. 1.2 MeV

Chapter 14

14-1. 1.11×10^8 alpha particles/sec; 3 mCi

14-2. 1.126×10^{14} alpha particles/sec; 3.045×10^6 mCi

14-3. 1.11×10^{-6} gm

14-6. (a) 1.69×10^{-17} sec⁻¹; (b) 2.54×10^5 beta particles/sec; (c) 29.97 beta particles/sec

14-7. (a) 4.88×10^{-18} sec⁻¹; (b) 1.41×10^4 alpha particles/sec

14-8. (a) 2.098×10^{-6} sec⁻¹; (b) 5.6×10^{13} alpha particles/sec

14-9. 8.625×10^4 beta particles/sec

14-10. 91.2×10^3 beta particles/min

14-11. (c) 3.13 cm

14-15. (a) 5.40 MeV; (b) 0.102 MeV

14-16. (a) 10.74 MeV; (b) 282.9 MeV/c; (c) 0.20 MeV

14-17. (a) $\mathcal{E} = \dfrac{300(Z-2)e \times 2}{r}$, r in cm, $e = 4.8 \times 10^{-10}$ stcoul, $\mathcal{E}$ in eV; (b) 35.63 MeV; (c) 5.43×10^{-12} cm

14-18. (a) 4.869 MeV; (b) 4.782 MeV

14-19. (a) 624.75 keV: (b) 662.19 keV

14-20. (a) 545.67 keV: (b) 661.26 keV

14-21. (a) 47 keV/c; (b) 5.65×10^{-3} eV

14-22. (a) 2.62 MeV/c; (b) 17.7 eV

14-23. $<t> = 1/\lambda$

14-24. 0.978 gm

14-25. (a) 1.067×10^{12} alpha particles/day; (b) 3.416×10^{-7} coul

14-26. 0.135 decay

14-27. $p = m\lambda/2\sigma_B n$, $n =$ no. of atoms/cm³ in target

14-28. (a) $\frac{1}{2}\frac{(h\nu)^2}{mc^2}$; (b) 1.7 eV

14-29. (a) $\frac{1}{2}\frac{p^2}{M}$; (b) $\frac{e^2}{2}\frac{(BR)^2}{M}$; (c) 4 eV

14-30. 1.22×10^{-7} sec

14-31. 5.02×10^{-3} eV

14-32. (a) $\nu = \dfrac{\nu'[1 + (v/c)]}{\sqrt{1 - (v/c)^2}}$

14-33. 122.13 yr

14-34. $t = \ln(A_1/A_2)/(1/\tau_2 - 1/\tau_1)$ with $\tau = T/\ln 2 =$ mean life

Chapter 15

15-3. 3.361 MeV

15-4. 14.1 MeV

15-5. 8.00479 u

15-6. 0.627 MeV

15-7. 5.021 MeV

15-8. (b 1.0; (c) 8/9; (d) 0.284

15-9. 5.063 MeV

15-10. 3.367 MeV

15-11. (b) 10 min

15-12. (a) $^{37}_{17}\text{Cl}\ (n,\gamma) \rightarrow\ ^{38}_{18}\text{A} + \beta^- + \bar{\nu}$; (c) 36 min

15-13. 2.970 MeV

15-14. (a) -4.036 MeV; (b) 4.278 MeV

15-15. (a) -3.869 MeV; (b) 4.062 MeV

15-16. (a) 3.982 MeV; (b) 4.964 MeV; (c) 3.383 MeV; (d) 4.033 MeV; (e) -15.669 MeV; (f) -12.126 MeV

15-17. (a) $^{231}_{90}\text{Th}$; (b) $^{24}_{11}\text{Na}$; (c) $^{3}_{1}\text{H}$; (d) $\nu + ^{20}_{10}\text{Ne}$; (e) $^{17}_{8}\text{O}$; (f) α

15-19. (a) $dN = -\lambda N\,dt + R\,dt$; (b) $N = \dfrac{R}{\lambda}[1 - \exp(-\lambda t)]$

15-20. 1.113 MeV; 7.976 MeV; 8.790 MeV; 8.548 MeV; 8.445 MeV; 7.916 MeV; 7.570 MeV

15-23. (a) Cu; (b) Zn

15-24. $^{13}_{6}\text{C}$; $^{12}_{6}\text{C}$; $^{10}_{5}\text{B}$; $^{13}_{7}\text{N}$; $^{14}_{7}\text{N}$; $^{12}_{6}\text{C}$

15-25 63.54 u

Chapter 16

16-1. 193.5 MeV

16-2. Smaller mass

16-3. (a) 6.48×10^6 V; (b) 291.6 MeV

16-4. 171.8 MeV

16-5. (a) 213.6 MeV; (b) 260 MeV

16-6. 0.02585 eV; 1.2 keV; 8.617 keV

16-7. (a) 0.151 MeV; (b) 0.094 MeV

16-8. 1.465×10^{-12} cm; $1.27\ r^1$

16-9. (a) 39.97 MeV; (b) -24.15 MeV; (c) 15.82 MeV

16-10. (a) 71.06 MeV; (b) 3.25 MeV; (c) 74.31 MeV

16-11. (a) 96.62 MeV; (b) -11.64 MeV; (c) 84.98 MeV

16-12. About 17

16-13. 19.27 MeV

16-14. (b) 168.04 MeV

16-15. 3.125×10^{10} fissions/sec

16-16. (c) 0.192, 2.097

16-17. 0.809

16-19. (a) 4.56 millisec; (b) 0.721 microsec

16-20. $L \geq 9.8$ meters

16-22. 0.72 MeV

16-23. $C = \sqrt{2/\pi e}$

16-24. 1.385 percent

16-25. (a) 1.2 MeV; (b) 9.28×10^9 K

16-26. (a) 2.75×10^{-13} cm; (b) 3.145 MeV; (c) 3.65×10^{10} K

16-27. Eq. (16-39) 1.442 MeV
Eq. (16-40) 5.494 MeV
Eq. (16-41) 12.859 MeV
Eq. (16-42) 19.795 MeV

16-28. In the order of Eq. (16-44): 1.944 MeV; 2.220 MeV; 7.550 MeV; 7.293 MeV; 2.760 MeV; 4.964 MeV

Chapter 17

17-1. (a) 7.275 MeV; (b) 2.425 MeV

17-2. (a) ^{23}Na; (b) ^{23}Ne; (c) ^{20}F

17-3. (a) 1.802 Å; (b) 6.374×10^{-12} cm

17-4. (a) 11.84 MeV; (b) 23.67 MeV

17-5. (a) 61.7 f; (b) 54.3 f

17-6. (a) 57 eV; (b) 9.6 eV; (c) 3.96×10^6 cm/sec, 7.08×10^5 cm/sec

17-7.   (a) 7.0 eV; (b) $p_e = 0.160$ MeV/c; $p_\nu = 0.13$ MeV/c; $p_R = 0.177$ MeV/c

17-8.   (a) 1.42 keV; (b) $p_e = 1.94$ MeV/c, $p_\nu = 2.0$ MeV/c, $p_R = 0.48$ MeV/c

17-9.   (b) 0.862 MeV; (c) 57.7 eV

17-10.   (b) 1.710 MeV; (c) 4700 gauss cm

17-11.   (a) 0.1806; (b) 1.88×10^{-7} eV; (c) 0.393 cm/sec

17-12.   (b) -0.102; (c) 1.07×10^{-8} eV; (d) 2.24×10^{-2} cm/sec

17-16.   (a) 380 MeV; (b) 0.207 MeV

17-17. 1.24×10^{36}

17-19.   0.04

17-22.   (a) ^{28}Si

Chapter 18

18-1.   (a) 954 MeV/c; (b) 2.14×10^8 m/sec

18-2.   (a) 0.282 c; (b) 31 MeV/c; (c) 20.6 cm

18-3.   (a) 0.36 c; (b) 54 MeV/c; (c) 30.0 cm

18-4.   (a) 157 MeV/c; (b) 259.8 MeV

18-5.   (b) 14.3 meters, 6.5 meters

18-6. $K^\circ \rightarrow \pi^+ + \pi^-$

18-7.   (a) 1.09×10^{-12} cm, 4.36×10^{-12} cm; (b) 1.04 MeV

18-8.   (a) 4.8×10^{-12} cm, 19.2×10^{-12} cm; (b) 45.6 keV

18-9.   (a) 1.4×10^{-12} cm, 5.7×10^{-12} cm; (b) 152 keV

18-10.   (a) 2.92×10^{-13} cm; (b) 0.41

18-11. ^{105}Mo

18-12.   (a) $\pi \rightarrow e + \nu$; (b) 139.07 MeV

18-13.   (a) 214.67 MeV; (b) 75.09 MeV; (c) 388.16 MeV

18-14.   (a) 5.62 GeV; (b) 4.03 GeV

18-15. 3.04×10^{17} cm; 0.323 light years

18-16.   29.9 MeV/c; 4.5 MeV

18-18.   52.5 MeV

18-20.   6.8 eV

18-22.   (a) 870 MeV

18-27.   (a) C and T; (b) P and T

18-28.   None

18-31.   (a) 303.5 MeV/c; (b) 1236 MeV

18-33.   (a) $l = 1$; (b) $l = 1$, (c) $l = 0$

18-36.   (a) $Q_\lambda = -\frac{1}{3}$, $Q_P = \frac{2}{3}$, $Q_n = -\frac{1}{3}$; (b) ppn; (c) $n\lambda\lambda$

18-37.   (a) K^+; (b) $p\bar{n}$

18-38.   (a) 0.121 Å; (b) $\frac{1}{2}$

18-41.   (a) 75 MeV; (b) 4.15×10^{-14} cm

Index

Date Due

JY 07 '92			
OB 3 '95			
OCI 0 '95			
10-5-00			

DATE DUE	BORROWER'S NAME	
JY07'92	Dan Lewis	AL
00 3 '95	Mark Garrison	04540
OC10'95 AM	Mark Garison	4540
10/5/00	Robinett	FAL

DATE DUE	BORROWER'S NAME	

PHYSICAL CONSTANTS FOR RAPID CALCULATION
(for better values see Appendix I, p. 649)

Speed of light	$c = 3 \times 10^8$ m sec^{-1}
Electron charge	$e = 1.6 \times 10^{-19}$C
Planck's constant	$h = 6.63 \times 10^{-34}$joule sec
	$\hbar = 1.05 \times 10^{-34}$joule sec
Gas constant	$R = 8.31 \times 10^3$joule kmole^{-1}K^{-1}
Avogadro's number	$N_0 = 6.02 \times 10^{26}$kmole^{-1}
Boltzmann's constant	$k = 1.38 \times 10^{-23}$joule K^{-1}
Gravitational constant	$G = 6.67 \times 10^{-11}$ newt m^2 kg^{-1}

CLUSTERS OF CONSTANTS

Electron charge-to-mass ratio	$e/m = 1.76 \times 10^{11}$ C kg^{-1}
Bohr radius	$\hbar^2/mke^2 = 5.29 \times 10^{-11}$ m
Electron Compton wavelength	$h/mc = 2.43 \times 10^{-12}$ m
Classical electron radius	$ke^2/mc^2 = 2.82 \times 10^{-15}$ m
Bohr magneton	$e\hbar/2m = 9.27 \times 10^{-24}$ joule tesla^{-1}
	$= 5.79 \times 10^{-9}$eV gauss^{-1}
Fine-structure constant	$\alpha = ke^2/\hbar c = 1/137$
	$\hbar c = 1.97 \times 10^{-11}$MeV cm